1st International Scientific Meeting on Engineering and its Vocational Education (ISMEVD 2019)

IOP Conference Series: Materials Science and Engineering
Volume 885

Makassar, Indonesia
31 October 2019

ISBN: 978-1-7138-3621-6
ISSN: 1757-8981

Printed from e-media with permission by:

Curran Associates, Inc.
57 Morehouse Lane
Red Hook, NY 12571

CURRAN ASSOCIATES INC. proceedings .com

Some format issues inherent in the e-media version may also appear in this print version.

Institute of Physics
Dirac House, Temple Back
Bristol BS1 6BE UK

Phone: 44 1 17 929 7481
Fax: 44 1 17 920 0979

techtracking@iop.org

Additional copies of this publication are available from:

Curran Associates, Inc.
57 Morehouse Lane
Red Hook, NY 12571 USA
Phone: 845-758-0400
Fax: 845-758-2633
Email: curran@proceedings.com
Web: www.proceedings.com

TABLE OF CONTENTS

ELECTRO

ADDITIONAL LOCATION SCENARIOS OF GENERATION CAPACITY WITH ETAP APPLICATION 1
Hestikah Eirene Patoding, Matius Sau, Rombe

APPLICATION OF FUZZY C-MEANS ALGORITHM FOR BASIC SCHOOL CLUSTERING IN SAMARINDA CITY BASED ON MINIMUM EDUCATIONAL SERVICE STANDARD INDICATORS 7
Karyo Budi Utomo, P Arbain, Suminto

THE ROLE OF BIG DATA IN IMPROVING E-LEARNING TRANSITION 15
Karim Moharm, Muhammed Eltahan

ELECTRONIC PRACTICUM ASSESSMENT SYSTEM BASED ON ANDROID AT POLITEKNIK ATI MAKASSAR 22
J Habibuddin, T Suhaebri, A Ainuddin, M F Azis, Mutmainnah

BLDC MOTOR DESIGN BY APPLYING THE CONTROL OF DIRECT CURRENT AND ANALYSIS OF LOADING MECHANISM SOFT STARTING 26
M Wirandi, Safril, F Sumasto, M Agus, F Imansuri

BLIND PEOPLE'S DIRECTION SUPPORT SYSTEM USING ULTRASONIC, COLOR SENSOR WITH FUZZY LOGIC 34
Nur Yanti, Taufik Nur, Qory Hidayati

UNIT COMMITMENT DIRECT CURRENT OPTIMAL POWER FLOW USING MIXED-INTEGER LINEAR PROGRAMMING 43
M F Azis, J Habibuddin, Mutmainnah, T Muchtar, D Purwanto

SOLAR-DIESEL HYBRID POWER PLANT BATTERY CHARGING SYSTEMS 49
Matius Sau, Hestikah Eirene Patoding, Agustina Kasa

POWER FACTOR CORRECTION OF THE INDUSTRIAL ELECTRICAL SYSTEM DURING LARGE INDUCTION MOTOR STARTING USING ETAP POWER STATION 56
Mutmainnah, Muhammad Fadli Azis, Julianti Habibuddin

CHEMISTRY

STUDY OF MICROWAVES ENERGY CONVERSION INTO ELECTRIC CURRENT ON SEMICONDUCTOR MATERIAL NAXCUYTIZ 61
Regna Tri Jayanti, Sri Haryati, M. Djoni Bustan

THE EFFECT OF ACTIVE CARBON ADSORBENTS FROM SOME WASTES IN REDUCING FREE FATTY ACIDS AND ACID NUMBER TO IMPROVE VCO QUALITY 67
Dimas Frananta Simatupang, Jenny Tarigan, Mansyur

EFFECT OF MIXED ADDITIONS IN RESIDUAL OIL TO IMPROVE THE QUALITY OF SUB-BITUMINOUS COAL FROM SANGATTA, EAST KALIMANTAN 72
Flaviana Yohanala Prista Tyassena, Widya Ashari Amin

QUALITY TEST OF LIQUID TOFU WASTE IN ONE OF THE INDUSTRIES IN BANTAENG DISTRICT 76
Neny Rasnyanti M Aras

THE EFFECT OF BOILER SIZE OF DISTILLATION TO THE CONCENTRATION OF BIOETHANOL AND YIELD FROM FERMENTATION OF SWEET SORGHUM JUICE 83
Rosalia Sira Sarungallo, A Mellolo, L Melawaty, M Djonny

ENZYMATIC HYDROLYSIS OF PROTEINS FROM SNAKEHEAD-FISH (CHANNA STRIATA) 87
H Hariyati, M Mahendradatta, A B Tawali, J Langkong

MACHINE

DEVELOPMENT OF POLISHING MACHINE FOR PREPARATION METALLOGRAPHIC SPECIMEN WITH RE-MANUFACTURING 92
Apollo, R Nur, M A Suyuti, M Madjid

APPLICATION OF AIR CONDITIONER (AC) AUTOMATION SYSTEM ON ARDUINO FLATFORM-BASED VEHICLE 97
Randis, S H Wijaya

EFFECT OF TYPICAL LUBRICATION ON THE PRESSURE RATIO THROUGH OPERATING THE EFI SUZUKI ENGINE 104
R Sihombing, Hidayat, A Fatmawati, R Nur

THE RELATIONSHIP OF TEMPERATURE AND COMPRESSIVE STRENGTH ON GEOPOLYMER MORTAR USING FLY ASH-BASED 109
E Bachtiar, I Marzuki, Erdawaty, M F Junaedy, F Ramadhan, A M Setiawan, F Rachim

THE INFLUENCE OF SHOOTING CONDITIONS DURING SHOT PEENING OF STAINLESS STEEL ON SURFACE ROUGHNESS 118
M. Iqbal Mukhsen, Rusdi Nur, M. Alif AF Yusuf, Christo Rakka

THE SIMULATION BUOY SHAPE BULLET OF SEA WAVE ENERGY ABSORPTION BASED ON DIAMETER VARIATIONS 123
Jufri, Daeng Paroka, Jalaluddin, Rustan Tarakka

COMPUTATIONAL STUDIES OF PRESSURE DISTRIBUTION OF FLOW THROUGH INLINE DIMPLED PLATE 132
M Setiawan Sukardin, Nasaruddin Salam, Rustan Tarakka, Jalaluddin Haddade, Muhammad Ihsan

INFORMATION TECHNOLOGY

THE DEVELOPMENT OF PREMARITAL MOBILE LEARNING VERSION 3.0 WITH JQUERY TECHNOLOGY TO BE LEARNING FLEXIBILITY PREMARITAL COURSE APPLICATION 139
Herlinah, Baso habibi, Abd Rochman

AUTOMATIC BUOY SYSTEM FOR POSITION CONTROL BASED ON GLOBAL POSITIONING SYSTEM (GPS) 146
Z Akbar, Z B Hasanuddin, A E U Salam

DESIGNING A ROAD TOOL FOR BLIND PEOPLE USING THE COMPUTER MINIMUM SYSTEM 151
AR Hakim, F Metandi

INDUSTRY

SIMULATION OF GENERATE VALUE FOR DEVELOPMENT OF BIOMASS PLANTS IN INDONESIA: A FEASIBILITY STUDY 159
Safril, F Sumasto, M Wirandi, F Imansuri

SUSTAINABLE DEVELOPMENT IMPACT OF IMPLEMENTING ELECTRIC TAXIS IN JAKARTA: A COST-BENEFIT ANALYSIS 165
F Sumasto, F Imansuri, M Agus, Safril, M Wirandi

PRODUCTION CAPACITY PLANNING USING RCCP METHOD WITH CPOF APPROACH: A CASE STUDY IN AN AUTOMOTIVE INDUSTRY 172
Siti Aisyah, Humiras Hardi Purba, R F Dewarani

PRODUCTION CAPACITY PLANNING IN MOTORCYCLE ASSEMBLY LINE USING CRP METHOD AT P T XYZ 177
Humiras Hardi Purba, Siti Aisyah, R F Dewarani

THE IMPLEMENTATION OF PREVENTIVE MAINTENANCE USING MACHINE DAMAGE ANALYSIS: A CASE STUDY OF POWER PLANT MACHINE 184
Sri S Islam, Tika Lestari, Anisa Fitriani, Dilla Ayu Wardani

DECISION SUPPORT SYSTEM FOR SELECTION OF TOURISM IN TANA TORAJA USING TECHNIQUE FOR ORDER PREFERENCE METHOD BY SIMILARITY TO IDEAL SOLUTION (TOPSIS) 189
Anton Topadang, Agus Triyono

THE LINK MODEL OF ERP-USAGE, ABSORPTIVE-CAPACITY, AND TASK-TECHNOLOGY-FIT, TO TASK-INNOVATION IN ERP USER 198
A Velahyati Baharuddin

ERGONOMIC RISK ANALYSIS OF LECTURE CHAIRS AT THE ENGINEERING FACULTY, HASANUDDIN UNIVERSITY 203
Mulyadi, Andi Nurwahidah, Dedi Nismar Satria

KNOWLEDGE MANAGEMENT AS A SOURCE OF INNOVATION AND BUSINESS PROCESS DEVELOPMENT TO IMPROVE THE COMPETITIVENESS OF SMALL AND MEDIUM ENTERPRISES (SMES) IN THE MADURESE HANDICRAFT SOUVENIR SECTOR (CASE STUDY: XYZ SOUVENIR) 211
D G Ramadhan, Sofiyanurriyanti

APPLICATION OF LINE BALANCING USING THE HEURISTIC METHOD TO EQUALIZE THE PRODUCTION LINE AT PT.BOGATAMA MARINUSA MAKASSAR 219
Ahmad Sawal, Ainun Jariah Hamzah

WORKING POSTURE ANALYSIS ON MUSCULOSKELETAL DISORDER TO WORKERS IN THE SACK WAREHOUSE PT MAKASSAR TENE MAKASSAR 227
Arminas, M Basri, I R Aulia

FLOOD HANDLING WITH PRIORITY SCALE DETERMINATION IN SAMARINDA SEBERANG FROM EAST KALIMANTAN PROVINCE INDONESIA 232
SSN. Banjarsanti, P. Kukuh, Theta Margaritifera

ANALYSIS OF THE IMPACT OF CONVECTION INDUSTRIAL WASTE 240
Nilda, Mulyadi, Andi Nurwahidah

MIXED MATRIX MEMBRANE OF ALGINATE ZEOLIT CROSSLINKED BY PVA FOR SEPARATION OF CO2/CH4 244
Syardah Ugra Al Adawiyah

WORKER FATIGUE IN SILO TOWER DUE TO HOT CLIMATE 250
A Nurwahidah, I Bakri, Mulyadi, N Ester Kelah

DETERMINATION OF LEAD TIME ARRIVAL OF RAW MATERIALS OPTIMAL VANNAMEI SHRIMP AT PT. XYZ MAKASSAR 256
Puadi Haming, Rizqa Awaliyah

THE EFFECT OF TOP ROLL DEFECTS ON DYH5000 TOYODA HARA FINISHER DRAW FRAME ON SLIVER UNEVENNESS AT PT XYZ 263
W A Subakdo

EFFECT OF GOVERNMENT REGULATION ON LABOR COMPETITIVENESS: STUDY OF NINE MAJOR INDONESIAN COMPANIES 267
A A Aris, S Asmal

THE MANUFACTURE OF AUTOMATIC CHICKEN FEED MAKING MACHINES AS AN ECONOMICAL SOLUTION FOR CHICKEN FARMERS IN INDONESIA 273
Amrin Rapi, Abdul Nasser Arifin, Asiyanthi Tabran Lando

EFFECT OF TYPES OF CONTROL MATERIALS IN A MODIFIED SEDIMENTATION EQUIPMENT BATCH LABORATORY SCALE 281
Mustafa, Masing, M Irwan

COMPARISON OF FREEZE DRYING AND FOAM MAT DRYING EFFECTS ON CHARACTERISTICS OF INULIN FROM GEMBILI (DIOSCOREA ESCULENTA) 289
N Indah, Zainal, D Ganesa

EFFECT OF KNOWLEDGE MANAGEMENT ON PERFORMANCE: A CASE STUDY AT PT. EASTERN PEARL FLOUR MILLS MAKASSAR 293
A Rapi, A V Baharuddin, M P Shabrina

PRELIMINARY ANALYSIS OF ELECTRICITY ENERGY EFFICIENCY IN SMES 297
Y S Akil, H Miyauchi, S Mangngenre, K Amar

BIOHYDROGEN PRODUCTION FROM PALM OIL MILL EFFLUENT BY DARK FERMENTATION USING CLOSTRIDIUM BUTYRICUM 304
M Pasaribu, F S P Purba, E Y R Sitompul, E A Nainggolan

ANALYSIS OF FMEA AND FTA METHODS IN IDENTIFYING THE CAUSES OF DECREASED QUALITY OF REFINED SUGAR PRODUCTS (CASE STUDY: PT. MAKASSAR TENE) 309
N Khaerani B, Arminas, R Hendrawan

APPLICATION OF QUALITY FUNCTION DEPLOYMENT FOR THE DEVELOPMENT OF SULAM TUMPAR MOTIF DESIGN 316
Dwi Cahyadi, Etwin Fibrianie Soeprapto, Andi Farid Hidayanto, Ditha Nizaora

IMPLEMENTATION OF THE ROUGHT CUT CAPACITY PLANNING (RCCP) METHODS FOR THE PLANNING OF BOTTLED DRINKING WATER PRODUCTION CAPACITY: A STUDY CASE 324
A Rapi, I S Muthalib, S Asmal

MODEL DEVELOPMENT OF RISK ASSESSMENT OF RISK-TAKING RELATIONSHIP WITH INTEGRATED APPROACH 329
A B R Indah, S Mangngenre, R D Mudiastuti, N I Syamsul, Nilda

A HYBRID FIREFLY AND LOCAL SEARCH ALGORITHMS TO SOLVE ORIENTEERING PROBLEMS 339
C Sitompul, R Pramudi, F Setiawan

REDUCING VARIABILITY IN PROCESS AND THE RISK OF PRODUCT DEFECTS: A CASE STUDY IN SME'S 344
A Darmawan, R Hanafi, A Asqarye

LOGISTIC REGRESSION MODEL FOR USER PREFERENCE OF ONLINE PARKING: A STUDY CASE OF SIDOARJO 353
E Widodo, R N Harnaningrum, Suparno, A Santoso

RECOMMENDATIONS OF ERGONOMIC CHECKPOINTS AND TOTAL ERGONOMICS INTERVENTION IN THE PEMPEK KEMPLANG PALEMBANG INDUSTRY 360
H Setiawan, M Rinamurti

DRYING OF EXTRACTED BETACYANIN FROM RED DRAGON FRUIT (HYLOCEREUS COSTARICENSIS) AS NATURAL COLORANT BY USING SPRAY DRYER IN VARIOUS OPERATING CONDITION 369
I Susanti, F Hasanah, H Wijaya, S Heryani, E S Hartanto

ANALYSIS OF TIME ACCELERATION USING CRITICAL PATH METHOD (CPM) TO INCREASE MOTORCYCLE MAINTENANCE IN AUTHORIZED SERVICE STATION 377
I Y Marit, E Nursanti, P Vitasari

HYGIENE FACTORS OF OUTSOURCING WORKERS: A LITERATURE REVIEW 382
I Ramdhani, I Bakri

OVERALL EQUIPMENT EFFECTIVENESS (OEE) ANALYSIS AND FAILURE MODE AND EFFECT ANALYSIS (FMEA) ON PACKER MACHINES FOR MINIMIZING THE SIX BIG LOSSES - A CEMENT INDUSTRY CASE 392
I S. Muthalib, M Rusman, G L Griseldis

INTERNATIONAL JOINT VENTURE (IJV) IN EMERGING ECONOMIES: CASE STUDY OF AUTOMOTIVE INDUSTRY 398
K Nadiyah, V Vojisavljevic, K Amar

ANALYSIS OF TOTAL PRODUCTIVE MAINTENANCE (TPM) AND FAILURE MODE AND EFFECT ANALYSIS (FMEA) TO IMPROVE MACHINE EFFECTIVENESS: A STUDY ON INDONESIA'S SUGAR MILLS 405
Lukmandono, R Prabowo, E Sulistyowati

FOOD SAFETY ASSURANCE THROUGH THERMAL PROCESS ON CANNED CORNED BEEF 414
M M Rohaman, N C Siregar

IDENTIFICATION OF PRODUCTION PLANNING AND SUPPLY OF GLASS BOTTLES AT THE BOTTLED TEA COMPANY 423
Nurulinzany, Amrin

BASIC PROPERTIES EVALUATION OF ALKYL PALMITATE TO BE USED AS OIL INSULATION 431
A Rajab, J S Ramadhani, R Kurniawan, K Qibran, L Melda, M I Hamid

ANALYSIS OF SOCIAL DEMOGRAPHIC FACTORS AND ITS EFFECT ON WORK-RELATED MUSCULOSKELETAL DISORDERS IN FRONT SHOVEL OPERATORS OF A MINING COMPANY 438
R A Imran, I Bakri, Mulyadi, A E U Husain

QUALITY IMPROVEMENT USING PDCA METHODOLOGY IN THE BEVERAGE INDUSTRY 448
Raodah, W Astutik, A A Aris, S Bahri

IDENTIFYING CYBERSICKNESS WHEN WEARING A HEAD-MOUNTED DISPLAY THROUGH HEART RATE VARIABILITY DATA 455
N O Setiowati, T Wijayanto, F Trapsilawati

Author Index

IOP Conf. Series: Materials Science and Engineering **885** (2020) 011001 doi:10.1088/1757-899X/885/1/011001

PREFACE OF ISMEVD 2019

Research and Community Service Unit Politeknik ATI Makassar is proud to present The 2019 International Scientific Meeting on Engineering and Its Vocational Education (ISMEVD) 2019 held on 31 October 2019. The event was also attended by Director of Politeknik ATI Makassar Ir. Amrin Rapi, ST., MT., Assistant Director I Windi Mudriadi, ST., MT., Assistant Director II Herlina Rahim, ST., M.Sc., and Assistant Director III Ir. Muhammad Basri, MM. This event was also attended by Head of Research and Service Unit, Merla Madjid, SS., M.Hum., Head of Professional Certification Unit Dr. Sariwahyuni, SP. M.Sc., Head of Teaching Factory Dr. Eng. Abdul Nasser Arifin, ST., M. Eng., Head of Business Incubator Widya Hastuti A., S.Si.T., MM., Ph.D., Head of Internal Quality Assurance Unit Drs. Haruddin. MM, as well as the academic community, lecturers, and researchers from polytechnic, other Ministry of Industry work units, and more than 100 participants and seminar invitations. The event was opened by the master ceremony and continued with the report of the committee chairman Dr. Eng. Abdul Nasser Arifin, ST., Men, followed by welcome and introduction for invited participants, guest speakers, and seminar participants from the Director of Polytechnics ATI, Ir. Amrin Rapi, MT. Dr.Ir. H. Masjono, M. Eng., Head of Quality Assurance and the moderator of the event, then gave insights on the theme of this seminar. The main event of the seminar was speeches from guest speakers. The first speaker was Prof. Dr. Isa Setiyasah Toha, M.Sc., who has gladly accepted the committee's invitation to become one of the main speakers at this seminar. He is a Professor of ITB Industrial Engineering and currently serves as Director of Politeknik Industri Logam Morowali and an academic senate member of Institut Teknologi Nasional Bandung. Currently, he is also an editor/reviewer in the ITENAS journal, ITS technology journal, Gema Teknik UNS, and ITB industry engineering and management journal. In his speech, he emphasized the role of a polytechnic in process technology. He discussed how to increase the role of engineering technology education in polytechnics, especially in the field of process technology, materials, and process engineering technology. After the tea break, the event continued with the second guest speaker, Prof. Kuan Yew Wong. He is a professor of Industrial Engineering and leads the Industrial and Systems Engineering Research Group at the Faculty of Engineering, Universiti Teknologi Malaysia (UTM), and working in close collaboration with industries and other stakeholders to develop and implement solutions for operational improvement. Prof. Wong discusses his research on the development of an integrated sustainability assessment method for the food manufacturing industry.

All the expert talks were quite engaging, and knowledge filled. The participants were actively involved and appreciative of the quality of the expert sessions delivered.

Merla Madjid, SS. M.Hum
Chair - ISMEVD 2019
Department of Agro-Industry Manufacturing Engineering
UPPM Politeknik ATI Makassar, Indonesia

Conference Committee

Scientific Advisory Board / Reviewer
Prof. Dr. Anton Satria Prabuwono
Prof. Ravindra S. Goonetilleke, Ph.D.
Assoc. Prof. Tan Kay Chuan, Ph.D
Asst. Prof. Aldy Gunawan, Ph.D.
Asst. Prof. Hendry Raharjo, Ph.D.
Assoc. Prof. Dr. A. F. M. Saifuddin Saif
Dr. Mohammad Shojafar
Prof. Doaa Abdel Nabi Abdel Aziz
Dr. Zain Anwar Ali
Dr. Xiao Huang
Prof. Cheng-Hsing Hsu
Prof. Jianmei He
Dr. Arfan Bukhtiar
Dr. Mohamed M. S. Wahsh
Dr. Firman Mangasa Simanjuntak
Assoc. Prof. Akawut Siriruk, Ph.D.
Assoc. Prof. Dr. Anas Maruf
Assoc. Prof. Yassierli, Ph.D.
Prof. Dr. Ali Altway
Prof. I Nyoman Pujawan, Ph.D.
Asst. Prof. Budi Hartono, Ph.D.
Assoc. Prof. Markus Hartono
Prof. Joniarto Parung, Ph.D.

Advisory Committee
Dr. Haris Munandar N., MA (Ministry of Industry, Indonesia)
Drs. Mujiyono, MM (Ministry of Industry, Indonesia)
Prof. Dr. Isa S. Toha, M.Sc (Institut Teknologi Bandung, Indonesia)
Prof. Dr. Wong Kuan Yew (Universiti Teknologi Malaysia, Malaysia)

Steering Committee
Assoc. Prof. Jhon Leonard, DEA
Prof. Dr. Syamsul Bahri
Prof. Dr. Salama Manjang
Prof. Dr. Anshar Suyuti
Assoc. Prof. Muhammad Arsyad
Amrin Rapi, ST., MT
Assoc. Prof. Ilham Bakri
Dr. Farid Mardin

IOP Conf. Series: Materials Science and Engineering **885** (2020) 011002 doi:10.1088/1757-899X/885/1/011002

Organizing Committee

Chair : Dr.Eng. Abdul Nasser Arifin, ST., M.Eng
Secretary : A. Velahyati Baharuddin., ST., MT
Treasurer : A. Nurwahidah. ST., MT

Conference Committee

Ir. Muhammad Basri, ST., MM
Windi Mudriadi, ST., MT
Hj. Merla Madjid, SS., M.Hum
Widya Hastuti Afris, SST., MM., Ph.D
Della Ginza Ramadhan, ST., MT
Muhammad Fadli Azis, ST., M.Sc
Syamsuddin, S.Pdi
Amiruddin, ST
Samsul
Flaviana Yohanala Prista Tyassena, S.ST., MT
Muh, Farid Hidayat, S.ST
Lilian Putri Dwi Intan Sekar Sari, A.Md
Flourensia Irene R A
Feri Fadli
Fadliah
Mahadiah
Radinal Razak
Andi Andri Al-Ghazali Putra
Belyana Audrey Ni Abbas
Beatrix J Kailem
Muh. Dirhamsyah
Hizkia Adityo
Syarifah infoasari
Siti niril alyssa kahar
Ichwan dzulkaidah
Verayani Bahri
Dwi Indayani Salam
Ricky Triputra
Muh. Fachry Syawari
Muhammad Fahmi Suhardi
Rapitha Ayu Greazya
Fadhlyna Mikhrudi Thahirah

IOP Conf. Series: Materials Science and Engineering **885** (2020) 011003 doi:10.1088/1757-899X/885/1/011003

Peer review statement

All papers published in this volume of *IOP Conference Series: Materials Science and Engineering* have been peer reviewed through processes administered by the proceedings Editors. Reviews were conducted by expert referees to the professional and scientific standards expected of a proceedings journal published by IOP Publishing.

IOP Conf. Series: Materials Science and Engineering **885** (2020) 012001 doi:10.1088/1757-899X/885/1/012001

Additional location scenarios of generation capacity with ETAP application

Hestikah Eirene Patoding[1], Matius Sau[2], and Rombe[3]

[1,2,3]Department of Electrical Engineering, Faculty of Engineering, Universitas Kristen Indonesia Paulus Makassar, South Sulawesi, Indonesia. 90245

Email: hestika@ukipaulus.ac.id; matiussau@ukipaulus.ac.id; rombe87@yahoo.co.id

Abstract. This study aims to determine the distribution of electrical power by considering the addition of generating capacity and the addition of networks using ETAP applications. The simulation results show that for the first scenario, the addition of generating capacity is carried out at Barru Bus without the addition of transmission networks, the voltage on bus 36 (Barawaja) stabilizes from 0.92 pu to 1.02 pu, with losses of 17.218 MW and 30.171 MVAR. This shows that power losses have decreased from 18.853 MW and 41.291 MVAR or decreased by 1.635 MW from power losses before the addition of generation in Barru, and the second scenario is the addition of generation at Bontoala and the addition of an express network. The system voltage that previously dropped became stable at the existing tolerance limit. Besides that, there were power losses when adding a generator at the Barru Bus. Thus, the distribution of electricity will be better when there is an additional generation at Bontoala Bus and the addition of an express network in the Sulselbar system.

1. Introduction

The magnitude of the need for electrical energy with different types of loads (varies) causes more complex network structure and operation of a power system. This can cause problems in meeting the required requirements [1,2].

The development and utilization of electrical energy are based on comprehensive and integrated energy policy by taking into account the increasing needs and existing generation capabilities. Generation capacity development can be done by combining Diesel with alternative energy such as Solar Power [3,4].

Planning for generation development in the Sulselbar electricity system 2015-2025 for conditions in 2017 is located in Jeneponto, Barru, Bakaru and Sengkang. This condition requires a generation placement scenario so that electricity supply to consumers becomes reliable and meets system security standards in terms of power losses and system stability. Electricity needs in Indonesia, especially in South Sulawesi, continue to increase along with development in various sectors. This has an impact on the electrical energy crisis resulting in rotating blackouts; however, these conditions have been minimized by the addition of generating capacity.

The study of power flow can be analyzed regarding the continuity of the distribution of electrical power, security and system stability. A system is said to be continuity in its distribution if the power supply from generation to the load has not failed. This depends on the ability of a plant and the total load to be supplied. Talking about continuity cannot be separated from a power system reliability which is expressed by the ability of an electric power system to supply electrical energy to a continuous load within a certain period of time and the electrical energy must be available when needed. In connection with the continuity of power distribution, it is necessary to analyze the security and stability of the system.

1.1. Security Approach

A power system can be said to have a safe system if there are no violations of the security constraints; in this case, it can be a static security constraint or dynamic security constraints. Security constraints that are considered in this study are static security constraints when peak loads occur [2].

Static security constraints are operating limits that must be met in the operation of the power system. The obstacles that must be met are:

a. Voltage. Operating limits that must be met with stress in each load bus (PQ bus) are: $V_i^M \leq V_i \leq V_i^M$ with V_i^m and V_i^Meach of which is the minimum voltage and maximum voltage allowed on the bus i.
b. Power flow in the lines. Operating limits that must be met by the power flowing through the T line are:
   $-T_L \leq S_T \leq T_L$ with S_T the total power flowing in the T line while T_L the thermal operation limit of the T line
c. Active power generation. Operating limits for active power generation are: $P_k^m \leq P_k \leq P_k^M$with P_k^m and P_k^M each of which is the minimum power and maximum power of the generator on the bus-k.
d. Reaktiv power generation. Operating limits for reactive power generation are $Q_k \leq Q_k \leq Q_k^M$ with Q_k^m and Q_k^M each of which is the minimum power and maximum power of the generation on the k-bus.

1.2. System stability

A system is said to state if the voltage on each bus does not exceed or less than the existing standard that has a nominal voltage not exceeding the voltage limit set at ± 5% of the normal voltage. To find out that the system is in a stable condition, a power flow study is needed. Voltage control of the system can be done by installing capacitors [5,6]. The results of this power flow will show buses that can experience an increase or decrease in voltage from the existing tolerance limits.

2. Research methodology

The implementation of the research was carried out in several stages

a. Load data, generating data, line data
b. Using the Newton Raphson Method for power flow analysis
c. Analyze simulation results in determining the stability and security of the system
d. Repeating the point b to get a stable system condition and security in terms of the voltage and power loss

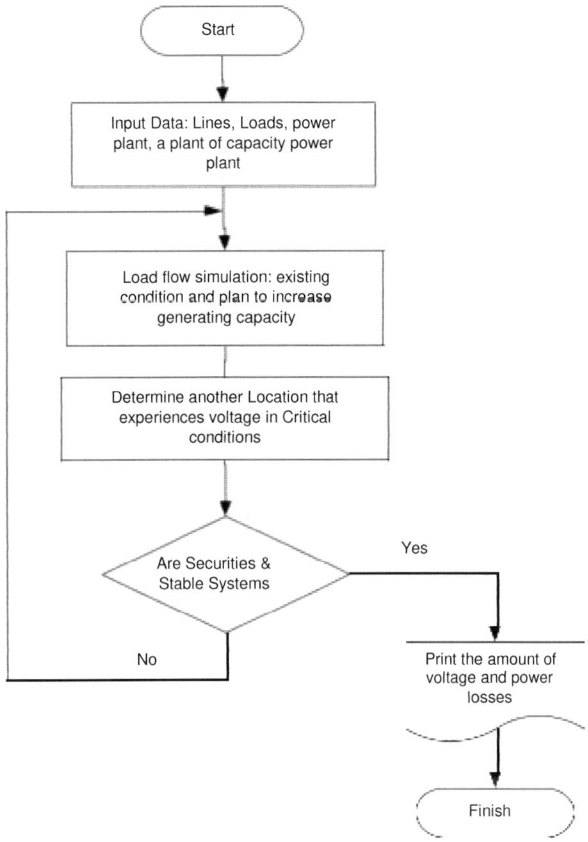

Figure 1. Research Flow Chart

3. Results And Discussion

3.1. Scenario Before Adding Generating Capacity
The simulation results of the power flow before the addition of the generator shows that the total peak load is 1050 MW and the total capable capacity is 1240 MW in accordance with the operating data of 2017, and the total losses are 18,853 MW and 41,291 MVAR. This value is still below the tolerance limit for power losses of 10% of the total power generated.

3.1.1. Analysis of system security
Voltage values indicate that on bus 33 and bus 36, the voltage is below the allowable tolerance limit of ± 5% of the nominal voltage. The power flow in each line shows that the power that is passed to each line is still within the tolerance limit that is dependent on the maximum capacity the power is passed. Active power and reactive power generated by each generator is still at the limit of the capable power, meaning that the engine is operated in normal conditions because it has not exceeded the power limit of the generator [7].

3.1.2. System Stability Analysis
From the simulation results, obtained the magnitude of the voltage experienced a drop on bus 36 (Brawaja) and bus 33 (Bontoala) at night peak load conditions.

IOP Conf. Series: Materials Science and Engineering **885** (2020) 012001 doi:10.1088/1757-899X/885/1/012001

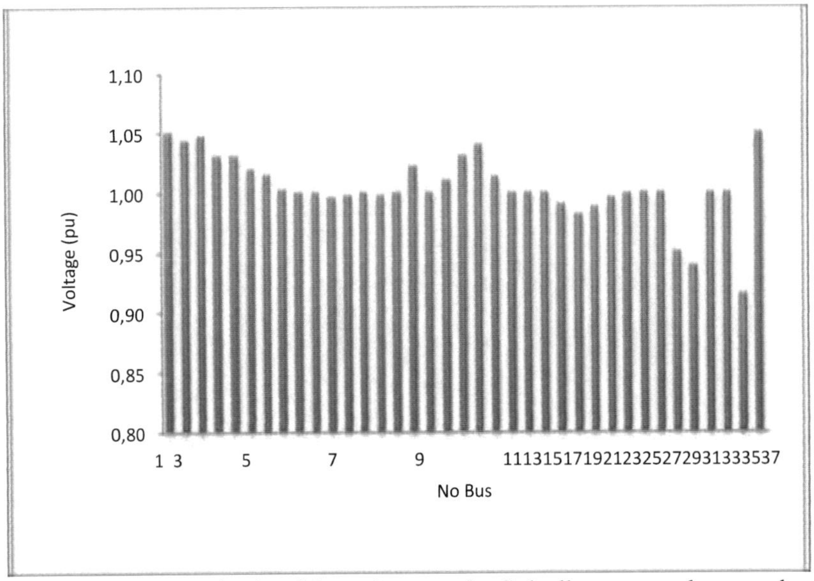

Figure 2. The magnitude of the voltage to the Sulselbar system bus number

3.2. Scenario when adding capacity on the Barru Bus without the Express Network

The generation increase in Barru is 50 MW, the condition of bus 32 and bus 33 is still below the tolerance limit, however, the voltage on bus 36 (Barawaja) has stabilized from 0.92 pu to 1.02 pu, with losses of 17,218 MW and 30,171 MVAR. This shows that power losses have decreased from 18,853 MW and 41,291 MVAR or decreased by 1,635 MW (8.7%) from power losses before the addition of generation in Barru. The power flow in each line shows that the power passed on each line is still within the capacity of the existing line.

Active power and reactive power generated by each generator are still at the limit of the capable power of each generator, meaning that the engine is operated under normal conditions because it has not exceeded the power limit of the generator. From the simulation results, obtained the amount of voltage experienced a drop on bus 32 (Tallo lama 70 kV) and bus 33 (Bontoala) in peak load conditions at night, this can be seen in Figure 3

IOP Conf. Series: Materials Science and Engineering **885** (2020) 012001 doi:10.1088/1757-899X/885/1/012001

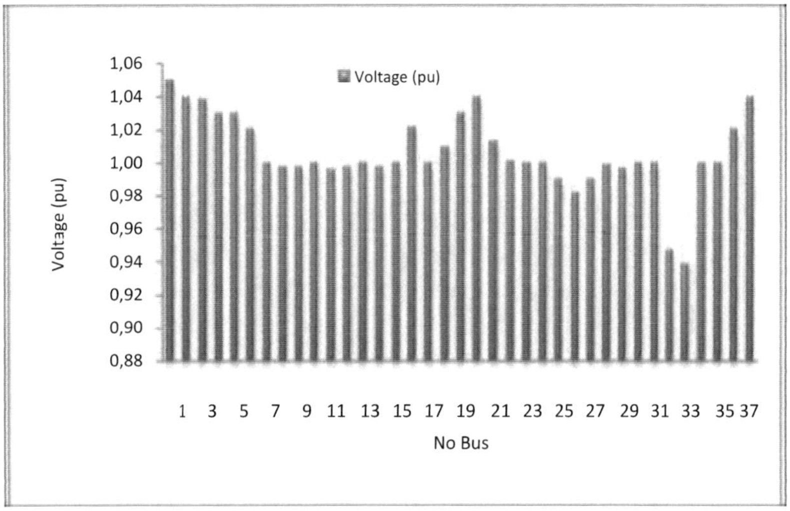

Figure 3. The magnitude of the voltage against the Sulselbar system bus number after the addition of 50 MW capacity

3.3. Capacity addition scenarios at Bontoala Bus and the addition of the Express Network

When the location of an additional generation is placed on bus 33 (Bontoala bus) and the addition of the express network middle line (Bus Sidrap with Sungguminasa Bus), the system voltage that previously dropped becomes stable at the existing tolerance limit.

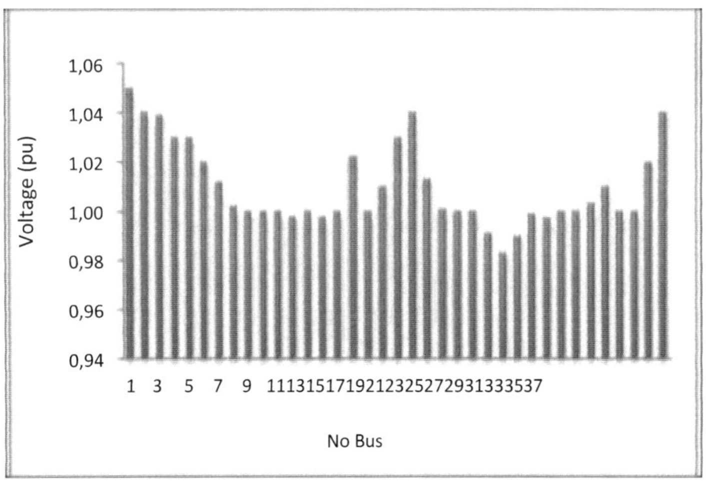

Figure 4. The magnitude of the voltage against the Sulselbar system bus number at Optimum conditions

The three results of the sucked simulation above can be seen in Figure 4 comparison of the system voltage in the three conditions.

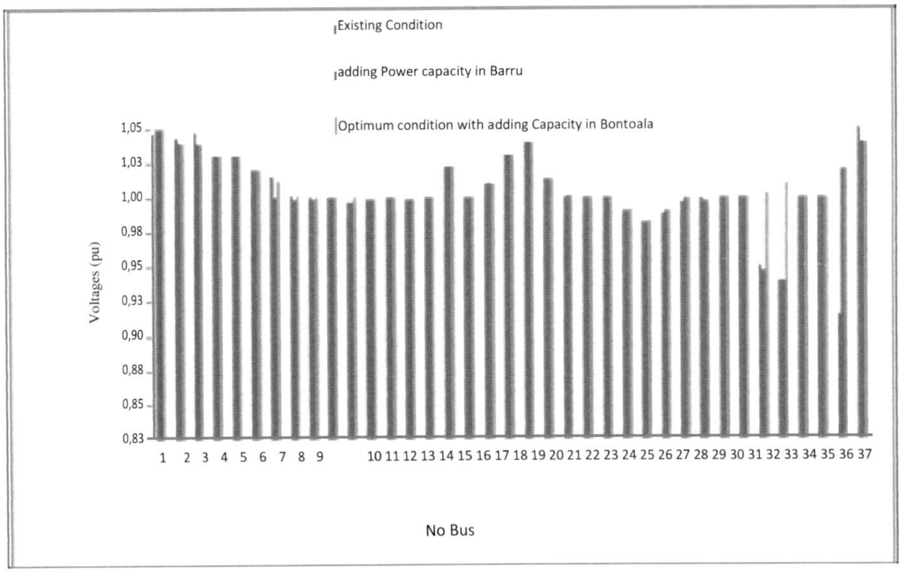

Figure 5. System voltage comparison for three conditions.

The amount of existing power losses is 15,682 MW and 11.68 MVAR. This value is less than the power losses when adding a power plant on bus 7 (Barru).

4. Conclusion

Based on the simulation results, it can be concluded that the quarantine of the distribution of electrical power to consumers in the Sulselbar system is better (increasing) when the addition of generating capacity compared to the conditions before the addition of generation, Electricity and stability of electricity in the Sulselbar system is better than before the addition of the generator both in terms of voltage and power losses that occur. The location of the optimum generation addition is shown at the Bontoala location (bus 33) in simulation in terms of system security and stability.

Acknowledgment

Thanks to the Ministry of Research, Technology and Higher Education Directorate General of Learning and Student Affairs and the Directorate of Research and Community Development that funded this Research.

Reference

[1] Feinberg E A and Genethliou D 2005 Load forecasting *Applied mathematics for restructured electric power systems* (Springer) pp 269–85
[2] Xifan W and McDonald J 1994 Modern power system planning *London Mcgraw-hillb. Co.*
[3] Sau M, Patoding H E and Kasa A 2019 Model design of solar-diesel hybrid power system with Homer Pro **4** 103–8
[4] Megantoro P, Prabowo I E and Shomad M A 2020 Development of Maximum Power Point Tracking Solar Charge Controller for 120 Volt Battery System at Pandansimo Hybrid Power Plant *J. Phys. Conf. Ser.* **1471**
[5] Wood A J and Wollenberg B 1996 Power generation operation and control–Second Edition–John Wiley & Sons
[6] Setiadi H, Krismanto A U, Mithulananthan N and Hossain M J 2018 Modal interaction of power systems with high penetration of renewable energy and BES systems *Int. J. Electr. Power Energy Syst.* **97** 385–95
[7] Saadat H 1999 *Power system analysis*

IOP Conf. Series: Materials Science and Engineering **885** (2020) 012002 doi:10.1088/1757-899X/885/1/012002

Application of fuzzy C-means algorithm for basic school clustering in Samarinda city based on minimum educational service standard indicators

Karyo Budi Utomo[1], Arbain P[2], and Suminto[3]

[1]Information and Technologi Enginering, Politeknik Negeri Samarinda, Indonesia
[2]Electrical Enginering, Politeknik Negeri Samarinda, Indonesia
[3]Bussiness Administration, Politeknik Negeri Samarinda, Indonesia

Email: kbu@polnes.ac.id

Abstract. Quality of service in schools determines the level of success of education service. The minimum service standard or better known as (SPM) in education is one of the references for evaluating the performance of education services. There are six categories in SPM of education, namely facilities and infrastructure, educators and education staff, curriculum, education quality assurance, education assessment, and school management. Each category has an achievement indicator to be used as a reference to assess the achievement that has been done by each school. School grouping based on the Minimum Service Standards (SPM) of Education is expected to be a reference for the government in making policies and references for Education organizers to know their position and then take strategic steps to enter the best groups. One of these groupings is achieved using the Fuzzy C-Means clustering method which is a non-hierarchical Clustering algorithm. The results of the study are expected to obtain a model that stakeholders can use in conducting elementary school groupings based on minimum service standard (SPM) in Education which are divided into 3 (three) categories, namely: Less, Medium, and Good.

1. Introduction

The 1945 Constitution Article 31 paragraph (1) mandates that every citizen has the right to education. The 1945 Constitution Article 31 paragraph (2), every citizen is required to attend basic education and the government is obliged to finance it [1]. To improve the quality of adequate education, a Minimum Service Standard (SPM) is needed. Regulation of the Minister of Education and Culture of the Republic of Indonesia, Number 23 of 2013 Article 6 paragraph (1), Education Minimum Service Standards (SPM) education, is a reference in program planning and budgeting achievement of targets in each regency/city [2].

Data from Samarinda City Education Office in 2015 showed that there were 184 public primary schools in Samarinda. The government has declared that every school must meet the SPM. There are 6 (six) categories of service types that become the minimum standard reference, namely: 1) Facilities and Infrastructure, 2) Educators and Education Personnel, 3) Curriculum, 4) Educational Quality Assurance, [1] Educational Assessment, and 6) School Management [2]. Each category has an achievement indicator to be used as a reference to assess the achievement of MSS that have been carried out by each school.

A large amount of data can slow down the clustering process of primary schools with good minimum service standards. The clustering of public elementary schools will be divided into three clusters, namely: poor, medium, and good.

The method used is fuzzy c-means, fuzzy c-means is a data clustering technique where the existence of each data point of a cluster is determined by the value of membership. The membership value will include real numbers at intervals of 0-1 [3,4]. The advantage of the fuzzy c-means method is that it is capable of grouping irregularly distributed data [5,6]. The cluster formed is influenced by the input of several inputs in the fuzzy c-means process, such as the number of iterations, ranks, and the smallest error, but is not affected by the objective function and the initial iteration [5].

2. Review of Theory

2.1. Minimal Service Standards (SPM)
According to the Regulation of the Minister of Education and Culture of the Republic of Indonesia Number 23 of 2013, Article 6 [2]:
(1) SPM education is a reference in program planning and budgeting the achievement of targets in each district/city.
(2) Program planning and budgeting for education Minimum service Standard as referred to in paragraph (1) shall be carried out in accordance with established technical guidelines/standards.
(3) The target of achieving basic services in education must be achieved by the end of 2014.
SPM education has six categories of service types, as described above.

2.2. Data Mining
Data mining is a method of processing data to find hidden patterns from the data. The results of data processing with this data mining method can be used to make decisions in the future. Data mining is data processing on a large scale, so data mining has an important role in the fields of industry, finance, weather, science and technology [7]. Data mining can also be done in various types of databases and information storage, but the type of pattern that will be found is determined by various data mining functions such as class/concept description, association, correlation analysis, classification, prediction, cluster analysis and others. Data mining uses the application of certain algorithms to extract patterns from data. Where this process will automatically look for simple patterns of large data using a particular analysis. Data mining also uses sophisticated mathematical algorithms to segment data and evaluate the possibility of some results set by the user [8].

2.3. Clustering
Clustering is the process of grouping objects based on information obtained from data that explains the relationships between objects with the principle of maximizing the similarity between members of one class and minimizing similarities between classes or clusters [9]. There are two stages that must be done in cluster analysis, namely: 1) Deciding whether the number of clusters is determined or not, and 2) Determine the algorithm to be used in clustering. To decide how many clusters to be formed [10]. In conducting clustering analysis can choose one of two approaches, namely 1) Hard Clustering or 2) Soft Clustering. The choice of the approach used depends on the type of data to be grouped. Hard clustering is used if the data is crisp, while soft clustering is used when the data is fuzzy [11].

2.4. Fuzzy C-Means
The Fuzzy C-Means algorithm was first proposed by Dunn in 1973 and then updated by Bezdek in 1981. This algorithm is one of the most popular soft clustering techniques using a data point approach where the cluster center point will always be updated according to the membership value of existing data and besides the fuzzy c-means algorithm is also an algorithm that works using the fuzzy model so that it allows all data from all group members to form with different degrees of membership between 0 and 1 [12][13]. Following is the elaboration of the FCM algorithm [14]:

- Enter data to be clustered in the form of a matrix X size n x m (n = number of sample data and m = number of variables per data). Xij = i-th sample data (i = 1,2, ..., n), the j-th variable (j = 1,2, ..., m).
- Determine:
  - Number of clusters to be formed (c)
  - The weighting device (w)
  - Maximum iteration (MaxItr).
  - The smallest error
  - The initial objective function (P0 = 0)
  - Initial iteration (t = 1)

- Generating the initial partition matrix Unxc = [μ_{ik}], μik is a random number representing a degree of membership
- Calculate the center of the k-th cluster (Vkj) with k = 1,2, ..., c; and j = 1,2, ..., m as follows:

$$V_{kj}=\frac{\sum_{i=1}^{n}(\mu_{ik})^{w}.X_{ij}}{\sum_{i=1}^{n}(\mu_{ik})^{w}} \quad (1)$$

- Calculate the objective function on the t-iteration, Pt, which describes the amount of data distance to the center of the cluster.

$$V_{kj}=\sum_{i=1}^{n}\sum_{k=1}^{c}\left(\left[\sum_{j=1}^{m}(X_{ij}-V_{kj})^{2}\right]\mu_{ik}{}^{w}\right) \quad (2)$$

Where:
Pt = objective function;
Xij = element X row i, column j;
Vkj = cluster center

- Fix the degree of partition matrix membership:

$$\mu_{ik}=\frac{\left[\sum_{j=1}^{m}(X_{ij}-V_{kj})^{2}\right]^{\frac{-1}{w-1}}}{\sum_{i=1}^{n}\left[\sum_{j=1}^{m}(X_{ij}-V_{kj})^{2}\right]^{\frac{-1}{w-1}}} \quad (3)$$

Where:
i = 1,2, ..., n
k = 1,2, ..., c
Xij = i-th sample data, j-th variable
Vkj = k-cluster center for the jth variable
w = weighting rank

- Check the stop condition

If you know (t> MaxIter) then stop
If not: t = t + 1, repeat step 4

2.5. Cluster Validity Testing

The results of clustering were tested for their level of validity by using the Modified Partition Coefficient measurement method to measure the level of cluster validity using values of μic membership degrees [15]. The main purpose of cluster validity is to evaluate the quality of the cluster and determine how well the data is represented by the selected cluster [16]. MPC is an improvement of the Partition Coefficient (PC) method. The PC method proposed by Bezdek in [17] is defined by Equation (4).

IOP Conf. Series: Materials Science and Engineering **885** (2020) 012002 doi:10.1088/1757-899X/885/1/012002

$$\text{PC(c)}=\frac{1}{N}\sum_{i=1}^{c}\sum_{j=1}^{N}(\mu_{ij}{}^{2}) \quad (4)$$

Where:
c = number of clusters
N = amount of data
μij = degree of k-j data membership in the i cluster
PC (c) = PC index value in the cth cluster

The partition coefficient tends to experience a monotonous change in various values of c (number of clusters) [18]. Modifications from the PC (Modified Partition Coefficient / MPC) index can reduce these monotonous changes. The MPC value is within the limit of $0 \leq$ PC (c) ≤ 1. Generally, the most optimal number of clusters is determined from the highest MPC value (max2 max c≤n-1 PC (c)). Following is the MPC method algorithm: [18].

$$MPC\ (c) = 1 - \frac{c}{c-1}\,(1 - PC(c)) \quad (5)$$

Where:
c = number of clusters
MPC (c) = MPC index value in the c-cluster

3. Research Methods

The research methods that will be carried out can be grouped into several sections, including:

3.1. Initial research

Studying matters related to the research topic. The main parts that need to be studied and explored include minimum education service standards (SPM), the basics of clustering, as well as an understanding of the Fuzzy C-Means method.

3.2. Data Collection

The data collected for this study is an indicator of the achievement of SPM (Minimum Service Standards) at Elementary Schools sourced from the Samarinda City Education and Culture Office

3.3. Data analysis

Data collected, then analyzed based on research needs, among others, by observing the data structure and looking for patterns of data adjustment based on research needs. In each minimum service standard (SPM) category, it has several achievement indicators; the clustering process is carried out based on the SPM category so that the values on each achievement indicator are added up to then become a category value.

3.4. Application of clustering

At this stage, the clustering process is carried out using the fuzzy c-means method, the steps in the clustering process as previously described.

3.5. Evaluation

At this stage, the results of clustering are tested for validity using the Modified Partition Coefficient measurement method to determine the optimal number of clusters. Cluster segmentation results obtained from the results of the Fuzzy C-means clustering algorithm are compared with other period datasets to measure the performance of the mining process.

4. Results and Discussion

As explained earlier, the number of public primary schools to be clustered in 184 schools and the number of clusters is 3. The test in this study was carried out with a number of scenarios that were carried out by changing the input on several inputs in the fuzzy c-means process. The input that is changed is the maximum iteration, rank, and smallest error. All inputs are carried out with the same number of clusters, the number of clusters is an assessment of the results of grouping, which in this study uses three clusters, namely: good, moderate and less.

4.1. Testing based on maximum iteration

Testing the maximum iteration is done with the aim to determine the effect of iteration on the results of the cluster. In this test, the number of clusters is 3, with rank weights is two and the smallest error is 10-7 (0.0000001), and the initial objective function is 0. The testing process is carried out 10 times, by taking random iterations which include iteration 1; 5; 10; 15; 20; 30; 40; 45; 50; and iteration 58.

Table 1. Comparison of test results with different maximum iterations

	Result of Cluster		
Iteration number	Good	Moderate	Less
1	38	77	69
5	47	69	68
10	68	61	55
15	75	64	45
20	75	67	42
30	73	66	45
40	73	66	45
45	73	66	45
50	73	66	45
58	73	66	45

Table 1 shows that from 10 scenarios, the maximum iteration data collection, at the maximum iteration 1; 5; 10; 15; and 20 shows the data distribution of cluster membership is still not definitive or still changing with each other. At a maximum of 30 iterations; 40; 45; 50; and 58 indicate that the cluster membership distribution level is definitive, and there is no change to the smallest error limit of 10-7 on the 58th iteration.

4.2. Testing based on rank

Testing of rank values is carried out in order to determine the effect of rank on cluster results. In this test, the number of clusters is 3, with a maximum iteration is 100, the smallest error is 10-7 (0.0000001), the initial objective function is 0 with the initial iteration 1. The testing process is carried out 10 times, with rank values including, 1; 2; 3; 4; 5; 6; 7; 8; 9; 10.

Table 2. Comparison of test results with different ranks

	Result of *Cluster*		
Instrument	Good	Moderate	Less
1	0	109	75
2	73	66	45
3	75	68	41
4	77	78	29
5	9	126	49
6	27	79	78
7	37	97	50
8	37	99	48
9	37	100	47
10	56	66	62

Table 2 shows that of the 10 data collection scenarios based on rank, starting from rank 1 to rank 10, the value of the members in each cluster varies.

4.3. Testing Based on the Smallest Error

Testing the smallest error is done with the aim to determine the effect of the smallest error on the results of the cluster. In this test the number of clusters is 3, with a maximum iteration is 100, the initial objective function is 0 and with an initial iteration one and with a rank value 2. The testing process is carried out seven times, including the smallest error 0.1; 0.01; 0.001; 0.0001; 0.00001; 0.000001; and 0.0000001.

Table 3. Comparison of test results with the smallest of different errors

The smallest error	Result of Cluster		
	Good	Moderate	Less
0.1	75	67	42
0.01	73	66	45
0.001	73	66	45
0.0001	73	66	45
0.00001	73	66	45
0.000001	73	66	45
0.0000001	73	66	45

Table 3 shows the seven scenarios for taking the smallest error data, the smallest error value of 0.1 occurs in the 20th iteration, then in the smallest error condition 0.01 until the smallest error, 0.0000001 condition distribution of member values in each cluster does not change.

4.4. Clustering Validity Testing

In testing the validity of clustering, the number of clusters is 3, with a maximum iteration of 100, the initial objective function is 0, with an initial iteration of 1 and with a rank of 2. and with the smallest error of 10-7 (0.0000001). The test is carried out using the Partition Coefficient method and the Modified Partition Coefficient method. The testing mechanism is carried out with variations of 10% data (18 data), 25% (46 data), 50% (92 data), 75% (138), and 100% (184) of a total of 184 data. With random data collection.

Table 4. Testing Clustering Validity with Comparison of data variations

Number of Data set	Method of Test	
	PC	MPC
100 % (184)	0.4797	0.2195
75% (138)	0.5152	0.2729
50% (92)	0.5252	0.2879
25% (46)	0.5303	0.2955
10% (18)	0.5764	0.3646

Table 4 shows with all the scenarios the amount of data variation, in the Partition Coefficient method, the validity of the clustering has a tendency of optimal validity. All data variations worth more than 50% are close to 1. While the Modified Partition Coefficient method of clustering validity has a tendency of suboptimal validity because all data variations are worth less than 50% close to 1 (one), meaning that all data variation values are closer to 0. In this test, both using the Partition Coefficient method and the Modified Partition Coefficient method show that with a smaller number of data sets, the potential for validity is optimal.

4.5. Analysis of Testing Results

Based on testing to determine the performance of the fuzzy c-means algorithm for clustering the following results are obtained:

- Input in some inputs in the fuzzy c-means process affects the output of the cluster. The scanned input to be changed is the maximum iteration, rank and smallest error. The maximum value of iteration and smallest error with different input scenarios affects the members of each cluster but still produces the number of clusters as planned. Whereas in rank testing scenarios in addition to affecting the number of members of each cluster also affect the number of clusters, it happens when the rank value is 1.
- Testing based on the maximum iteration, the results vary. The maximum iteration 1 was 38.3 good schools, 77 schools and 69 schools were lacking. The number of iterations is too small, so the results are not right because the center of the cluster has not moved to the right position. Likewise, iterations in 5, 10 and 20 still obtain results that are varied in each cluster, in the iteration not yet obtained maximum results because the iteration is too little and does not meet the conditions according to the rules of the fuzzy c-means algorithm (t> MaxIter). Whereas in the 30, 40, 45, 50 and 58 iterations, the number of clusters and the number of members of each cluster is the same, namely the good category of 73 schools, 66 schools, and 45 schools less; this is because the iteration will stop if the objective function value during iteration to t and the objective function during the previous iteration (| Pt - Pt-1 | <ξ) the difference is smaller with the smallest expected error. So the exact center position of the cluster center is between the 20th and 30th iterations so that in more than 30 iterations, the same results will be obtained.
- Testing is based on the number of ranks, from 10 test scenarios, on testing with rank 1 produces 2 clusters even though it is conditioned by using 3 clusters. Whereas the testing with rank values 2 to 10 produces 3 clusters with different members of each cluster.
- At the test, based on the smallest error, the results vary. Only the smallest error value of 0.1 has a different number of members in each cluster, which is a good category of 75 schools, 67 schools and 42 schools less. While the smallest error is 0.01 to 0.0000001, the number of members in each cluster has similarities; that is, 73 schools are good, 66 schools are moderate and 45 schools are less.
- In testing the validity of clustering with 3 cluster scenarios tested based on the number of varied data sets divided into five variations of the amount of data, namely 10%, 25%, 50%, 75% and 100% of the total data of 184 schools, and conducted by using the Partition Coefficient and Modified

Partition Coefficient methods. Testing uses the Partition Coefficient method; the validity of the clustering has a tendency for optimal validity. All data variations worth more than 50% are close to 1. While the Modified Partition Coefficient method of clustering validity has a tendency of suboptimal validity because all data variations are worth less than 50% close to 1 (one), meaning that all data variation values are closer to 0. In this test, both using the Partition Coefficient method and the Modified Partition Coefficient method show that with a smaller amount of data set, the potential validity will be increasingly optimal.

5. Conclusions

Based on the results of the discussion, the following conclusions are obtained:

- The fuzzy c-means method can be applied in clusters, variations in the input of several inputs in addition to affecting the value of each cluster member also affect the number of clusters formed
- The rank value affects the number of clusters, even though the number of clusters has been determined as many as 3 clusters, in the rank value 1, there are 2 clusters, which are medium and low clusters.
- The more maximum iterations and the smaller the error value of the cluster center will be in the right position
- Test the validity of the Partition Coefficient method, cluster validity has a tendency for optimal validity, whereas the Modified Partition Coefficient method has cluster validity that tends to be less than optimal.
- The less amount of data, both using the Partition Coefficient method and the Modified Partition Coefficient method, the potential for validity will be more optima

Reference

[1] Anon 1945 *The 1945 Constitution of the Republic of Indonesia* (Republic of Indonesia)
[2] Kemdikbud 2013 *[2]. Regulation of the Minister of Education and Culture Number 23 of 2013 concerning Amendments to the Regulation of the Minister of National Education Number 15 of 2010 Minimum Service Standards for Basic Education in Regencies / Cities*
[3] Megawati N, Mukid M A and Rahmawati R 2013 Segmentasi Pasar Pada Pusat Perbelanjaan Menggunakan Fuzzy C-Means (Studi Kasus: Rita Pasaraya Cilacap) *J. Gaussian* **2** 343–50
[4] Priandini N, Zaman B and Purwanti E 2017 Categorizing document by fuzzy C-Means and K-nearest neighbors approach *AIP Conference Proceedings* vol 1867 (AIP Publishing LLC) p 20012
[5] Ahmadi A and Hartati S 2013 Application of Fuzzy C-Means in Decision Support Systems for the Determination of PNPM-MPd Community Direct Assistance Recipients (BLM) (Case Study of PNPM-MPd Ngadirojo District Pacitan) mencanangkan Program Nasional Pemberdayaan Mas *Berk. MIPA* **23** 264–73
[6] Kustiyahningsih Y and Suprajitno H 2017 Development of an interval value fuzzy number based on MCGDM model by hybrid AHP and TOPSIS methods *2017 International Conference on Sustainable Information Engineering and Technology (SIET)* (IEEE) pp 62–7
[7] Ong J O 2013 Implementasi algoritma k-means clustering untuk menentukan strategi marketing president university
[8] Ndehedehe C, Ogunlade S and Akwaowo E 2013 Spatial image data mining using K-Means analysis: a case study of Uyo Capital City, Nigeria *Int. J. Adv. Res.* **1** 6–15
[9] Agusta Y 2016 Clustering
[10] Kusumadewi S and Purnomo H 2010 Aplikasi Logika Fuzzy untuk pendukung keputusan *Yogyakarta Graha Ilmu* 33–4
[11] Gan G, Ma C and Wu J 2007 *Data clustering: theory, algorithms, and applications* vol 20 (Siam)
[12] Bora D J, Gupta D and Kumar A 2014 A comparative study between fuzzy clustering algorithm and hard clustering algorithm *arXiv Prepr. arXiv1404.6059*
[13] Sanmorino A 2012 Clustering batik images using fuzzy c-means algorithm based on log-average luminance *Comput. Eng. Appl. J.* **1** 25–31
[14] Andriyani T ., Linawati L and Setiawan A 2013 Application of Fuzzy C-Means (FCM) Algorithm in Determining the Location of Establishment of the Salatiga PDAM Water Payment Counter in Proceedings of the SWCU *Sci. Educ. Sci. VIII Salatiga*
[15] Amirah M M A, Widodo A W and Dewi C 2017 Pengelompokan Lagu Berdasarkan Emosi Menggunakan Algoritma Fuzzy C-Means *J. Pengemb. Teknol. Inf. dan Ilmu Komput. e-ISSN* **2548** 964X
[16] Sandhir R P, Muhuri S and Nayak T K 2012 Dynamic fuzzy c-means (dFCM) clustering and its application to calorimetric data reconstruction in high-energy physics *Nucl. instruments methods Phys. Res. Sect. A Accel. spectrometers, Detect. Assoc. Equip.* **681** 34–43
[17] Maimon O and Rokach L 2005 Data mining and knowledge discovery handbook
[18] Nugraheni Y 2013 Data Mining Using Fuzzy Method for Customer Relationship Management in Retail Industry *Lontar Komput. J. Ilm. Teknol. Inf.*

The Role of Big Data in Improving E-Learning Transition

Karim Moharm[1], and Muhammed Eltahan[2,3,4]

[1]Electrical Engineering, Alexandria University, Egypt
[2]Aerospace Engineering Department, Cairo University, Cairo, Egypt
[3]Now at Institute of Bio-Geosciences (IBG-3, Agrosphere), Forschungszentrum Jülich GmbH, 52425 Jülich, Germany
[4]Centre for High-Performance Scientific Computing, Geoverbund ABC/J, 52425 Jülich, Germany.

Email: karim.i.moharm@gmail.com, m.eltahan@fz-juelich.de

Abstract. Educational organizations are operating in a dynamic competitive environment. Online learning, education, emerging technologies, and the number of learners are growing very fast and produce big data that contain meaningful information to be extracted and processed. The idea of applying the big data in e-learning supports developing and analyzing the students as well as the educational organization's stakeholders. A big data framework architecture, methods, and tools in the e-learning platform are introduced. The paper studies briefly the opportunities of big data deployment in the educational sector and the value that the students and organizations gain. However, it also highlights the challenges and future research directions as the tool's selection and value extraction from complex educational data sets.

1. Introduction

Educational organizations are operating in a very dynamic and highly competitive environment. The educational institutes have to have precise decisions to sustain its competitiveness. The traditional data warehouses in the educational field are going beyond the capabilities of traditional processing techniques for structured, semi-structured, and unstructured data sets and extracting the value beyond data sources as videos, audio, etc.

E-learning concept first aroused in 1990 and had no standard definition [1]. However, the American Society for Training and Development suggests that anything delivered using electronic technology for learning purposes [2,3].

The main obstacle for e-learning spreading is the active need for big educational data processing to have decisions based on analysis and not only intuitive or experience. Big data generated by learners, teachers, universities, institutions, etc. offer interesting applications in the educational field. Big data is described by the six V's (volume, velocity, variety, veracity, value, and volatility). The volume refers to that big data processes large amounts of data that was impossible to process via traditional techniques. Data from thousands of learners who are enrolled in hundreds of courses in several organizations is a high-volume data issue. The velocity refers to the rate of obtaining the data is very high. For example, the data processing of a learner interacting with an educational platform that stores and analyses the streaming data gathered from the learner. The variety refers to the information that can be structured (learner profile data stored in rational databases), semi-structured (emails), or unstructured (free text, PDFs, videos, images, web-pages, RSS feeds, Facebook feeds, etc.). The veracity represents the truthfulness of the data and ensuring the data is reliable. The value indicates that the processed data is useful to end-users. The volatility shows that gathered data importance changes over time.

Institutes are gathering large amounts of data for each student to track performance and progress. Big data employment in the educational sector results in more academic achievements, less drop-out rate of students through continuous feedback and course adaptation, and effectiveness for the student and the teachers as well. The effective performance analysis of the learner can also help in an effective recommendation for the student, along with the behavioral and historical data.

The remaining of this paper is organized as follows. Section 2 presents the big data sources in the educational sector and section 3 studies the background of big data tools and techniques. Section 4 discusses the proposed big data architecture layered in the e-learning, section 5 discusses big data opportunities and related work while section 6 proposes the challenges of implementing the big data in the education sector. Finally, the conclusion is drawn in section 7.

2. E-Learning Data Sources

Data in educational sectors is generated by several sources. Educational data sources include the academic data sets collected by the institutions, online public available data sets, and the communication data between different e-learning platforms. Other sources can be smartphones, web, media, audio, and blogs. Sensors such as wearable electronics, student attention sensors, and student electrical activity measurement sensors like NeuroSky [4] head tilt angle for learner engagement measurement are generating important data about the education process. Traditional data sets gathered by assessments and class drop rates are important data sources to be processed and analyzed. Online data sources through simulation and educational games [5] can contain important information about the learners. Educational environment interaction logs include MOODLE logs [6] which contain data about user ID, number of log-in trials, total time spent, user clicks, lessons progress, course information, and dated accessing of course materials [7]. Personal data sources include social media data such as university advertisements, online forums, student health status (disabled, diseases, medical reports, etc.), memory and concentration levels measurements, address, and financial situations. Historical data about the learner which involves Daily activities as workshop attendance, extracurricular activities as competitions participation, seminars participation, previous detailed grades, and previous schools' history [8]. Classroom specific data such as the classroom cleaning status and bus routes are hiding big data that affect student learning [9].

3. Big Data Techniques & Technologies

3.1. Tools

Different tools are used for big data analysis, including data mining, analysis, and processing. The most used tools in the big data processing of e-learning are Apache Hadoop and Apache Spark. Apache Hadoop [10,11] is an open-source ecosystem for big data. It can be used as a cloud service. It contains several components such as Hadoop Distributed File System (HDFS), Yet Another Resource Negotiator (YARN), Pig and MapReduce. HDFS divides and replicates the data across multiple nodes. MapReduce [12] is responsible for mapping and reducing processes that are splitting tasks into chunks for hardware resources processing. Then, combining the outputs of the processed chunks of data. MapReduce programs can be programmed using Apache Pig and Pig Latin programming language [13]. Apache YARN [14] can replace the MapReduce process. Apache Mahout [15] is a scalable open-source machine learning library used for collaborative filtering, clustering, and classification. Mahout can be used with Hadoop. Apache Spark [16–18] is an open-source computing platform based on data analysis and computational memory capabilities for real-time and batch data processing. It is used by Yahoo, eBay, etc. It supports machine learning libraries as MLlib and visualization tools like GraphX.

3.2. Techniques

Several analytical methods are applied for the big data analysis and discovering value behind the data as the association rule mining, regression, classification (as decision trees, ANN, Bayesian networks) and statistical methods.

4. Design of Layered Big Data Framework in E-learning

Several analytical methods are applied for the big data analysis and discovering value behind the data as the association rule mining, regression, classification (as decision trees, ANN, Bayesian networks) and statistical methods. According to [19,20], the general framework for big data analytics in the e-learning can be through the following layers:

- Presentation layer: for visualization and statistical views. Output of the layer can be reports and graphs.
- Analysis layer: through MapReduce algorithms of the big data.
- Data pre-processing layer: data merging and data cleaning processes.
- Big data storage layer: through HDFS and HBase of heterogeneous data as learning platform data, card records, course videos, etc.

5. Opportunities

The output of big data analysis can be statistical analysis or visual presentation for the educational process and progress. Deploying the big data in the learning sector can enhance e-learning market and result in [9]:

- Dynamic tracking effectively the current students' progress and performance per each subject using data sources such as personal data (ex. grades, hobbies, residence, activities, preferences), historical, behavior, and student interaction.
- Expecting the students who will fail and drop out of the course, especially in online courses [21].
- Preparation of personalized student development plan based on the student model, traits, behavioral analysis, and target of the learner whether to master the subject or learn an overview
- The digital textbook is emerging in a big data system. The digital textbook can be adapted based on the big data analytic of the student and learning style.
- Student interaction and engagement analysis with the class materials and provide modifications and enhancements, leading to student satisfaction [22].
- Marketing research about students and future market directions to attract more outstanding students.
- Improving student progress measurement and adaptive learning [23]. The traditional education system relies only on exam results to identify student progress. However, big data could generate progress metrics based on which questions are not correct and the meaning behind them, how long the student was thinking about the questions and selecting a wrong answer, and how many trials it took the student to get the correct answer.
- Providing relevant previous experiences from previous learners who faced the same issue that the current learner is struggling with. For example, big data analytic can recommend to the current learner different education lessons or insights that worked with previous learners with similar issues [24,25].
- Keeping the student in an energetic state through noticeable immediate actions implemented by big data analytic.
- Identification of each student learning depth of each concept.
- Courses enhancements through big data analytics of learners' behaviors. If there are some very hard questions or materials that many students could not master, more explanation and modification of learning objects can solve the problem [22].
- Management and decision support for administration decisions as admissions, student profiles, alumni, strategic planning, industrial connections, campus management, and security monitoring [26].

5.1. Big Data Potential in E-Learning: Case Studies

5.1.1. Big Visualization, Dashboards, and Decision-Making Support: proposed Hadoop based system based on the student performance data warehouses. The system produces a visual presentation through graphs that represent the student academic progress for supporting top management decision making and statistical visualization dashboard for institution management.

5.1.2. Student Progress Tracking: suggested a random forest algorithm applied on big data for student grade forecasting based on Spark to monitor the student progress based on stored data in HBase based on different sources of data such as student personal data, academic history, absence rates, and previously detailed grades.

5.1.3. Educational Organization Quality Monitoring System: [27] designed a Spark-based big data solution for analysis of the institution's big data such as daily instructional activities, student activities, academic results, etc. to monitor and evaluate the quality of the institution teaching and learning process and adjusting teaching strategies accordingly.

5.1.4. Performance Prediction: Big data enables the prediction of student performance and test grades based on the analysis of student information [28]. This can help student's selection of best fit classes such as the application Degree Compass introduced by [29] which recommends the best class for the student to enroll.

5.1.5. Choosing Future Career and Pathways: Big data can support the future student decision for the future career. Zhenyu [30] proposed HDFS for data storage and MapReduce with Mahout to analyze exam results, student behavior data, and student life information for helping the student in Holland in the optimal selection of the future profession.

5.1.6. Improved Personalized learning, Instruction, and Assessment: Big data can adapt the e-learning learning objects. Big data analytics in the e-learning also reveals the power for providing personalized e-learning materials, student learning style identification, and adaptive recommending of learning objects [25,31]. Big data analytics can rapidly analyze learning difficulties and provide an opportunity to gradually improve and challenge the learner. [32] proposed a Spark-based system in C++ class to detect the hardiest questions to students relying on the question expected difficulty and number of attempts to solve. [33] proposed a big data system for elective course recommendation using analyzing the grades of other subjects for Indian school students based on the Mahout machine learning method built on the Hadoop ecosystem. [34] proposed a Hadoop and MapReduce big data collaborative filtering system for resources recommendation such as videos, classes, papers, etc. for the learners based on click streams and web accesses of the users. [23] proposed adaptive course recommendation system based on student profile and teacher's objective from the class. The system based on genetic algorithm implemented on HDFS and MapReduce on Hadoop big data ecosystem based on analysis of social media data and emotional learner state.

5.1.7. Instant Feedback: Big data deployment in e-learning can give the institution and stakeholders instant feedback of which learning style and patterns work and which does not work per each learner [31,35]

5.1.8. Student Behavior Monitoring: Provided obtaining the student permission for behavioral data collection (includes data about emails, calls, photos, payments, notes, social media interaction, family, friends, habits, and interests), [36] suggested big data-based system for monitoring the attitudes and behaviors of the student. Behavioral analytics detects the tendency of the student towards violence and terrorism.

6. Discussion, Challenges and Future Research

6.1. Technical Challenges
Data integration among the data generated by different sources with different structures and formats is a technical challenge for big data analytics in the e-learning platforms. Data collection and handling of data when not all sources are present as missing some data important for processing is another technical challenge that needs to be addressed. The literature in the e-learning big data analytics misses a holistic case study that merges the historical data, behavior data, and other data sources. Yet, acceptance of big data in the e-learning systems is still in a primary phase [26]. The big data analytics in the e-learning is also facing difficulties in providing databases for big data research and analysis in the e-learning field.

6.2. Big Data Analytics Implementation challenges
Hadoop has over 190 configurations to affect big data analysis performance. Superdeep and Hadoop configuration tuning need to be investigated to achieve the optimal configuration in the e-learning big data analytics platforms. There is a need for the adoption of big data novel solution by traditional educational institutes. Staff at universities and institutes have to be trained in innovative data management skills. In addition, the cost with the holistic, big data system (storage warehouses, tools, etc.) is still high and to be further optimized and decreased. However, Data standard can be found in

[37] it is not fully applied to the educational data. Data analytics in the education sectors lacks the applicable standard measures and indicators to represent educational variables such as educational progress estimation. Big data analytics researchers should ensure a careful analysis of historical data. Past performance must be used carefully for the only estimation and not for grading. The student may not get under-estimated for previous good or bad performance. Also, similar behavior patterns for several learners may not conclude the similarity between the learning styles and preferences.

6.3. Political Issues
Data privacy and protection, such as the huge information gathered from the learners, can be mistreated and expresses risk for the learners. Data of the learner must have legal protection for its intended use only and must be protected. Also, full consideration of the big learning data and numbers, regardless of the human perspective, represents a challenge. Numbers only as a learning measurement, maybe not sufficient for student assessment.

7. Conclusion
Big data is evolving in the educational sector. Learners and educational organizations are generating large amounts of data to hide important information. The integration of big data technologies in the educational sectors reveals viable information for all involved education process stakeholders. There are several benefits of applying big data in the education field. Educational institutions will benefit from enhanced market analysis, improved adapted learning environment, analyzed career consulting based on student data, minimize students drop-out from classes, support decision making and administration works. Nevertheless, the big data analysis is having several challenges as well, such as data privacy, security constraints, limitation of data integration among different data sources, and the relatively high analysis cost of the holistic, big data solution. Moreover, new future research areas are urging like deployment of other platforms such as Superdoop in the e-learning platforms, standardized data protocols that ensure privacy and security, and holistic case studies that combine historical and different data sources within optimized Hadoop configuration. In summary, e-learning will be boosted with the growing big data penetration in the e-learning.

References
[1] Yanhui W 2018 Language E-learning based on learning analytics in big data era *Proceedings of the 2018 International Conference on Big Data and Education* pp 106–11
[2] Fee K 2009 Delivering e-learning: a complete strategy for design *Appl. Assess.*
[3] Kwary D A and Fauzie S 2018 Students' achievement and opinions on the implementation of e-learning for phonetics and phonology lectures at Airlangga University *Educ. e Pesqui.* **44**
[4] Bhat A Z and Ahmed I 2016 Big data for institutional planning, decision support and academic excellence *2016 3rd MEC International Conference on Big Data and Smart City (ICBDSC)* (IEEE) pp 1–5
[5] Santoso L W 2017 Data warehouse with big data technology for higher education *Procedia Comput. Sci.* **124** 93–9
[6] Dougiamas M and Taylor P 2003 Moodle: Using learning communities to create an open source course management system *EdMedia+ Innovate Learning* (Association for the Advancement of Computing in E ducation (AACE)) pp 171–8
[7] West D M 2012 Big data for education: Data mining, data analytics, and web dashboards *Gov. Stud. Brookings* **4** 1–10
[8] Alblawi A S and Alhamed A A 2017 Big data and learning analytics in higher education: Demystifying variety, acquisition, storage, NLP and analytics *2017 IEEE Conference on Big Data and Analytics (ICBDA)* (IEEE) pp 124–9
[9] Kochetkov O T and Prokhorov I V 2017 The research of approaches of applying the results of big data analysis in higher education *AIP Conference Proceedings* vol 1797 (AIP Publishing LLC)

p 20008
[10] Hadoop A 2016 Apache hadoop yarn
[11] White T 2012 *Hadoop: The definitive guide* (" O'Reilly Media, Inc.")
[12] Dean J and Ghemawat S 2004 MapReduce: Simplified data processing on large clusters
[13] Fuad A, Erwin A and Ipung H P 2014 Processing performance on apache pig, apache hive and MySQL cluster *Proceedings of International Conference on Information, Communication Technology and System (ICTS) 2014* (IEEE) pp 297–302
[14] Vavilapalli V K, Murthy A C, Douglas C, Agarwal S, Konar M, Evans R, Graves T, Lowe J, Shah H and Seth S 2013 Apache hadoop yarn: Yet another resource negotiator *Proceedings of the 4th annual Symposium on Cloud Computing* pp 1–16
[15] Ingersoll G 2009 Introducing apache mahout: scalable, commercial-friendly machine learning for building intelligent applications *IBM Corp.*
[16] Zaharia M, Xin R S, Wendell P, Das T, Armbrust M, Dave A, Meng X, Rosen J, Venkataraman S and Franklin M J 2016 Apache spark: a unified engine for big data processing *Commun. ACM* **59** 56–65
[17] Apache Software Foundation 2018 Apache Spark™ - Unified Analytics Engine for Big Data *Apache Spark*
[18] Chen J, Tang J, Jiang Q, Wang Y, Tao C, Zhang X and Liao J 2017 Research on architecture of education big data analysis system *2017 IEEE 2nd International Conference on Big Data Analysis (ICBDA)(* (IEEE) pp 601–5
[19] Zhang W and Jiang L 2018 Algorithm Analysis for Big Data in Education Based on Depth Learning *Wirel. Pers. Commun.* **102** 3111–9
[20] Dandan H, Yajuan Z, Junfeng L, Chen L, Mo X and Zhihai S 2017 Research on Centralized Data-Sharing Model Based on Master Data Management *MATEC Web of Conferences* vol 139 (EDP Sciences) p 195
[21] Liang J, Yang J, Wu Y, Li C and Zheng L 2016 Big data application in education: dropout prediction in edx MOOCs *2016 IEEE Second International Conference on Multimedia Big Data (BigMM)* (IEEE) pp 440–3
[22] Chen G, Wen S, Yang J and Song G 2014 Building an Experimental Platform for Cloud and Big Data Education *2014 International Conference on Education Reform and Modern Management (ERMM-14)* (Atlantis Press)
[23] Madani Y, Bengourram J, Erritali M, Hssina B and Birjali M 2017 Adaptive e-learning using genetic algorithm and sentiments analysis in a big data system *IJACSA) Int. J. Adv. Comput. Sci. Appl.* **8** 394–403
[24] Anna N E V and Mannan E F 2020 Big data adoption in academic libraries: a literature review *Libr. Hi Tech News*
[25] Drigas A S and Leliopoulos P 2014 The use of big data in education *Int. J. Comput. Sci. Issues* **11** 58
[26] Hanapiyah Z M, Hanafi W N W and Daud S 2018 Issues, Challenges and Opportunities of Big Data Management in Higher Education Institutions in Malaysia *Indian J. Sci. Technol.* **11**
[27] Li Y, Li P, Zhu F and Wang R 2017 Design of higher education quality monitoring and evaluation platform based on big data *2017 12th International Conference on Computer Science and Education (ICCSE)* (IEEE) pp 337–42
[28] Yu X and Wu S 2015 Typical applications of big data in education *2015 International Conference of Educational Innovation through Technology (EITT)* (IEEE) pp 103–6
[29] Denley T 2013 Degree compass: A course recommendation system *Educ. Rev. Online* 1–5
[30] Zhenyu C 2017 The application of big data in higher vocational education based on Holland vocational interest theory *2017 International Conference on Industrial Informatics-Computing Technology, Intelligent Technology, Industrial Information Integration (ICIICII)* (IEEE) pp 37–40
[31] Maresca P and Molinari A 2016 Is e-learning ready for big data? And how big data would be useful to e-learning? *DMS* pp 34–40
[32] AL-Taie R R K and Abu-Alsaad H A 2018 Using Big Data Technology for Prediction of Quiz Difficulty Level in E-learning Systems *Iraqi J. Inf. Technol.* **8** 164–81

[33] Dwivedi S and Roshni V S K 2017 Recommender system for big data in education *2017 5th National Conference on E-Learning & E-Learning Technologies (ELELTECH)* (IEEE) pp 1–4
[34] Shaikh A A Big Data Recommendations Algorithm For E-Learning Using Click Stream Data
[35] Secades V A and Arranz O 2016 Big data & eLearning: A binomial to the future of the knowledge society *IJIMAI* **3** 29–33
[36] Baig A R and Jabeen H 2016 Big data analytics for behavior monitoring of students *Procedia Comput. Sci.* **82** 43–8
[37] Rankin J G 2016 *Standards for reporting data to educators: What educational leaders should know and demand* (Routledge)

Electronic practicum assessment system based on android at politeknik ATI makassar

J Habibuddin[1], T Suhaebri[2], A Ainuddin[2], M F Azis[1], and Mutmainnah[1]

[1]Machining System Automation, Politeknik ATI, Makassar, Indonesia
[2]Akademi Komunitas Industri Manufaktur Bantaeng

Email: juliantihabibuddin@atim.ac.id

Abstract. The Practicum assessment conducted conventionally is gradually replaced with an Android-based system. It aims to support the government program of Making Indonesia 4.0. In this research, the electronics practicum assessment is designed using the App Inventor application. The attendance assessment and initial assignment are designed to obtain an objective grade. Attendance assessment is conducted by making a standard assessment form, determining points obtained by looking at the time of arrival. Assessing the initial task was done by making ten multiple-choice questions for each experiment. The initial assignment assessment form was created in Microsoft Excel to make it easier to import, export data to .csv (comma separated values) to enter the database system. The results obtained from 6 types of initial assignment assessment experiments conducted using the application. There were 5 trials that were successfully run, and there was still one error trial.

1. Introduction

Assessment is an important component of teaching. With the assessment, lecturers can measure the ability of students. Practical assessment using a manual system is sometimes ineffective and not yet objective. This is because the preliminary tasks given when entering the practicum are still in the form of oral questions. As technology develops, there are many ways to assess students. One of them is by using an application on Android [1]

The design of the appraisal using an android application can be made using android studio software or the inventor app. App Inventor is an open-source web application originally provided by Google, which is currently managed by the Massachusetts Institute of Technology (MIT). It makes it possible to create software applications for the Android operating system (OS) [2]. The MIT version was launched in March 2012. On December 6, 2013, MIT released App Inventor 2, renaming the original version of "App Inventor Classic". App Inventor 2 makes it easy for teachers who are not programming specialists or lack programming experience to be able to develop Android applications in the teaching process [3–5].

App inventor is block-based programming, which means programmers don't have to remember code or syntax, and more importantly, users don't have to worry about syntax errors [6]. App Inventor 2 has a user interface with component designer components and block editors. Component designers are visual objects in the display to adjust and change the size and color and general appearance of the application. The block editor determines what application is created and manages the application under certain conditions [3,5].

From the description above, this research discusses practicum assessment using an android application with Inventor 2 app software on the electronics practicum at the ATI Makassar polytechnic.

2. Methods

In designing the electronic practicum assessment, the steps taken before the planning system are the analysis of laboratory conditions when the electronics practicum is taking place. The electronics practicum in its grading system uses a control card. Students before entering the laboratory must print a control card. When entering the practicum, the lecturer fills in the attendance column. After that, giving the questions as preliminary tasks carried out in writing or orally. Students answer the preliminary questions, and the lecturer grades the preliminary assignment column contained on the control card. The attendance assessment and this preliminary task will be made by the android application to make it easier for the lecturer to assess.

2.1. *Planning System*

The planning in this system can see in figure 1.

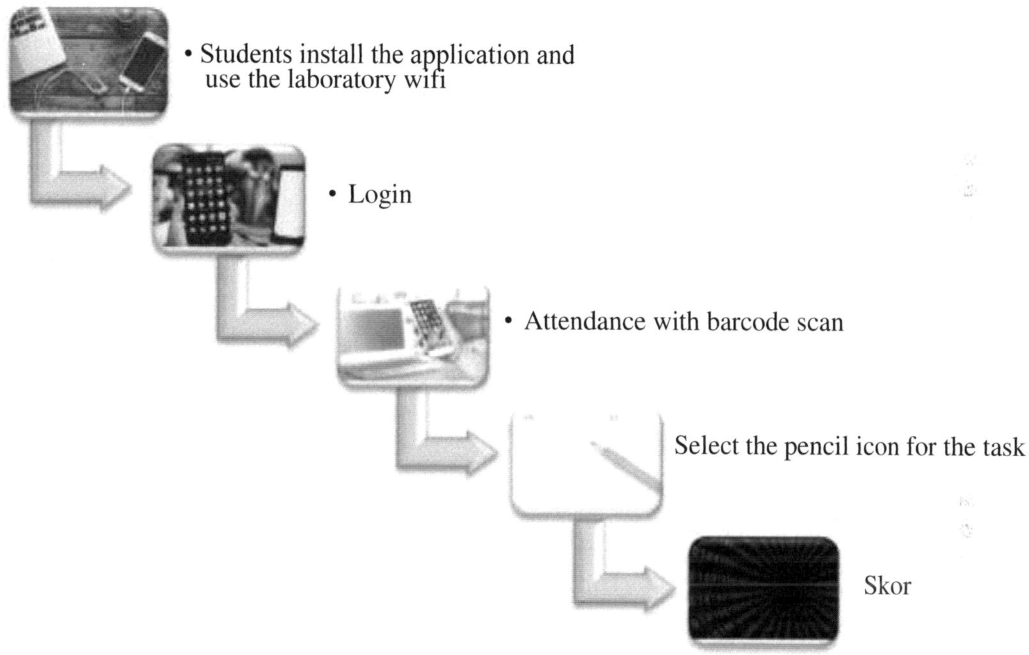

Figure 1. Planning of system

2.2. *Design System*

Design the system using App Inventor. The initial step is to create a database on Microsoft Excel. Then imported into .csv, the system database can be seen in Figure 2

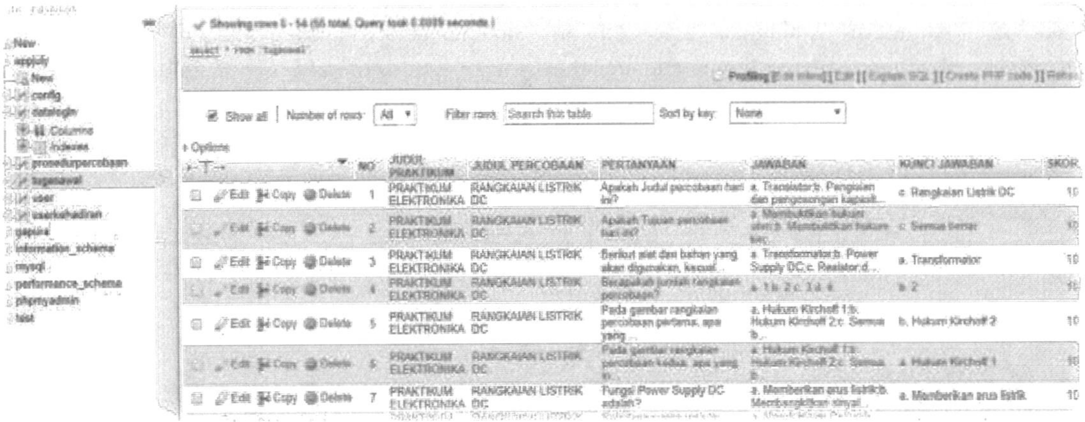

Figure 2. Design of system

3. Result and Discussion

3.1. System Analysis

System analysis is performed when students use the practicum assessment application found on smartphones. Attendance assessment is done by selecting the key icon on the application, then scanning the barcode available on the laboratory computer. The results obtained there were no errors in the attendance assessment; this was seen when observing students scanning barcodes and all data stored in the database. Assessment of the initial task in 6 experiments, there was one trial that error was the initial task in the transistor experiment.

Figure 3. Application Display

3.2. Problem Analysis
In attendance assessment, when students scan barcodes, there are students who have difficulty when scanning barcodes, this is because the position of the smartphone is not parallel to the position of the computer laboratory. Next assessment of the initial task error that occurs when working on the problem, when you want to proceed to the next problem, there is an error warning, so it needs to be checked again in the App Inventor software.

Figure 4. System Implementation

3.3. Discussion results
In its application in electronics practicum, this application still uses the localhost. Student smartphones still need to be connected to the wifi signal available in the laboratory, so it is less efficient.

4. Conclusions
In designing an Android-based practicum assessment still needs development. This design still uses the localhost network. So that future development requires a system that already uses the internet network and is able to integrate with the campus academic information system.

References
[1] Setiadi B R, Setuju S, Ratnawati D, Suparmin S, Sugiyono S and Widatmoko A 2018 E-Monitoring in The Vocational Process-Based Assessment *Int. J. Eng. Technol.* **7** 78–81
[2] Singh A, Pal A and Rai B 2015 GSM based home automation, safety and security system using android mobile phone *Int. J. Eng. Res. Technol.* **4**
[3] Kang H, Cho J and Kim H 2015 Application study on android application prototyping method using app inventor *Indian J. Sci. Technol.* **8** 1
[4] Harinath R and Santhi S 2015 GSM based Home Automation System using app-inventor for android mobile phone *Int. J. Comput. Sci. Mob. Comput.* **4** 158–67
[5] de Moura Oliveira P B 2015 Teaching automation and control with App Inventor applications *2015 IEEE Global Engineering Education Conference (EDUCON)* (IEEE) pp 879–84
[6] Tharishny S, Selvan S and Nair P 2016 Android based smart house control via wireless communication *Int. Journanl Sci. Eng. Technol.* 323–5

IOP Conf. Series: Materials Science and Engineering **885** (2020) 012005 doi:10.1088/1757-899X/885/1/012005

BLDC motor design by applying the control of direct current and analysis of loading mechanism soft starting

M Wirandi, Safril, F Sumasto[1], M Agus, and F Imansuri

Department of Automotive Industrial Engineering, STMI Polytechnic of Jakarta, Ministry of Industry Republic of Indonesia, DKI Jakarta, Indonesia

Email: wirandi13@kemenperin.go.id

Abstract. In this research, there will be a design driver brushless DC (BLDC), which on the input voltage of the motor driver applying direct current controllers and the implementation mechanism of soft starting the operator of BLDC, which uses fuzzy logic controllers. Testing is done with a view of controlling the application of direct current to the BLDC to many conditions, namely the amount of value given fuzzy logic and the load that is given to the mechanics. The design of the driving BLDC has successfully created and implemented by applying a direct current control and the results obtained in the form of efficiency that is better than the previous design. The soft starting mechanism that controls using the fuzzy logic method possible produces a low inrush current and generates steady conditions that are better on BLDC. For values that are applied to the load, the greater the torque load placed on the BLDC, the higher the initial current needed to achieve transient time equally and opposite that the smaller the torque load placed on the BLDC, then the lower torque the load required to achieve the transient time.

1. Introduction

In the use of BLDC, one that must be considered is the large wave in the initial start of the motor because if it does not take into account things like this, it will cause some good problems that result in damage to the BLDC driver and have an impact on energy efficiency problems [1]. Several studies use current protection to control BLDC using both hardware and software. The application by using hardware a method that can protect the wave in a classic way that is by using a rheostat, how to slowly increase the resistance value to the maximum resistance value so that the initial inrush current can be minimized. But this method has several disadvantages; namely, the rheostat still uses a manual shear system, so that the current that is retained is discharged into heat, and the price of the rheostat is still relatively expensive. More conventional hardware current protection by using a start and trip system for BLDC [2].

The induction motor, when turned on, can immediately draw 5 to 7 times the full load current. This large current can result in a decrease in voltage and disrupt other devices on the same channel, so motors with power above 30 or 50 horsepower are not recommended to be turned on immediately. In the research carried out, to achieve the nominal current needed 3 seconds, because the setting at the start and trip time is 3 seconds, if it exceeds the time limit set, then the trip circuit will function. A time tolerance setting of fewer than 3 seconds can be set on IC 555 via the potentiometer. So that the delay system, which consists of a timer configuration start and the trip, will reduce the current, and others are expected to control the current to the motor gradually by following the desired settings.

Current protection research using the software by looking at the BLDC current profile for several variations, namely the height of steps, a time delay between steps, and mechanical load provided by changing the resistor resistance at the BLDC terminal. The greater the time delay between input steps, the lower the maximum start-up wave that occurs at BLDC. When the given delay time is greater than the BLDC transient time for each step, the addition will not decrease the maximum value of the start-up wave but will extend the time needed for the BLDC to reach steady-state [3]. Referring to several references, the researcher got an idea to develop BLDC current protection by using fuzzy logic and made a closed loop to control the current entering BLDC.

2. Literature Review

2.1. Three-phase BLDC motor
The drive is the main part of a vehicle. One part of the drive is the Brushless Direct Current Motor (BLDC). BLDC has many advantages from DC motors, one of which is the commutation that can be released, namely the commutation of 2 connected phase windings and the commutation of 3 connected phase windings. Allowing Intermediate power to occur when 2 phase coil commutations are connected, and maximum power occurs when 3 phase coil commutations are connected [4].

2.2. Hall sensor
The hall-effect sensor was implemented for driving asynchronous three-phase motor with a permanent magnet in the rotor side through the three-phase inverter. In censored driver, hall-effect sensors are generally of three pieces mounted on the stator and separated by 120 electrical degrees or 60 electrical degrees. Noisy hall-effect based rotor position sensor could result in erratic switching patterns if not properly addressed [5]

2.3. Fuzzy logic
Basically, the fuzzy set theory is an extension of classical set theory (crisp). The existence of an element in a set, A, will only have two possible memberships, which are members of A or not members of A [6,7]. The reasoning with the Sugeno method is almost the same as the Mamdani reasoning, it's just that the (consequent) output of the system is not a fuzzy set, but a constant or linear equation. The stages in the Sugeno method are as follows: the formation of the fuzzy set, application function implications, and defuzzification.

2.4. Current sensor
The ACS712 current sensor is a hall-effect based current sensor. This type of sensor can be used to measure AC or DC currents and is commonly used to control motors, detect electrical loads, and detect overload. This sensor has a reading with high accuracy because, in the sensor, there is a series of the low-offset linear hall with a path made of copper. The workings of this sensor are by reading the current through the copper pathway to produce a magnetic field that is captured by the integrated hall and converted to a proportional voltage [8].

3. Method

3.1. Design scheme
The architecture and initial data of the system to be used can be seen in Figure 1. System design in this research consists of 5 main parts, i.e., Selection of microcontroller, BLDC switching design, current sensor design, speed calculation design, and software design.

IOP Conf. Series: Materials Science and Engineering **885** (2020) 012005 doi:10.1088/1757-899X/885/1/012005

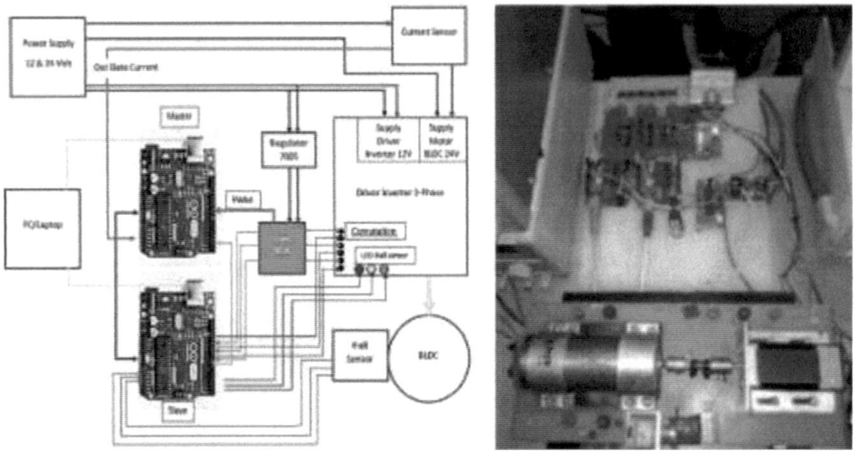

Figure 1. System hardware configuration.

3.2. Selection of microcontroller

This research uses two Arduino Uno microcontrollers which are used as a master and as a slave where the microcontroller "used as a master is used to acquire data from the sensor and run the controller algorithm" that has been made while the microcontroller used as a slave is "used to regulate the combination of commutations based on the rotor position". By using the master and slave configuration, it is expected to function to optimize the performance of the microcontroller and to accommodate the limitations of clock speed, number of timers, and the number of ports needed in the system.

3.3. BLDC switching design

BLDC uses a switching circuit consisting of MOSFET transistors with type IRF3205, which has the advantage of a large current rating and is used as an electronic switch to represent the phase of BLDC and uses IC IR2110 which is used as a commutation to trigger the gate as controlling MOSFET. Each IC has two outputs that are used to trigger the activation of the MOSFET.

BLDC drivers require three voltage inputs: 5V input, is used to supply the hall sensor, 12V input, is used to supply the IR2110 IC as an electronic component to trigger the gate, or as a MOSFET driver, and 24V input is used to supply power to the BLDC motor. This research uses a combination of two connected switches, where only two phases are supplied for each combination of commutation, and one other phase is not connected.

3.4. Current sensor design

How to find the value of the current on the current sensor is to convert the voltage (sensor output) into an electric current. The form of a mathematical equation for conversion from the sensor output voltage to the current can be written:

$$vout = 2.5 + (0.185\ lout) \tag{1}$$

The sensor output current equation is:

$$lout = \frac{512 - data\ ADC}{1023}\ x\ \frac{5}{0.815} \tag{2}$$

The conversion calculation above is done by the master controller by adding an average algorithm to reduce the ripple of current measurement data. Noteworthy in the algorithm is the determination of the ADC value to represent a voltage of 2.5 volts.

3.5. Speed calculation design

To calculate the speed of a BLDC motor, the first thing to do is find out the rotational speed information of the hall sensor. In this research, the motor used had four pairs of poles, and the smallest speed measurement obtained was 50 rpm or 12.5% of the maximum speed of the BLDC.

3.6. Software design

In making an algorithm, the program is inseparable from the design of the algorithm. Therefore, the form of the overall system algorithm can be seen in Figure 2. In the implementation of the BLDC speed algorithm, the calculation is performed by the master controller, and the input bit position from the Hall Sensor is carried out by the slave controller. Every time an interrupt occurs, the number of counters will increase, and each time interval that has been determined, the counter value is taken to calculate the BLDC speed.

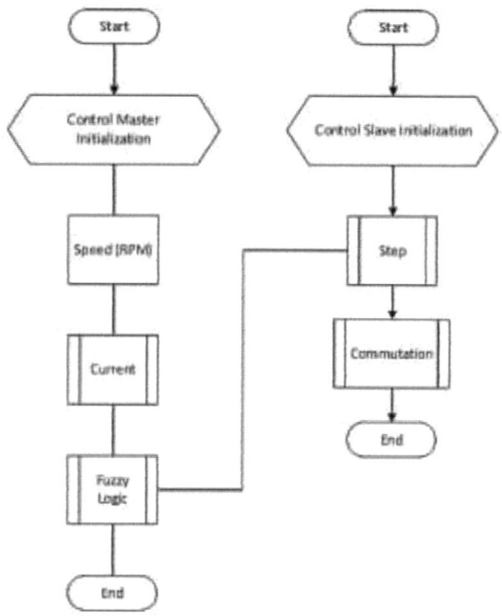

Figure 2. Implementation of the overall system algorithm.

In the implementation of 2 phase switching connected to be used to play BLDC, the switching model used by the slave controller is used, whereas the order of commutation logic will always be the same for all types of BLDC. In general, the design of fuzzy logic controllers is divided into three stages, namely the fuzzification stage, the stages determining the implication function (fuzzy rule base), and the defuzzification stage (fuzzy inference).

a. BLDC current without load and load is used as the first fuzzy input.
b. Delta's current BLDC is used as the second fuzzy input.
c. The step value is used as a fuzzy output.

3.6.1. Fuzzification

Figure 3. Shows the steps that must be made in the form of determining the Input and Output functions, where the input function is the current and output values in the form of the current delta value of the BLDC and the output function in the form of step values. After fuzzification, the next process is determining the implication function. As previously designed, the output of the fuzzy logic controller is in the form of adding or subtracting step values. Fuzzy inputs and outputs are linked based on rules called if-then rules.

IOP Conf. Series: Materials Science and Engineering **885** (2020) 012005 doi:10.1088/1757-899X/885/1/012005

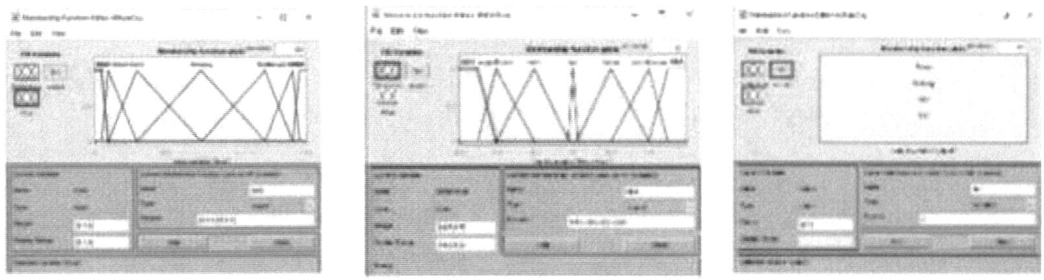

(a) (b) (c)

Figure 3. From left to right: (a) function enter the current value, (b) enter the delta current function, (c) step output Function implications

3.6.2. Defuzzification

The method used during defuzzification is the Sugeno method. To implement the fuzzy logic, a program is created on the Arduino Uno master microcontroller, where the flow diagram is used to control the speed with fuzzy logic, which can be seen in Figure 4.

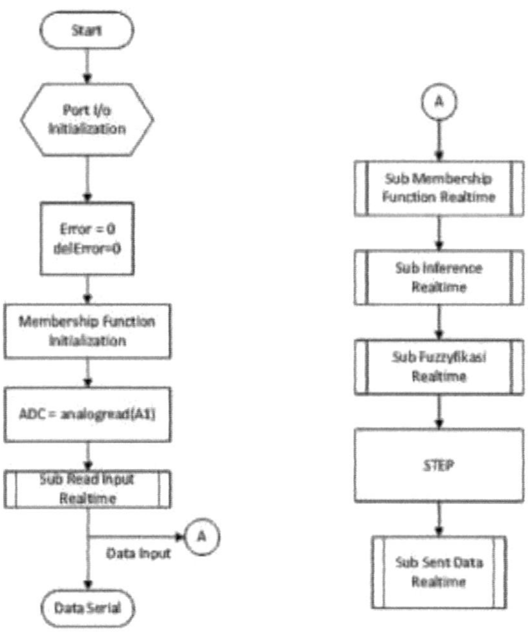

Figure 4. Current chart of the fuzzy logic control program

4. Result and Discussion

This test is done by combining the whole system and testing, which basically, each block is assembled into one to see the results of the system that is made to run as it should. Testing and data collection conducted in this study are aimed at seeing as current protection (soft starting) using a microcontroller which contains a fuzzy logic method that can run well and compare with the response of the system without using current protection so that it is expected that this control method can be used as a reference for the use of BLDC soft starting.

4.1. System response without overcurrent (soft starting)

In this section, the system is tested without current protection and uses current protection. This test is carried out using 24v and 6A currents sourced from the power supply and with a sampling time of 50ms. This test is also carried out in several stages, namely using load and without load and with different PWM values. The results of testing this system can be seen in Figures 5 and 6.

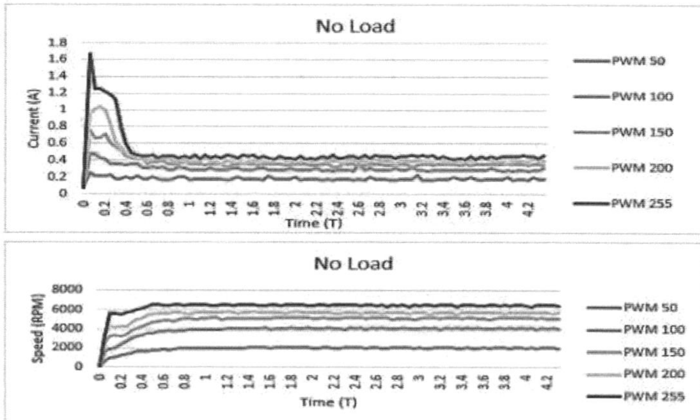

Figure 5. System response current and speed without using a load.

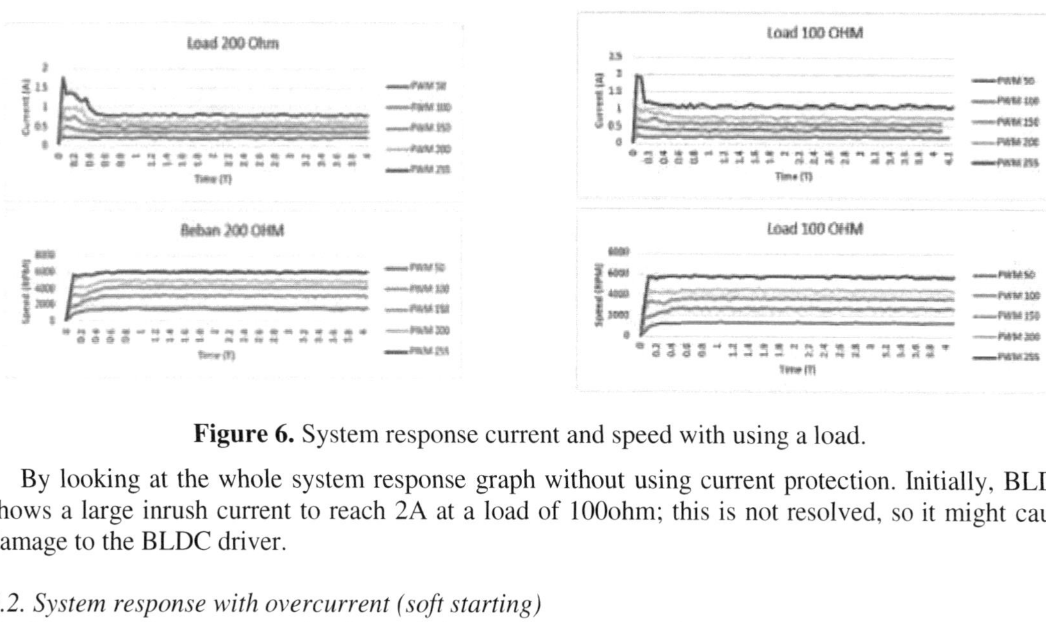

Figure 6. System response current and speed with using a load.

By looking at the whole system response graph without using current protection. Initially, BLDC shows a large inrush current to reach 2A at a load of 100ohm; this is not resolved, so it might cause damage to the BLDC driver.

4.2. System response with overcurrent (soft starting)

At this stage, the system testing is carried out using current protection. In this test, the fuzzy logic system is applied to protect the inrush current with the input in the form of current and delta current obtained from the sampling steps that have been taken previously and the output of the input in the form of steps whose data has been previously taken. This test is also carried out in several stages, namely by using a load and no-load and with different PWM values. The results of testing this system can be seen in Figures 7 and 8.

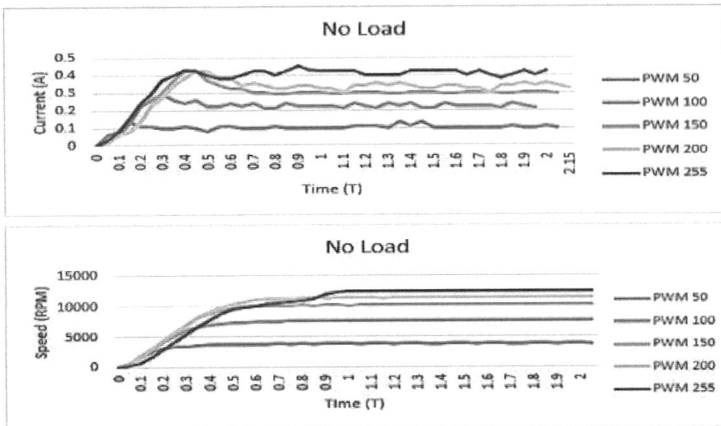

Figure 7. System response current and speed without using a load.

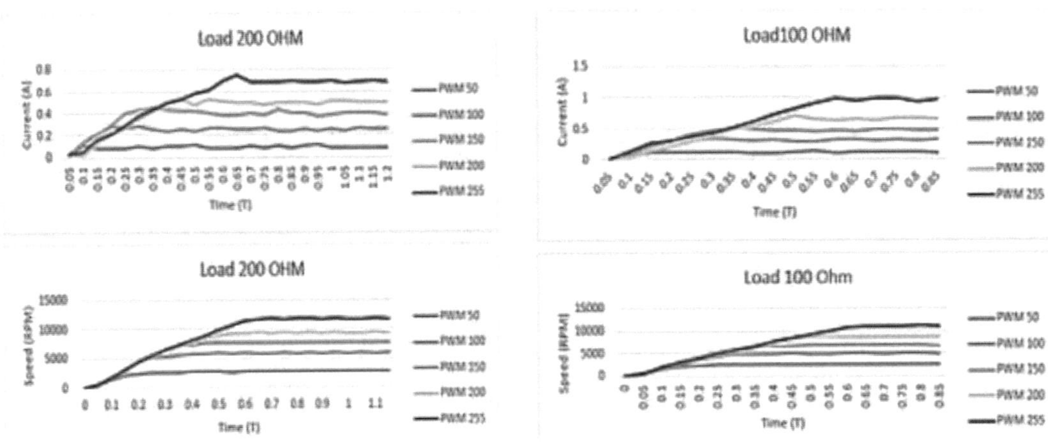

Figure 8. System response current and speed with using a load.

By looking at the system response graph using current protection. The current at the start of BDLC shows a very small surge in current and by using several types of load current can be protected properly and minimize the impact of damage that occurs in the BLDC driver.

5. Conclusion

After testing and analysis, BLDC speed without using fuzzy logic in protecting currents reaches a steady-state faster than using fuzzy logic. When given PWM 255 without a current protection speed of 0.95 seconds is obtained, and a current protection speed of 1.1 seconds is obtained to reach a steady-state so that the difference obtained is 0.15 seconds.

Before using fuzzy logic for current protection with PWM 255, obtained a wave of 1.67A and after using a damped current to 0.43A to obtain a difference of 73.53%. By using fuzzy logic to protect the current with PWM 255 and given a load of 100 Ohm, an attenuation current of 50.25% from 1.97A and 200Ohm can be obtained from the current attenuation of 57.15% from 1.75A.

Acknowledgments

This research is supported and funded by the Directorate of Research and Community Engagement (P2M), STMI Polytechnic of Jakarta, Number: SP DIPA-019.10.2.579350/2019.

References

[1] Lü Z and Sun F 2011 Soft-start circuitry for DC motor *2011 International Conference on Electrical and Control Engineering* (IEEE) pp 3151–3

[2] Tukananto A Rancang Bangun Sistem Proteksi Arus Lebih Motor 3 Fasa dengan Timer Start dan Trip *J. Tek. Elektro Univ. Tanjungpura* **2**
[3] Irsyadi F 2016 *Controller design and analysis of the influence of variations in input ladder and loading on the current profile on the soft starting motor mechanism of direct current without brush* (LAPAN)
[4] Amalia S 2018 Implementation of 2 phase winding and 3 phase winding connected to voltage in brushless direct current (BLDC) outdoor rotor motors with ANOVA analysis *J. Ipteks Terap.* **12** 167–76
[5] Wicaksono R, Firmansyah E and Isnaeni B S M 2016 Current control and protection of block commutated for 3 phase inverter BLDC motor *2016 8th International Conference on Information Technology and Electrical Engineering (ICITEE)* (IEEE) pp 1–6
[6] Nauck D, Klawonn F and Kruse R R 1997 *Foundations of Neuro-Fuzzy Systems*
[7] Lastomo D, Widodo W and Setiadi H 2018 Optimal power flow using fuzzy-firefly algorithm *Proceeding Electr. Eng. Comput. Sci. Informatics* **5** 210–5
[8] Microsystems A 2017 ACS712 Fully Integrated, Hall-Effect-Based Linear Current Sensor IC with 2.1 kV-RMS isolation and a low-resistance current conductor Datasheets

Blind people's direction support system using ultrasonic, color sensor with fuzzy logic

Nur Yanti [1], Taufik Nur [2] and Qory Hidayati [3]

[1,3] Electrical Engineering Department, Politeknik Negeri Balikpapan, Jln. Soekarno Hatta KM.8 Balikpapan 76129, Indonesia
[2] Industrial Engineering Department, Faculty of Industrial Technology-Universitas Muslim Indonesia, Jl.Urip Sumoharjo KM.5, Panaikang, Kec. Panakkukang, Kota Makassar, Sulawesi Selatan 90231, Indonesia

E-mail: nur.yanti@poltekba.ac.id

Abstract. Someone who has impaired vision or blindness has a condition that is susceptible to dangers that are around him such as the presence of animals, vehicles or obstructions. People with visual impairments carry out their activities based on estimates of their feelings. One obstacle that is often faced by blind people is that if they are in a new environment, they must recognize patterns around the environment such as roads, rooms, or certain important places.his research in the form of blind shoe design using ultrasonic, color sensor and Arduino Uno microcontroller, with fuzzy logic, aims to help blind people, especially children who are educated in extraordinary schools so that in carrying out their learning activities do not experience obstacles in recognizing the school environment, such as study rooms or classrooms, toilets and roads that are in the school environment. The method used is experimental, the color sensor can read input in the form of red, blue, green and yellow which is the blind pointing pathway, and ultrasonic can read the distance of obstacles in front of the shoe using an intelligent fuzzy logic system.A pair of shoes uses Bluetooth communication, output in the form of sound through earphones as a guide to the classroom or toilet.

1. Introduction

Vision is one of the most important five senses, functioning as the recipient of information received directly by way of seeing. Visual impairment in the form of blindness causes blind people who are not free to move or do daily activities so that a tool designed to help blind people to facilitate their activities, especially when walking, and reduce the risk of persons with danger around them.

As a direction when walking the blind shoes are designed using color sensors, ultrasonic sensors, Arduino microcontrollers, Bluetooth and earphones. The color sensor is used to detect an object, objects, colors or the surrounding environment. The use of color sensors as exploration and mapping of autonomy, the determination of octree-based localization in indoor robots [1], [2]. Another advantage of the color sensor is that it can track objects for pizza chef robots with RGB-D [3], Offline programming using RGB on cell dispensing robots [4]. In this design, ultrasonic sensors are used to detect distances that function as identification of obstacles or hazards, other uses as detection of underwater objects [5], as detection of structural changes on the bridge [6], to detect traffic participants in low-speed traffic situations [7]. Fuzzy logic engineering is needed for the classification of RGB colors as input systems [8] The fuzzy logic approach is very precisely applied to tracking

IOP Conf. Series: Materials Science and Engineering **885** (2020) 012006 doi:10.1088/1757-899X/885/1/012006

multiple visual objects [9], for fast and efficient color enhancement based on fuzzy logic and histograms [10].

2. System Design

The design of a pointing system for people with visual impairment in (figure 1) includes a series of ultrasonic sensors, TCS230 color sensors, fuzzy logic and programming of an Arduino microcontroller.

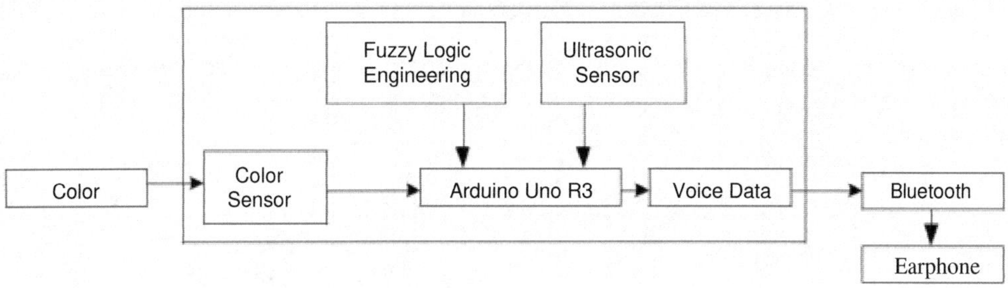

Figure 1. System design

2.1. Color sensor

TCS230 color sensor is a color sensor that is often used in microcontroller applications to detect an object or the color of an object being monitored. The TCS230 color sensor can also be used as a motion sensor, where the sensor detects the movement of an object based on the color changes received by the sensor.

Figure 2. TCS230 color sensor

Color sensor with input in the form of red, green, blue and yellow. The sensor is exposed to the color and the reading results will be seen on the serial monitor with the provisions of the RGB value.

Table 1. Classification of red

No	Color	R,G,B Output Value
1		2,15,10
2		5,21,15
3		9,37,30

4		12,43,38
5		15,52,46
6		25,70,85
7		30,74,80
8		33,77,70
9		38,85,76
10		43,89,80

The (table 1) shows the RGB value of each red color that is detected as a value of each in pink R values tend to be lower than red because of the lack of red content contained in the object being detected.

Table 2. Classification of green

No	Color	R,G,B Output Value
1		5,3,9
2		7,3,8
3		10,7,17
4		15,7,17
5		22,17,26
6		32,23,29
7		33,22,25
8		33,26,30
9		35,30,33
10		39,28,35

The (table 2) shows the RGB value of each red color detected has a value of each in light green color G values tend to be lower than the green color because of the lack of green content contained in the object being detected.

Table 3. Classification of blue

No	Color	R,G,B Output Value
1		14,10,2
2		21,14,2
3		28,18,3
4		32,25,9
5		45,31,15
6		45,33,16
7		48,31,17
8		51,35,20
9		57,40,22
10		60,44,28

Table 3 shows the RGB value of each red color detected has a value of each in the light blue color B values tend to be lower than the blue color because of the lack of blue content contained in the object being detected.

2.2. Ultrasonic Sensor

Ultrasonic sensors can detect the distance of an object by emitting ultrasonic waves with a frequency of 40 KHz, then detect the reflection. The HC-SR04 Ultrasonic Sensor is faced perpendicular to an obstacle in the form of the paper with a smooth surface at a certain distance. Figure 3 shows an ultrasonic sensor.

Figure 3. Ultrasonic Sensor

2.3. Fuzzy logic

The process of the fuzzy logic method is divided into three, namely: fuzzification, rule evaluation, and defuzzification. (figure 4) is the process of the fuzzy method of blind shoes. The fuzzification stage is divided into 2 groups, namely color sensor fuzzification and ultrasonic sensor fuzzification.

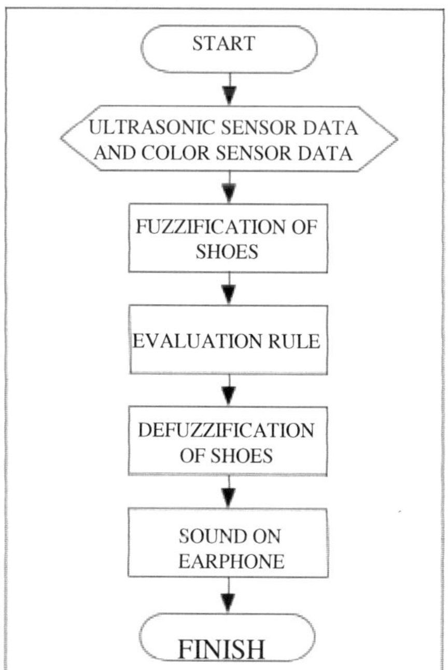

Figure 4. Fuzzy logic process

The color sensor has 3 Red, Green and Blue (RGB) output conditions, each of which has an output value. Figures 5, 6 and 7 show the membership function of the input colors Red, Green and Blue on fuzzy-MATLAB.

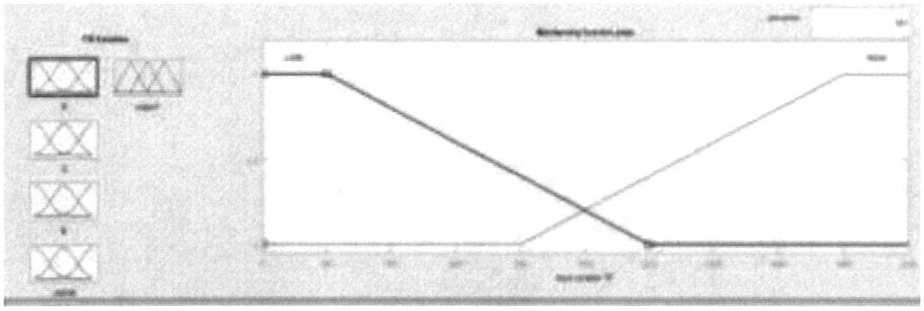

Figure 5. Membership input is red

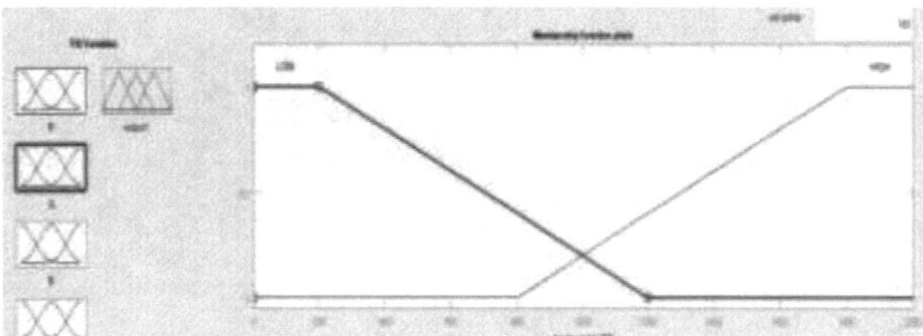

Figure 6. Membership input is green

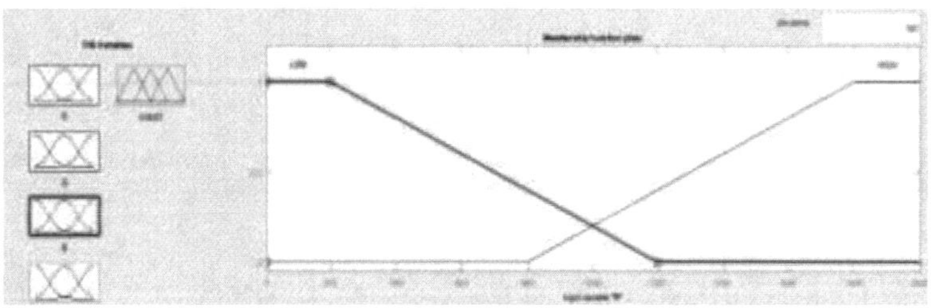

Figure 7. Membership input is Blue

Ultrasonic sensors make the distance as a membership to function as a barrier detection. Obstacles are determined with a range of 0 - 200 cm, if the person is at a distance of 40 cm then that is said to be safe or is far from the obstacle, but if the person is at a distance of 15 cm to 22.5 cm then it is said that the distance is not safe or in a condition very close to the obstacle.

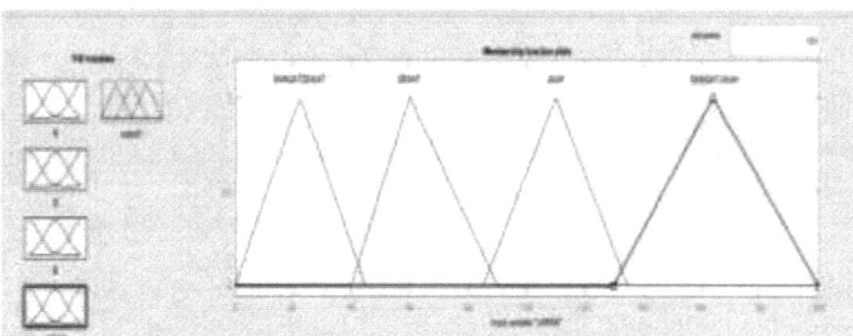

Figure 8. Distance input membership

From the rule evaluation we will know the output conditions of the RGB value and distance. At (table 4) shows the evaluation rule.

Table 4. Rule base

No.	Color			Distance	Output
	R	G	R		
1	Low (15)	High (50)	High (40)	Very close (30 cm)	WAV2
2	Low (15)	High (50)	High (40)	Close (40 cm)	WAV2
3	Low (15)	High (50)	High (40)	Far (60 cm)	WAV1
4	Low (15)	High (50)	High (40)	Very far (70 cm)	WAV1
5	High (23)	Low (18)	High (25)	Very close (30 cm)	WAV2
6	High (23)	Low (18)	High (25)	Close (40 cm)	WAV2
7	High (23)	Low (18)	High (25)	Far (60 cm)	WAV4
8	High (23)	Low (18)	High (25)	Very far (70 cm)	WAV4
9	High (34)	High (23)	Low (7)	Very close (30 cm)	WAV2
10	High (34)	High (23)	Low (7)	Close (40 cm)	WAV2
11	High (34)	High (23)	Low (7)	Far (60 cm)	WAV3
12	High (34)	High (23)	Low (7)	Very far (70 cm)	WAV3
13	Low (24)	Low (12)	High (65)	Very close (30 cm)	WAV2
14	Low (24)	Low (12)	High (65)	Close (40 cm)	WAV2
15	Low (24)	Low (12)	High (65)	Far (60 cm)	-
16	Low (24)	Low (12)	High (65)	Very far (70 cm)	-

Defuzzification produces firm values in the form of sound output produced according to the conditions of the blind shoe. There are five conditions, namely active1 functions when the shoe is outside the path of the blind (red), active2 is the condition where the shoe is facing close to the obstacle, active3 is the condition of the shoe is in the classroom or towards a blue classroom, active4 is the condition of the shoe is located in the toilet room or green, and the last one is active0 ie the shoes are in a condition in the lane, so the sound does not come out of earphone. All sounds will sound according to the condition of the shoes are.

2.4. Arduino programming

The blind shoe program in Arduino has been combined with an ultrasonic sensor program, useful when the blind shoe is faced with obstacles. The blind shoe output is sound. This sound is stored on the sd-card module in the way format, which is the sound recording format that will be issued when the shoe gets a response or input.

IOP Conf. Series: Materials Science and Engineering **885** (2020) 012006 doi:10.1088/1757-899X/885/1/012006

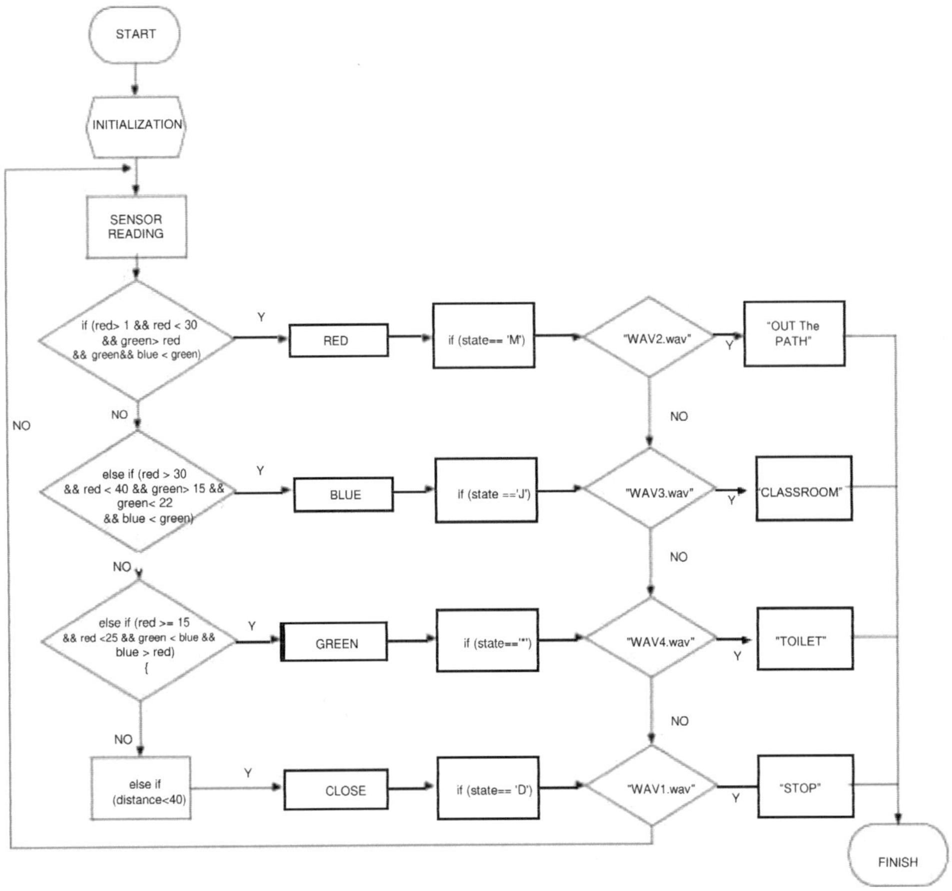

Figure 9. Arduino programming flowchart

3. Result

Figure 10 shows the color sensor and ultrasonic sensor when reading input in the form of color and obstruction. Shoes that have been designed as a direction for people with visual impairment can read red, blue, green, and yellow on the track plan. Shoes can also read distances, so obstacles can be identified. Communication between the right and left shoes using Bluetooth and output in the form of voice information through earphones.

Figure 10. Pointing shoes

The implementation of shoes as a direction is used for children with visual impairment in Special Schools (SLB). The path used is the path that is given color according to the sensor reading. The lane plan has 4 colors, each of which functions as a sign of location and direction, where children with visual impairment will walk on the yellow lane as a blind lane and the red lane as a marker or warning when leaving the lane.

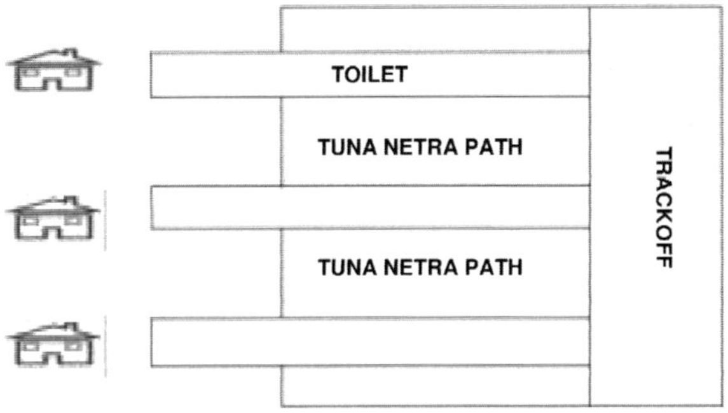

Figure 11. Plan direction shoe lane

Table 5 shows information related to the path position, obstacles or identification of hazards given in the form of voice messages via earphones.

Table 5. Sound, color and distance functions

No	Color and distance	Voice	Function
1	Red	"Out the Path"	Information when people with visual impairment are out of line
2	Blue	"Classroom"	Information when the blind are in front of the classroom
3	Green	"Toilet"	Information when people with visual impairment are in front of the bathroom (toilet)
4	Distance = <40	"Stop"	Information when visually impaired people are facing obstacles (hazard identification)

4. Conclusion

Shoes that are designed as a direction for the blind can function well, the color sensor can read input in the form of red, blue, green, and yellow according to the colors found in the lane plan. Ultrasonic sensors can read distances as identification of obstacles or hazards of a maximum of 300 cm or 3 meters. Fuzzy logic engineering is used in making comparisons to bring up the desired color depending on the output R, G, B values of the color sensor and depending on the distance between the sensor and the color object detected. Development can be done by adding an ultrasonic sensor that is pointing down in order to detect holes, and completing it with GPS as a location detector when people with visual impairment get lost, making it easy to find.

References

[1] Yu N and Wang S 2019 Enhanced Autonomous Exploration and Mapping of an Unknown Environment with the Fusion of Dual RGB-D Sensors *Engineering* **5** 164–72

[2] Martín F, Matellán V, Rodríguez F J and Ginés J 2019 Octree-based localization using RGB-D data for indoor robots *Eng. Appl. Artif. Intell.* **77** 177–85

[3] Petit A, Lippiello V, Fontanelli G A and Siciliano B 2017 Tracking elastic deformable objects with an RGB-D sensor for a pizza chef robot *Rob. Auton. Syst.* **88** 187–201

[4] Maiolino P, Woolley R, Branson D, Benardos P, Popov A and Ratchev S 2017 Flexible robot sealant dispensing cell using RGB-D sensor and off-line programming *Robot. Comput. Integr. Manuf.* **48** 188–95

[5] Bai X, Hu M, Gang T and Tian Q 2019 An ultrasonic sensor composed of a fiber Bragg grating with an air bubble for underwater object detection *Opt. Laser Technol.* **112** 467–72

[6] Chakraborty J, Katunin A, Klikowicz P and Salamak M 2019 Embedded ultrasonic transmission sensors and signal processing techniques for structural change detection in the Gliwice bridge *Procedia Struct. Integr.* **17** 387–94

[7] Li G, Li S E, Zou R, Liao Y and Cheng B 2019 Detection of road traffic participants using cost-effective arrayed ultrasonic sensors in low-speed traffic situations *Mech. Syst. Signal Process.* **132** 535–45

[8] Hassan M F A, Yusof Y, Azmi M A and Mazli M N 2012 Fuzzy Logic Based Intelligent Control of RGB Colour Classification System for Undergraduate Artificial Intelligence Laboratory *Proceedings of the World Congress on Engineering* vol 2 pp 4–6

[9] Liang-qun L, Xi-yang Z, Zong-xiang L and Wei-xin X 2018 Fuzzy logic approach to visual multi-object tracking *Neurocomputing* **281** 139–51

[10] Raju G and Nair M S 2014 A fast and efficient color image enhancement method based on fuzzy-logic and histogram *AEU-International J. Electron. Commun.* **68** 237–43

IOP Conf. Series: Materials Science and Engineering **885** (2020) 012007 doi:10.1088/1757-899X/885/1/012007

Unit commitment direct current optimal power flow using mixed-integer linear programming

M F Azis[1], J Habibuddin[2], Mutmainnah[3], T Muchtar[4], and D Purwanto[5]

[1,2,3,4]Electrical Engineering, Politeknik ATI Makassar
[5]Electronic and Computer Engineering, National Taiwan University of Science and Technology

Email: fadli@kemenperin.go.id, fadli@atim.ac.id

Abstract. This paper presents unit commitment direct current optimal power flow (UCDCOPF) using mixed-integer linear programming (MILP) for daily power system generation planning. This method is used to solve optimum operation in an electric power system by scheduling generator status and calculating the optimal power output of each generator in a power system. The calculation of this method considers several constraints in the power system. A modified IEEE 24 bus system is used as a test system to show the ability of the proposed approach.

1. Introduction

The main objective of power system operation is to minimize the operating cost. In achieving this, the optimization must not violate all constraints involved in the system. There are some operation level methods which are applied in the power system. Those operations are unit commitment (UC), economic dispatch (ED), and optimal power flow (OPF) [1].

Unit commitment (UC) is one of the power system operations to schedule all generators optimally in the system to meet the electricity demand. There are several operational constraints of the power system that must be accounted for UC. Those are minimum uptime, minimum downtime, startup cost, and shutdown cost. Minimum uptime is the minimum operation time of the unit to operate, whereas minimum downtime is the minimum operation time to not operate during the operation. The startup cost is an additional cost which is added to the unit when it is operated for the first time, whereas shutdown cost is an additional cost which added to the unit when it is committed from online to the offline condition [2].

Another power system operation is an economic dispatch (ED). This optimization is used to adjust the output of each online generator optimally. The constraints which are considered in ED are minimum stable generation, maximum generation, and power system balance. Minimum stable and maximum generation is a condition in which the online unit must inject the power with the value between both conditions. Power system balance is a condition in which the total supplied power to the system is equal to the system demand. Also, optimal power flow (OPF) is a development of ED by adding a transmission constraint [3].

Many studies related to power system operation has been conducted. R.S. Wibowo et al. [4–6] developed a dynamic DC optimal power flow using quadratic programming. This method solved OPF with multi-load levels by adding ramp rate constraint. V.S. Borra and K. Debnath [7] analyzed the comparison between dynamic programming and particle swarm optimization for solving unit commitment problems. Their simulation results demonstrated that the PSO technique is more accurate than DP technique in solving UC problems. T. Kim, G. Won and Y. Chung [8] designed an optimal power dispatch integrated with the unit using dynamic programming.

According to the existing result, in this paper, an integrated unit commitment direct current optimal power flow (UCDCOPF) will be calculated using a high-performance solver, mixed-integer linear programming (MILP), to achieve a better optimal result. To show the ability of the proposed approach, a modified IEEE 24 bus system is used as a test system.

2. Unit Commitment Direct Current Optimal Power Flow (UCDCOPF) Formulation

2.1. Objective Function

The objective of UCDCOPF is formulated as the minimization of $f(x)$ as presented in the equation below [9,10].

$$f(x) = f_p(p) + f_{uc}(v, w) \tag{1}$$

Where,

-power dispatch cost:

$$f(p) = \sum_{t \in T} \sum_{i \in I^t} [C_P^{ti}(p^{ti})] \tag{2}$$

t : Index of periods
T : Set of indices of periods
i : Index of units
I : Set of indices of units
C_P^{ti} : Cost function of real power injection
p^{ti} : Real power injection

-startup and shutdown cost:

$$f_{uc}(v, w) = \sum_{t \in T} \sum_{i \in I^t} (C_v^{ti} v^{ti} + C_w^{ti} w^{ti}) \tag{3}$$

C_v^{ti} : Startup cost
C_w^{ti} : Shutdown cost
v^{ti} : Binary startup state
w^{ti} : Binary shutdown state

2.2. DCOPF Constraints

The constraint of DCOPF consists of power balance and transmission limit. The power balance defines that the total load and the total generation must be equal to zero, while the transmission limit defines that the power flow in the branch must be below the line capacity limit. The formulation of these constraints can be seen as follows [1,2].

-power balance

$$g^t(p^t, \theta^t) = P_{Gn}^t + P_{Ln}^t + P_{INn}^t = 0 \tag{4}$$

$$P_{INn}^t = F_{nm}^t = \frac{1}{x_{nm}}(\theta_n^t - \theta_m^t) \tag{5}$$

-transmission limit

$$h^t(p^t, \theta^t) \leq 0 \tag{6}$$

$$F_{nm}^j \leq F_{nm}^{max} \tag{7}$$

Where,

g^t : DC power flow equations
h^t : Transmission limit
p^t, θ^t : Active power injection and voltage angle
P_{Gn}^t : Power generation at bus *n*
P_{Ln}^t : Power demand at bus *n*
P_{INn}^t : Sum of power flows at transmission lines connected to bus *n*
F_{nm}^j : Power flow at branch *nm*
F_{nm}^{max} : Line capacity of branch *nm*
x_{nm} : Reactance of branch *nm*
θ_n^t : Voltage angle at bus n
θ_m^t : Voltage angle at bus m

2.3. Unit Commitment Constraints

The purpose of unit commitment is to schedule all generations optimally in the power system for a given period. Several constraints of unit commitment, including maximum and minimum power generation, minimum uptime, minimum downtime, startup and shutdown condition, are considered as presented in the equation below [1,2].

-maximum generating capacity, minimum stable generation and commitment

$$u^{ti} P_{min}^{ti} \leq p^{ti} \leq u^{ti} P_{max}^{ti} \tag{8}$$

-minimum up and downtime

$$u^{ti} - u^{(t-1)i} = v^{ti} - w^{ti} \tag{9}$$

-startup and shutdown condition

$$\sum_{y=t-\tau_i^+}^{t} v^{yi} \leq u^{ti}, \quad \sum_{y=t-\tau_i^-}^{t} w^{yi} \leq 1 - u^{ti} \tag{10}$$

Where,
u^{ti} : Binary commitment state
v^{ti} : Binary startup state
w^{ti} : Binary shutdown state
P_{min}^{ti} : Minimum real power injection
P_{max}^{ti} : Maximum real power injection
τ_i^+ : Minimum uptime
τ_i^- : Minimum downtime

2.4. Intertemporal Constraints

The intertemporal constraints in this system define the upward and downward ramp rate of the unit in each period as presented in the equation as follows [1,2].

-ramp rate limit

$$0 \leq \delta_+^{ti} \leq \delta_{max+}^{ti} \tag{11}$$

$$0 \leq \delta_-^{ti} \leq \delta_{mux-}^{ti} \tag{12}$$

Where,
δ_+^{ti} : Upward load-following ramping reserve
δ_-^{ti} : Downward load-following ramping reserve
δ_{max+}^{ti} : Maximum upward load-following ramping reserve
δ_{max-}^{ti} : Maximum downward load-following ramping reserve

3. Study Simulation

The UCDCOPF problem is a mixed-integer linear programming (MILP) problem. The definition of MILP is a problem with conditions as follows [11].

$$\min_{x} f^T x \text{ subject to } \begin{cases} x \text{ are integers} \\ A \cdot x \leq beq \\ Aeq \cdot x = beq \\ lb \leq x \leq ub \end{cases}$$

1. The linear objective function, $f^T x$, where f is a column vector of constants, and x is the column vector of unknowns.
2. Bounds and linear constraints, but no nonlinear constraints.
3. Restrictions on some components of x to have integer values.

The problem formulation of UCDCOPF is implemented using MATLAB. Several toolboxes are used, including MATPOWER Optimal Scheduling Toolbox (MOST) and CPLEX to perform the MILP problem. The use of MOST in this simulation is to program the UCDCOPF. On the other hand, CPLEX is one of the high-performance solvers added in MATLAB to perform MILP calculation.

In this simulation, a modified IEEE 24 bus system is used [12]. This system consists of 12 thermal generation units and six wind farms. To show the effectiveness of UCDCOPF method, its result will be compared with dynamic DCOPF which uses linear programming for optimization and ignores the unit commitment constraints.

4. Result and Analysis

The main results of UCDCOPF are generation scheduling, power generation units and operation cost which can be seen in Table 1, Figure 1 and Figure 3. On the other hand, the main result of dynamic DCOPF are only power generation unit and operation cost which can be seen in Figure 2 and Figure 3. This is because the dynamic DCOPF method ignores the unit commitment constraints so that all the generation units are online over the 24 hours.

Table 1. Generation scheduling result of UCDCOPF

Unit-i	Period (h-i)																						
	1	2	3	4	5	6	7	8	9	10	11	12	13	14	15	16	17	18	19	20	21	22	23	24
Gen1	ON	ON	ON	ON	ON	ON	ON	ON	ON	ON	ON	ON	ON	ON	ON	ON	ON	ON	ON	ON	ON	ON	ON	ON
Gen2	ON	ON	ON	ON	ON	ON	ON	ON	ON	ON	ON	ON	ON	ON	ON	ON	ON	ON	ON	ON	ON	ON	ON	ON
Gen3	OFF	OFF	OFF	OFF	OFF	OFF	OFF	OFF	OFF	OFF	OFF	OFF	OFF	OFF	OFF	OFF	OFF	OFF	OFF	OFF	OFF	OFF	OFF	OFF
Gen4	ON	ON	ON	ON	ON	ON	ON	ON	ON	ON	ON	ON	ON	ON	ON	ON	ON	ON	ON	ON	ON	ON	ON	ON
Gen5	OFF	OFF	OFF	OFF	OFF	OFF	OFF	OFF	OFF	OFF	OFF	OFF	OFF	OFF	OFF	OFF	OFF	OFF	OFF	OFF	OFF	OFF	OFF	OFF
Gen6	ON	ON	ON	ON	ON	ON	ON	ON	ON	ON	ON	ON	ON	ON	ON	ON	ON	ON	ON	ON	ON	ON	ON	ON
Gen7	OFF	OFF	OFF	OFF	OFF	OFF	OFF	OFF	OFF	OFF	OFF	OFF	OFF	OFF	OFF	OFF	OFF	OFF	OFF	OFF	OFF	OFF	OFF	OFF
Gen8	ON	ON	ON	OFF	ON	OFF	ON	ON	ON	ON	ON	ON	ON	ON	ON	ON	ON	ON	ON	ON	ON	ON	ON	OFF
Gen9	ON	ON	ON	ON	ON	ON	ON	ON	ON	ON	ON	ON	ON	ON	ON	ON	ON	ON	ON	ON	ON	ON	ON	ON
Gen10	ON	ON	ON	ON	ON	ON	ON	ON	ON	ON	ON	ON	ON	ON	ON	ON	ON	ON	ON	ON	ON	ON	ON	ON
Gen11	ON	ON	ON	ON	ON	ON	ON	ON	ON	ON	ON	ON	ON	ON	ON	ON	ON	ON	ON	ON	ON	ON	ON	ON
Gen12	ON	ON	ON	ON	ON	ON	ON	ON	ON	ON	ON	ON	ON	ON	ON	ON	ON	ON	ON	ON	ON	ON	ON	ON

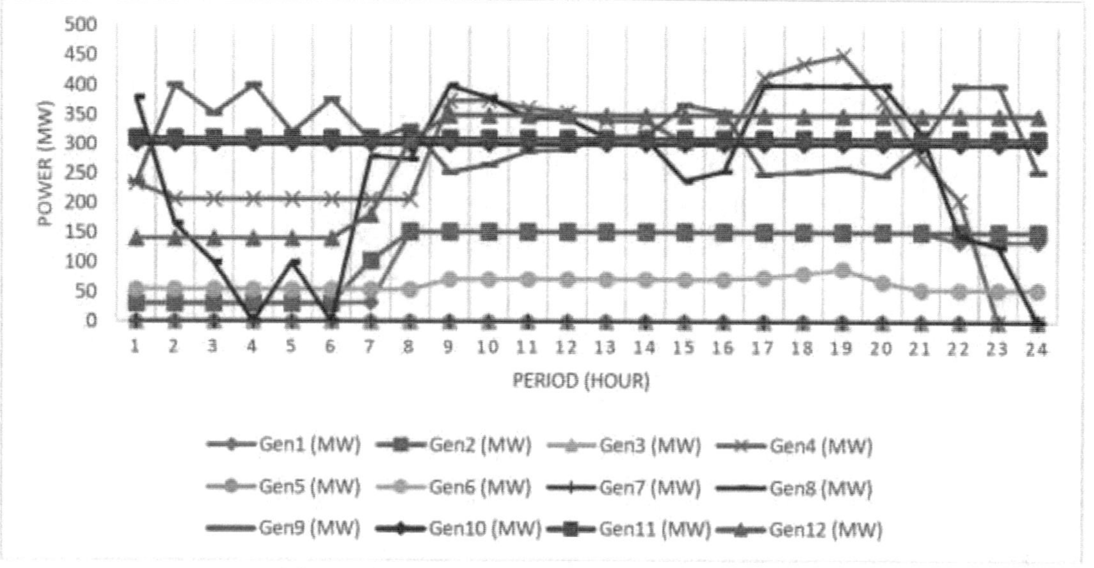

Figure 1. Power generation units of UCDCOPF

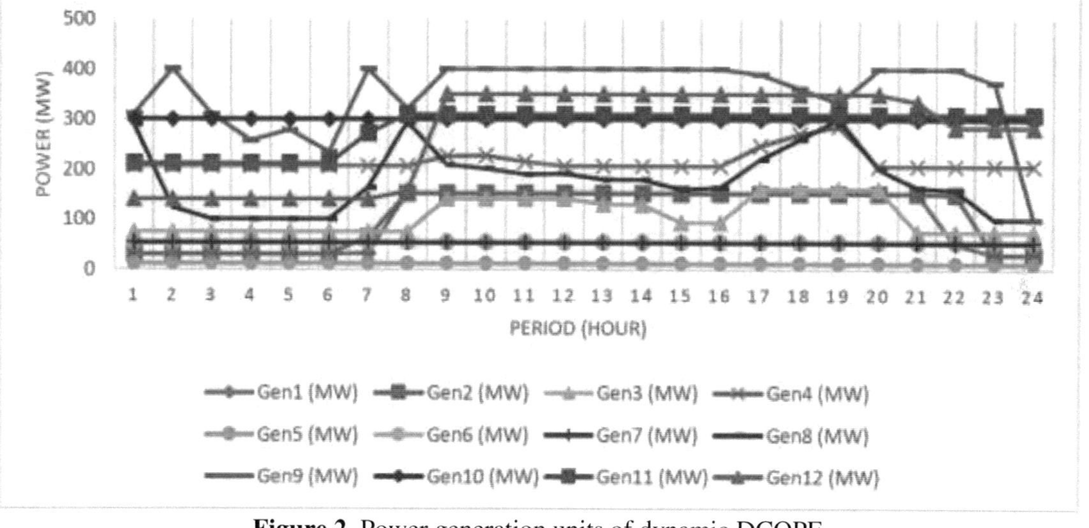

Figure 2. Power generation units of dynamic DCOPF

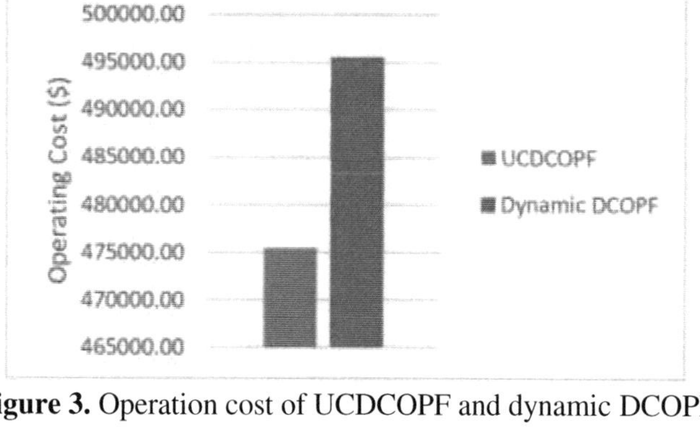

Figure 3. Operation cost of UCDCOPF and dynamic DCOPF

Based on the result, it can be seen that the UCDCOPF and dynamic DCOPF have different simulation results on generation scheduling, power generation units, and operation cost. This is because the UCDCOPF considers more constraints than the dynamic UCDCOPF, specifically the unit commitment constraints such as startup cost, shutdown cost, minimum uptime and minimum downtime. As a result, this will affect the value of power generation units and operation costs.

The most important result in this simulation is the operating cost, respectively $475392.30 for the UCDCOPF and $495465.90 for dynamic DCOPF. Based on Figure 3, it can be seen that the UCDCOPF has a lower cost than dynamic DCOPF with a difference of $20073.60.

5. Conclusions

In this paper, a unit commitment direct current optimal power flow (UCDCOPF) has been simulated and calculated by using the mixed-integer linear programming. Based on the simulation result, it can be concluded that the UCDCOPF has a more optimal operation cost than the dynamic DCOPF.

References

[1] Wood A J, Wollenberg B F and Sheblé G B 2013 *Power generation, operation, and control* (John Wiley & Sons)

[2] Kirschen D S and Strbac G 2018 *Fundamentals of power system economics* (John Wiley & Sons)

[3] Saadat H 2009 Power System Analysis,(2nd) *McGraw-Hill High. Educ.*

[4] Wibowo R S, Soeprijanto A and Penangsang O 2013 Dynamic DC optimal power flow using quadratic programming *2013 International Conference on Information Technology and Electrical Engineering (ICITEE)* (IEEE) pp 360–4

[5] Setiadi H and Jones K O 2016 Power system design using firefly algorithm for dynamic stability enhancement *Indones. J. Electr. Eng. Comput. Sci.* **1** 446–55

[6] Setiadi H, Krismanto A U, Mithulananthan N and Hossain M J 2018 Modal interaction of power systems with high penetration of renewable energy and BES systems *Int. J. Electr. Power Energy Syst.* **97** 385–95

[7] Borra V S and Debnath K 2019 Comparison between the dynamic programming and Particle Swarm Optimization for solving unit commitment problems *2019 IEEE Jordan International Joint Conference on Electrical Engineering and Information Technology (JEEIT)* (IEEE) pp 395–400

[8] Kim T-Y, Won G-H and Chung Y 2018 Optimal Dispatch and Unit Commitment Strategies for Multiple Diesel Generators in Shipboard Power System Using Dynamic Programming *2018 21st International Conference on Electrical Machines and Systems (ICEMS)* (IEEE) pp 2754–7

[9] Zimmerman R D, Murillo-Sánchez C E and Thomas R J 2010 MATPOWER: Steady-state operations, planning, and analysis tools for power systems research and education *IEEE Trans. power Syst.* **26** 12–9

[10] Murillo-Sánchez C E, Zimmerman R D, Anderson C L and Thomas R J 2013 Secure planning and operations of systems with stochastic sources, energy storage, and active demand *IEEE Trans. Smart Grid* **4** 2220–9

[11] Wolsey L A and Nemhauser G L 1999 *Integer and combinatorial optimization* vol 55 (John Wiley & Sons)

[12] Ordoudis C, Pinson P, González J M M and Zugno M 2016 An Updated Version of the IEEE RTS 24-Bus System for Electricity Market and Power System Operation Studies.

IOP Conf. Series: Materials Science and Engineering **885** (2020) 012008 doi:10.1088/1757-899X/885/1/012008

Solar-diesel hybrid power plant battery charging systems

Matius Sau[1], Hestikah Eirene Patoding[1], and Agustina Kasa[2]

[1]Electrical Engineering, Paulus Christian University of Indonesia, Makassar, Indonesia
[2]Mechanical Engineering, Paulus Christian University of Indonesia, Makassar, Indonesia

Email: sau_tpsdp@yahoo.com

Abstract. This study aims to examine the charging current of a solar energy hybrid generator with a Genset / Diesel and the time used to fill the accumulator in conditions without load and load conditions. The results showed that the solar energy hybrid generator and Genset / Diesel prototypes were able to produce electrical power at 08.00 - 17.00 in the average sunny weather of 290.7 Wp without loading and at the time of load of 236.4 Wp. At night, the energy stored in the battery is converted into electrical energy to supply the load. Thus it can save electrical energy.

1. Introduction

Today, the source of electrical energy that is of concern is solar energy and vegetable energy because it is easy to obtain and environmentally friendly. This situation helps the government in saving electric energy, especially in offices, hospitals, housing, and so on, because the electricity produced does not use fuel oil.

Indonesia's large enough area separated by land and sea is one of the factors of inequality in PLN electricity services so that the archipelago uses electricity generation separate from the interconnection system. Thus, PLN must build a PLTD or GENSET in serving customers or people who live in rural areas, especially in the islands. This situation is very influential in the development of current technology and rural economic growth. Seeing this condition, the available electric power capacity certainly cannot be sufficient anymore, so that another power plant is needed that is able to serve the needs of the community. However, the current conditions are no longer possible because of the investment costs of generation and the price of fuel oil, which is high enough so that the parties concerned are trying to save electricity.

1.1. The working principle of photovoltaic solar cells

Solar or in the international world, better known as a solar cell or photovoltaic cell, is a semiconductor device that has a wide surface and consists of a series of diodes of type p and n, which can convert the energy of sunlight into electrical energy [1,2].

Solar cells are arranged by combining p and type n silicon. P-type silicon is positive silicon due to lack of electrons, while n-type silicon is negative silicon due to excess electrons when it receives (charged) solar radiation (in the form of photons) in both (silicon p and n) forms positive (holes) and negative (electron). This causes polarization where the hole moves toward silicon type n. by connecting both types of silicon (type p and type n) through an external conductor, there is a potential difference between the two and direct current flows. This illustration is presented in Figure 1.

The definition of photovoltaic itself is the process of converting light into electrical energy [1,3]. Therefore the field of research related to solar energy is often also known as photovoltaic research [4].

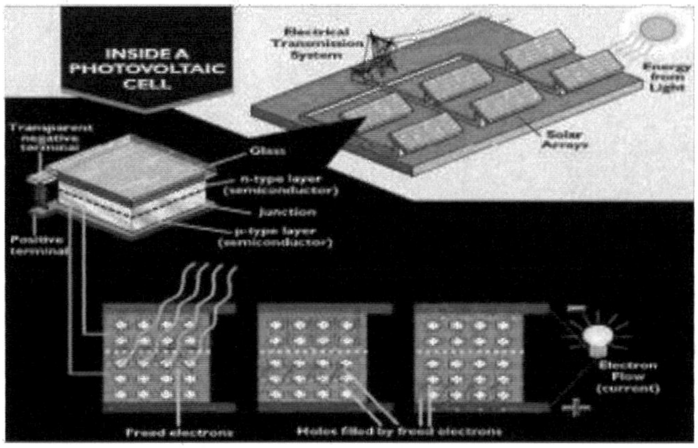

Figure 1. Working principle of photovoltaic cells.

1.2. Diesel / Genset Power Plant

The Diesel Power Plant in this study is stated as a generator set that works using fuel oil. Genset is a device that produces electrical power, which is obtained from the conversion of mechanical energy into electrical energy. Genset consists of two main devices, namely engine, and generator (G1), and is supported by a starter motor (M1), battery, and other control devices, as shown in Figure 7 [4,5]. If the start button is pressed, the starter motor M1 works and starts the engine, if the engine is working then the start button is released, then about 10 seconds the generator (G1) works then the MCB can be turned on to supply power. When the stop button is pressed, the engine will stop (Off).

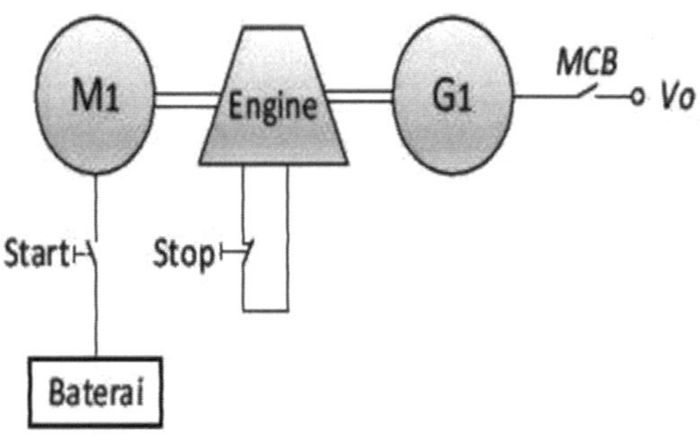

Figure 2. Single line Genset

1.3. Hybrid PLTS – Genset / Diesel (PLTD)

The term Hybrid is defined as the use of 2 or more power plants with different energy sources, generally used for captive generators, so that a synergy that provides economic and technical benefits means the reliability of the supply system.

The main purpose of a hybrid system is basically to try to combine two or more energy sources (generating systems) so that they can cover each other's weaknesses and can achieve supply reliability and economic efficiency on certain load profiles.

Type load (load profile) is an important keyword in a hybrid system. For each different load profile, a hybrid system with a certain composition is needed, so that the optimum system can be achieved. Therefore, system design and sizing systems play an important role in achieving the target of a hybrid system. For example, a relatively constant load profile for 24 hours can be supplied efficiently and economically by the generator (with the appropriate capacity), but the load profile where electricity usage during the day differs greatly compared to night time, will make the generator use not optimum. Hybrid PV-Genset [6] combination will reduce the generator operating hours (for example from 24 hours per day to only 4 hours per day at peak load only) so that O & M costs can be more efficient, while PLTS is used to supply base loads, so no initial investment is needed big. Thus Hybrid PV-Genset will be able to save O & M costs, reduce inefficiency in the use of generators, and at the same time, avoid the need for large initial investments.

Hybrid system modeling can be expressed in the form of a centralized AC and DC relationship, as shown in Figure 3.

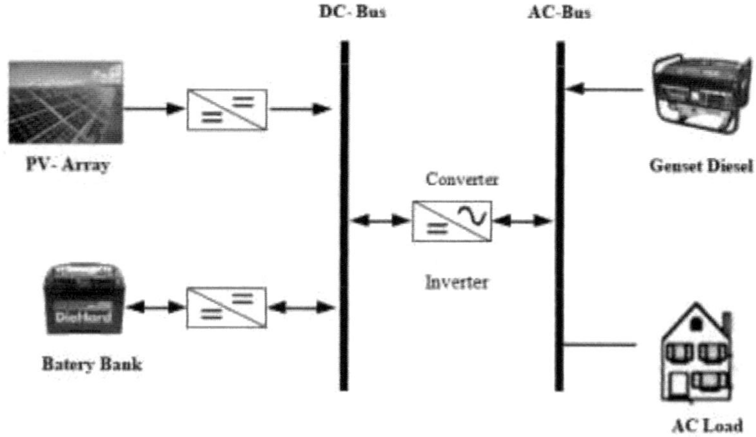

Figure 3. Centralized AC and DC coupled hybrid generating systems (Centralized AC and DC-coupled Hybrid Power Systems)

2. Research Method

This research was carried out in 3 stages:

2.1. Hybrid system

The making phase of a hybrid system prototype Experimental prototype for hybrid PLTS system with PLTD as in Figure 4 below:

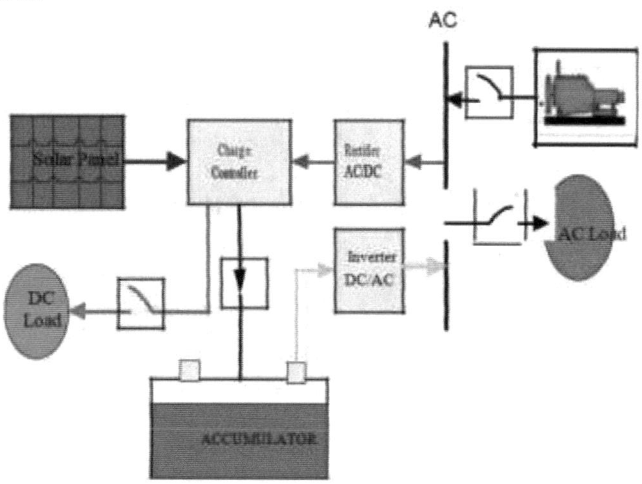

Figure 4. Hybrid system scheme for PLTS and PLTD/Genset

2.2. Testing and measurement phase

Based on the prototype of the solar energy hybrid generator and Genset / PLTD that has been made, it is tested with two conditions as follows:

IOP Conf. Series: Materials Science and Engineering **885** (2020) 012008 doi:10.1088/1757-899X/885/1/012008

a. Hybrid system between PLTS and Genset (PLTD) is operated in no-load conditions. Tests are carried out during the day by
   1. Solar energy is converted into electrical energy and then stored in the Accumulator while the electrical energy from the Genset is aligned using an AC-DC converter and dihybrid on the charge controller.
   2. DC load and AC load in the OFF position
   3. Measuring current and voltage from solar energy (PLTS and from Genset / PLTD 4. Measure the hybrid current and voltage in the charge controller
b. Hybrid systems between PLTS and Genset (PLTD) are operated in heavy conditions
   1. Solar energy is converted into electrical energy and then stored in the Accumulator while the electrical energy from the Genset is aligned using an AC-DC converter and hybridized on the charge controller
   2. DC load and AC load are ON
   3. Measuring current and voltage from solar energy (PLTS and from Genset / PLTD
   4. Measure the hybrid current and voltage in the charge controller. Measure current and voltage in the AC clutch

2.3. Analysis phase

At this stage, the results of the tests that have been carried out are then analyzed to find out the total power supplied and the total load that can be served by the hybrid system.

3. Results and Discussion

Prototype of a hybrid solar power generator and generator / PLTD system as shown in Figure 5.

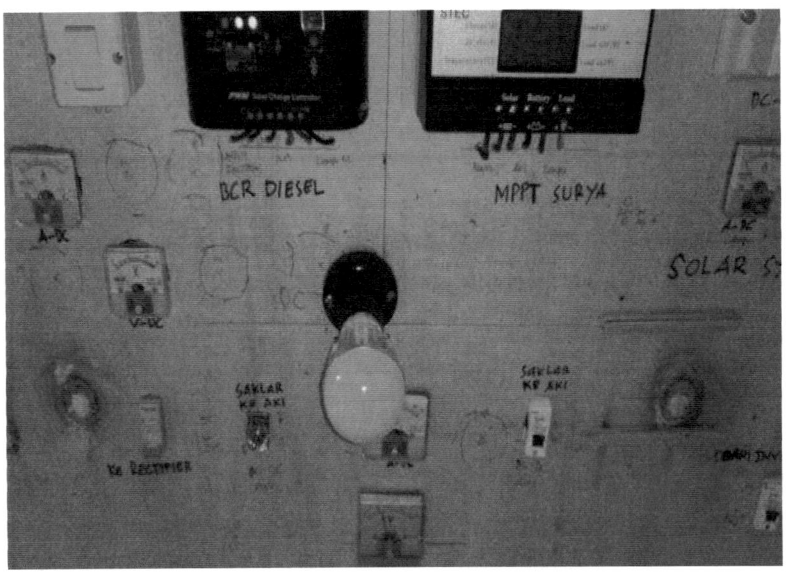

Figure 5. Prototype of hybrid PLTS system testing with Genset

3.1. Testing the Hybrid (PLTS – Genset) on Conditions without Loads

Table 1. The Current and Voltage of Hybrid PLTS systems with Genset for time during sunny days

Times	V Hybrid (V)	I Hybrid (A)
08.00-09.00	19	8.0
09.00-10.00	19.2	10.4

10.00-11.00	19.4	16.8
11.00-12.00	19.5	18.4
12.00-13.00	20.2	19.2
13.00-14.00	20	18.6
14.00-15.00	19.1	18.2
15.00-16.00	19	12.8
16.00-17.00	19	12.6

Based on the test results data in table 1, it can be seen that the relationship between the voltage and current of the Hybrid system is a no-load condition. At 12.00 - 13.00 West Indonesia Time, seen from the current Hybrid PLTS system output with the largest generator. This happens because the solar panel is perpendicular to the sunlight it absorbs

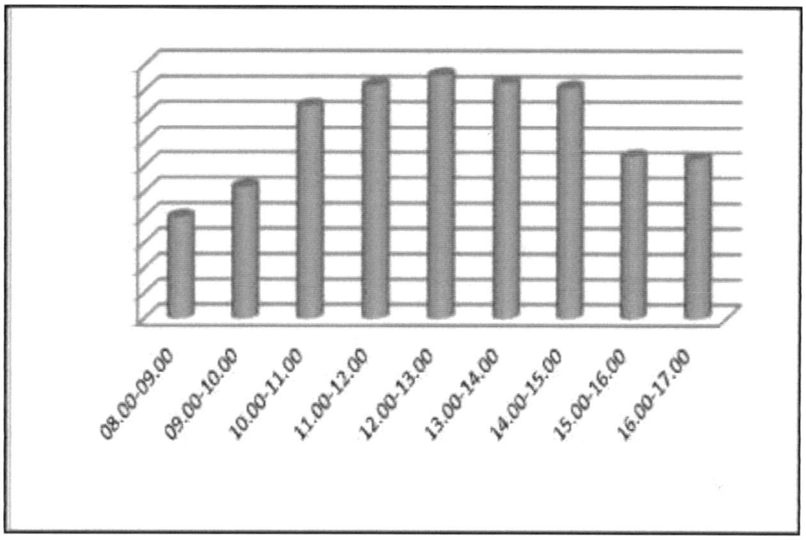

Figure 6. The current of the Hybrid PLTS system with Genset

3.2. Testing the Hybrid (PLTS – Genset) on Load Conditions

Table 2. The Current and Voltage of Hybrid PLTS systems with Genset for time during sunny days

Times	V Hybrid (V)	I Hybrid (A)
08.00-09.00	19	6.0
09.00-10.00	19.2	8.2
10.00-11.00	19.4	14.6
11.00-12.00	19.5	15.4
12.00-13.00	20.2	17.0
13.00-14.00	20	15.2
14.00-15.00	19.1	14.4
15.00-16.00	19	10.0
16.00-17.00	19	9.0

Based on the test results data in Table 2 shows the current and voltage of the Hybrid system are directly proportional. In overloaded conditions, there is a decrease in hybrid currents, this is because part of the current flows into the load. Thus, the charging time to the accumulator will be longer than the no-load condition.

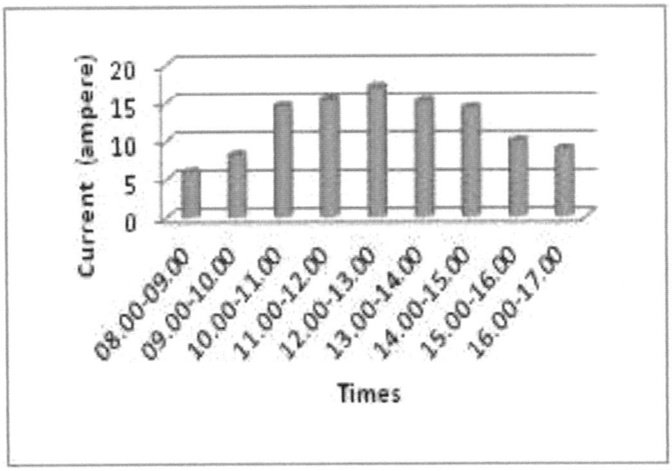

Figure 7. The current of the PLTS Hybrid system with Genset

3.3. Discussion

On average, the voltage for charging a hybrid system of PLTS with Genset for the time period 08.00 to 17.00 is 19.4Volt. The average for the hybrid system output of PLTS with Genset for the time range 08.00 to 17.00 is 15 Ampere. If this power is used to turn on the lamp load of 7 x 18 Watt or 126 Watt it can last for 2.31 hours while the average for charging voltage of hybrid systems with Genset PLTS for the time range 08.00 to 17.00 uses a load of 7 x 18 Watt of 19.4Volt, the average for the output current of the hybrid PLTS system with Genset for the time range 08.00 to 17.00 is 12.2 Ampere. If this power is used to turn on the lamp load of 7 x 18 Watt or 126 Watt, it can last for 1,876 hours. So charging from a Solar Power Plant and Genset for one hour produces an average voltage of 19.4 Volts with a current of I = 12.2 Ampere, capable of turning on a 126 Watt lamp load for ± 1,876 Hours Both conditions above show that the power stored in the accumulator at no-load conditions will longer turn on the lamp with 126 Watts of power compared to the conditions when charging accumulator while being loaded with a power of 126 Watt.

The test results show that the Genset directly supplies to the load and simultaneously fills the battery, while the PLTS only fills the battery; this is in accordance with the function of the tool made. So the batteries function as backup energy and store electricity from the two plants. If at any time the supply from the generator (Genset) goes out, then the source of supply to the load uses the battery.

4. Conclusion

Based on the results of prototype testing for conditions without and with a load it can be concluded that the prototype solar energy hybrid generator and Genset / Diesel are able to produce electrical power at 08.00 - 17.00 on a sunny average of 15 Ah at a voltage of 19.4 Volts or 290.7 Wp without loading and at an average load of 12.2 Ah at a voltage of 19.4 Volts or 236.4 WP. At night, the energy stored in the accumulator is converted into electrical energy to supply the load. Thus it can save electrical energy.

Acknowledgments

Thanks to the Ministry of Research, Technology and Higher Education Directorate General of Learning and Student Affairs and the Directorate of Research and Community Development that funded this Research.

IOP Conf. Series: Materials Science and Engineering **885** (2020) 012008 doi:10.1088/1757-899X/885/1/012008

References

[1] Commission C E 2001 A guide to photovoltaic (PV) system design and installation *Energy Technology Development Division* vol 1516

[2] Setyawati H, Hadi M, Darmokoesoemo H, Murwani I, Jaya Permana A and Rachman F 2020 Modification of Methyl Orange dye as a light harvester on solar cell *IOP Conference Series: Earth and Environmental Science* vol 456 p 12010

[3] Firdaus A A, Yunardi R T, Agustin E I, Nahdliyah S D N and Nugroho T A 2020 An improved control for MPPT based on FL-PSo to minimize oscillation in photovoltaic system *Int. J. Power Electron. Drive Syst.* **11** 1082–7

[4] El-wakil M M 1984 *Powerplant Technology* (Singapore: Mc Graw-Hill Book Company)

[5] Watson G O 2013 *Marine Electrical, Practice* (Butterworths: Elsevier)

[6] Bertheau P, Cader C, Müller H, Blechinger P, Seguin R and Breyer C 2014 Energy storage potential for solar based hybridization of off-grid diesel power plants in Tanzania *Energy Procedia* **46** 287–93

IOP Conf. Series: Materials Science and Engineering **885** (2020) 012009 doi:10.1088/1757-899X/885/1/012009

Power factor correction of the industrial electrical system during large induction motor starting using ETAP power station

Mutmainnah, Muhammad Fadli Azis and Julianti Habibuddin

Electrical Engineering, Makassar ATI Polytechnic, South Sulawesi, Indonesia

E-mail: innahrafrin@gmail.com

Abstract. This paper presents the power factor correction of the electrical system during the induction motor starting using ETAP power station. There are some problems faced by the electrical system when the motor is started, including voltage drop, high insulation current, and low power factor. The starting motor method used in this paper is starting with VFD. The aim of this research is to improve the power factor during the starting time. The system with a capacitor resulted in a low power factor that lowers the quality of electrical power. By adding capacitor 30 kVAR with each step 0.1 kVAR in the system, the power factor increased to 95.5%.

1. Introduction

Induction motors are the most widely used electric motors in the industry; therefore induction motors are the most widely produced throughout the world and their numbers continue to increase significantly. One of the benefits of an induction motor is that the starting motor does not need an external circuit for starting. However, this starting effect on the utility and on the motor itself where it cannot be ignored. When a voltage is applied to the motor, most of the voltage is taken from the utility which causes a decrease in the system voltage. Due to the highly inductive nature of the motor, the power factor at the start is quite low, usually decreasing by 10-20%. Therefore the most effective component when starting a motor is reactive power to improve the system power factor.

During the last few years, the starting motor method has become the focus of research. Various methods of induction motor starting are known, including the Direct On-Line (DOL) method, Starting Y-delta, Auto-transformer, soft starter, and VFD. Several studies have discussed how to overcome the decrease in power factor when starting a motor by using supporting devices such as capacitor and SVC. These devices are widely explained to overcome the problem of voltage drop during the starting of a large power induction motor and without affecting the dynamics of the induction motor. In the research of J. Williams [1] capacitor can improve electrical system performance during the process of starting large capacity motors and the research conducted by Ho-Chiao Chuang et al. [2] recalls the efficiency of induction motors and improvement of power factors by improving frequency and voltage using SCR.

IOP Conf. Series: Materials Science and Engineering **885** (2020) 012009 doi:10.1088/1757-899X/885/1/012009

2. Starting Analysis of Induction Motor

During the motor starting period, Induction motors is at rest, & it appears just like a short-circuited transformer and if connected to the full supply voltage, draw a very high current known as the "Locked Rotor Current." the Locked Rotor Current (LRC) is a function of the terminal voltage of the motor and the motor design. The current is about six times the motor rated current, which therefore results in voltage dip in the system and poses disturbances to the normal operation of other system loads. The magnitude of this dip is proportional to the magnitude of the surge and the impedance of the system. Because of the highly inductive nature of the motor circuit at rest, the power factor of the surge is quite low, usually on the order of 10 to 20 percent. As the motor accelerates to rated speed, the surge decays and the system voltage recovers. One of the starting methods is known as VFD.

VFD is an electronic electric motor starting aid device (commonly applied to pumps, air handlers, etc.). It functions by electronic power principles in varying the frequency of the input power to the motor, thereby controlling the motor speed. This gives reduced motor starting current, reduction in thermal and mechanical stresses on motor and belts during starting, etc [3].

3. Power Factor Correction

The cosine of the angle between voltage and current in an AC circuit is known as the power factor. In an AC circuit, there is generally a phase difference Ø between voltage and current. The term cos Ø is called the power factor of the circuit. If the circuit is inductive, the current lags behind the voltage and the power factor is referred to as lagging. However, in a capacitive circuit, the current leads the voltage and the power factor is said to be leading.

The analysis of the power factor can be made in terms of power drawn by the AC circuit. The power triangle OAB showed (figure 1). Where OA = VI cos Ø and represents the active power in watts of kW, AB = VI sin Ø and represents the reactive power in VAR of kVAR, OB = VI cos Ø and represents the apparent power in VA of kVA.

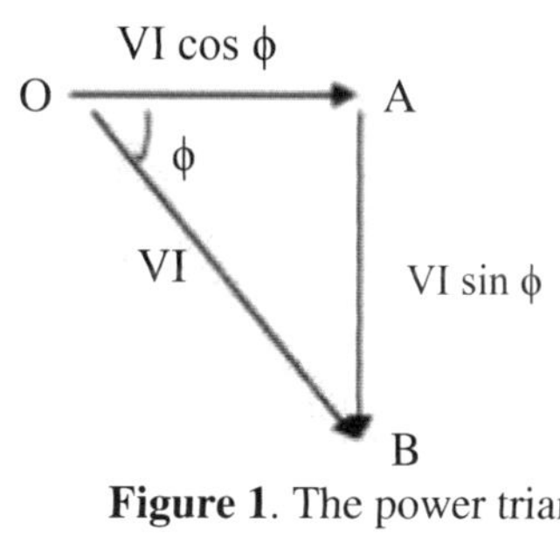

Figure 1. The power triangle

$$OB_2 = OA_2 + AB_2 \quad (1)$$

(Apparent power)2 = (active power)2 + (reactive power)

$$(kVA)2 = (kW)2 + (kVAR)2 \quad (2)$$

Cos ϕ = OA/OB, cos ϕ = power factor

IOP Conf. Series: Materials Science and Engineering **885** (2020) 012009 doi:10.1088/1757-899X/885/1/012009

The low power factor is mainly due to the fact that most of the power loads are inductive and, therefore, take lagging currents. In order to improve the power factor, some devices taking leading power should be connected in parallel with the load. One of such device can be a capacitor. The capacitor draws a leading current and partly or completely neutralizes the lagging reactive component of load current. This raises the power factor of the load.

The use of capacitors as starting aids to electrical motors are aimed at reducing the voltage dip experienced by neighboring loads during the in-rush moment of these electrical motors by connecting them to either the motor bus or the motor terminals [4], this is because these capacitors generate reactive power. They are known as capacitor banks or reactive power banks. One major merit of this starting aid method is its ability to improve the power factor of electrical machines when connected in parallel to the electrical machines [5,6]

4. Study Simulation

In this case, the single system line has two induction motors, 100 HP and 50 HP 0.4 kV, VFD (Variable Frequency Drive), Capacitor bank (0.1 Mvar), Static load 5 kVA, Dynamic load 10 kVA. Single line diagram is shown in (figure 2).

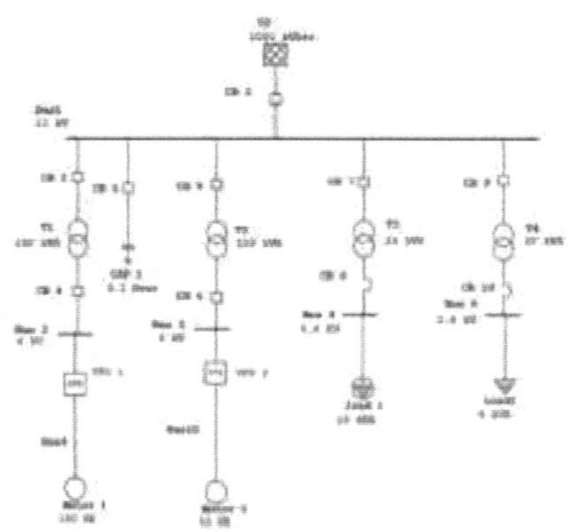

Figure 2. Single line diagram

5. Result and Analysis

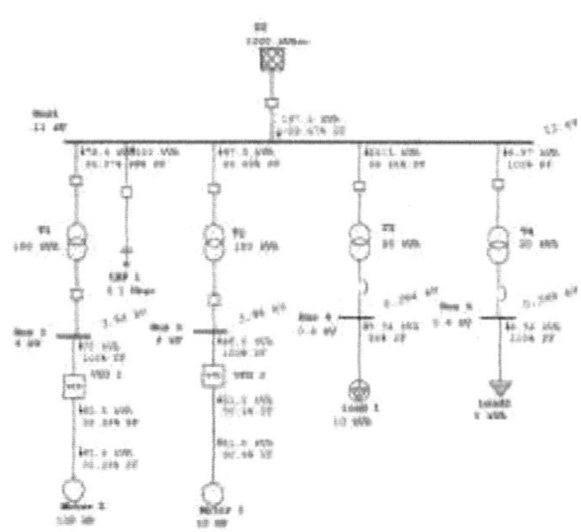

Figure 3. System load flow without capacitor

The case study for VFD starting adopts the system configuration shown in figure 3 with a low-voltage drive. The VFD is rated at VA and 0.4 kV at 50 Hz. The step-up transformer is rated at 100 kVA with multiple secondary taps, and its impedance is 5.2%. The required surface voltage under normal operating conditions is achieved by adjusting the secondary tap settings of the step-up transformer for various system configurations. Four starting frequencies, namely, 7, 8, 9, and 10 Hz, are applied to the VFD. Before installing the capacitor, the power factor of the motor is 84.5% as shown in (figure 3).

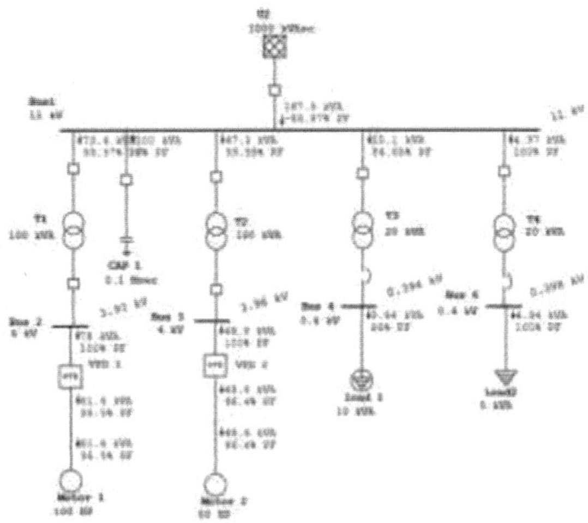

Figure 4. System load flow with capacitor

The figure shows the load flow of the circuit with a capacitor. The installed capacitor is a capacitor with a capacity of 30 kVAR. It shows that the power factor of the motor is improved to 95.5%. (Figure 5) shows the bus voltage of the motor and figure 6 shows the terminal voltage of the motor base.

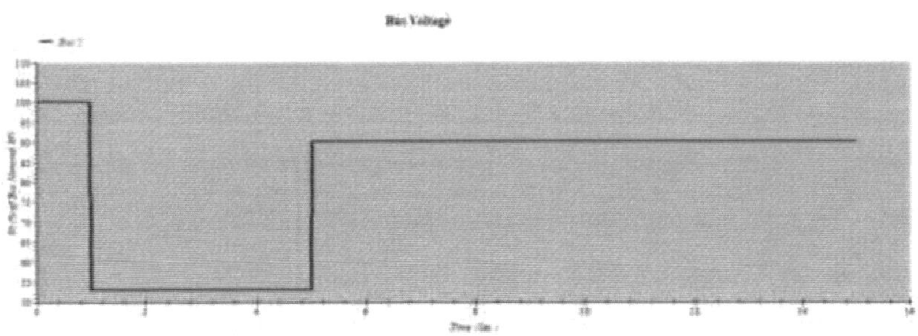

Figure 5. The bus voltage

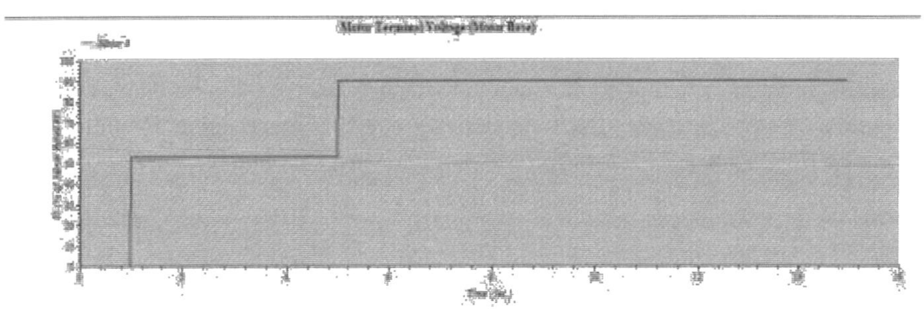

Figure 6. The terminal voltage of motor base

6. Conclusion

In this paper, the power factor of the system that consists of high-power induction motors can be improved by adding the capacitor bank in the system. To improve power factor from 84.5% to 95 % and to minimize the dip in system voltage, the system is installed with capacitor bank 30 kVAR with each step 0.1 kVAR. VFD is added to the system to reduce the suddenly high current in starting the induction motor.

References

[1] Williams A J and Griffith M S 1978 Evaluating the effects of motor starting on industrial and commercial power systems *IEEE Trans. Ind. Appl.* 292–305
[2] Chuang H-C and Lee C-T 2019 The efficiency improvement of AC induction motor with constant frequency technology *Energy* **174** 805–13
[3] Prihandana G S, Mahardika M, Hamdi M and Mitsui K 2009 The current methods for improving electrical discharge machining processes *Recent Patents Mech. Eng.* **2** 61–8
[4] Jabbar R A, Al-Dabbagh M, Akmal M and Lafit K 2002 Reducing Voltage dip on switching of heavy loads on power systems *IEE Trans. energy Convers.* **3** 147–52
[5] John W 2006 Power Factor correction and improvement on power systems *IEEE Trans. Ind. Appl.* **26** 719–25
[6] Setiadi H, Krismanto A U, Mithulananthan N and Hossain M J 2018 Modal interaction of power systems with high penetration of renewable energy and BES systems *Int. J. Electr. Power Energy Syst.* **97** 385–95

Study of microwaves energy conversion into electric current on semiconductor material $Na_xCu_yTi_z$

Regna Tri Jayanti[1], Sri Haryati[2], and M. Djoni Bustan[2]

[1]Departement of Chemical Mineral Engineering, Politeknik Industri Logam Morowali
[2]Departement of Chemical Engineering, Sriwijaya University

E-mail: regnatj@gmail.com

Abstract. Converting microwaves energy into electric current has been studied through co-precipitation and sintering process on the semiconductor material. Most of the application of these methods still had a challenge of lowering the resistivity value of semiconductor material. Some researches indicated by adding molten salt into semiconductor material can reduce the resistivity value. In this study, the synthesis process by combining the co-precipitation and sintering method of transition metal (titanium and copper) and alkali metal (Sodium) shown an improvement in lowering resistivity problems. Composition ratio, temperature, and distance of radiation were the variable of measurement in this research. Three semiconductor materials $Na_xCu_yTi_z$ were synthesized by using 60 ml of hydrochloric acid (HCl) and 80 ml of ammonium hydroxide (NH_4OH) as a reagent in co-precipitation method and sintering at temperature 500°C. An Olympus microscope was used to analyze the microstructures of all three materials. Fractions x, y, and z in the materials were determined by analysis of X-ray spectrograph model 2501. The resistivity values of three materials, namely as $Na_{0.52}Cu_{0.37}Ti_{0.11}$, $Na_{0.26}Cu_{0.39}Ti_{0.35}$, and $Na_{0.26}Cu_{0.67}Ti_{0.07}$ were obtained by varying the temperature of measurement. The microwave radiation was measured by observing the output voltage and current along with the radiation at a variable distance (60, 90, 10, 150, 180, and 210 cm) through the semiconductor material. The experiment showed that material $Na_{0.52}Cu_{0.37}Ti_{0.11}$ had low resistivity and small bandgap energy as 0.875 eV and able to convert microwave energy from 1.624 x 10^{-24} Watt to 9.137 Watt at 60 cm of microwaves radiation distance.

1. Introduction

Energy harvesting technologies seem developed to deliver green energy from diverse types of sources such as solar, thermal, kinetic, and electromagnetic by converting them into electric current [1]. Microwave as a part of an electromagnetic wave is one of alternative for efficient power transmission [2,3]. Microwave radiation can be directed to any desired location, can be collected and converted back to electricity, also can pass unimpeded through clouds and precipitations [4]. A number of methods have been used to produce semiconductor material in order to absorb microwaves and convert them into electric current such as sol-gel, solvothermal, sintering, and co-precipitation [5–7]. Most of the application of these methods still had a challenge to lowering sintering temperature and the resistivity value of semiconductor material [8,9]. Some researches indicated by adding molten salt into semiconductor material can reduce the resistivity value [10,11]. This study intended to make semiconductor material $Na_xCu_yTi_z$ in low resistivity and convert microwave energy to electrical current by adding molten salt (NaCl).

2. Materials and Methods

The Synthesis of semiconductor material $Na_xCu_yTi_z$ has been performed in material R1, R2, and R3 by co-precipitation and sintering method. The co-precipitation method was conducted by mixed the basic compound of material in order to get the pure homogeny composite. Whereas the sintering method was performed in purpose to obtain the compact semiconductor material.

2.1. Co-Precipitation of $Na_xCu_yTi_z$

Some materials Titanium dioxide (TiO_2), Sodium Chloride (NaCl), Copper dichloride ($CuCl_2$), Hydrochloric acid (HCl) 32%, and Ammonium Hydroxide (NH_4OH) 25% were prepared from Merck. Synthesis preparation started by determining composition ratio of semiconductor material. TiO_2 nanoparticles were prior synthesized by precipitation method of 80 ml Hydrochloric acid to make similar form as a variable of composition ratio with other (molten salt) compound NaCl and $CuCl_2$. Then, all three compounds (R1, R2, and R3) are synthesized by co-precipitation method by adding Sodium hydroxide (NH_4OH) and heating at the temperature 50°C along 30 minutes. The deposits were precipitated and rinsed to clean the remaining acid and base. The precipitation of $Na_xCu_yTi_z$ filtered and dried in the oven at temperature 100°C for two hours. The product of this co-precipitation process was powder $Na_xCu_yTi_z$.

Table 1. Variaton of Material Composition

Variation	R1	R2	R3
Material	%	%	%
NaCl	50	25	25
$TiCl_4$	25	50	25
$CuCl_2$	25	25	50

2.2. Sintering of $Na_xCu_yTi_z$

In order to make compact semiconductor material, material powder $Na_xCu_yTi_z$ was milled in the mortar, weighed to obtain the appropriate volume in the cast (15.07 cm^3) and pressed with the pressure 0.106 N/m^2. The material was sintered at temperature 500°C for 4 hours. Size of semiconductor material $Na_xCu_yTi_z$ was proposed to validate the effect of microwave distribution that converted to electric current. In this research, main material was designed with the diameter 8 cm and the thickness 0.3 cm and was put in the cast made from Aluminium (Al) with the diameter 10 cm and the thickness 1 cm.

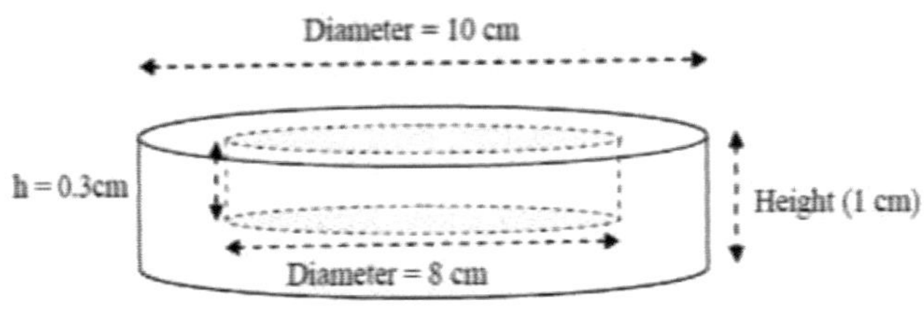

Figure 1. Design of the semiconductor material $Na_xCu_yTi_z$.

2.3. Measurement of Resistivity

The electrical and optical properties of a semiconductor are dependent upon the energy gap [12]. Measurement of resistivity material at various temperatures was intended to get the value of energy bandgap. Material is considered to be a semiconductor if energy band gap E_g is small (0 – 4 eV), and material is considered to be an insulator if E_g large (4 - 12 eV). Energy band gap for every semiconductor material is determined by the temperature [13]. The measurement of resistivity was obtained by measurement of output voltage and current. The measurement was applied three times with variation of temperature 303, 308, 313, 318, 323 ^{0}K, average data, and deviation standard used for calculation. The input voltage for resistivity measurement were 9 and 12 Volt. If current (I) flow through the material with the area (A) and length (L) which is connected with input voltage (V), the resistance can be obtained by the equation R = V/I. where the relationship between resistance (R) and resistivity (ρ) is written as follow:

$$\rho = R\frac{A}{L} \quad (1)$$

Because the resistivity (ρ) inversely proportional to conductivity (σ), so conductivity value could be achieved, and the energy bandgap (Eg) of material could be obtained by the equation (3).

$$\sigma = \frac{1}{\rho} \tag{2}$$

$$\ln \sigma = \ln \sigma_0 - \left(\frac{E_g}{2k_B}\right)\frac{1}{T} \tag{3}$$

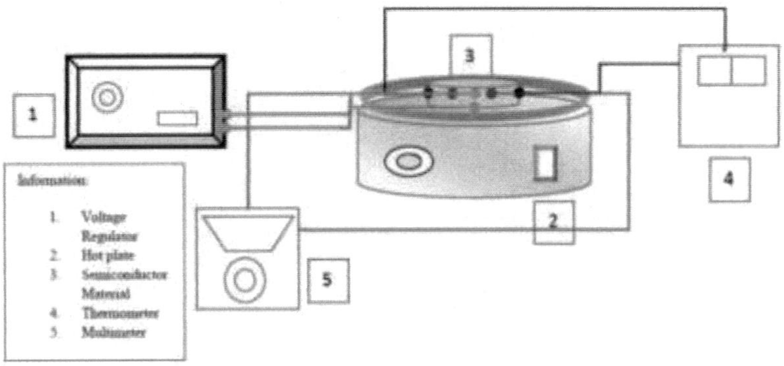

Figure 2. Scheme of Resistivity Measurement

2.4. Measurement of Microwave Radiation

Measurement of microwave radiation of semiconductor material $Na_xCu_yTi_z$ was proposed to determine the amount of electric current that could be converted for every second of microwave radiation. Microwave radiation measurement is done by observing the output voltage and current on digital multimeter KW0600276 along with the microwave radiation at variation distance 60 cm, 90 cm, 120 cm, 150 cm, 180 cm, and 210 cm. The energy source from microwave generator is 1.014 x 10^{-5} eV or similar to 1.624 x 10^{-24} Watt/s with 12.24 cm wavelength. This value was known from the input frequency of microwave generator used in this measurement (2.45 GHz). Microwave power generated by the microwave generator, then microwave power stabilized by stabilizer and transmitted to the magnetron. In the receiving side, the receiver 1 and receiver 2 reflected the microwave radiation in order to reduce radiation loss. The semiconductor material $Na_xCu_yTi_z$ transmitted the microwave power and converted it into direct current shown by the fluorescence of the light bulb. Then, voltage and electrical current of fluorescence of light bulb measured by a multimeter.

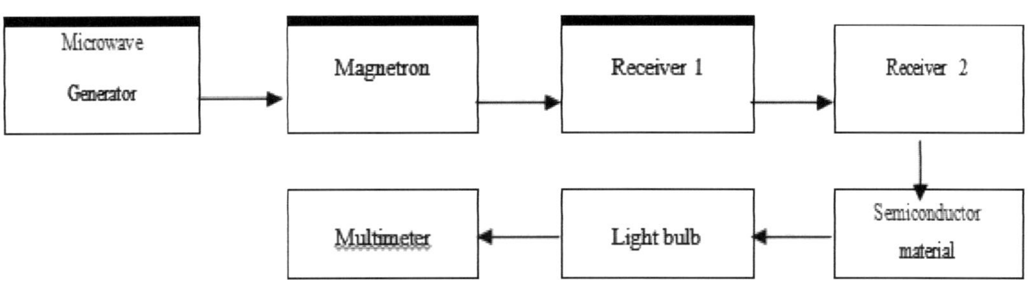

Figure 3. Block Diagram of Microwave Radiation

3. Result and Discussion

3.1. Synthesis of $Na_xCu_yTi_z$

Microstructure of semiconductor material $Na_xCu_yTi_z$ was observed by Olympus microscope D-20 with 11 times magnification.

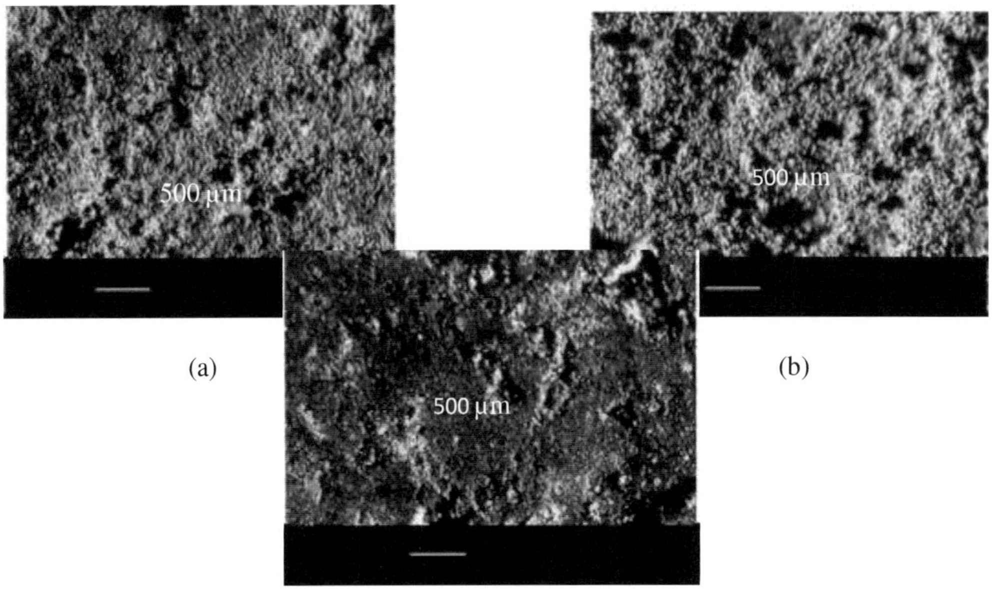

(a) (b)

(c)

Figure 4. (a) Microstructure of material R1, (b) Microstructure of material R2, and (c) Microstructure of material R3.

As shown in figure 4 (a), (b), and (c), the Synthesis of material $Na_xCu_yTi_z$ by co-precipitation and sintering at 500^oC affected the grains in compact material. Semiconductor material R1 with a high composition of molten salt (NaCl) has better particles grown than R2, and R3 where the porous of material formed in such a diameter around 30.9 µm, 43.6 µm, and 45.09 µm. All of the three materials seem colored light blue because of the reaction between copper chloride ($CuCl_2$) and ammonium hydroxide (NH_4OH). Value of fraction x, y, and z of the semiconductor material Na $_xCu_yTi_z$ was measured by X-Ray spectrograph model 5201. The fraction value of semiconductor material R1, R2, and R3 were Na0.5Cu0.36Ti0.11, Na0.25Cu0.38Ti0.34, and Na0.25Cu0.64Ti0.07.

3.2 Band Gap Energy of Semiconductor Material $Na_xCu_yTi_z$

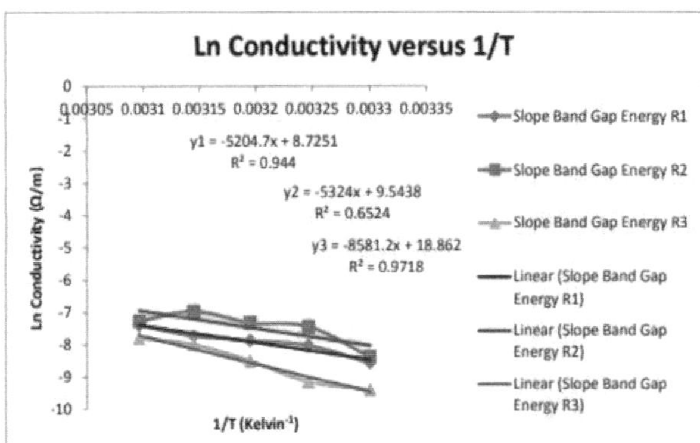

(a)

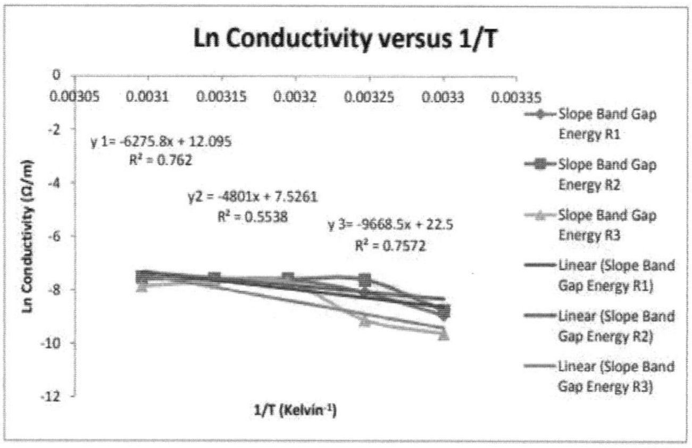

(b)

Figure 5. The slope of bandgap energy with input voltage (a) 9 Volt and (b) 12 Volt.

The graph of ln σ versus 1/T linear equation of y = ax + b. Where, ln σ = y, ln σ_0 = a, $-Eg/(2k_BT)$ = b, and 1/T = x. Slope b refers to bandgap calculation, while k_B is Bolztmann constant = 8.617 x 10^{-5} eV/K. R^2 is correlation coefficient. For both input voltages (9 and 12 volt), increasing the temperature tends to decrease the resistivity of materials R1, R2, and R3. As the conductivity is inverse of resistivity, the value of band gap energy of materials R1, R2, and R3 for each input voltage 9 and 12 were 0.875, 0.985, and 1.575 eV.

3.3 Converting Microwave into Electrical Current

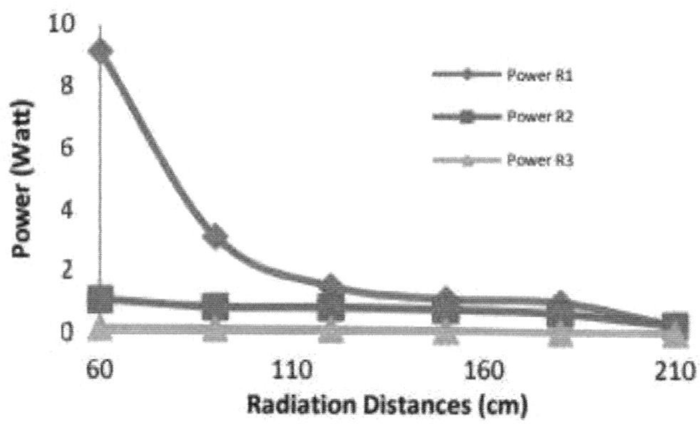

Figure 6. Microwave Power versus Radiation Distance

According to the graph of microwave power conversion to radiation distance in figure 6, the production power of material R1, R2, and R3 were decreasing as the farther distance of microwave radiation. Material R1 converted microwave energy from 1.624 x 10^{-24} watt to 9.137 watts at a distance of microwave radiation 60 cm while material R2 and R3 converted it around 1.09 watt and 0.168 watts.

4. Conclusions

The Synthesis of material $Na_xCu_yTi_z$ by using co-precipitation and sintering methods could produce semiconductor material in order to convert microwave energy into the electrical current with low resistivity and bandgap energy. Material $Na_{0.5}Cu_{0.36}Ti_{0.11}$ was a semiconductor material with the highest composition of molten salt (NaCl) produced electrical power as 9.137 watts in radiation distance 60 cm.

References

[1] Niotaki K, Collado A, Georgiadis A, Kim S and Tentzeris M M 2014 Solar/electromagnetic energy harvesting and wireless power transmission *Proc. IEEE* **102** 1712–22

[2] Sonawane J and Benare S 2017 Wireless Power Transmission Using Interational *J. Innov. Res. Comput. Commun. Eng* **5** 5143–5148

[3] Reddy M V, Hemanth K S and Mohan C H V 2013 Microwave power transmission–a next generation power transmission system *IOSR J. Electr. Electron. Eng.* **4** 24–8

[4] Satavekar M S G 2014 Solar Power Satellites and Microwave Wireless Power Transmision Technology *Adv. Electron. Electr. Eng* **4** 193–200

[5] Abbas S I, John H T and Fraih A J 2017 Preparation of nano crystalline zinc–ferrite as material for micro waves absorption by sol-gel methods *Indian J. Sci. Technol.* **10** 1–6

[6] Rashad M M, Ibrahim I A, Osama I and Shalan A E 2014 Distinction between SnO 2 nanoparticles synthesized using co-precipitation and solvothermal methods for the photovoltaic efficiency of dye-sensitized solar cells *Bull. Mater. Sci.* **37** 903–9

[7] Zabotto F L, Gualdi A J, Eiras J A, Oliveira A J A de and Garcia D 2012 Influence of the sintering temperature on the magnetic and electric properties of NiFe2O4 ferrites *Mater. Res.* **15** 428–33

[8] Nofianti R D and M Darminto S 2008 Nanopartikel, N. Zn, X. F. O. Dengan, and M. Kopresipitasi, Sintesis Nanopartikel Ni1-xZnxFe2O4 dengan Metode Kopresipitasi *Indones. J. Mater. Sci.* 225–227

[9] Deng L, Zhao Y, Xie Z, Liu Z, Tao C and Deng R 2018 Magnetic and microwave absorbing properties of low-temperature sintered BaZr x Fe (12− x) O 19 *RSC Adv.* **8** 42009–16

[10] Rattanaphan S and Kanlahasuth P 2016 Enhanced Electrical Conductivity by Modifying LiCl-KCl Mole Fraction at High Temperature *Int. J. Adv. Sci. Eng. Technol.* **4** 201–206

[11] Kolay P K, Burra S G and Kumar S 2018 Effect of salt and NAPL on electrical resistivity of fine-grained soil-sand mixtures *Int. J. Geotech. Eng.* **12** 13–9

[12] Strehlow W H and Cook E L 1973 Compilation of energy band gaps in elemental and binary compound semiconductors and insulators *J. Phys. Chem. Ref. Data* **2** 163–200

[13] Jorena J 2009 Menentukan Energi Gap Semikonduktor Silikon Melalui Pengukuran Resistansi Bahan pada Suhu Beragam *J. Penelit. Sains* **12**

IOP Conf. Series: Materials Science and Engineering **885** (2020) 012011 doi:10.1088/1757-899X/885/1/012011

The effect of active carbon adsorbents from some wastes in reducing free fatty acids and acid number to improve vco quality

Dimas Frananta Simatupang[1], Jenny Tarigan[1], and Mansyur[2]

[1] Department of Chemical Engineering, Politeknik Teknologi Kimia Industri Medan, 20142 North Sumatera, Indonesia
[2] Department of Mechanical Engineering, Politeknik Teknologi Kimia Industri Medan, 20142 North Sumatera Indonesia

Email: difratas@ptki.ac.id

Abstract. The use of several wastes such as durian outer skin, cassava outer skin and corncobs could be used to improve VCO quality. The wastes were converted into activated carbon as an adsorbent that was activated using sodium carbonate in purifying VCO in terms of reducing free fatty acid (FFA) and acid number. VCO used was oil that has made directly in the laboratory of oleo chemistry PTKI Medan and tested to reduce level of FFA and acid numbers. A total of 5 ml VCO samples were used to determine reducing in level of FFA and acid numbers by a titrimetric method with three times repetition using 0.1 N NaOH and previously treated with activated carbon adsorbent from durian outer skin, cassava outer skin and corncobs as much as 1 gr, 2 gr, and 3 gr. Based on the results of the examination, it was found that the level of FFA and acid number of VCO had a significant decrease with the best-activated carbon adsorbent using waste from cassava outer skin where level of FFA decreased by 0.16% and acid number decreased by 0.448.

1. Introduction

Virgin Coconut Oil (VCO) is a kind of vegetable oil which become more popular nowadays because of its functions related to human health and high-value material for the health food market [1,2]. Many studies related to the benefit of VCO were done, such as people who consume VCO have fewer incidences of the digestive complaint, cancer, prostate and cardiovascular diseases [3]. Generally, the content of coconut (*Cocos nucifera*) fruit is 51.7% kernel, 9.8% water, and 38.5% shell [4]. Mature coconut meat contains 65% saturated fatty acids with medium-chain or called as medium-chain saturated fatty acids (MCFAs) and the chemical content of VCO such as phenolic compounds, volatile compounds and act as antioxidant activity made this oil become widely used for food and industrial purposes [5-7].

VCO could be obtained using several methods. VCO is extracted from mature coconut meat and processed through physical methods without any contamination by chemical materials [8]. In addition, VCO is not only extracted through physical method but involved chemical method using microorganisms for fermentation process with heat or without heat and centrifugation method [9-10].

Research on VCO quality showed that the highest oil yield was through the centrifugation method, but through freezing and thawing were the most preferred in sensory properties [11]. The quality of VCO influenced by many factors such as quality raw materials, processing techniques, packaging method and storage conditions. One parameter of quality control for oil is free fatty acids (FFA) and acid number. The number of FFA might be increased because of several factors; therefore it delivered bad quality of VCO. One technique to reduce FFA and AN is using activated carbon to adsorb oils. Adsorption is a technical operation widely used in industrial applications. Carbon with excellent surface characteristics and specific functionalities should be developed to create a high affinity for the adsorption process. Activated carbon has been extensively used as material for purification contaminated or uncontaminated oil and for increasing oil quality [12].

In this research, we manufactured activated carbon from some local wastes such as durian and cassava outer skin and corncobs. This study aims to reduce FFA and AN in VCO by adsorption of modified activated carbon.

2. Experimental Methods

VCO directly produced in Laboratory of Oleochemistry Politeknik Teknologi Kimia Industri (PTKI) Medan through fermentation method using commercial tempe's yeast. Durian outer skin, cassava outer skin and corncobs were obtained from local wastes.

2.1. Production of Virgin Coconut Oil (VCO)

Mature coconut fruit was cut and separated coconut meat and water. Then coconut meat was grated and obtained 940 gram. As much 940 mL of coconut water was mixed with coconut meat, then it was squeezed using filter woof and produced coconut milk. Coconut milk was placed in separating funnel and let it until formed three layers. The very bottom layer was discarded and poured as much 2.5 gram of yeast and fermented for 24 hours. Oil and coconut milk formed after 24 hours and collected oil (VCO) and centrifuged with 3000 x g for 5 min.

2.2. Preparation of Activated Carbon

Durian outer skin, cassava outer skin, and corncobs as local wastes were cleaned then cut into small pieces. All sample was dried by sunlight for three days and placed them in a container. It was burnt for 3 hours until formed charcoal or carbon. Furthermore, carbon was placed in desiccator and mashed up until 100 mesh. Carbon was activated with sodium carbonate 0.1 N with ratio 1:2 (m:v) for 30 min and over-heated in furnace for 30 min then cooled down in desiccator. The activated carbon was washed with aquadest to remove activator and heated in oven to remove water content.

2.3. Adsorption Technique

An experiment was performed in 250 mL Erlenmeyer flask with 50 mL VCO and mixed with 1 gram, 2 gram, and 3 gram activated carbon, respectively. It was shaked with bar stirrer for 24 hours. The supernatant was used to determine FFA and AN.

2.4. Determination of Free Fatty Acid (FFA) and Acid Number (AN)

The titration method was performed to determine FFA and AN for VCO in condition previous and after adsorption process with activated carbon. As much, 5 gram of VCO was placed in Erlenmeyer flask and mixed with 50 mL ethanol p.a. Then it was heated for 10 min and shaked vigorously to dissolve FFA. After it was cooled, then titrated with 0.1 N KOH in addition to phenolphthalein until reach permanent faint pink color. The amount of FFA and AN were calculated using this following formula

Where C is the concentration of potassium hydroxide for standard titration (N), V is titration volume of potassium hydroxide (mL), Mw is the molecular weight of fatty acid (gram/mole), M is the mass of VCO (gram), and CF is the conversion factor. VCO contains most of lauric acid as major fatty acid composition [13]. Therefore this study used lauric acid with conversion factor 2.8.

3. Results and Discussion

VCO, which directly produced through fermentation method was first analyzed the FFA and AN as one parameter to know the VCO quality. %FFA for VCO before conducted adsorption process was 0.4% and the acid number was 1.12, as shown in Table 1. The value of %FFA from this sample was higher than 0.2% for commercial VCO and not suitable for Indonesia's National Standard for VCO (max 0.2%). It was included as non-consumable VCO because of many factors such as some contamination of other microbes or not steril and not in good condition of raw materials. Oxidation reaction in VCO might have occurred while in storage conditions [14]. Oxidation could be delivered free fatty acids because it attacked unsaturated fatty acids in the oil and formed peroxide [15]. FFA might be formed because of hydrolysis reaction. This reaction could damage oil because of the amount of water in oil, which accelerated by heat, acidity, and impurities.

Table 1. %FFA and acid number of VCO before immersion with activated carbon.

Sample	%FFA	Acid Number
VCO	0.4	1.12

Three different sources of carbon from local wastes named durian outer skin, cassava outer skin, and corncobs were chosen as candidate for manufacturing activated carbon as bio-adsorbents. All sample was collected from local waste and easy to find them in large amount. Utilization of local waste is one way to reduce waste in Indonesia, especially in North Sumatera, Medan, and recycling them into a new form of a good product. Based on the result of titration, the %FFA of VCO, which treated with activated carbon in various mass was shown in Figure 1.

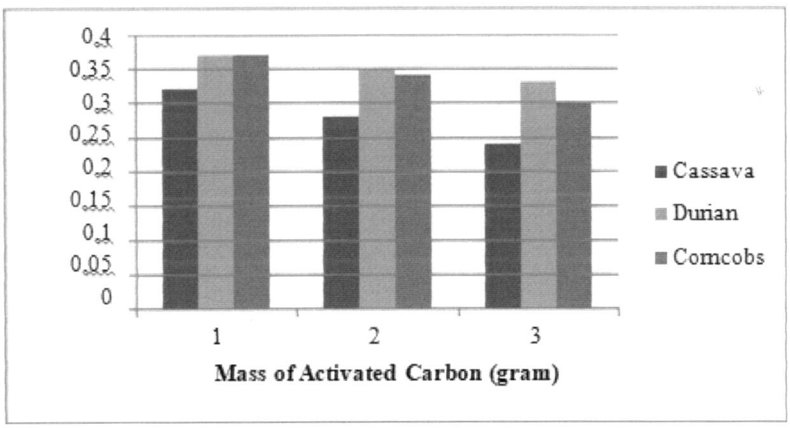

Figure 1. Profil of %FFA content in VCO.

In Figure 1, the profile of %FFA content in VCO after immersion with various mass (1-3 gram) of activated carbon for 24 hours. The activated carbon has successfully reduced %FFA in VCO with various ranges. Activated carbon made from cassava outer skin was the best one to reduce %FFA by using 3 gram and %FFA was reduced from 0.4% to 0.24%. It was obtained that activated carbon made from cassava outer skin could reduce %FFA as much 0.16%. This value of %FFA was closer to Indonesia's Standard National with max 0.2% of FFA in VCO. Based on Figure 1, it was obtained that activated carbon made from durian outer skin and corncobs has also reduced %FFA from VCO, but the value of %FFA decrease after immersion was not higher than using activated carbon made from cassava outer skin. Activated carbon made from durian outer skin could be reduced %FFA of VCO as much 0.07% and from corncobs as much 0.1%.

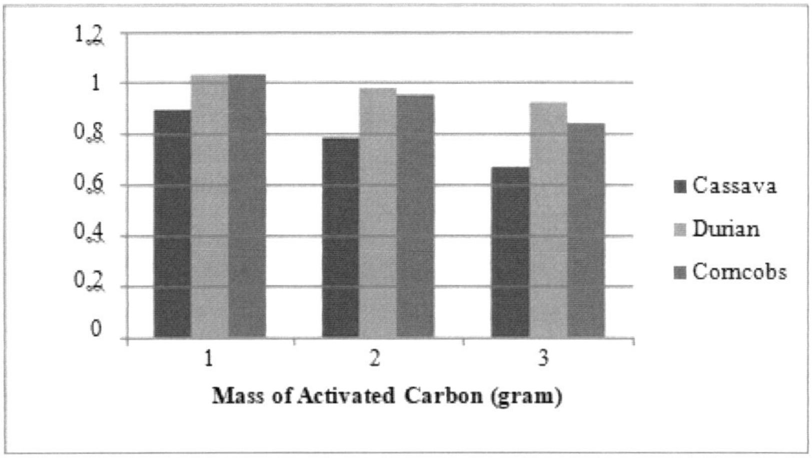

Figure 2. Profil of acid number in VCO.

The result of acid number of VCO could be determined directly with the same method, and the result was shown in Figure 2. The profile of acid number of VCO was same with the profile of %FFA. Activated carbon made from outer cassava skin was able to reduce the highest value of acid number of VCO as much 0.448. All activated carbon source was a good candidate to reduce %FFA of VCO but different ability in the adsorption process. Durian outer skin and corncobs were able to reduce %FFA of VCO as much 0.1% using mass of 1 gram, but it was different using mass of 2 gram which corncobs was better to reduce %FFA of VCO. More mass of activated carbon used, more %FFA of VCO reduced.

Activated carbon was one material that could be reduced %FFA and acid number of VCO and made better VCO quality. Local wastes were modified into carbon or charcoal and activated using chemicals; therefore it has ability to adsorb oil. This activated carbon called bio-adsorbent has a wide surface area to contact with the sample and increased adsorption capacity.

4. Conclusion

This study presents production of VCO in lab-scale using fermentation method, synthesized activated carbon made from some local wastes such as cassava outer skin, durian outer skin, and corncobs and improve VCO quality using activated carbon through determination of %FFA and acid number. The result of this study delivered that outer cassava skin was the best carbon source for synthesized activated carbon and able to reduce the highest value of %FFA and acid number of VCO. More mass of activated carbon used, more %FFA, and acid number of VCO reduced.

References

[1] Agarwal RK and Bosco SJD 2017 *MOJ Food Process Technol.* **4(2)** 54–56
[2] Erawati T, Retnowati, Firdaus A W and Soeratri W 2017 *Int. J. Drug Deliv. Technol.* **7** 42–4
[3] Marina MA, Chen-Man YB, Amin A. 2009 *Trends in Food Sci. and Technol* **20(10)** 4817
[4] Patil U and Benjakul S. 2018 *J. Food Science* 1–9
[5] Debmandal M and Mandal S. 2011 *Asian Pac. J. Trop. Med.* **4(3)** 241–247
[6] Mulyadi AF, Schreiner M and Dewi IA. 2018 *Proc. The 8th Annual Basic Science Int. Conf.* 2021
[7] Ghani NAA, Channip AA, Chok Hwee Hwa P, Ja'afar F, Yasin HM and Usman A. 2018 *Food Science and Nutrition* **6(5)** 1298–1306
[8] Mulyadi AF, Schreiner M and Dewi IA. 2019 *Proc. Int. Conf. Biology and Applied Science* 2120

[9] Tagle FR. 2018 *Proc. ICEAST* **192** 01045 0–3
[10] Mansor TST, Che Man YE, Shuhaimi M, Abdul Afiq MJ and Ku Nurul FKM. 2012 *Int. Fd. Rsrch. Jour.* **19(3)** 837–845
[11] Ndife J, Obot D and Abasiekong K. 2019 *Asian Food Science Journal* **8(4)** 1–10
[12] Sivakumar V, Asaithambi M and Sivakumar P. 2012 *Advances in Applied Science Research* **3(1)** 219–226.
[13] Asmoro N, Afriyanti, Widyastuti R and Ndrudu J. 2018 *Proc. Int. Conf. Appl. Sci. Eng.* **175** 74–77
[14] Rahayu S, Supriyatin and Bintari A. 2018 *Proc.The 9th Int. Conf. Global Resource Conservation and AJI from Ritsumeikan Univ.* 050004
[15] Lee J, Lee S, Lee H, Park K and Choe E. 2002 *J. Agr. Food Chem.* **50** 5664-5669

IOP Conf. Series: Materials Science and Engineering **885** (2020) 012012 doi:10.1088/1757-899X/885/1/012012

Effect of mixed additions in residual oil to improve the quality of sub-bituminous coal from sangatta, East Kalimantan

Flaviana Yohanala Prista Tyassena and Widya Ashari Amin

Department of Mineral Chemical Engineering, Politeknik ATI Makassar, Sulawesi Selatan, Indonesia

Email: flaviana.yohanala@atim.ac.id

Abstract. Indonesia is well-known for its abundant coal reserves but is included in low-rank coal with a total moisture content of 40%. In the combustion process, this water content will reduce the calorific value so that the use of coal becomes inefficient and not environmentally friendly. The purpose of this study was to determine the effect of added additives to residual oil to improve the quality of sub-bituminous coal originating from Sangatta, East Kalimantan. Variations of additives used include NaOH, HCl, and Pertalite, mixed with residual oil. The results in this study indicate an increase in calorific value in each composition of mixed additives. The best results were obtained for residual oil mixtures with pertalite, which obtained coal with a calorific value of 7204 cal/g, moisture content 16.9%, ash content 3.5%, carbon content 40.7%, sulfur content 0.41%, and flying substance content 38.9%.

1. Introduction

The amount of volcanic activity makes Indonesia rich in mineral content, one of which is coal. In 2009 coal production in East Kalimantan was 123,256,163 tons, in 2010 it rose to 140,753,374 tons and in 2011 production had reached 204,989,756 tons [1]. However, coal reserves in Indonesia are generally low-rank coal. Based on BP Statistical Review of World Energy data, June 2012, Indonesia has 5,529 billion tons of coal resource reserves, with the percentage of high-quality coal and low-quality coal, respectively 27.49%.

The type of coal is influenced by the process of its formation. One type of coal that is commonly found in Sangatta, East Kalimantan, is the Sub Bituminus type. Sub Bituminous contains little carbon and a lot of water content. In the combustion process, water content will reduce the calorific value and the amount of coal used is much greater. It makes a negative environmental impact. Therefore sub-bituminous coal is a less efficient source of heat compared to bituminous, which has higher calorific value. Sub-bituminous coal has calorific value for about 4500 – 5000 cal/g, while bituminous coal has a calorific value of more than 7000 cal/g [2].

Used oil is a class of B3 waste because used oil can cause the soil to become barren and lose its nutrients. While its nature, which is not soluble in water, can cause water pollution, besides that oil is also flammable [3]. The nature of flammable used oil can be utilized to improve the quality of coal.

Improving the quality of coal can be done by upgrading. Several studies that have been done previously show that with the upgrading process, the calorific value of coal will increase so that it will be more efficient to be used as an energy source, for example, in steam power plants. Several studies have been carried out, among others, by Arisandy et al. (2017), the process of improving the quality of sub-bituminous coal using a process by adding residual oil and premium that is heated at a temperature of 150-200°C results in water content, and sulfur content decreases, while ash content, fly levels increased [2]. The disadvantages of this research are that the levels of ash and flying matter have increased, and this will reduce the heating value and environmental pollution. Therefore in this study, additional NaOH, HCl, and pertalite are used in combination with used oil and are expected to increase the calorific value of coal without increasing sulfur and ash content.

2. Experimental Methods

2.1. Size Reduction
Sub-bituminous coal from Sangatta, East Borneo was first crushed using jaw crusher then sieved to get particle below 3 mm.

2.2. Desulphurisation
Coal that has been crushed before then dried in dry sheet for about 18 hours at 40°C. After obtaining a constant weight, coal then mixed by each solvent. For the desulphurisation method, coal was mixed with used oil and hydrochloride acid in a ratio of 2: 1: 1. The mixture was put into a beaker glass then heated in hotplate at temperature 200°C for 75 minutes. The mixture was filtered at room temperature to get the coal particle. And then the coal was put back into the dry sheet for further drying.

2.3. Demineralisation
For demineralization, coal, which has been crushed before, was mixed by used oil and sodium chloride in a ratio of 2: 1: 1. As in desulphurization method, the mixture was put into a beaker glass then heated in hotplate at temperature 200°C for 75 minutes. The mixture was filtered at room temperature to get the coal particle. And then the coal was put back into the dry sheet for further drying.

2.4. Pertalite Addition
The third method is addition method. In this method, some solvent that has high calorific value was added into the coal. For addition method, pertalite was chosen as additives. As the other methods, coal, which has been crushed before, was mixed by used oil and pertalite in a ratio of 2: 1: 1. As in desulphurization method, the mixture was put into a beaker glass then heated in hotplate at temperature 200°C for 75 minutes. The mixture was filtered at room temperature to get the coal particle. And then the coal was put back into the dry sheet for further drying.

2.5. Analysis
The filtered coal sample was dried in dry sheed for several days. After it was really dry, the coal was analyzed. Analysis conducted includes inherent moisture, ash content, volatile matter, fixed carbon, total sulfur, and calorific value. Inherent moisture, ash content, volatile matter, and fixed carbon analysis using laboratory instrument and chemical apparatus. Total sulfur was analyzed using a sulfur analyzer. Furthermore, the calorific value was analyzed using a bomb calorimeter.

3. Results and Discussion
In improving the quality of coal three methods can be used, namely: desulphurization, demineralization, and the addition of a solution (Pertalite), all three methods are used in this study to improve the quality of coal. The principle of the desulfurization method is to improve the quality of coal by reducing the sulfur content of coal. The principle of the demineralization method is to improve the quality of coal by reducing ash content, while the last method is to add a solution containing high calorific value.

In this research, several compositions of mixed materials were used, namely used oil as an adsorbate, HCl as a solvent in the desulphurization method, NaOH as a solvent in the demineralization method, and pertalite as a solvent in the method of adding solutions.

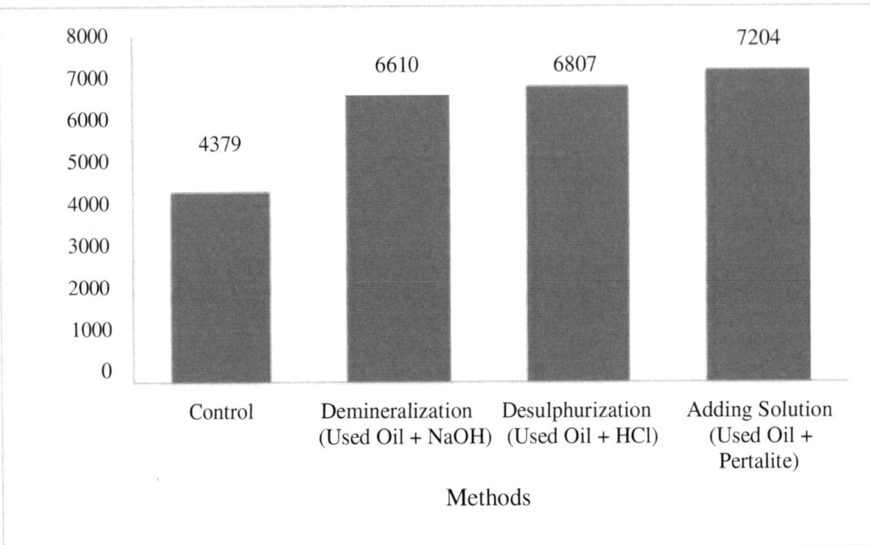

Figure 1. Effect of method variation in improving calorific value of coal.

Based on the results of the study as shown in Figure 4.1, for a combination of used oil and NaOH (demineralization method) has a calorific value of 6,610 cal / g, for a combination of used oil and HCl (desulfurization method) of 6,807 cal / g, and combination used oil and pertalite (method of adding solution) is 7,204 cal / g.

The combination of used oil and pertalite can increase the calorific value of coal from 4,379 cal / g to 7,204 cal / g and provides the highest value compared to the other two solvents. This increase in calorific value is due to the high calorific value of pertalite, which is 44,260 kJ / kg [4]. while used, oil will facilitate the absorption of pertalite into coal. This combination of used oil and pertalite can make this sub-bituminous coal has the same calorific value as bituminous coal (more than 7000 cal/g), the highest rank of coal.

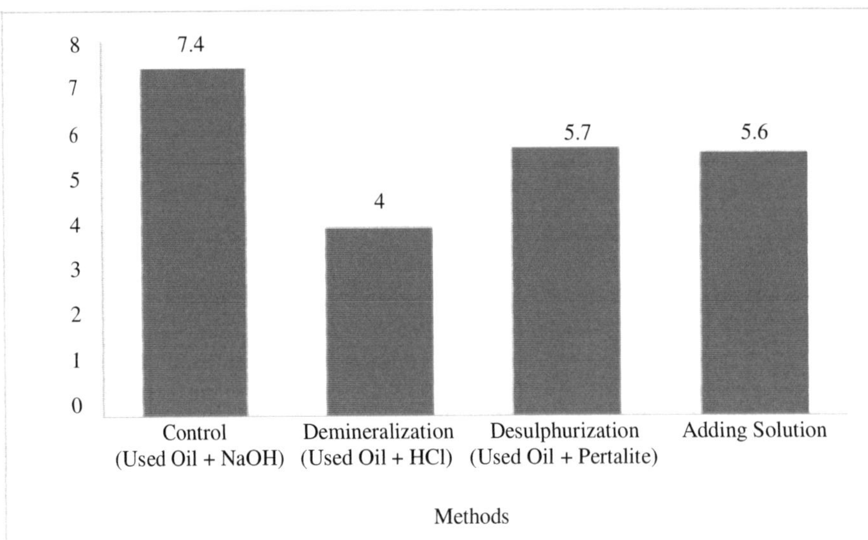

Figure 2. Effect of method variation in reducing sulphur content of coal.

Besides calorific values, sulfur content and ash content are also important parameters in determining coal quality. Low levels of coal ash and sulfur content are desirable, so coals will not cause adverse effects on the environment. The combination of used oil solvents and pertalite can increase the calorific value without increasing the sulfur content and ash content of coal. Figure 2 shows the sulfur content before and after treatment. Sulfur content from untreated sub-bituminous coal was 0.44%. While the sulfur content after being treated was 0.43%. Meanwhile, Hydrochloric Acid can reduce the sulfur content from 0.44% to 0.37%.

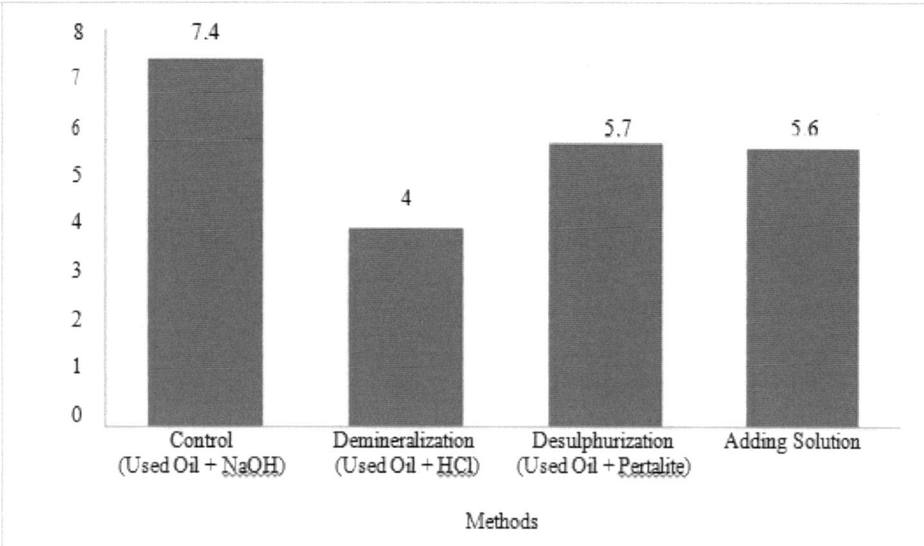

Figure 3. Effect of method variation in reducing ash content of coal.

Figure 3 shows the ash content of sub-bituminous coal before and after treatment. The combination of used oil solvents and pertalite can also reduce the ash content of coal from 7.4% to 5.6%. However, Sodium Hydroxide is the best additive to reduce ash content. Sodium hydroxide can dissolve some minerals content in coal, such as Si, Al, Fe, Cr, Ti, Mn, Mg, Na, and K [5].

4. Conclusion

The combination of used oil and pertalite can increase the calorific value of sub-bituminous coal from Sangatta, East Kalimantan, without increasing sulfur and ash content. The calorific value obtained was 7,204 cal / g with sulfur content of 0.43% and an ash content of 5.6%. This result gives the highest calorific value compared to the combination of used oil + NaOH and used oil + HCl.

References

[1] Badan Pusat Statistik 2012 Statistik Daerah Provinsi Kalimantan Timur 2012 *Badan Pus. Stat.*
[2] Arisandy A A and Uzaimi A 2017 Peningkatan Kualitas Batubara Sub Bituminous Menggunakan Minyak Residu di PT. X Samarinda, Kalimantan Timur *J. Teknol. Miner. FT UNMUL* **5**
[3] Mukhlisoh 2016 *Dampak Oli Bekas Terhadap Lingkungan* (Jakarta: Gramedia Pustaka Utama)
[4] Hafiz 2016 Spesifikasi Petralite Uji Eksperimental Nilai Kalori Bahan Bakar
[5] Sukandarrumidi 1995 *Batubara dan Gambut* (Yogyakarta: University Gadjah Mada Press)

Quality test of liquid tofu waste in one of the industries in Bantaeng district

Neny Rasnyanti M Aras

Akademi Komunitas Industri Manufaktur

Email: neny.rasnyanti@gmail.com

The tofu industry is growing rapidly, along with the increasing number of people in Indonesia. On the other hand, industrial liquid waste contains very high organic compound that potentially pollutes the environment. Without a good handling process, tofu liquid waste can cause a variety of negative impacts such as water pollution, unpleasant odor, and lower aesthetic to the environment. Most tofu industries are run by small and medium industries so that the disposal of its waste is less attention. Liquid waste from the tofu industry can be used to produce biogas through an anaerobic process if it is well processed. The aim of this research is to measure the physical and chemical parameters of liquid waste in one of the tofu industries in Bantaeng by taking samples at three different points. Based on the physical parameters obtained temperature, TDS, TSS, smell and color from three points for Point I are 29.6°C, 1932 mg/L, 7690 mg/L, smelly and white-colored, point II are 27.8°C, 540 mg/L, 1646 mg/L, smelled, murky-colored, point III temperature of 29°C, 510 mg/L, 288 mg/L. Based on chemical parameters, the tofu liquid waste pH, BOD (Biochemical Oxygen Demand), and COD (Chemical Oxygen Demand) for the point I, II, and III in a row are pH (7-7.8); BOD 7.5 mg/L, 3.3 mg/L, 9.6 mg/L; COD 1405 mg/L, 930 mg/L, 370 mg/L. From the parameters that have been measured. It be can be concluded that some parameters for industrial tofu liquid waste is still unqualified to the requirements of the quality standards set by KEPMENLH No. 5 in 2014. The unqualified standards are the TSS parameter and COD so that further research is required in order for the industrial liquid waste can be disposed of to the environment.

1. Introduction

Wastewater is water that comes from solids, both soluble and insoluble, which comes from human activities that contain hazardous substances and can damage the health of the environment and humans [1]. Wastewater from industries, especially household industries, has not been well managed before being discharged into water flow. Around 50 m^3 /ha/day of domestic waste produced that does not undertake a wet process.

One of the home industries that produce a lot of liquid waste is the tofu industry [2]. According to Darsono's report in 2007, the number of tofu industry in Indonesia reached 84,000 business units, with a production capacity of more than 2.56 million tons per year. Wastewater discharged into the environment is around 20 million cubic meters per year. Wastewater consists of water (99.9%), and the remainder consists of dissolved solids and suspended solid particles of 0.1%. Solid particles of inorganic substances (± 30%) and organic substances (± 70%) of this 70 % are protein (± 65%), carbohydrates (± 25%), fat (± 25%) [3].

Tofu waste contains high organic matter [4], because the raw material for making tofu (soybean) contains protein up to 40 - 60%. Tofu waste has 8640 mg / L COD, 297.5 mg / L total Nitrogen, COD: N ratio is 203: 7 [5].

The presence of organic compounds causes the liquid waste of tofu industry to contain high Biochemical Oxygen Demand (BOD), Chemical Oxygen Demand (COD), and Total Suspended Solid (TSS). High TSS causes several disadvantages, including blocked drainage due to sludge formation. Sedimented mud will cause odors because the sun is blocked from entering the water. This directly reduces the amount of dissolved oxygen due to the disrupted photosynthesis process [6].

The amount of dissolved oxygen is also related to the BOD and COD parameters. Biochemical Oxygen Demand (BOD) is the amount of dissolved oxygen needed by microorganisms (usually bacteria) to decompose organic matter under aerobic conditions [7,8]. Whereas COD or Chemical Oxygen Demand is the amount of oxygen needed to decompose all organic matter in water [9].

According to the Regulation of the Minister of Environment of the Republic of Indonesia Number 5 of 2014 concerning Wastewater Quality Standards, the TSS of wastewater from soybean process is 200 mg/L. Most households, industrial wastes such as liquid waste from tofu factories, do not become aware of wastewater treatment technology (IPAL) due to expensive operational costs and the complexity of the operating system.

One of the liquid waste's utilization is as fertilizer and pesticide and even organic fungicide with additional other materials, including using empon-empon or herbal plants through fermentation. Tofu liquid waste contains a lot of residual protein and vinegar so that it can boost the effectiveness of fermentation [10].

2. Materials and Methods

2.1. Research Place

This research was conducted at the Laboratory of Basic Testing Laboratory of AK-Manufacturing Bantaeng, Chemical Analysis Study Program, and the sampling of tofu waste in the Tala-tala, Bontoala Village, Kec. Bissapu, Kab. Bantaeng at the three points, the first point (T1) taking directly at the disposal source, the second point (T2) is about 20 meters from the tofu waste disposal stream and the third point (T3) at the flow to paddy fields

2.2. Tools and Materials

Materials: tofu wastewater in the Tala-tala tofu factory, Kelurahan Bontoala, Kec. Bissapu, Kab. Bantaeng, washing solution and BOD, COD analysis reagent with grade pro analysis, Whatman filter paper and reagents - reagents according to SNI procedures.

Tools: buckets, dipper, plastic jerry cans, sample bottles, furnaces, grab samplers, pH meters, thermometers, cool boxes, and laboratory equipment for BOD-COD analysis.

Methods: After the samples are taken, then measured the temperature and pH. After that proceed with a physical observation of wastewater. The samples were then analyzed with the BOD5, COD and TSS parameters and compared with the wastewater quality standard according to the Regulation of the Minister of Environment of the Republic of Indonesia Number 5 of 2014 [11].

2.3. Testing Methods for Physical and Chemical Parameters

2.3.1. Physical Parameters

2.3.1.1. Observation of color, odor, and temperature. Odor color observations are carried out using the organolepic test. While the temperature is done with thermometers.

2.3.1.2. Chemical Parameters

a. Determination of pH Value
The pH value is determined by using Germany-Original pH paper

b. Determination of TDS / TSS
TDS and TSS testing is carried out in accordance with the procedures in SNI 06-6989.3-2004

Formula : $\text{mg/L TSS} = \frac{(A - B) \times 1000}{\text{mL sample}}$ (1)

Annotation:
A = Weight of filter paper containing suspended substances (mg)
B = Weight of empty filter paper (mg)

c. Determination of BOD
A number of test samples were added to an oxygen-saturated diluent solution that had been added with a nutrient solution and microbial seedlings, then incubated in a dark room at 20°C ± 1°C for 5 days. The BOD is calculated based on the difference in dissolved oxygen concentration in 0 (zero) days until 5 five days. This analysis method is in accordance with SNI 6989.72: 2009 [12].

d. Determination of COD
Pipette the volume of the test sample, add a digestion solution, and add an adequate sulfuric acid reagent solution to the tube or ampoule. Cover the tube and shake gently until it is homogeneous. Put the tube on the heater that has been heated at 150 °C, reflux for 2 hours, and this analysis method is in accordance with SNI 6989.73: 2009 [13].

3. Discussions

Sampling was taken at three locations in the village of Tala-tala. The following are the pictures of the location of the tofu wastewater sampling place in the Tala-tala area, Kelurahan Bontoala, Kec. Bissapu, Kab. Bantaeng. Locations 1, 2 and 3 written as T1, T2, and T3

Figure 1. Location of the sampling points at T1, T2, T3

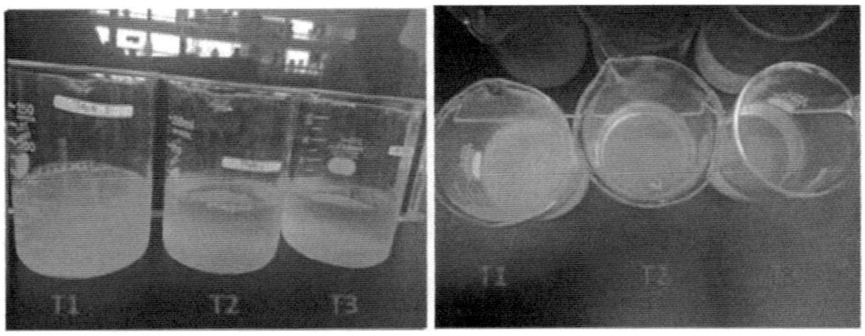

Figure 2. Samples from different location point

From figure 2 above, the samples are murky. The further the point sampling, the less murky of the sample is. Besides the color, the odor of the samples is also tested. The sampling point I (T1) has a unique odor, which more like smelly soya. At the T2 and T3 sampling point the odor of the waste tofu liquid is foul. These are because samples from points II and III mixed with domestic waste. Analysis of tofu wastewater in the Tala-tala area, Kelurahan of Bontoala, Kec. Bissapu, Kab. Bantaeng tested both physical and chemical properties, including temperature, pH, BOD5, COD, and TSS. The results of the analysis are shown in Table 1 below.

Table 1. Physical parameters of tofu liquid waste

Parameters	Units	Location			Max conc.
		I	II	III	
Temperature	^{o}C	29.6	27.8	29	38
TDS	mg/L	1932	540	510	2000
TSS	mg/L	7690	1646	288	200

* Max levels taken from KEPMENLH No. 5 of 2014 concerning Wastewater Quality Standards

The results of the analysis showed that the condition of tofu wastewater in the Tala-tala area, Kelurahan of Bontoala, Kec. Bissapu, Kab. Bantaeng at locations 1, 2 and 3 for temperature and TDS parameters have met the quality standards for wastewater. Temperature of water is influenced by the temperature of the air ambient and the intensity of exposure of sunlight entering the body of water, the intensity of sunlight is influenced by cloud, season [14].

In the TSS parameters, the results obtained are very far beyond the quality requirements of wastewater. According to Tarigan and Edward (2003), TSS (Total Suspendes Solids) are all solid substances (sand, mud, and clay) or particles suspended in water and living components (biotic) such as microorganisms or dead components (abiotic) [15]. The high TSS level at T1 presumably due to a large amount of waste from the remaining waste given that the tofu water waste discharged into the environment was not treated beforehand. As for the next pick up point, tofu water waste has been mixed with the flow of domestic discharges around the plant so that the concentration of solids is reduced. As a whole, from T1 to T3 the parameter values have almost decreased. This is, as mentioned in Hendrasari and Cahyarini 2010, that in each body of water has a self-purification, so the farther the distance from the source of pollutants, the lower waste content [16].

High TDS levels generally indicate hard water, which can cause crust in the pipe, and filter, reduce performance and increase system maintenance costs. This effect can be seen in aquariums, swimming pools, and "reverse osmosis" water treatment systems. From the hydroponics and aquaculture sectors, TDS parameters are useful for creating an environment that can increase crop productivity. For hydroponic use, total dissolved solids are considered to be one of the best indicators of nutrient availability for developed aquatic plants. TDS is a solute in the form of inorganic anions or cations, the low value of TDS indicates the inorganic content in wastes is fairly low. Tofu waste, as previously reported more dominantly, contains organic substances because tofu is made from soy, which is an organic material. From the results of TDS measurements, it is found that at three points palace, the TDS value of tofu waste is still relatively lower than the standard requirement except at point I, which is concentrated waste due to it does not mix with domestic waste around the tofu factory.

Besides physical parameters, the chemical parameters of tofu waste also need to be monitored. The value of these chemical parameters can be seen in Table 2. Fluctuations in pH values are influenced by the presence of organic and inorganic waste discharges into the river [17]. The difference in pH values can be influenced by the addition of organic and inorganic that enters along with the domestic waste flow. Overall at all points (T1, T2, T3), the pH parameters are fulfill the requirement standard of wastewater specified by class I i.e. pH ranges between 6-9.

Table 2. Chemical parameters of tofu liquid waste

Parameters	Units	Results			Max Conc.
		I	II	III	
pH	-	7	7	7.8	6-9
BOD	mg/L	7.5	3.3	9.6	50
COD	mg/L	1405	930	730	100

* Max levels taken from KEPMENLH No. 5 of 2014 concerning Wastewater Quality Standards

Decomposition of organic substances is natural. If a body of water is polluted by organic substances, aerobic bacteria will use the oxygen to decompose organic matter presents in the waste. Lack of oxygen supply impacts on the death of fish and other biotas in the waters, and the condition being anaerobic then can cause a foul odor to the water. The amount of organic matter pollution that can be decomposed by bacteria is measured by the BOD parameter.

The principle of measuring BOD is basically quite simple, the initial dissolved oxygen content (DOi) of the sample measured immediately after sampling, then measuring the dissolved oxygen content of the sample that has been incubated for 5 days under dark conditions at a constant temperature (20°C) which is often referred to as DO5. The difference between DOi and DO5 (DOi - DO5) is the BOD value expressed in milligrams of oxygen per liter (mg / L). Biochemical oxidation is a slow process. Within 20 days, the oxidation of carbon organic matter reached 95-99%, and within 5 days, around 60-70% of the organic material had been decomposed [8]. Five days of incubation are general agreements in determining BOD. Based on this, the calculation of BOD levels starts from DOi to DO5.

Oxygen measurements are carried out using a device called a DO meter which is equipped with a special probe. During incubation, the waste contained in the Winkler bottle needs to be incubated in a dark room to prevent photosynthesis. The process of photosynthesis obscures DO values because photosynthesis produces oxygen. The measured DO value is expected to be oxygen, which only comes from waste then is decomposed by microorganisms. DO5 should not be zero so that BOD can be determined.

The temperature of 20°C in incubation is also a standard temperature. The temperature of 20°C is the average temperature of slow-flowing rivers in temperate regions (Metcalf & Eddy, 1991), where this BOD theory originated [8]. Tropical waters temperatures generally range between 25 - 30 °C with a lower incubation temperature, the activity of decomposing bacteria may also be lower and not optimal as expected. BOD gives a description of the amount of oxygen used by bacteria to decompose the material in the waste during the next five days. In other words BOD can determine the burden of pollution due to wastewater from residents and industries. If the BOD value is obtained, then biological treatment systems for the polluted water can be designed. Furthermore, by comparing the value of BOD to COD, it will also show how much the amount of organic material that persistent in the waters.

Based on the observation of the physical condition of wastewater from the point I to point II, which is approximately 20 meters away, there is a muddy puddle of water accompanied by a bad odor, so this condition needs to get serious treatment. This is possible because of the high value of BOD, as said by Fajri et al. 2017 that the Biological Oxygen Demand (BOD) or the amount of oxygen needed by bacteria to decompose or oxidize almost all dissolved organic substances and some organic substances that are suspended in water [18]. Based on the results obtained BOD values from three sample points are still relatively below the threshold required.

While another parameter, COD, still has a higher value than the maximum allowable standard. As explained by Puspayana and Damayanti 2013, that tofu wastewater has characteristics such as COD (Chemical Oxygen Demand), BOD (Biological Oxygen Demand), which still exceeds the wastewater quality standard [19]. The high BOD and COD values indicate the high organic content contained in the waste. Waste with high COD value is very dangerous for the environment because it can reduce the content of dissolved oxygen in water [20]. While a high TSS value will inhibit the entry of light from the surface into the water and will cause disruption of photosynthesis [21].

4. Conclusion

Based on the TSS and COD values obtained, the researchers suggest that further waste treatment be done to reduce the levels of both of them so that they can be safely disposed of in water bodies.

References

[1] Azwar A 2000 *Pengantar Ilmu Kesehatan Lingkungan* (Jakarta: Mutiara Sumber Jaya)
[2] Sakinah N, Rahmatullah L, Kuncoro E and Oktavitri N 2019 Performance of sequencing batch reactor (SBR) of treated tofu wastewater: variation of contact time and activated sludge sources *IOP Conf. Ser. Earth Environ. Sci.* **259** 12017
[3] Djabu U 1991 *Pedoman Bidang Studi Pembuangan Tinja dan Air Limbah pada Sanitasi Lingkungan* (Jakarta: Depkes RI Pusat Pendidikan Tenaga Kesehatan)
[4] Darmokoesoemo H, Kuncoro E, Supriyanto G, Manuhara W and Wijaya L 2019 Development of solid waste of tofu industry immobilized on silica surface as adsorbent for the adsorption of Cd (II) using method of continuous flow column *IOP Conf. Ser. Earth Environ. Sci.* **259** 12014
[5] Myrasandri P 2009 *Degradasi senyawa organik limbah cair tahu dalam* (Institut Teknologi Bandung)
[6] Soemirat J 2004 *Kesehatan Lingkungan* (Yogyakarta: Gadjah Mada University Press)
[7] Umaly R C and Cuvin M 1988 *Limnology: Laboratory and field guide, Physico-chemical factors, Biological factors* (National Book Store Philippines)
[8] Tchobanoglous G, Metcalf A K, Eddy A C, Tchobanoglous G, Metcalf M E M, Metcalf J L and Eddy H P 1991 *Wastewater Engeneering: Treatment, Disposal and Reuse* (New York: McGraw-Hill,Inc)
[9] Boyd C E 1990 *Water quality in ponds for aquaculture* (Alabama: Alabama Agricultural Experiment Station)
[10] Lashanta 2011 *Memanfaatkan limbah Tahu Menjadi Pestisida dan Pupuk Organik Cair* (Gerbang Pertanian)
[11] Hidup K L 2014 *Peraturan menteri lingkungan hidup republik indonesia nomor 5 tahun 2014 tentang baku mutu air limbah* (Jakarta: Kementerian Lingkungan Hidup Republik Indonesia)
[12] Standar Nasional Indonesia 2009 *Air dan air limbah–Bagian 72: Cara uji Kebutuhan Oksigen Biokimia (Biochemical Oxygen Demand/BOD)* vol 6989
[13] Standar Nasional Indonesia 2009 *Air dan Air Limbah – Bagian 73 : Cara Uji Kebutuhan Oksigen Kimiawi (Chemical Oxygen Demand / COD) dengan refluks tertutup secara spektrofotometri*
[14] Agustiningsih D 2012 *Kajian Kualitas Air Sungai Blukar Kabupaten Kendal Dalam Upaya Pengendalian Pencemaran Air Sungai* (Program Magister Ilmu Lingkungan Undip)

[15] Tarigan M S 2010 Kandungan total zat padat tersuspensi (total suspended solid) di perairan Raha, Sulawesi Tenggara *Makara J. Sci.* **7** 110–1
[16] Hendrasarie N 2010 Kemampuan Self Purification Kali Surabaya, Ditinjau dari Parameter Organik Berdasarkan Model Matematis Kualitas Air *J. Ilm. Tek. Lingkung.* **2** 1–11
[17] Yuliastuti E 2011 *Kajian Kualitas Air Sungai Ngringo Karanganyar dalam Upaya Pengendalian Pencemaran Air* (Program Magister Ilmu Lingkungan)
[18] Fajri N, Hadiwidodo M and Rezagama A 2017 Pengolahan Lindi dengan Metode Koagulasi-flokulasi Menggunakan Koagulan Aluminium Sulfat dan Metode Ozonisasi untuk Menurunkan Parameter Bod, Cod, dan Tss (Studi Kasus Lindi Tpa Jatibarang) *J. Tek. Lingkung.* **6** 1–13
[19] Puspayana D R and Damayanti A 2013 Pengolahan Limbah cair tahu menggunakan membran nanofiltrasi silika aliran cross flow untuk menurunakan kadar nitrat dan amonium *J. Tek. ITS* **2** D87–91
[20] Tchobanoglous G, Burton F L and Stensel H 2003 *Wastewater Engineering: Treatment and Reuse. 4th edn, 983–1027* (New York: McGraw Hill Inc)
[21] Estikarini H D, Hadiwidodo M and Luvita V 2016 Penurunan Kadar Cod Dan Tss Pada Limbah Tekstil Dengan Metode Ozonasi *J. Tek. Lingkung.* **5** 1–11

IOP Conf. Series: Materials Science and Engineering **885** (2020) 012014 doi:10.1088/1757-899X/885/1/012014

The effect of boiler size of distillation to the concentration of bioethanol and yield from fermentation of sweet sorghum juice

Rosalia Sira Sarungallo, A Mellolo, L Melawaty, and M Djonny

Department Chemical Engineering, Universitas Kristen Indonesia Paulus, Perintis Kemerdekaan Street Km. 13, Daya, Makassar, Indonesia

Email: rosalia_sira@ukipaulus.ac.id

Abstract. Bioethanol from sorghum juice can be obtained by the fermentation process involving microorganisms. The distillation process is then carried out to increase the bioethanol concentration. The purpose of this study was to determine optimal of the boiler size from a distillation device based on the concentration of bioethanol final and yield. The bioethanol production research method consists of three stages, namely preparation of raw materials covers cleaning and pressing sorghum stem to get sweet sorghum juice, removal of phenolic compounds by the adsorption method using activated charcoal. Furthermore, the fermentation of sorghum juice is carried out in a batch system. The last stage, the main process of this research is the distillation process at column temperature 70 ^{O}C, with variations in the size of boiler A (diameter 7.5; height 29 cm), B (diameter 10 cm; height 20,5 cm), C (diameter 13; height 10 cm). The bioethanol concentration parameters were analyzed using Gas chromatography. The yield calculation is based on the comparison of the volume of bioethanol distillate with the volume of sweet sorghum juice. The results of the optimum bioethanol production research on boiler size C, with final bioethanol concentration (distillate) 75%, final volume of pure bioethanol 16.5 ml, and yield 6.6%.

1. Introduction

Bioethanol is a biochemical fluid from the process of glucose fermentation using the help of microorganisms and to increase the concentration of bioethanol, followed by the process of distillation. There are various types of plants that can be used as a source of bioethanol raw materials, one of which is the most potential developed in Indonesia is a sweet sorghum plant.

Sweet Sorghum plant is a multipurpose plant, ranging from seeds, leaves, stems, and roots all parts of sorghum useful, namely as a source of food, livestock feed and raw materials of various industries. Specifically, on the stem sorghum can be obtained juice used as a source of bioethanol, the content of sugar Brix stem ranging from 7,67-12,67%, on average, 10.02% [1]. Cultivation of sweet sorghum (Sorghum bicolor L Moench) is generally done to obtain Juice as raw material of ethanol as well as liquid sugar [2,3]. Sorghum stem extracts that form juice can be fermented and distilled to be used as ethanol. The ethanol can be used to fuel the motorbike with a mixture of as much as 20 percent without revamping its motor system [4]. Through the process of fermentation, juice sorghum can produce bioethanol. The fermentation batch system is widely applied in the bioethanol industry because it can produce high levels of bioethanol, but its weakness, which requires a long overall operation time [5].

The distillation process is performed to increase the concentration of bioethanol using the distillation apparatus. The problem faced at this stage is that it is not easy to increase the purity of bioethanol up to 85-95%; it requires high-performance distillation tools, which can cost high. To overcome this, a thorough study is required to modify bioethanol-based sweet sorghum distillation tools. According to the study of Ninik A et al. (2013), about the design of terraced distillation tools that the more number of columns and the number of shavings will result in higher levels and amounts of alcohol [6].

The research is aimed at increasing the purity of bioethanol sourced from the sweet sorghum juice. Data to improve the performance of distillation tools based on variations boiler size will be studied influence in increasing the concentration of bioethanol and bioethanol yield.

2. Methods

2.1. Materials and tools

The main ingredient used is the stem of the sweet sorghum with the varieties of Super 1. The supporting ingredients are yeast (*Saccharomyces cerevisiae*), active charcoal, alcohol 95%, cotton, and aquades. The main tool used is a unit of distillation equipment with different boiler sizes. The supporting tools used are beaker glass, measuring cup, Erlenmeyer, thermometer, electric stove, pump, press machine, fermentor, hand refractometer, and place for bioethanol shelter.

2.2. Experimental

2.2.1. Preparation of raw materials. Sorghum stems obtained from the experimental Garden of Bontobili Balitserealia Gowa. Preparation of raw materials begins with cleaning the sweet sorghum stems freshly harvested. Furthermore, the sorghum rods are squeezed using a pressing tool to get the juice of sweet sorghum stem. Analysis of the concentrated juice of the sweet sorghum using a hand refractometer.

2.2.2. Fermentations. Yeast (*Saccharomyces cerevisiae*) for inoculation, is grown in an Erlenmeyer 250 ml containing sweet juice sorghum with a sugar rate of 18% Brix. After incubation for 48 hours at room temperature, *Saccharomyces cerevisiae* is inoculated into the culture media to begin fermentation.

The fermentation process is carried out by taking juice sorghum sweet 750 ml inserted into the fermentor bottle with a volume of 1 liter and pasteurized at a temperature of 75 ^{o}C for 30 minutes then cooled. Next, add an Inoculum of 75 ml to the fermentor bottle aseptically, pH Medium 5.0. The fermentation process is discontinued after 30 days. Analysis of the concentration sugar of juice sweet sorghum and bioethanol levels using a hand refractometer.

2.2.3. Distillation process. The distillation process is done by inserting a solution of fermentation (crude) as much as 250 ml into the boiler, then install the fractionation column and condenser then run the water pump. At this stage, heating is held at a column temperature of 70 ^{o}C to vaporize bioethanol from crude. Variation of the size of the boiler used is the boiler A (diameter 7.5; height 29 cm), boiler B (diameter 10 cm; height 29 cm), and boiler C (diameter 13; height 10 cm). The upper result of the distillation process is analyzed of its bioethanol levels using the GC, which is hereinafter referred to as the final bioethanol concentration.

3. Results and Discussion

3.1. Effect of boiler size against final bioethanol concentration

Distillation is carried out to purify Bioethanol, the purification process begins with heating crude at 70^{0}C at a pressure of 1 Atm. Bioethanol turns the phase into vapor and passes through the column, and subsequent vapors are condensed into bioethanol fluid with a higher concentration of crude.

During the distillation process, the vapor leading to the condenser and has condensed into a liquid at the top of the column, a portion of the liquid is returned to the column as reflux, and the remainder is pulled as distillate [7].

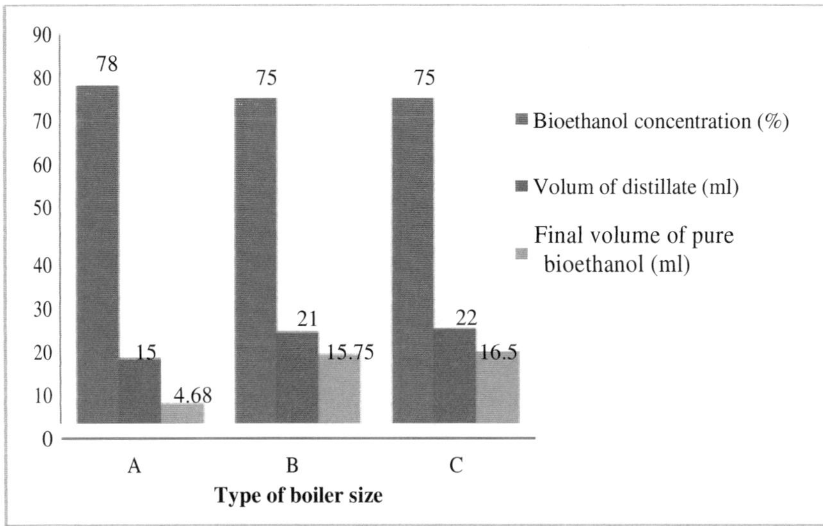

Figure 1. Influence of boiler size based on bioethanol concentration and distillate volume

The results of fermented sweet sorghum juice (*crude*) with Brix bioethanol concentration of 22%. High distillate concentrations were obtained at 78%, concentrations resulting from the distillation process using type A boilers, but the lowest distillate volume charge of was 15 ml. Based on the concentration bioethanol of the distillate and the volume distillate, it is estimated that the final pure bioethanol volume is approximately 4.68 ml. Whereas for the high distillate volume of 22 ml obtained from a type C boiler with a bioethanol concentration of 75%, so the final pure bioethanol volume is estimated at around 16.5 ml. For type B boilers, it can produce 21 ml distillate volume, and distillate concentration 75% with estimated final pure bioethanol volume 15.75 ml.

3.2. Influence of boiler diameter size on bioethanol yield

The volume of distillate bioethanol is the volume of bioethanol per 250 ml of sweet sorghum juice, which is fermented and then distilled. Calculation of bioethanol yield is based on the volume of final pure bioethanol compared with the volume of distilled sweet sorghum juice.

Table 1. The diameter of the boiler effect on the yield of distillation

Type boiler	diameter boiler (cm)	Yield (%)
A	7.5	4.68
B	10	6.30
C	13	6.60

Table 1 shows a boiler diameter 13 cm C type yielding the greatest yield (6.60%), but the yield value of a diameter 10 cm type boiler B approaches the value of a type C boiler yield (6.30%). The yield of type A boilers with a value of 4.68% provide fewer yield than boiler type B and C. It is explained that the more surface area of the boiler, then the yield percentage will also increase.

4. Conclusion

The results of bioethanol distillation research from optimal sweet sorghum juice fermentation on boiler C size, with a final bioethanol concentration (distillate) 75%, final volume of pure bioethanol 16.5 ml, and yield 6.6%.

References

[1] Pabendon M B, Sarungallo R S and Mas'ud S 2015 Pemanfaatan nira batang, bagas, dan biji sorgum manis sebagai bahan baku bioetanol *J. Penelit. Pertan. Tanam. Pangan* **31** 180–7

[2] Andrzejewski B, Eggleston G and Powell R 2013 Pilot plant clarification of sweet sorghum juice and evaporation of raw and clarified juices *Ind. Crops Prod.* **49** 648–58

[3] Agung I G, Sardiana I K, Diara I W and Nurjaya I G 2013 Adaptation, biomass and ethanol yields of sweet sorghum (Sorghum bicolor L. Moench) varieties at dryland farming areas of Jimbaran Bali, Indonesia *J. Biol. Agric. Healthc.* **3** 111–5

[4] Irawan B and Sutrisna N 2016 Prospek pengembangan sorgum di Jawa Barat mendukung diversifikasi pangan *Forum Penelitian Agro Ekonomi* vol 29 pp 99–113

[5] Puspita M, Silviana H and Ismail T 2010 Fermentasi Etanol dari Molasses dengan Zymomonas mobilis A3 yang diamobilisasi pada κ-karaginan *Proceeding Seminar Rekayasa Kimia Dan Proses*

[6] Agustin N, Wahyuningrum L and Harjunowibowo D 2011 *Rancang Bangun Teknologi Destilasi Bioetanol untuk Bahan Bakar Terbarukan* (Universitas Sebelas Maret)

[7] Patil K, Choudhary P and Bhatia T 2009 Distillation Operations: Methods, Operational and Design Issues *Conference on Advances in Heat & Mass Transfer, FAMT, Ratnagiri,* vol 2 p 3

IOP Conf. Series: Materials Science and Engineering **885** (2020) 012015 doi:10.1088/1757-899X/885/1/012015

Enzymatic hydrolysis of proteins from snakehead-fish (*Channa striata*)

Hariyati H, M Mahendradatta, A B Tawali, J Langkong

Departement of Food Science and Technology, Faculty of Agriculture, Hasanuddin University, Indonesia

Email: metamahendradatta@gmail.com

Abstract. The purpose of this research is to study the characteristics and bioactive components of the production of snakehead fish (*Channa striata*) protein hydrolysates through enzymatic methods. The enzymes used in this study were the papain enzyme and without the enzyme as a comparison. Substrate used is fresh snakehead fish (*Channa striata*) obtained from fish reservoirs. Analysis carried out in the form of total degrees of hydrolysis and nitrogen content in protein. The results showed that the use of the papain enzyme was excellent in producing snakehead fish (*Channa striata*) protein hydrolyzate. The papain enzyme helps break down fish protein into peptides or amino acids.

1. Introduction

Fisheries resources as one of the marine products potential in Indonesia which have long been used by the community. Community needs for fishery products continue to increase every year. Abundant fishery products are an opportunity to be utilized as processed products to meet people's food needs. One type of fishery product in South Sulawesi that is favored by the community is snakehead fish. It is known that snakehead fish has a high protein content [1]. The amount of fresh snakehead fish protein reached 21.98% [2], and after further processing, it became a protein concentrate using additional solvents, the protein content increased to 62.9% [3].

Snakehead fish, which has a high protein content, has the opportunity to be used as protein hydrolyzate. Fish protein hydrolyzate is a product produced from the breakdown of proteins into peptides and amino acids through the hydrolysis process by acids, bases, enzymes, or fermentation. Protein hydrolyzate can be obtained by the hydrolysis process, where the protein is broken down by the enzyme protease. The enzyme that will be used is called the papain enzyme[4]. Papain is a commercially available proteolytic enzyme that can be isolated from the sap of the papaya plant (*Carica papaya*).

The use of enzymes in producing fish protein hydrolyzate is expected to produce quality hydrolyzate. Protein hydrolyzate has functional properties as a nutritional supplement that acts as a bioactive compound [5]. Therefore this research was conducted to find out more about the nature and characteristics of snakehead fish protein hydrolysates using the papain enzyme.

IOP Conf. Series: Materials Science and Engineering **885** (2020) 012015 doi:10.1088/1757-899X/885/1/012015

2. Material and Methods

2.1. Preparation of samples
Snakehead fish are used in research this is a fresh snakehead fish that is obtained from the reservoirs District of Antang, South Sulawesi. Snakehead fish unit weight is used around 1.4kg per fish, and then it is weeded to remove from the attached dirt. Next, the fish is fillet to taken about 100g of meat. The Meat is then washed to clean it with flowing water. Furthermore, the enzyme that is used is the papain enzyme at a concentration of 5% (w/v) brands PAYA with the activity of the enzyme specific for 1.0593 u/g.

2.2. Production of protein hydrolyzate
The procedures were performed, namely the snakehead fish meat substrate as much as 100g homogenization with solvent distilled water 1:1 (w/v) and heated with a hotplate stirrer (Stuart) for 2 minutes. Furthermore, the papain enzyme is added as much as 5% to achieve a pH of 7. The sample was hydrolyzed using the tool water bath for 6 hours at a temperature of 55°C and inactivated for 20 minutes at a temperature of 90°C. Furthermore, the sample centrifuged (Heraeus Labofuge A) at a temperature of -4 ° C at the speed of 5000 rpm to separate the supernatant and natan. The supernatant obtained from the centrifuge is then dried using a freeze dryer. Furthermore, the hydrolyzate in the form of a supernatant was further analyzed.

2.3. Analysis of samples
In this study, several analyzes were conducted to determine the effect of the papain enzyme in the production of snakehead fish protein. Analysis was conducted, which analyzes the content of nitrogen in proteins and the degree of hydrolyzate in the protein hydrolyzate.

2.3.1. Determination of nitrogen content in protein. Nitrogen content was analyzed by the Kjeldahl standard method using protein analyzer (KjelMaster K-375) [6].

2.3.2. Determination of degree of hydrolyzate in protein hydrolyzate. The degree of hydrolysis was calculated according to percent of trichloroacetic acid (TCA) ratio method [5,7]. After hydrolysis, 20 ml of protein hydrolysate was added to 20 ml of 20% (w/v) TCA to produce 10% TCA soluble material. The mixtures were left to stand for 30 min to allow precipitation, followed by centrifugation (7800 x g for 15 min) (High-Speed Refrigerated Micro Centrifuge, TOMY, MX-305). The supernatant was analyzed for protein content by Kjeldahl method. Sample from the hydrolysate was also analyzed for protein content.

3. Results and discussion
In the research, measurements are done to determine the degree of hydrolysis and nitrogen in the protein hydrolyzate. Material that is used is a substrate of snakehead fish100g and the papain enzyme 5% (w/v), these things are done to see the influence of the degree of hydrolysis to produce the hydrolyzate protein. The results of the study can be seen in Figure 1 for nitrogen levels in proteins and Figure 2 for the degree of hydrolysis.

In Figure 1 shows the levels of nitrogen in the protein for the 5% papain enzyme in the amount of 0.081% and without the enzyme in the amount of 0.008%. It is shown that the addition of the enzyme is very influential to the results of the analysis of nitrogen in proteins. In addition to the time of incubation are over 6 hours will increase the levels of nitrogen and the degree of hydrolysis. The longer the incubation time, the easier it is for enzymes to break the peptide bonds of proteins into simple compounds.

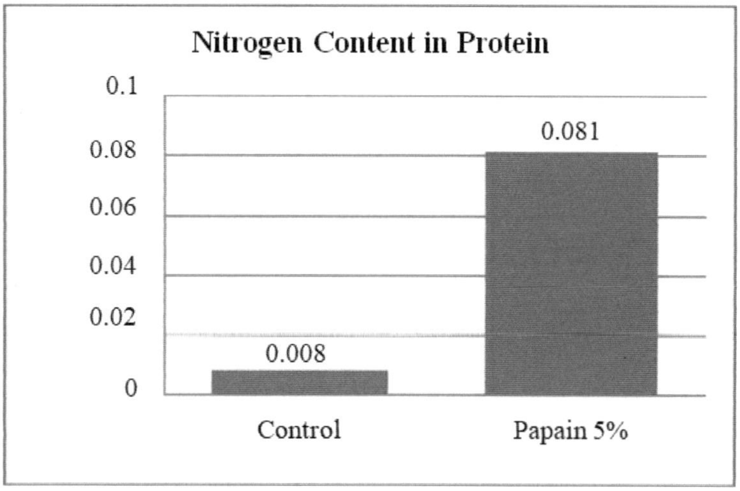

Figure 1. Nitrogen Content in Protein

Hydrolysis Enzymatic with the papain enzyme 5% indicates the percentage that is different by without using enzymes. The results in Figure 1 percentage of 5% papain enzyme with a value of 0.081%, showed the occurrence of the hydrolysis process due to the addition of the enzyme. This increasing shows the reaction of proteolysis to generate the levels of nitrogen dissolved in the hydrolysis of proteins during the process of hydrolysis takes place [8]. Results of nitrogen dissolved in the protein together with the results of the degree of hydrolysis are produced.

The degree of hydrolysis is an important factor in the making of the hydrolyzate of protein by using enzymes. Unknown enzymes greatly determine the magnitude of the degree of hydrolysis. In this study, the papain enzyme was compared with control. The results showed that with the addition of the papain enzyme, as much as 5% indicates the degree of hydrolysis is getting higher as compared with no addition of enzyme. Increased concentrations of the enzyme produce a correlation that is linear to the increase in the degree of hydrolysis [9].

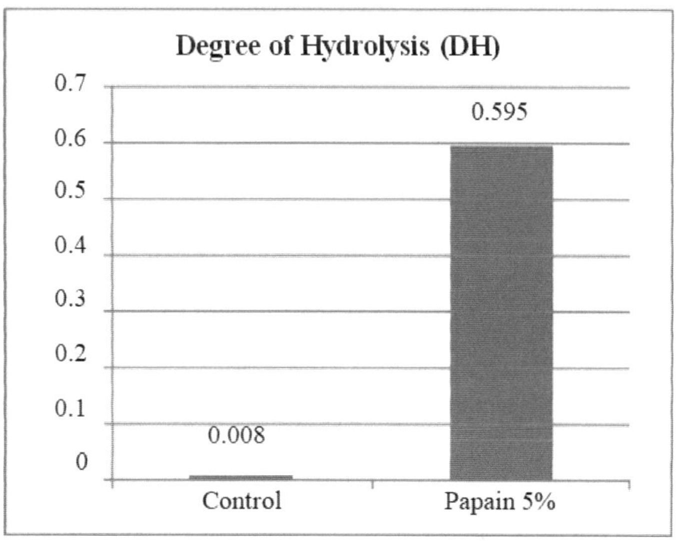

Figure 2. Degree of Hydrolyzate

The value in the degree of hydrolysis is very influential in producing a good protein hydrolyzate. Hydrolyzate will give certain functional properties. It is known during the process of hydrolysis, the value of the degree of hydrolysis will increase along with the time of hydrolysis. In the study, this time, hydrolysis was carried out for 6 hours at pH 7 at a temperature of 55°C and inactivated for 20 minutes at a temperature of 90ºC.

In Figure 2 the use of the 5% papain enzyme was obtained by the degree of hydrolyzate 0.595% and control 0.008%. It is caused during the process of hydrolysis, the process breaking the bonds of peptides from the protein into compounds more simple. Furthermore, the process of hydrolysis is aided by adding the papain enzyme, which is a class of protease that can break down the molecules of the protein to hydrolyze the bond of the peptide into a compound that is more simple. Furthermore, the value of the degree of hydrolysis as the rate of proteolysis, which refers to the percentage of binding peptides that split [8].

In determining the value of the degree of hydrolysis using the addition of 20% TCA. The use of TCA (*trichloroacetic acid*) as a medium to measure the degree of hydrolyzate. This analysis is known as the SN-TCA method. The advantage of the method is that the process of analysis is relatively more rapid and practical as compared with methods other [10].

The technique of analysis is done by measuring samples that do not precipitate in the centrifugation process, which is then carried out by measuring the levels of nitrogen dissolved in the TCA by using a Kjeldahl. The results of TCA showed the value of the degree of hydrolysis of proteins using the addition of the enzyme papain 5% is better than with no enzyme. The increase in the degree of hydrolysis is caused by an increase in peptides and amino acids dissolved in TCA due to termination of peptide bonds during protein hydrolysis [11]. By because it is with the addition of the enzyme is better to produce a product of hydrolyzed protein.

4. Conclusion

The use of enzymes in a yield of hydrolyzed protein highly influential. The addition of enzymes is better than control. The use of enzymes can produce protein nitrogen and a good degree of hydrolysis to increase the functional value of snakehead fish protein hydrolyzate.

References

[1] Fitriani E N, Rozi, Arief M and Suprapto H 2019 Prevalence and intensity of ectoparasites in gabus fish (Channa striata) at Cangkringan Fishery Cultivation Technology Development Center, Sleman, Yogyakarta *IOP Conf. Ser. Earth Environ. Sci.* **236**

[2] Amoo I A, Adebayo O T and Oyeleye A O 2006 Chemical evaluation of winged beans (Psophocarpus Tetragonolobus), Pitanga cherries (Eugenia uniflora) and orchid fruit (Orchid fruit myristic a) *African J. Food, Agric. Nutr. Dev.* **6**

[3] Mahendradatta M, Tawali A B, Asfar M, Suryani and Nurpudji 2012 *Optimalisasi Ekstraksi Albumin Ikan Gabus (Ophiocephalus striatus) sebagai Food Supplement Optimization of Albumin Exctraction on Snakehead Fish (Ophiocephalus striatus) as Food Supplement* (Hibah Strategis Nasional DIKTI)

[4] Harahap N, Satyantini W and Sudarno 2020 Effect of enzyme papain against natural flavor of raw meat waste laundering surimi *IOP Conf. Ser. Earth Environ. Sci.* **441** 12154

[5] Hoyle N T and Merritt J H 1994 Quality of fish protein hydrolysates from herring (Clupea harengus) *J. Food Sci.* **59** 76–9

[6] AOAC 1990 Official Methods of Analysis of AOAC International *Association of Official Analysis Chemists International*

[7] Fonkwe L G and Singh R K 1996 Protein recovery from mechanically deboned turkey residue by enzymic hydrolysis *Process Biochem.* **31** 605–16

[8] Amoo I A, Adebayo O T and Oyeleye A O 2006 Chemical evaluation of winged beans (Psophocarpus Tetragonolobus), Pitanga cherries (Eugenia uniflora) and orchid fruit (Orchid fruit myristic a) *African J. Food, Agric. Nutr. Dev.* **6**

[9] Mahendradatta M, Tawali A B, Asfar M, Suryani and Nurpudji 2012 *Optimalisasi Ekstraksi Albumin Ikan Gabus (Ophiocephalus striatus) sebagai Food Supplement Optimization of*

Albumin Exctraction on Snakehead Fish (Ophiocephalus striatus) as Food Supplement (Hibah Strategis Nasional DIKTI)

[10] Hoyle N T and Merritt J H 1994 Quality of fish protein hydrolysates from herring (Clupea harengus) *J. Food Sci.* **59** 76–9
[11] AOAC 1990 Official Methods of Analysis of AOAC International *Association of Official Analysis Chemists International*
[12] Fonkwe L G and Singh R K 1996 Protein recovery from mechanically deboned turkey residue by enzymic hydrolysis *Process Biochem.* **31** 605–16
[13] Wang J, Zhao M, Zhao Q, Bao Y and Jiang Y 2007 Characterization of hydrolysates derived from enzymatic hydrolysis of wheat gluten *J. Food Sci.* **72** C103–7
[14] Kamau S M and Lu R-R 2011 The effect of enzymes and hydrolysis conditions on degree of hydrolysis and DPPH radical scavenging activity of whey protein hydrolysates *Curr. Res. Dairy Sci* **3** 25–35
[15] Rutherfurd S M 2010 Methodology for determining degree of hydrolysis of proteins in hydrolysates: a review *J. AOAC Int.***93** 1515–22
[16] Hasnaliza H, Maskat M Y, Wan A W M and Mamot S 2010 The effect of enzyme concetration, temperature and incubation time on nitrogen content and degree of hydrolysis of protein precipate from cockle (Anadara granosa) meat wash water *Int. Food Res. J.***17** 147–52

IOP Conf. Series: Materials Science and Engineering **885** (2020) 012016 doi:10.1088/1757-899X/885/1/012016

Development of polishing machine for preparation metallographic specimen with re-manufacturing

Apollo[1], R Nur[1,2], M A Suyuti[1], and M Madjid[3]

[1]Mechanical Engineering Department, Politeknik Negeri Ujung Pandang Jl. Perintis Kemerdekaan Km. 10 Makassar 90245 South Sulawesi Indonesia

[2]Center for Materials and Manufacturing, Politeknik Negeri Ujung Pandang Jl. Perintis Kemerdekaan Km. 10 Makassar 90245 South Sulawesi Indonesia

[3]Politeknik ATI Makassar Jl. Sunu No. 220 Makassar 90211 South Sulawesi Indonesia

Email: rusdinur@poliupg.ac.id

Abstract. A metallographic specimen should be performed to determine the microstructure of a material. One of the processes of preparation of metallographic specimen is polishing process. The designing and manufacturing a polishing machine will be very helpful in the process of polishing the metallographic specimen. This paper was proposed to design and manufacture a polishing machine for laboratory scale. In this research, several stages of the process are performed, such as designing the model, preparing tools and materials used, making components, assembling and testing the polishing machine. the results obtained are polishing machine that works well and can be used for the metallographic test for students and researchers. it can be concluded that this polishing machine can produce the level of specimen preparation for a metallographic test.

1. Introduction

The nature of mechanically polished surfaces and the mechanism by which they are produced have been subjects of inquiry for a considerable period of time. All held the view that polishing is a cutting process in which asperities on the surface are cut away so that the topography of a polished surface consists of a series of fine grooves--the finer the polishing medium the finer the resultant grooves-until eventually a mirror-like specularly reflecting surface is obtained.

In more recent times, Samuels [1,2] has seriously questioned the concept of the Beilby layer, particularly in regard to surfaces of metals mechanically polished by the methods commonly used in the laboratory and in an industry that employ fine abrasives held by a comparatively soft backing. Moreover, Samuels obtained evidence supporting the earlier conclusions that polishing is usually a cutting process that differs from abrasion essentially only in degree. He has also concluded [2] that a polished surface is crystalline but plastically deformed and, moreover, that the severity of plastic deformation at the surface decreases progressively with increasing fineness of polish.

This paper was proposed to design and manufacture a polishing machine for laboratory scale. A particular objective was to investigate polished surfaces by a microscope.

IOP Conf. Series: Materials Science and Engineering **885** (2020) 012016 doi:10.1088/1757-899X/885/1/012016

2. Grinding and polishing process

In the process of metallography, polishing is the final process of the precipitating part of the test for obtaining a smooth surface of the surface by using a metallographic machine which is composed of rotating plates and in it using force [2]. Polishing is often used to enhance the glossy, smooth work piece, prevent the contamination of medical equipment, remove oxidation, or prevent corrosion of the pipe [3]. In metallography and metallurgy, polishing is used to make plates making the work surface free from defects so easily in the inspection of microstructures of metals with a microscope.

To ensure proper graphite retention, use of an automated grinder-polisher machine is recommended. This equipment makes possible, in comparison to manual specimen preparation, uniform control of the desired load on the specimens and the time for each preparation stage. Furthermore, this equipment ensures uniform specimen alignment against the preparation surface, influencing the graphite retention.

The general principle is that the number of grinding and polishing stages should be minimized to preclude the risk of pulling out of the graphite phase. Also, the loading of the specimen should be most favorable to achieve a correctly polished surface in the shortest time. Four specimens with five different types of graphite were chosen as examples to study an effect of grinding/polishing processes on graphite shape and [4,5] They are:

1. coarse, needle, "star shape" graphite combined with regular medium size flake graphite in a hypereutectic gray iron;
2. spheroidal graphite in ductile iron;
3. compacted graphite;
   temper graphite in malleable iron

3. Experimental Setup

3.1. Designing Method

The polishing machine is designed and manufactured for the purpose of improving the quality of the preparation of metallographic test specimens. This polishing machine is designed using Autodesk Inventor 2011 software, as shown in figure 1.

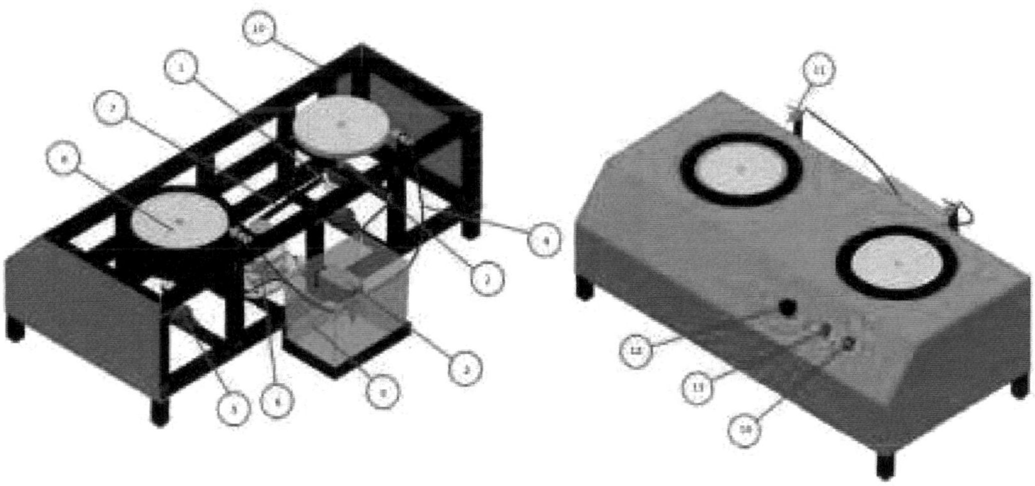

Figure 1. Design of polishing machine

Where :

1. Pulley discs	8. Discs
2. Axis	9. Water tank
3. Water pump	10. Machine frame
4. Hose	11. Water faucet
5. Bearing	12. Speed control button
6. AC motor	13. Water switch

IOP Conf. Series: Materials Science and Engineering **885** (2020) 012016 doi:10.1088/1757-899X/885/1/012016

7. Pulley motor 14. ON/OFF switch

3.2. Manufacturing Method

After the design process, then proceed with the process of making. In the manufacture of polishing machines, need to pay attention to the sequence or procedure either of the design to be made. The manufacturing process was performed in accordance with the work in groups of each component, so that at the time of assembly process easier. Polishing machine manufacturing procedure includes several components, namely:

1. Main components, such as a frame, polishing discs, and machine cover.,
2. Sub-components, such as electric motor, water pump, pulley, belt, axis, and bearing.

Assembly stage was the next stage after the making stage. The assembly process is a process of assembling or combining each component into a mutually supportive form so as to form a tool in accordance with the planned.

Testing of polishing machine is done in the mechanical laboratory of Mechanical Engineering Department PNUP. The materials used are steel with different hardness (VCL and VCN code). Variable of data that is researched is the result of polishing and time polishing. The result of polishing is obtained by varying the use of the type of sandpaper paper and limited by the time of polishing. The data collection result of polishing was taken at every time polishing has finished. The polishing process will be finished if the polishing results have been glossy, then microstructures are tested using a microscope and take pictures of the workpiece result.

4. Results and discussion

Based on the design concept, the manufacture of polishing machine has been produced. This polishing machine consists of the frame, transmission drive, 2 pieces of disc and body cover of the engine. The resulting assembly of polishing machine can be seen in the picture below.

Figure 2. The manufacturing result of polishing machine

A polished machine completed will be tested for steel material by using different types of sandpaper and polishing time. The result of polishing test on steel can be seen in figures 3-9.

(a)

(b)

Figure 3. Polishing result using a) VCL (360) and b) VCN (360) for 15 minutes

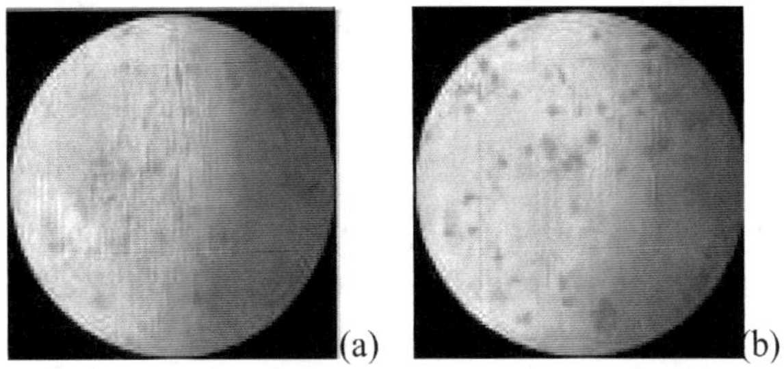

Figure 4. Polishing result using a) VCL (500) and b) VCN (500) for 15 minutes

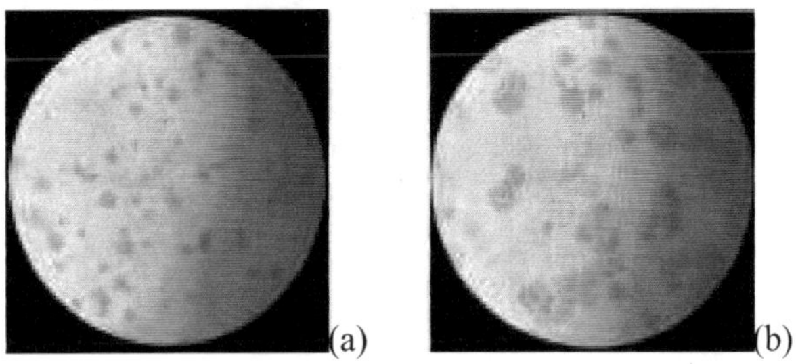

Figure 5. Polishing result using a) VCL (800) and b) VCN (800) for 15 minutes

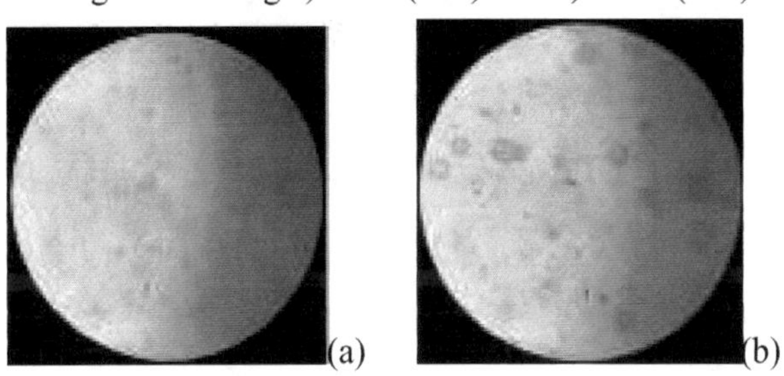

Figure 6. Polishing result using a) VCL (1200) and b) VCN (1200) for 15 minutes

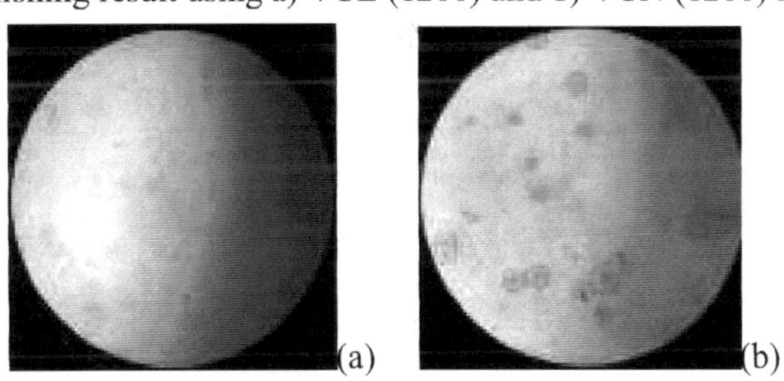

Figure 7. Polishing result using a) VCL (2000) and b) VCN (2000) for 15 minutes

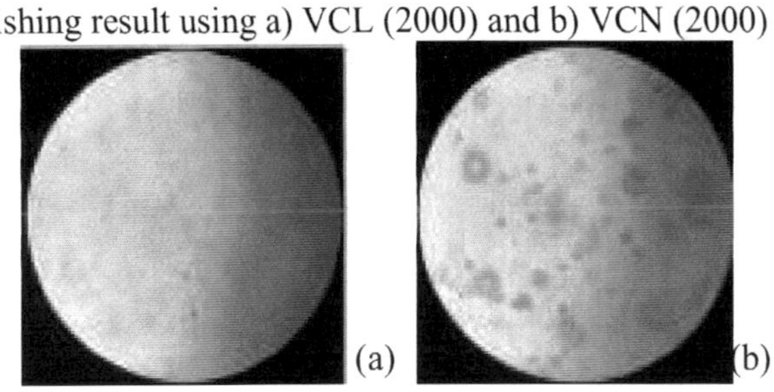

Figure 8. Polishing result using a) VCL (3000) and b) VCN (3000) for 15 minutes

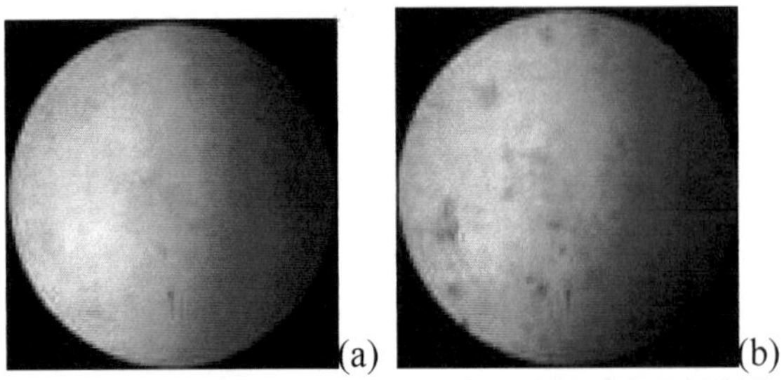

Figure 9. Polishing result using a) VCL (5000) and b) VCN (5000) for 15 minutes

The polishing results as shown in Fig. 3 - 9 show that the ability of the polishing machine has been reliable for use as a means of preparing a metallographic specimen. The results of this polishing are in line with the preparation of specimens performed by Unnikrishnan [6] to investigate the structural development of un-stabilized ultra-low carbon steel. Also in their experiment [7] using the polishing process to prepare the specimen of dental glass.

5. Conclusion

Stages of design, manufacture, and testing polish machine has been completed, it can be drawn conclusion as follows:

5.1. The polishing time can be reduced for approximately 24 minutes for the VCL sample and 22 minutes for the VCN sample.

5.2. Testing results prove that the length of time polishing and rotation speed of the disk will affect the level of fineness of the test specimen. While the level of coarse sandpaper roughness is directly proportional to the level of fineness of the test object polish.

References

[1] Samuels L E 1956 The nature of mechanically polished metal surfaces: the surface deformation produced by the abrasion and polishing of 70: 30 brass *J. Inst. Met.* **85** 51–62

[2] Samuels L E 2003 *Metallographic polishing by mechanical methods* (Asm International)

[3] Krejci I, Lutz F and Boretti R 1999 Resin composite polishing--filling the gaps *Quintessence Int. (Berl).* **30** 490–5

[4] Radzikowska J M 2005 Effect of specimen preparation on evaluation of cast iron microstructures *Mater. Charact.* **54** 287–304

[5] Liu P C, PC L I U and HK P 1980 Observations on the graphite morphology in cast iron *Trans AFS* 88–97

[6] Unnikrishnan R, Kumar A, Khatirkar R K, Shekhawat S K and Sapate S G 2016 Structural developments in un-stabilized ultra low carbon steel during warm deformation and annealing *Mater. Chem. Phys.* **183** 339–48

[7] Silva R M, Pereira F V, Mota F A P, Watanabe E, Soares S M C S and Santos M H 2016 Dental glass ionomer cement reinforced by cellulose microfibers and cellulose nanocrystals *Mater. Sci. Eng. C* **58** 389–95

Application of air conditioner (AC) automation system on arduino flatform-based vehicle

Randis[1] and S H Wijaya[2]

Heavy Equipment Study Program, Mechanical Engineering Department, Balikpapan State Polytechnic, Balikpapan Indonesia

Email: randis@poltekba.ac.id

Abstract. This study aims to design and create a safety device and automation system in the use of air conditioner (ac) use in-vehicle units, especially in four-wheeled vehicles based on Arduino UNO microcontroller Atmega 328. The safety device system created in this study used Arduino UNO as its microcontroller, a tool was equipped with a PIR sensor that was positioned in the cabin and a limit switch placed at the door of the vehicle. If there are humans inside the vehicle unit, the PI sensor will detect movement, and if the unit door is in the closed position, the limit switch will be depressed. The principle used the AND logic gate principle where both inputs both the PIR sensor and the limit switch must work together to turn on the AC, but if only one input works, then the AC will not turn on. The results of this study indicated the tool could work under the design that has been done, and the programming of the safety device can work well according to the desired working principles through the programming commands on the Arduino UNO microcontroller.

1. Introduction

The energy-saving behavior is being promoted by the government to ensure the fulfillment of energy achievements, especially electricity that is increasing. Saving electricity consumption is considered very important; research is developed in order to design and implement the system to be able to save energy consumption on the electrical devices used [1–6].

The role of the human is the biggest factor in the context of saving electrical energy waste. Generally, it is caused mostly by human factors and a small portion by technical factors. Examples of electrical energy waste that often occurs in the community are the use of Air Conditioners (AC), both the use of Air Conditioners on buildings and air conditioners on vehicles that are still turned on even though they are no longer needed.

Various ways and efforts have been made in order to save electrical energy, one of which is by saving electricity by turning off air conditioners and other electrical devices when they are no longer needed, but these savings are seen to be less effective because of the limitations of people who often forget or are lazy to turn off AC when the device is no longer used.

One alternative solution in the problem caused by forgetting or lazy someone turns off the AC in the vehicle by designing a system that can work automatically. With this system, the AC in the vehicle can be turned on or off by utilizing the Passive Infrared (PIR) sensor to detect human movements and limit switches placed installed on the vehicle door, if the door is closed, the AC will turn on and when the door is open the AC will turn off.

Systems built and developed using Arduino board microcontroller devices and Passive Infrared (PIR) sensors as sensors are considered capable of overcoming these problems because the use of Arduino and PIR sensors as a system capable of reducing the use of electrical energy and used as a security system has been developed before [7–12]. The tools in [7] use Arduino and PIR in order to conserve the use of electrical energy in the use of light lamps with the help of lithium tantalate (LiTaO3), while in research [8] Arduino and PIR sensors are used on their tools not to reduce the use of street lights with automation techniques. In the security system, the PIR sensor and the Arduino microcontroller are used in [9] to detect and give warnings of theft in a house. While in [10], the vehicle safety system uses Arduino to develop a warning and feedback system for children in vehicles.

The system to be built and developed is different from [1,2,7,8,11,12], with the application of tools only on simulations and prototypes, but in this study, the system will be applied directly to the four-wheeled vehicle units that use the most devices Air conditioner. The system will be developed using inputs in the form of PIR sensors and limit switches and using LCD devices and relays that are connected to the Air Condition (AC) / magnetic clutch to save electricity consumption in vehicle AC.

2. Research Method

2.1. Block Diagram

The system block diagram is shown in Figure 1. Simply put, this block diagram shows the working system where the 12-volt battery power supply will be reduced by a step-down device in accordance with the input voltage of Arduino. PIR sensor and Limit Switch as input to Arduino Uno before relay and magnetic clutch. Arduino UNO is a data processing center that will provide information to the output in the form of LCDs and relays that function to activate the magnetic clutch. This tool uses the AND gate, so the two inputs used must work together so that the AC is on.

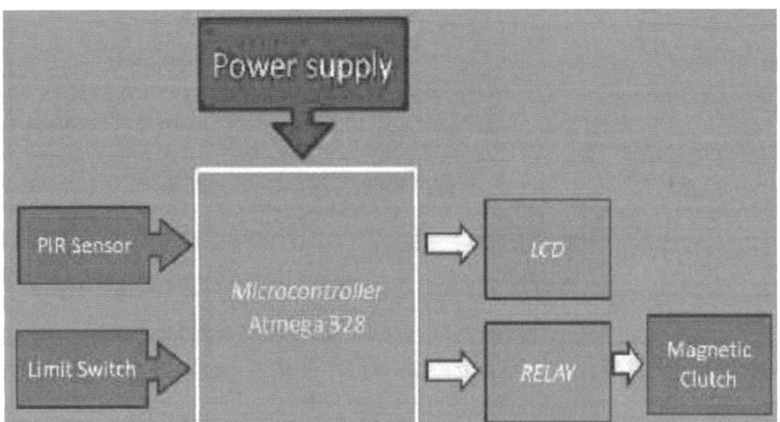

Figure 1. System Block Diagram

2.2. *Flow Chart*

The system flow chart is made for the procedures and work processes of tools and algorithms in making programs as in [13–17]. The flow chart of the instrument developed is shown in Figure 2, starting with the reading of the Arduino input variable. In the variable reading, there are two inputs, the first, the limit switch, and the second, the PIR sensor. Both inputs enter the AND gate variable, if all variables are 1 then the AC will turn on and the LED will turn off while the LCD gives a sign (closed-door / "AC On"), if one of the variables is 1 then the AC will turn off, and the LED lights will be on while the LCD gives a sign (the door is open / "AC Off"). When the second condition does not work, it will return to the initial process, if one of the variables does not work, the Ac is not alive. The difference between the two input variables includes the limit switch working when the door is closed while the PIR sensor works when detecting human movement.

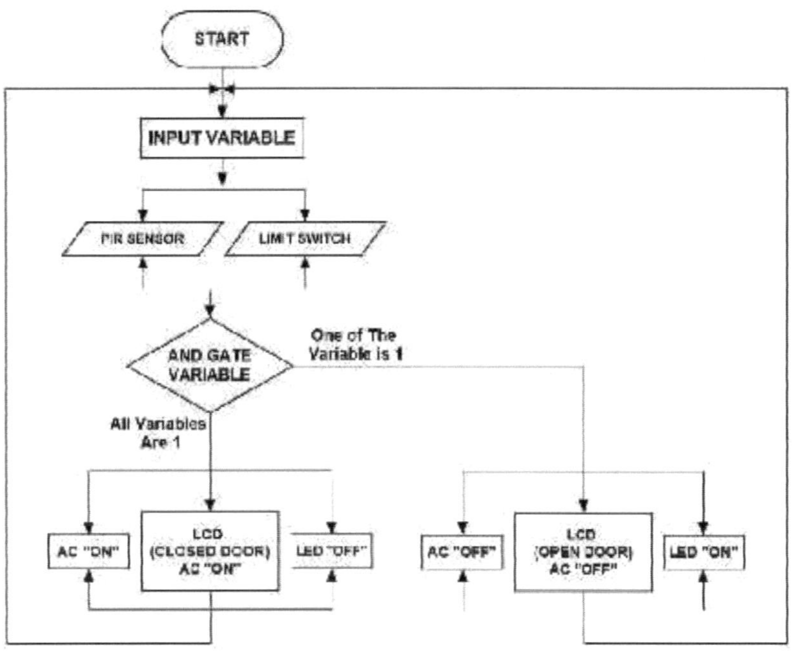

Figure 2. Flow Chart System

3. Result and Discussion

3.1. *Implementation of Electrical Systems*

The electrical system developed in this study is shown in figure 3. The electrical system shows the wiring and layout arrangement of all components used. Arduino Uno becomes the center for processing information from sensors and is processed into information so that it can be sent to the output to move the actuator. The application of the program to the Arduino microcontroller uses Sketch and C language as in [18–20]. Arduino board is a microcontroller device that gets input voltage from the battery to the vehicle, which is first lowered by the operating voltage step down the device. After Arduino gets power, the device will be active, then the PIR sensor and the limit switch as input will be supplied by the VCC and GND currents. The red and black lines indicate the VCC and Ground power connected in parallel to the device, and the yellow lines indicate that the input signal current on the microcontroller, if one of

IOP Conf. Series: Materials Science and Engineering **885** (2020) 012017 doi:10.1088/1757-899X/885/1/012017

the PIR sensors and the limit switch gets a signal, then the LED and the relay (magnetic clutch) will be activated based on the command received by the microcontroller.

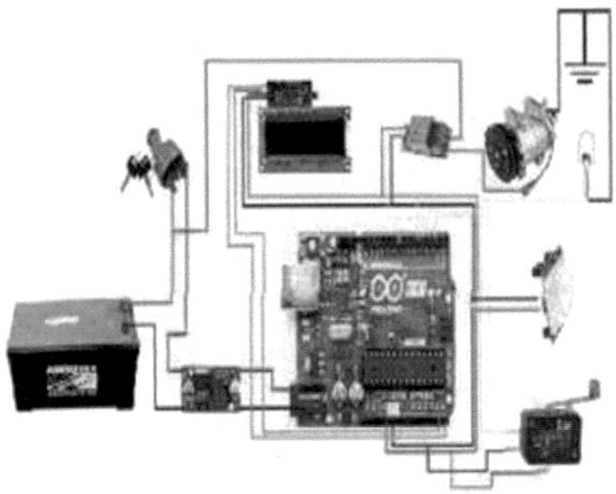

Figure 3. Implementation of Electrical system

3.2 Implementation of mechanical systems

The mechanical system shown in Figure 4, this system shows the placement of sensor components and tools that are developed and applied to four-wheeled vehicle units. Shown in the first picture is a description of the position of the PIR sensor in the cabin that will detect movement with an angle of 45° in the cabin area. The second picture is the position of the limit switch that exists between the cabin and the door, shown in Figure 5, the position of the limit switch. The third image description is the placement of the toolbox of the system tools, which contains a microcontroller and contains other components. The fourth image is a magnetic clutch (air conditioner) that will be connected to the relay in the toolbox. The relay is connected to an AC magnetic clutch via normally closed (NC) and COM.

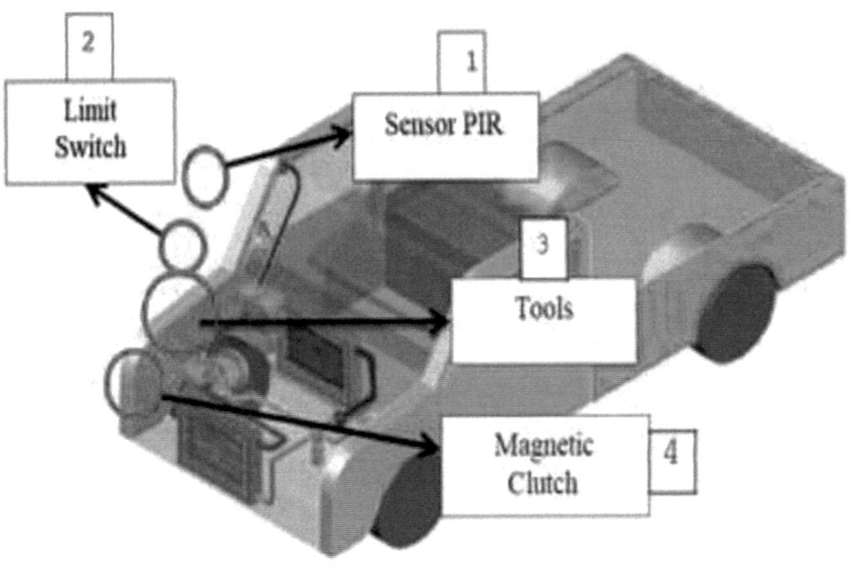

Figure 4. Implementation of Mechanical System

3.3. Sensor Testing

3.3.1. Limit Switch Testing. Limit switch testing aims to determine the limit switch can work well or not. The limit switch component is installed in the NO (Normally Open) position according to the needs of the device, where the limit switch is placed at the door in a depressed/closed door position so that the limit switch is connected.

.

Figure 5. Closed Door Limit Switches

Figure 5 shows the limit switch works when the door is closed because the limit switch is installed in NO (Normally Open) so that when pressed, the limit switch will close and send a signal to the microcontroller and the output on the LCD will give the sign "AC On."

Figure 6 shows the limit switch does not work because the door is open so that the limit switch is not depressed, which causes the limit switch not to be connected so that there is no signal to the microcontroller and the output on the LCD will give the sign "AC Off."

Figure 6. Open Door Limit Switches

3.3.2. PIR Sensor Testing

PIR sensor testing (Passive Infrared Receiver) aims to determine the distance from the PIR sensor that can detect human movement. Testing is done by connecting the PIR sensor to the PORT B 0 microcontroller and the output in the form of an LCD to display the display is human or not. If there is detected a human movement, then the LCD will display "H = 1" if no human movement is detected, then the LCD will display "H = 0". The experiments were conducted at a distance of 1-8 meters, and the results are shown in Table 1, while the placement of the PIR sensor (Passive Infrared Receiver) is shown in Figure 7. Overall the testing process, the use of this PIR sensor can be used well on this tool because the sensor reading distance can reach a distance of 4 meters while in the cabin of the vehicle, the maximum range is only 2 meters so that the system can work just as expected.

Table 1. PIR Sensor Testing

PIR and Human Sensor Distance (meters)	LCD Display
1	H=1
2	H=1
3	H=1
4	H=1
5	H=2

6	H=2
7	H=2
8	H=2

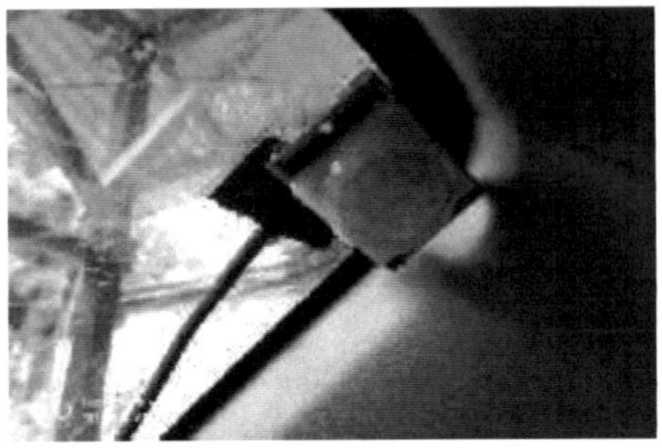

Figure 7. Placement of PIR sensors

3.4 . Overall component testing

After testing each input and output device, the whole system is tested, which is by combining all input and output devices into a system that can control electrical equipment automatically. Table 2 shows the results of testing the behavior of the electronic equipment controlling system automatically from a number of circumstances.

Table 2. Overall Componen Testing

No	PIR sensor	Limit Switch	LCD	Air Conditioner
1	Detect movement (Human)	On	Displays "AC on"	AC on
2	Detect movement (Human)	Off	Displays "AC off"	AC off
3	Does not detect movement (human)	On	Displays "AC off"	AC off
4	Does not detect movement (human)	Off	Displays "AC off"	AC off

4. Conclusion

This research succeeded in designing and applying an automation system on the use of air conditioners on four-wheeled vehicles to save electricity consumption in vehicles by using the Arduino UNO AT mega 328 as an information processing center and utilizing the PIR sensor as a sensor to detect movement and limit switches to identify door openings on the vehicle. Output in the form of character display on the LCD and relay connected to the magnetic clutch can work well with the unit with a 12-volt working voltage and is able to provide automation on the basis of AND logic gates that can save the use of electrical energy in vehicles, especially on four-wheeled vehicles.

Reference

[1] Alam S U, Ahmed R, Imam M S, Farshid M, Hossain M A and Islam M A 2019 Design and Implementation of Website based Energy Consumption Monitoring and Controlling *2019 International Conference on Computer Communication and Informatics (ICCCI)* (IEEE) pp 1–7

[2] Alghamdi H and Almawgani A H M 2019 Smart and Efficient Energy Saving System Using PDLC Glass *2019 Smart City Symposium Prague (SCSP)* (IEEE) pp 1–5

[3] Tamkittikhun N, Tantidham T and Intakot P 2015 AC power meter design based on Arduino: Multichannel single-phase approach *2015 International Computer Science and Engineering Conference (ICSEC)* (IEEE) pp 1–5

[4] Lee K 2015 Drivers and barriers to energy efficiency management for sustainable development *Sustain. Dev.* **23** 16–25

[5] Nasution H, Dahlan A A, Aziz A A, Azmi U, Zulkifli A H and Windiarti H 2016 Indoor Temperature Control And Energy Saving Potential Of Split-Type Air Conditioning System Using Fuzzy Logic Controller *J. Teknol.* **78**

[6] Ventura D, Casado-Mansilla D, López-de-Armentia J, Garaizar P, López-de-Ipina D and Catania V 2014 ARIIMA: a real IoT implementation of a machine-learning architecture for reducing energy consumption *International Conference on Ubiquitous Computing and Ambient Intelligence* (Springer) pp 444–51

[7] Irzaman, Siskandar R, Nabilah N, Aminullah, Yuliarto B, Hamam K A and Alatas H 2018 Application of lithium tantalate (LiTaO3) films as light sensor to monitor the light status in the Arduino Uno based

energy-saving automatic light prototype and passive infrared sensor *Ferroelectrics* **524** 44 55
[8] Ramli N L, Yamin N M, Ghani S A, Saad N M and Sharif S A M 2015 Implementation of passive infrared sensor in street lighting automation system *ARPN J. Eng. Appl. Sci* **10** 17120–6
[9] Surantha N and Wicaksono W R 2018 Design of smart home security system using object recognition and PIR sensor *Procedia Comput. Sci.* **135** 465–72
[10] Morsalin S, Islam A M J, Rahat G R, Pidim S R H, Rahman A and Siddiqe M A B 2016 Machine-to-machine communication based smart home security system by NFC, fingerprint, and PIR sensor with mobile android application *2016 3rd International Conference on Electrical Engineering and Information Communication Technology (ICEEICT)* (IEEE) pp 1–6
[11] David N, Chima A, Ugochukwu A and Obinna E 2015 Design of a home automation system using arduino *Int. J. Sci. Eng. Res.* **6** 795–801
[12] Baraka K, Ghobril M, Malek S, Kanj R and Kayssi A 2013 Low cost arduino/android-based energy-efficient home automation system with smart task scheduling *2013 Fifth international conference on computational intelligence, communication systems and networks* (IEEE) pp 296–301
[13] Bangali J and Shaligram A 2013 Design and Implementation of Security Systems for Smart Home based on GSM technology *Int. J. Smart Home* **7** 201–8
[14] Zeebaree S R and Yasin H M 2014 Arduino based remote controlling for home: power saving, security and protection *Int. J. Sci. Eng. Res.* **5** 266–72
[15] Huda A-S and ERÇELEBİ E 2017 Development of smart security system for remote control using small computer *Fırat Univ. Turkish J. Sci. Technol.* **12** 107–12
[16] Sahoo K C and Pati U C 2017 IoT based intrusion detection system using PIR sensor *2017 2nd IEEE International Conference on Recent Trends in Electronics, Information & Communication Technology (RTEICT)* (IEEE) pp 1641–5
[17] Ju S H, Lim Y H, Choi M S, Baek J-M and Lee S-Y 2011 An efficient home energy management system based on automatic meter reading *2011 IEEE International Symposium on Power Line Communications and Its Applications* (IEEE) pp 479–84
[18] Syed S N, Radhika S and Rani M N S 2015 Differential current protection of transformer using Arduino with voice alert *Int. J. Innov. Eng. Technol.* **6** 206–12
[19] Mobin M I, Abid-Ar-Rafi M, Islam M N and Hasan M R 2016 An intelligent fire detection and mitigation system safe from fire (SFF) *Int. J. Comput. Appl* **133** 1–7
[20] Gor M, Vora J, Tanwar S, Tyagi S, Kumar N, Obaidat M S and Sadoun B 2017 GATA: GPS-Arduino based Tracking and Alarm system for protection of wildlife animals *2017 International Conference on Computer, Information and Telecommunication Systems (CITS)* (IEEE) pp 166–70

IOP Conf. Series: Materials Science and Engineering **885** (2020) 012018 doi:10.1088/1757-899X/885/1/012018

Effect of Typical Lubrication on the Pressure Ratio Through Operating the EFI Suzuki Engine

R Sihombing[1], Hidayat[1], A Fatmawati[1], and R Nur[2]

[1]Mechanical Engineering Department, Politeknik Negeri Samarinda, Jl. Cipto mangunkusumo Kampus Gn. Lipan, Samarinda 75126, East Kalimantan, Indonesia
[2]Mechanical Engineering Department, Politeknik Negeri Ujung Pandang Jl. Perintis Kemerdekaan Km. 10 Makassar 90245 South Sulawesi Indonesia

Email: rusdinur@poliupg.ac.id

Abstract. Many types of lubricants was used as a lubricant in the Suzuki Baleno EFI engine. There is some questions in the community which better lubricants are used without prior research. This paper aims to determine the pressure ratio produced by the Suzuki Baleno EFI engine by using different lubricants. The experimental method was used in this study by testing two lubricants (i.e. SAE 30 and SAE 40) and comparing the ratio of the pressure produced. The results showed that the ratio of the pressure produced between lubricants of SAE 30 : SAE 40 was 1: 1.5 at 800 rpm; 1: 1 at 1500 rpm; 1: 1.06 at 2000 rpm and 1.1: 1 at 2500 rpm. T-test was used to compare the significance of the use of SAE 30 and SAE 40 as lubricants on the pressure ratio through operating the EFI Suzuki Engine. It can be concluded that the SAE 40 lubricant will produce a greater pressure compared to SAE 30 at a low speed of 850 rpm. However, the rotation of 1500 rpm and 2000 rpm will produce the same pressure, as well as at 2500 rpm with using SAE 30 and SAE 40.

1. Introduction

Demand on engine lubricant is to prevent or reduce wear as a result of direct contact between one metal surface with another metal surface continuously moving. The efficiency and effectiveness of motor vehicle engine performance are influenced by the condition of the lubricating oil. An essential condition and quality of the lubricant is its viscosity. Oil viscosity is a dimension of the internal cohesiveness, also known as a measurement of resistance to flow. The Society of Automotive Engineers (SAE) provides a global standard for grading engine oil according to its viscosity.

Many researchers on the influence of oil viscosity for lubrication to the machine have been reported. A number of study reported the effect of lubricant to mechanical purposes. Another study reported the effects of lubricant viscosity reduced mechanical losses. Low viscosity near mid-stroke and high viscosity near and strokes were considered [1]. Heavy-duty truck engine was used to investigate the effect of reducing oil viscosity on wear, friction and fuel consumption with conventional and low viscosity lubricants [2]. An experimental study reported the effect of lubricating oil temperature and pressure on compressor performance [3]. A comparison of two types of lubricants used in CNG buses to find out the fuel consumption is performed by the experiment [4]. The other study investigated the influence of lubricant properties on CEC L-54-96 standard engine fuel economy [5]. The performance and efficiency of needle bearing type turbocharger were reported by determined inlet and outlet temperatures, pressure and flow rates turbine and compressor [6]. An experimental test was used to evaluate piston ring and cylinder liner in realistic lubricants to support the development of new standards [7]. The variation of friction means effective pressure with viscosity provides a basis for comparing the effect of surface finish was reported [8]. Modern engines were developed for evaluating the fuel efficiency of automotive crankcase oils [9]. A study evaluated lube oil performance and engine friction at conditions typical for a fired engine. Commercial engine modeling software is used to predict the effect of lube oil formulations on the engine friction of the same engine [10]. However, there is no study reported the effect of typical lubrication on the pressure ratio through the specific engine. In the present study, the pressure ratio produced by a specific engine such as Suzuki Baleno EFI was determined with different lubricants. The lubricants were classified according to its viscosity and work temperature. Two types of lubricants, namely SAE 30 and SAE 40 were compared by the experimental method. Then, the independent samples t-test was carried out to find out the significance of the use of two types of lubricants on the pressure ratio through the operating EFI Suzuki engine. The pressure ratio was determined at 850 rpm, 1500 rpm, 2000 rpm and 2500 rpm, respectively.

2. Method

2.1 Engine Specification

The engine specification used in experiments in this study is shown in Table 1. The year manufacture of the Engine is 1998. The EFI Suzuki Engine model used in this work is the first generation Baleno, manufactured from 1995 to 2002.

Table 1. EFI Suzuki Engine Specification

Component	Specification
Engine	Inline, four-stroke, 16 valves
Fuel	Gasoline
Displacement [cc]	1,590
Compression ratio	9.5:1
Maximum power	67 kW at 6000 rpm
Maximum torque	127 Nm at 3000 rpm
Firing order	1-3-4-2
Fuel system	Multipoint Injection
Drive type	Front-Wheel Drive
Gearbox	Manual, five-speed

2.2 Experimental setup

Figure 1 shows the EFI Suzuki Engine SY416 Baleno DX. In this work, the two types of oil lubricants, namely SAE 30 and SAE 40 were used to compare the effect of lubricants on pressure ratio through operating engine EFI Suzuki Baleno. Firstly, install the pressure gauge to the oil switch indicator located in the oil pump. Set the engine speed constant at 850 rpm, then record the oil pressure on the switch indicator after 2 minutes. Furthermore, turned off the engine then check the temperature on the engine using an infrared thermometer, then point note the temperature results listed

IOP Conf. Series: Materials Science and Engineering **885** (2020) 012018 doi:10.1088/1757-899X/885/1/012018

on the thermometer. The experiment was carried out seven times with the engine speed varying from 850 to 2500 rpm. Then, an independent samples t-test was used to compare the significance of the use of SAE 30 and SAE 40 on the pressure ration through the operating system of the engine.

Figure 1. EFI Suzuki Engine SY416 Baleno DX

3. Results and Discussion

Figure 2 shows the effect of engine speed to the average pressure at the oil switch indicator. It is well known that the oil pressure will be increased as the horsepower output and engine rotary speed increase. As engine rotary speed demands increase, it is needed to have higher oil pressure between the shafts and bearings to prevent wear contact between spinning components. Moreover, when the oil pressure is increased, the flow rate also increases; hence, more lubricant is available to cool the engine system. It can be seen that the average pressure at a low engine speed of 850 rpm using SAE 40 is smaller than SAE 40, then at the high engine speed 1500, 2000 and 2500 rpm shows almost same pressure. The results showed that the ratio of the pressure produced between lubricants of SAE 30 : SAE 40 was 1: 1.5 at 800 rpm; 1: 1 at 1500 rpm; 1: 1.06 at 2000 rpm and 1.1: 1 at 2500

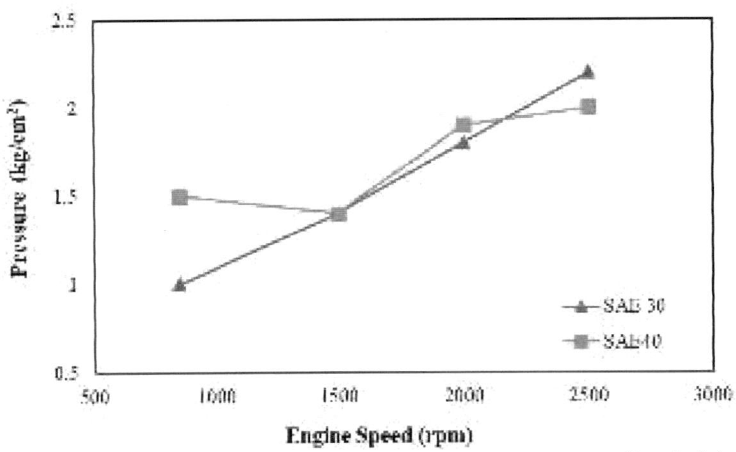

Figure 2. Effect of engine speed to the average pressure at oil switch indicator

In order to find out the significance of the average pressure by using SAE 30 and SAE 40 on all variations of engine speed starting from 850 rpm, 1500 rpm, 2000 rpm, and 2500 rpm, the independent samples t-test was carried out. The results of independent samples t-test for each engine speed is shown in Tabel 2.

Table 2. Independent samples t-test for the of oil lubricants SAE 30 and SAE 40

Engine Speed (rpm)		Levene's Test for Equality of Variances		t-test for Equality of Means						
		F	Sig	T	df	Sig. (2-tailed)	Mean Difference	Std. Error Difference	95% Confidence Interval of the Difference	
									Lower	Upper
850	Equal variances assumed	6.090	.030	-3.250	12	.007	-.51429	.15822	-.85902	-.16955
	Equal variances not assumed			-3.250	7.760	.012	-.51429	.15822	-.88112	-.14746
1500	Equal variances assumed	.000	1.000	.000	12	1.000	.00000	.05714	-.12450	.12450
	Equal variances not assumed			.000	12.000	1.000	.00000	.05714	-.12450	.12450
2000	Equal variances assumed	.000	1.000	.000	12	1.000	.00000	.07377	-.1603	-.16073
	Equal variances not assumed			.000	12.000	1.000	.00000	.07377	-.16073	-.16073
2500	Equal variances assumed	.000	1.000	.000	12	1.000	.00000	.05714	-.12450	.12450
	Equal variances not assumed			.000	12.000	1.000	.00000	.05714	-.12450	.12450

Table 2 shows the independent samples t-test for the use of oil lubricants SAE 30 and SAE 40. The value of F = 6.090 and significance value 0.030 from Levene's test for equality of variances means that the independent samples t-test at 850 rpm of engine rotary speed used the equal variances assumed. Furthermore, from the t-test for equality of means shows the value of T -3.250 and the significance 0.007 means that the differences of the oil pressure using two types of oil lubricants SAE 30 and SAE 40 are significant. Otherwise, as for the engine rotary speed at 1500 rpm, 2000 rpm and 2500 rpm, the result of the independent samples t-test shows the Levene's test which composes value of F = 0.000 and significance 1.000 means that the equal variances not assumed is used. Finally, the t-test for equality of means shows the value of T 0.000 and significance 1.000. It means that there are no differences in the oil pressure between the use of oil lubricants SAE 30 and SAE 40 due to the result of the t-test is not significant.

4. Conclusion

Based on these experimental results, it can be concluded that the use of two types of oil lubricants according to its viscosity, the SAE 40 has higher oil pressure than SAE 30 at engine rotary speed 850 rpm. Then, the result of t-test at 850 rpm is significant. However, there are no differences in oil pressure using SAE 30 and SAE 40 at engine rotary speed 1500 rpm, 2000 rpm and 2500 rpm.

References

[1] Takata R and Wong V W 2006 Effects of lubricant viscosity on ring/liner friction in advanced reciprocating engine systems *ASME 2006 Internal Combustion Engine Division Fall Technical Conference* (American Society of Mechanical Engineers Digital Collection) pp 281–93

[2] Carden P, Pisani C, Andersson J, Field I, Lainé E, Bansal J and Devine M 2013 The effect of low viscosity oil on the wear, friction and fuel consumption of a heavy duty truck engine *SAE Int. J. Fuels Lubr.* **6** 311–9

[3] Podevin P, Clenci A and Descombes G 2011 Influence of the lubricating oil pressure and temperature on the performance at low speeds of a centrifugal compressor for an automotive engine *Appl. Therm. Eng.* **31** 194–201

[4] Tormos B, Ramírez L, Johansson J, Björling M and Larsson R 2017 Fuel consumption and friction benefits of low viscosity engine oils for heavy duty applications *Tribol. Int.* **110** 23–34

[5] Maroto-Centeno J-A, Pérez-Gutiérrez T, Fernández-Ruíz-Morón L and Quesada-Pérez M 2016 Prediction of fuel economy performance of engine lubricants based on laboratory bench tests *Tribol. Int.* **94** 67–70

[6] Chuepeng S and Saipom S 2018 Lubricant thermo-viscosity effects on turbocharger performance at low engine load *Appl. Therm. Eng.* **139** 334–40

[7] Truhan J J, Qu J and Blau P J 2005 A rig test to measure friction and wear of heavy duty diesel engine piston rings and cylinder liners using realistic lubricants *Tribol. Int.* **38** 211–8

[8] Leong D K W, Shayler P J, Pegg I G and Murphy M 2007 Characterizing the effect of viscosity on friction in the piston assembly of internal combustion engines *Proc. Inst. Mech. Eng. Part J J. Eng. Tribol.* **221** 469–78

[9] Inoue K, Ogura H, Igawa S and Takeda K 1993 *Effects of lubricant composition on wear in methanol-fueled SI engines* (SAE Technical Paper)

[10] Rohr W F 2013 Experimental and Theoretical Investigations of Lube Oil Performance and Engine Friction *Tennessee Res. Creat. Exch. Exp. Theor. Investig. Lube Oil Perform. Engine Frict. Recomm.*

IOP Conf. Series: Materials Science and Engineering **885** (2020) 012019 doi:10.1088/1757-899X/885/1/012019

The Relationship of Temperature and Compressive Strength on Geopolymer Mortar using Fly Ash-Based

E Bachtiar [1], I Marzuki [2], Erdawaty[1], M F Junaedy[1], F Ramadhan[1], A M Setiawan[1] and F Rachim[1].

[1] Department of Civil Engineering, Faculty of Engineering, Fajar University, Makassar, Indonesia, 90231

[2] Department of Infrastructure & Environmental Engineering, Fajar University, Makassar, Indonesia, 90231

Email: erni@unifa.ac.id

Abstract. Infrastructure development in Indonesia has increased. Therefore, the use of cement has increased as well. For this reason, it is necessary to replace the replacement cement material with environmentally friendly material. Fly ash is one of the waste materials that can be used as a substitute for cement. The purpose of this paper is to determine the compressive strength values of geopolymer mortars using fly ash-based from different Thermal Power Station and to determine the relationship of temperature curing with the compressive strength on geopolymer mortars using fly ash-based. The fly ash used was obtained from two Thermal Power Station in South Sulawesi, Indonesia. The sample is in the form of a cube with 5 cm x 5 cm size, with variations in room temperature; 65°C, 85°C, and 105°C. Activators used were Na_2SiO_3 and NaOH with 10 M with a ratio of Na_2SiO_3 per NaOH of 2. Compressive strength tests were carried out at 28 days. From the results of the study that the compressive strength of geopolymer mortar based on fly ash-Thermal Power Station A is higher than the compressive strength value of geopolymer mortar based on fly ash - Thermal Power Station B. The relationship of compressive strength with curing treatment with geopolymer mortar temperature variations is very significant. The higher the temperature, the higher the compressive strength of the mortar.

1. Introduction

The development of construction that utilizes cement material has increased very rapidly. Cement production follows the increasing market demand as we all know that producing cement will have a negative effect, which is to create CO_2 gas. CO_2 gas will be delivered during the production of cement, which is very damaging to the air environment. Now it is necessary to find alternatives to cement replacement. Cement replacement is expected to be environmentally friendly and sustainable material.

Researchers have done a lot of research on materials that can replace part or all of cement as a binder in making concrete. As for some materials that have the same cement composition that has been studied by several previous researchers, namely rice husk ash, bagasse ash, and flies ash. Rice husk ash and bagasse ash can be used as cement substitution in concrete. Rice husk ash cannot be used as a substitute for whole cement, but rice husk ash can only replace part of the cement as much as it should not exceed 10% of the amount of cement [1]. Fly ash is a material that can be used as a substitute for cement entirely with a polymerization system [2,3]. Fly ash is an industrial waste that can be obtained from side products in thermal power plants that use coal as fuel. Fly ash has the same pozzolan properties as cement so that it can replace cement in concrete construction. Fly ash contains SiO_2 and Al_2O_3, which can react with calcium hydroxide, so fly ash can be used instead of cement [4]. The elements fly ash contained can react if given activator. Activators are single or combined alkaline solutions [5]. It is this alkaline activator that plays an essential role in reacting to the Al and Si elements contained in the fly ash so that it can produce strong polymeric bonds and speed geopolymers up the reaction. The activators commonly used are NaOH (sodium hydroxide), Na_2SiO_3 (sodium silicate), KOH (potassium hydroxide) with reactive aluminosilicate material [6–9]. The age and temperature variables of curing have an effect on the compressive strength of mortar or geopolymer concrete samples [10]. Besides this, the concentration of activator also has a considerable effect on strength rather than curing temperature and the amount of curing time [5,6,11,12]

Mixtures that have been prepared to make geopolymer materials can be cured at room/ room temperature [13]. However, according to Somna K et al. (2011) that the strength is low in geopolymer materials that use fly ash when using room temperature [9]. As research conducted by Bachtiar. E (2018) that concrete content that is not cured or in other words, concrete is placed following room temperature, the strength decreases [14]. Besides, there have been several previous studies which argued that materials made with the polymerization process could achieve maximum strength if curing at temperatures between 40-90°C or more [15–17].

From the previous studies above that fly ash can replace cement because it has the nature of pozzolan. However, due to the different characteristics of fly ash, it is essential to pay attention to the concentration of the activator used. Besides that, the temperature and the amount of curing time used for the production of geopolymer materials need attention. With this basis, researchers are interested in examining how the temperature relationship used in curing on mortar geopolymer is different at 28 days and 56 days.

2. Materials and methods

The material used in this study uses local materials in South Sulawesi, Indonesia. Local materials used are fly ash, sand, water, and activator. The necessary materials used in the manufacture of Geopolymers consist of coal fly ash waste originating from two different power plants in South Sulawesi, Indonesia. The fly ash used comes from two different Thermal Power Station, namely the Jenneponto Thermal Power Station and the Barru Thermal Power Station. The given name is distinguished, namely, fly ash from the Jenneponto power plant named fly Ash A and fly ash from the Barru power plant are named Fly Ash B.

This study is a continuation of previous research by Bachtiar, E et al. (2018) using the same fly ash [2]. The results of the examination of fly ash characterization used by X-Ray Fluorescence (XRF) analysis can be seen in Table 1.

Table 1 The chemical composition of fly ash [2]

Code of fly ash	Chemical Composition of Fly Ash (%)							
	SiO2	Fe2O3	CaO	Al2O3	K2O	Ti2O2	BaO	ZrO2
FA_A	35.880	29.200	23.520	9.230	1.030	0.710	0.270	0.054
FA_B	41.450	25.280	16.190	11.060	1.350	1.050	0.136	0.069

The activator used is a mixture of NaOH and Na_2SiO_3. The concentration of NaOH used is a mixture of NaOH and Na2SiO3. The concentration of NaOH used was 10 M; the ratio of Na2SiO3 to NaOH was 2. The ratio of the activator to the binder (fly ash) was 0.3. The ratio between fly ash and sand is 1: 2.75. Samples were made with variations in the type of fly ash and temperature variations during maintenance. Fly ash obtained from Jenneponto Thermal Power Station is coded FA. A and fly ash obtained from Barru Thermal Power Station is coded FA.B. Temperature variation consists of several temperatures, namely room temperature ± 25°C, 65°C, 85°C, and 105°C. Samples were made with 8 (eight) variations namely Geopolymer mortar using A fly ash-based which is a cure at an ambient temperature of ± 25°C (GM_FA.A_T25); Geopolymer mortar using A fly ash-based which is a cure at 65°C (GM_FA.A_T65); Geopolymer mortar using A fly ash-based which is cured with temperature 85°C(GM_FA.A_T85); Geopolymer mortar using A fly ash-based which is a cure with temperature 105°C (GM_FA.A_T105), Geopolymer mortar using B fly ash-based which is curing air with a temperature around ± 25°C (GM_FA.B_T25), Geopolymer mortar using B fly ash-based with temperature 65 (GM_FA.B_T65), Geopolymer mortar using B Fly ash-based with heat cure 85°C (GM_FA.B_T85), Geopolymer mortar using B Fly ash-based with temperature 105°C (GM_FA.B_T105).

The sample was made in the form of cubes 5 cm x 5 cm x 5 cm. Samples that have been made cured by being inserted in the oven in accordance with a predetermined temperature for 24 hours. Then removed in the oven and treated again with room temperature ± 25oC. Specifically, the samples GM_FA.A_25 and GM_FA.A_25 were not included in the oven, but since they were removed in the mold. They were directly treated with laboratory temperature. Compressive strength testing with the Unit Testing Machine (UTM) is done at the age of 7, 14, 28, and 25 days. Tests are carried out based on SNI 03-6825-2002.

3. Result and discussion

3.1. Development of Compressive Strengths on Geopolymer Mortars

The development of mortar/concrete compressive strength generally increases with age. As previous research on concrete using both seawater and freshwater as its mixing water, the development of compressive strength has a relationship with a very significant age. The relationship of compressive strength and age in self-compacting concrete that uses freshwater/seawater as its mixing water forms a nonlinear regression with a correlation of R close to 1 [18].

Compressive strength values obtained from the test results in the laboratory using the UTM (Unit Test Machine). Sample testing was carried out at 7, 14, 28, and 56 days. The results of research on the average compressive strength of three samples in geopolymer mortar using A fly ash-based and geopolymer mortar using B fly ash-based can be seen in figure 1 and figure 2.

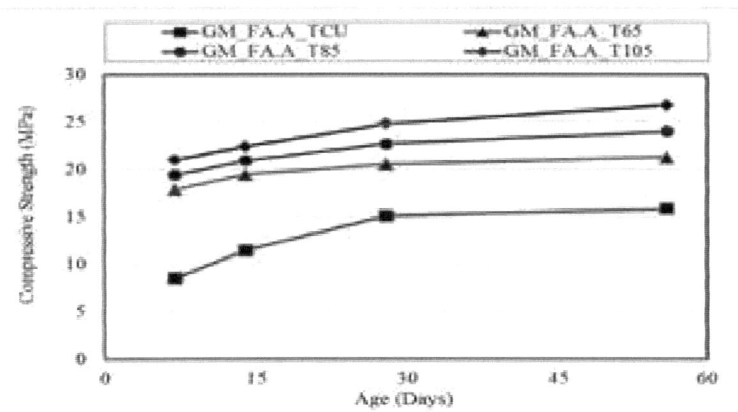

Figure 1. Compressive strength development of geopolymer mortar using A fly ash-based (GM_FA.A)

Figure 1 shows the development of the average compressive strength values of the three samples along with the age of the geopolymer mortar samples using A fly ash-based. The development of the 4 sample variations GM_FA.A_TCU, GM_FA.A_T65, GM_FA.A_T85, and GM_FA.A_T105 shows an increase in compressive strength often with increasing age of geopolymer mortar samples. Nonetheless, there are differences in the patterns of increase that vary.

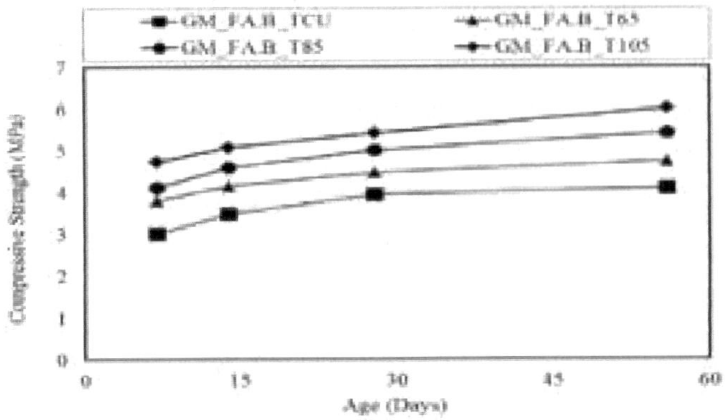

Figure 2. Age relationship to the compressive strength of geopolymer mortars using B fly ash-based (GM_FA.B)

Figure 2 shows the development of the average compressive strength values of the three samples along with the age of the geopolymer mortar samples using B fly ash-based. The development of the 4 sample variations GM_FA.B_TCU, GM_FA.B_T65, GM_FA.B_T85, and GM_FA.B_T105 shows an increase in compressive strength often with increasing age of geopolymer mortar samples. Nonetheless, there are differences in the patterns of increase that vary.

Geopolymer mortars that use A fly ash-based and B fly ash-based have in similarity in increasing compressive strength along with increasing mortar life. As in sameness in the development of mortar/concrete that uses cement, there is an increase along with the increasing age of the mortar/ concrete. This means that mortar/concrete that uses cement or flies ash as a binder has similarities in increasing compressive strength along with increasing mortar/ concrete age. As Erniati et al. (2018), that there was a significant relationship between age and compressive strength, and the increasing compressive strength of mortar, the higher the age of mortar [19].

3.2. Comparison of Compressive Strength between Geopolymer mortar using A Fly Ash-based and B Fly Ash-based

Comparison between the compressive strength of geopolymer mortar A fly ash-based and geopolymer mortar using B fly ash-based can be seen in figure 3, figure 4, figure 5, and figure 6. From the four pictures, they show how the compressive strength value resulting from the use of fly ash A and fly ash B in making Geopolymer mortars which are curing with variations in temperature and without curing.

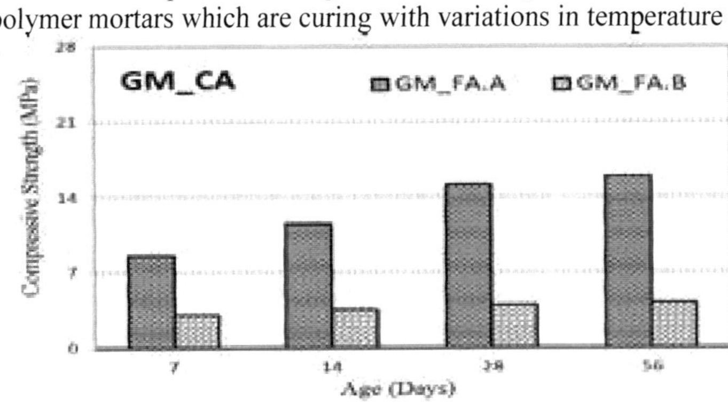

GM_CA : Geopolymer Mortar curing of air with the temperature at ±25oC
GM_FA.A, GM_FA.B : Geopolymer Mortar using A fly ash-based and B fly ash-based

Figure 3. Comparison of compressive strength values between geopolymer mortar using A fly ash-based and geopolymer mortar using B fly ash-based with cured of air temperature at ± 25°C.

Figure 3 shows the comparison of fly ash A (GM_FA.A) base Geopolymer mortar is greater than Geopolymer mortar using B fly ash based (GM_FA.B), which is curing with room temperature ± 25°C. It can be seen from the figure that since the age of 7, 14, 28, and 56 days, the compressive strength of GM_FA. A sample is greater than the GM_FA.B sample. The average percentage difference in the value of GM_FA.B compressive strength against GM_FA_A compressive strength is 70.56%.

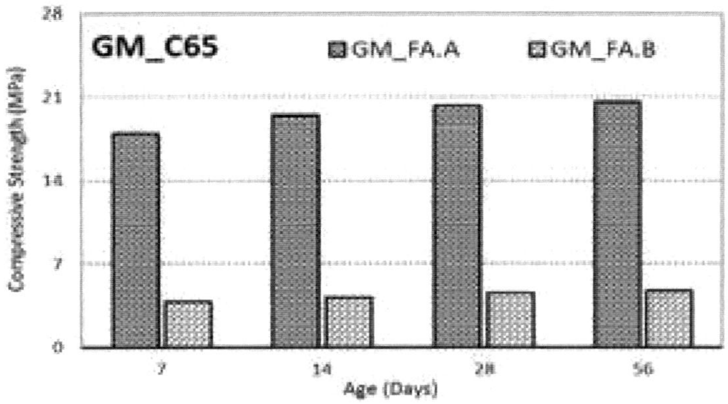

GM_C65 : Geopolymer Mortar cured of temperature at $65^{o}C$
GM_FA.A, GM_FA.B : Geopolymer Mortar using A fly ash-based and B fly ash-based

Figure 4. Comparison of compressive strength values between Geopolymer Mortar using A fly ash-based and B fly ash-based with cured of temperature at 65°C.

Figure 4 shows the comparison of geopolymer mortar using A fly ash-based (GM_FA.A) is greater than geopolymer mortar using B fly ash-based (GM_FA.B), which is curing at 65°C. Seen in the figure, that at 7, 14, 28 and 56 days old, the compressive strength of the GM_FA. A sample is greater than the GM_FA.B sample. The average percentage difference in the value of the compressive strength of GM_FA.B against the compressive strength of GM_FA_A is 78.31%.

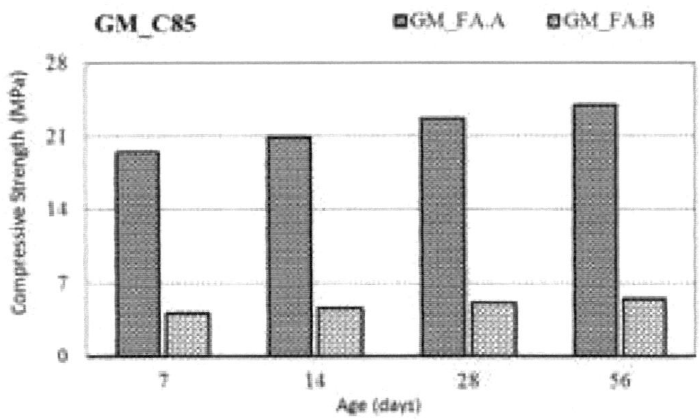

GM_C85 : Geopolymer Mortar, cured of temperature at 85°C
GM_FA.A, GM_FA.B : Geopolymer Mortar using A-fly ash-based and B fly ash-based

Figure 5. Comparison of compressive strength values between Geopolymer Mortar using A fly ash-based and Geopolymer Mortar using B fly ash-based with cured of temperature at 85°C.

Figure 5 shows the comparison of geopolymer mortar GE_FA. A is greater than geopolymer mortar GM_FA.B which is curing at 85°C. It can be seen from the figure that since the age of 7, 14, 28, and 56 days, the compressive strength of GM_FA. A sample is greater than the GM_FA.B sample.

The average percentage difference in the value of GM_FA.B compressive strength against GM_FA_A compressive strength is 78.04%.

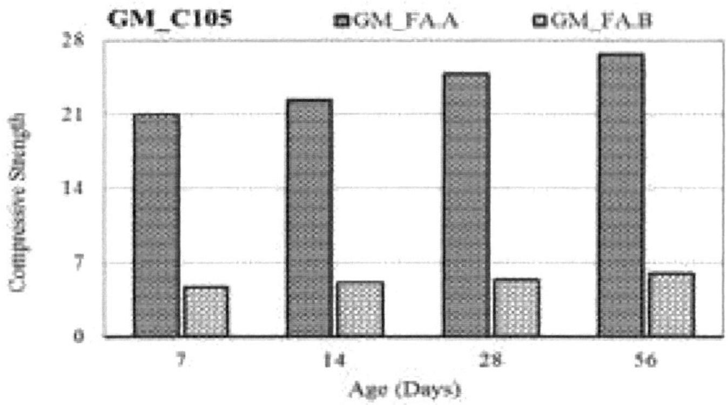

GM_C105 : Geopolymer Mortar, Geopolymer Mortar, cured of temperature at 105°C
GM_FA.A, GM_FA.B : Geopolymer Mortar using A fly ash-based and B fly ash-based

Figure 6. Comparison of compressive strength values between geopolymer mortar using A fly Ash based and Geopolymer Mortar B fly Ash-based with cured of temperature at 105°C.

Figure 6 shows the comparison of geopolymer mortar GE_FA. A is greater than geopolymer mortar GM_FA.B which is curing at 105°C. Seen in the figure, that since the age of 7, 14, 28 and 56 days, the compressive strength of the GM_FA. A sample is greater than the GM_FA.B sample. The average percentage is a difference in the value of the compressive strength of GM_FA.B against the compressive strength of GM_FA_A is 77.61%.

The results showed that the compressive strength value produced on geopolymer mortar A fly ash based is greater than B fly ash-based. This can be seen in Fig. 3, Fig. 4, Fig. 5, and Fig. 6. The compressive strength value of geopolymer mortar using A fly ash-based is greater than geopolymer mortar using B fly ash-based at all temperature variations ± 25oC, 65°C, 85°C, and 105°C. The average percentage is the difference in compressive strength of fly ash B against fly ash B due to differences in temperature variations is 76.13%. This difference is probably due to the material of the fly ash, as with the chemical composition of fly ash A and fly ash B, which are different, as in Table 1. Comparison of the amount of chemical composition between SiO2: Fe2O3: CaO: Al2O3 in A fly ash is 1: 1.23 : 1.53: 3.89, while for B fly ash is 1: 1.64: 2.56: 3.75. From this comparison, it can be seen that the CaO content in fly ash B is very large on the value of SiO2.

3.3. Relationship between Temperature and Compressive strength Value of Geopolymer mortar.

The relationship of temperature used in treating geopolymer mortar with compressive strength of geopolymer mortar produced can be seen in Fig. 7 and 8. Fig. 7 shows the temperature relationship used in curing geopolymer mortar and compressive strength of geopolymer mortar at 28 days. While Fig. 8 shows the relationship of the temperature used in curing geopolymer mortar and compressive strength of geopolymer mortar at 56 days.

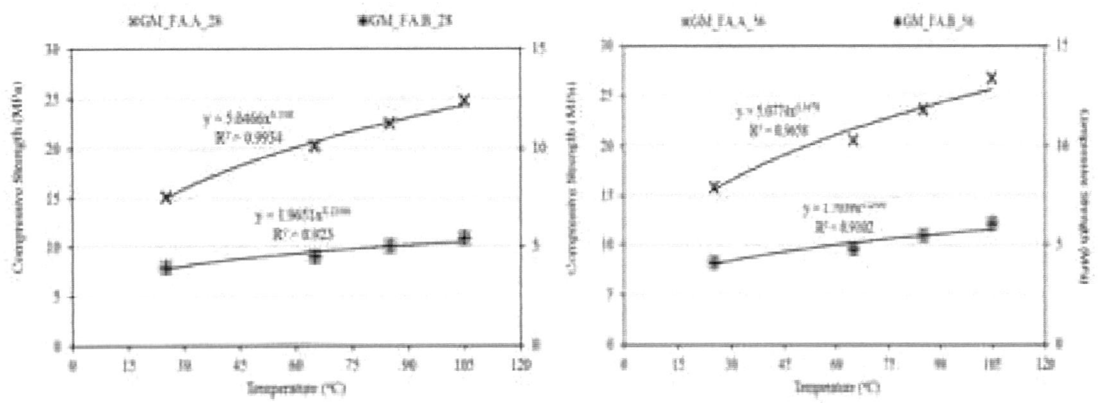

(a) (b)

Figure 7. The relationship of temperature with the compressive strength value of Geopolymer mortars at 28 days (a) and 56 days (b).

Fig. 7 (a) shows the relationship between the temperature used for curing geopolymer mortar with the compressive strength of at 28 days. Geopolymer mortar using A fly ash A-based (GM_FA.A_28) produces a nonlinear equation = 5.0466 $^{0.338}$ and the square root of the value ^{2,} which results in a correlation value of R = 0.9967. while the geopolymer mortar using B fly ash-based (GM_FA.B_28) produces a nonlinear equation = = 1.9651 $^{0.2106}$ and the square root of the value ^{2} which results in a correlation value of R = 0.9607. Fig. 7 (b) shows the relationship between the temperature used for curing geopolymer mortar with the compressive strength of geopolymer mortar at 56 days. For Geopolymer mortar using A fly ash-based (GM_FA.A_56) base produces a nonlinear equation with = 5.0774 $^{0.3479}$ and the square root of the value ^{2,} which results in a correlation value of R = 0.9828. while the geopolymer mortar using B fly ash-based (GM_FA.B_56) produces a nonlinear equation = 1.7939 $^{0.2496}$ and the square root of the value ^{2,} which results in a correlation value of R = 0.9540. From the two non-linear equations, the correlation value that is close to number 1 is obtained, so that the relationship between the temperature used for curing geopolitical mortar with the pressure of geopolymer mortar at 28 or 56 days is very significant. The higher the temperature used in curing the geopolymer mortar, the higher the compressive strength of the geopolymer mortar produced.

The curing of the material greatly influences the strength of the material. As the Research by Erniati et al. (2018) said that the self-compacting without curing produced the pressure decreased. A Geopolymer mortar has been removed from the mold. It needs nursing care, which is different from mortar, which uses cement as its cylinder. Generally, for mortar and concrete that use cement as a binder, the treatment is either soaked or watered with water. It is different from Geopolymer mortar. It acts as a radiant heat [14]. As a study by Gökhan Görhan and Gökhan Kürklü (2014) have conducted studies on the effect of curing NaOH concentrations at temperatures of 65°C and 85°C, with the result that NaOH concentrations have a clear effect on the mortar properties cured at 85°C [13]. Optimal NaOH concentrations obtained from research by Gökhan Görhan and Gökhan Kürklü (2014) were concentrated in NaOH 6 M, until treatment with temperatures of 85°C for 24 hours [13]. From this, it can be seen that the temperature used in geopolitical mortar curing has a great influence on the pressure of geopolymer mortar greatly. Curing both the mortar and the concrete influences the microstructure of a mortar material. The polymerization process is not maximized in geopolitical mortar working without treatment with the required temperature. The same thing with mortar/ concrete without curing, the hydration process is not maximized so that the structure of the material is not good, and microstructure is too influential in its mechanical properties, porosity, and absorption [20,21].

4. Conclusion

Geopolymer mortar by using fly ash which chemical composition is different, it will produce different strengths of geopolymer mortar as well. The development of compressive strength in mortal geopolymer was along with age. The more increasing age of the geopolitical mortar has, the more increased the height of the pressure. The relationship between the temperature used in geopolymer mortar treatment with the geopolymer mortar compressive strength is very significant, that is, the higher the treatment temperature, the higher the compressive strength of mortar geopolymer.

Acknowledgment
Thank you to the Ministry of Research, Technology, and Higher Education Research. Directorate General of Research and Development Strengthening, Fiscal Year 2019 that has funded the research of this Basic Research.

References

[1] Ahmad I A, Pertiwi N and Taufieq N A S 2018 Reliability of rice husk ash as substitution of Portland composite cement producing green concrete *Ecol. Environ. Conserv* **24** S56–63
[2] Davidovits J 1999 Chemistry of geopolymeric systems, terminology *Geopolymer* vol 99 (sn) pp 9–39
[3] Van Jaarsveld J G S, Van Deventer J S J and Lukey G C 2002 The effect of composition and temperature on the properties of fly ash-and kaolinite-based geopolymers *Chem. Eng. J.* **89** 63–73
[4] Rattanasak U and Chindaprasirt P 2009 Influence of NaOH solution on the synthesis of fly ash geopolymer *Miner. Eng.* **22** 1073–8
[5] De Vargas A S, Dal Molin D C C, Vilela A C F, Da Silva F J, Pavao B and Veit H 2011 The effects of Na2O/SiO2 molar ratio, curing temperature and age on compressive strength, morphology and microstructure of alkali-activated fly ash-based geopolymers *Cem. Concr. Compos.* **33** 653–60
[6] Bakharev T 2005 Geopolymeric materials prepared using Class F fly ash and elevated temperature curing *Cem. Concr. Res.* **35** 1224–32
[7] Brough A R and Atkinson A 2002 Sodium silicate-based, alkali-activated slag mortars: Part I. Strength, hydration and microstructure *Cem. Concr. Res.* **32** 865–79
[8] Fernández-Jiménez A and Palomo A 2005 Composition and microstructure of alkali activated fly ash binder: Effect of the activator *Cem. Concr. Res.* **35** 1984–92
[9] Somna K, Jaturapitakkul C, Kajitvichyanukul P and Chindaprasirt P 2011 NaOH-activated ground fly ash geopolymer cured at ambient temperature *Fuel* **90** 2118–24
[10] Andi Arham Adam H 2014 The Effect of Temperature and Duration of Curing on the Strength of Fly Ash Based Geopolymer Mortar *Procedia Eng.* **95** 410–4
[11] Hu M, Zhu X and Long F 2009 Alkali-activated fly ash-based geopolymers with zeolite or bentonite as additives *Cem. Concr. Compos.* **31** 762–8
[12] Sitoresmi D A N and Pursetyo K T 2020 The effect of concrete tanks for the breeding technique of the sand sea cucumber (Holothuria scabra) To *IOP Conf. Ser. Earth Environ. Sci.* **441** 1–10
[13] Görhan G and Kürklü G 2014 The influence of the NaOH solution on the properties of the fly ash-based geopolymer mortar cured at different temperatures *Compos. part b Eng.* **58** 371–7
[14] Bachtiar E 2018 The Self Compacting Concrete (Scc) Using Seawater As Mixing Water Without Curing
[15] Chindaprasirt P, Chareerat T and Sirivivatnanon V 2007 Workability and strength of coarse high calcium fly ash geopolymer *Cem. Concr. Compos.* **29** 224–9
[16] Swanepoel J C and Strydom C A 2002 Utilisation of fly ash in a geopolymeric material *Appl. geochemistry* **17** 1143–8
[17] Panias D, Giannopoulou I P and Perraki T 2007 Effect of synthesis parameters on the

mechanical properties of fly ash-based geopolymers *Colloids Surfaces A Physicochem. Eng. Asp.* **301** 246–54

[18] Erniati, M. W. Tjaronge, R. Djamaluddin V sampebulu 2015 Compressive Strength and Slump Flow of Self Compacting Concrete Uses Fresh Water and Sea Water *ARPN J. Eng. Appl. Sci. Asian Res. Publ. Netw.* **10** 2373–7

[19] Marzuki I, Bachtiar E, Chaerul M and Fatmawaty R 2018 The Development Of Compressive Strength On Geopolymer Mortar Using Fly Ash As Based Material In South Sulawesi

[20] Tjaronge M W and Djamaluddin R 2014 Microstructure Characteristics of Self Compacting Concrete using Sea Water *Int. J. Appl. Eng. Res.* **22** 18087–95

[21] Tjaronge M W and Irfan U R 2015 Porosity, pore size and compressive strength of self compacting concrete using sea water *Procedia Eng.* **125** 832–7

IOP Conf. Series: Materials Science and Engineering **885** (2020) 012020 doi:10.1088/1757-899X/885/1/012020

The influence of shooting conditions during shot peening of stainless steel on surface roughness

M. Iqbal Mukhsen[1], Rusdi Nur[1,2], M. Alif AF Yusuf [1], and Christo Rakka[1,3]

[1]Mechanical Engineering Department, Politeknik Negeri Ujung Pandang Jl. Perintis Kemerdekaan Km. 10 Makassar 90245 South Sulawesi Indonesia

[2]Center for Materials and Manufacturing, Politeknik Negeri Ujung Pandang Jl. Perintis Kemerdekaan Km. 10 Makassar 90245 South Sulawesi Indonesia

[3]Student of Manufacturing Engineering Study Program, Politeknik Negeri Ujung Pandang Jl. Perintis Kemerdekaan Km. 10 Makassar 90245 South Sulawesi Indonesia

Email: rusdinur@poliupg.ac.id

Abstract. Stainless steel AISI 316L is a suitable material used for orthopedic purposes. In order to improve the hardness of stainless steel, the heat-treatment process cannot be done, but with a shot peening process. This study aims to determine the roughness of the workpiece surface and investigate the effect of shooting conditions (i.e., angle and time) on the surface roughness during shot peening of Stainless steel AISI 316L. The shooting parameters used are 2, 4, and 6 minute for the duration, and angle of 0°, 15°, 30° and 45°. The ball material is a type of steel with 0.6mm in diameter. The experimental results presented that the increased shooting time will result in lower surface roughness, and vice versa. Therefore, it can be concluded that the use of varying shooting conditions will greatly affect the quality of the surface produced but needs to be compared with other responses, namely hardness.

1. Introduction

Material stainless steel grade AISI 316L is often used for medical equipment is biomaterial [1]. This biomaterial is expected to be resistant to corrosion, low toxicity, and the ability of a material to be easily formed. Some steel materials that have such characteristics are very difficult to obtain, so they need to look for materials that are close to their properties. In the case of connecting broken bones using implant plates, the material often used by orthopedic doctors is titanium or stainless steel. The implant material has a higher resistance to corrosion due to direct contact with the human body. Human body fluids have many aggressive ions that cause corrosion in the implant material.

Stainless steel is a material that have resistant to corrosion, but the hardness is below that of titanium material, so it needs to do an additional treatment. Stainless steel has low carbon content making it difficult to do heat treatment. Thus it was more suitable for cold treatment. The treatment processes carried out by cold processes are machining processes such as sandblasting [2–4], shot peening, acid etching [2,5], anodizing, and surface mechanical attrition treatment commonly abbreviated as SMAT [2,6,7]. Stainless steel is also a material that is more easily formed than titanium material, so the cost of its formation can be cheaper.

Previously, Liu et al. present that the structure of changes due to the shot peening shooting process results in an increase in deformation thickness and grain increase as the pressure and duration of the shot peening increase [8]. In Jiang's research, it was shown changes in the structure of the layer due to the collision of material [9]. The aims of this work are to study the effect of shooting conditions during the shot peening process of stainless steel 316L on the surface roughness.

2. Shot Peening Process

Shot peening is a method mostly used to treat surface mechanical. This method has some major advantages, i.e., easy surface cleaning, high processing quality, adjustability of the strengthening effect, and able to establish in the industry. The shot peening process starts with the pressurized air flowing from the compressor with 4-7 bar air pressure flowing into the spray-gun, the steel shot in the spray box will be flowed through the spray-gun due to the vacuum in the flowing steel.

The shot peening process produces a notch on the surface of the material caused by small balls firing on the surface of the material.

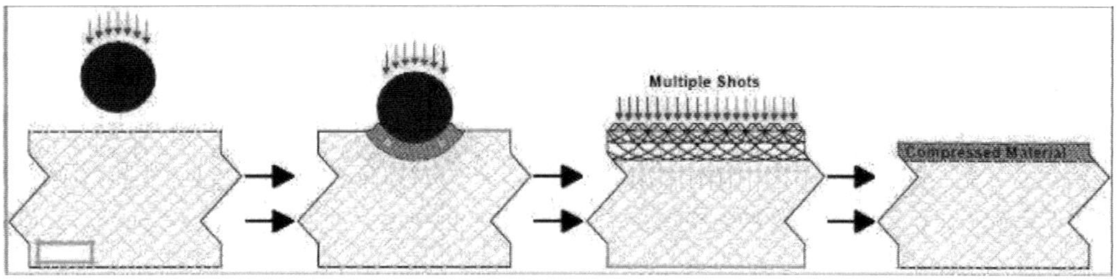

Figure 1. Illustration of shot peening

Relatively small steel balls are fired at high speed on the workpiece surface repeatedly. This process will create compressive residual stress on the entire surface of the object of the same size and is evenly distributed, and changes with depth or thickness of the object. This residual stress is useful for increasing material resistance to loads. The schematic of the shot peening process was shown in Figure 2.

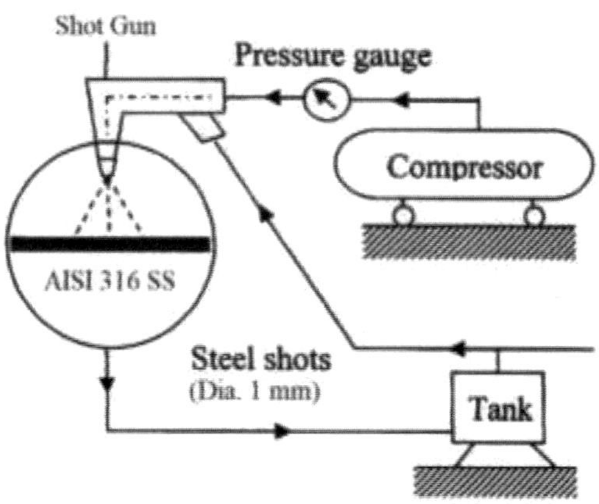

Figure 2. Schematic of the shot peening process

3. Experimental Setup

3.1. Materials

AISI 316 stainless steel of 5 mm thickness was used as a workpiece material. The composition of AISI 316 stainless steel is shown in Table 1.

Table 1. Chemical composition of AISI 316 SS

Composition	C	Mn	Si	Cr	Ni	Mo	P	S	N	Fe
Weight (%)	0.08 max	2.00 max	0.75 max	16-18	10-14	2 - 3	0.045 max	0.03 max	0.10 max	Bal.

3.2. Shooting Parameters

In this study, the steel shot used was ball type with a diameter of 1 mm. The experimental setup was using the shooting parameters as follow:

Table 2. Parameters for the shot peening process

Parameters	Range
Shooting Time, S_T (minute)	2, 4, and 6
Shooting Angle, S_A (degree)	0, 15, 30, and 45

Surface roughness testing was performed using surface roughness tester SURFTEST SJ-310. Data on Surface roughness was measured five times for each position.

4. Results and Discussion

The surface roughness after the shot peening process is shown in Table 3.

Table 3. The result of surface roughness

S_T	S_A	Surface Roughness (μm)					
(minute)	(degree)	Ra_1	Ra_2	Ra_3	Ra_4	Ra_5	Average
2	0	1,320	1,649	1,923	1,979	1,954	1,765
2	15	1,801	1,494	1,715	1,786	1,67	1,693
2	30	1,517	1,578	1,445	1,619	2,077	1,647
2	45	1,965	2,15	1,928	1,789	1,836	1,934
4	0	1,65	1,601	1,601	1,902	1,727	1,696
4	15	1,807	1,446	1,473	1,452	1,55	1,546
4	30	1,608	1,65	1,724	1,437	1,689	1,622
4	45	1,235	1,235	1,235	1,235	1,235	1,235
6	0	1,649	1,368	1,232	1,312	1,409	1,394
6	15	1,652	1,532	1,552	1,599	1,743	1,616
6	30	1,663	1,675	1,673	1,779	1,417	1,641
6	45	1,278	1,278	1,278	1,278	1,278	1,278

In order to gain a clear understanding, the experimental results data can be illustrated in Figures 3 and 4.

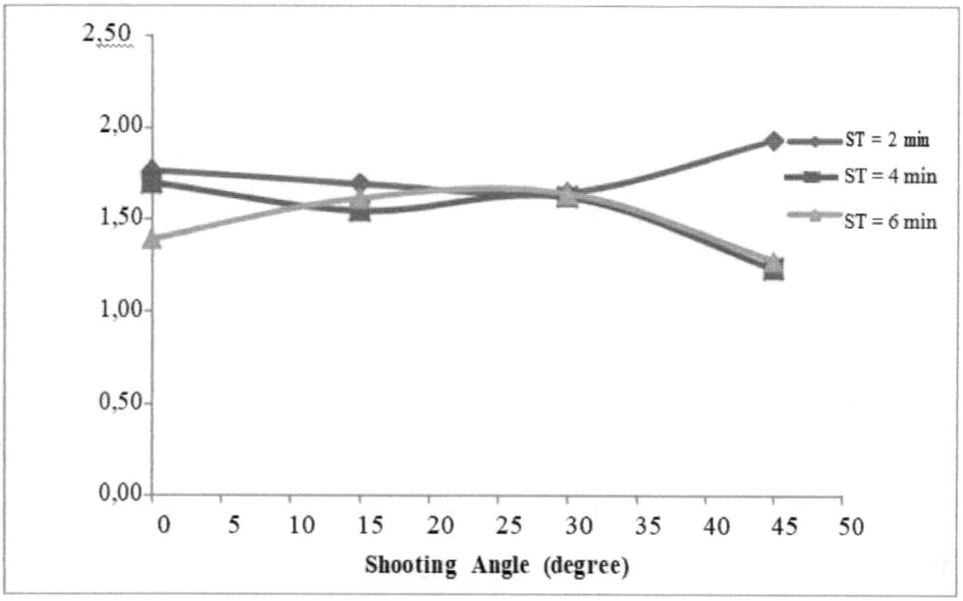

Figure 3. Data on surface roughness for different shooting time (S_T)

Figure 3 presents that surface roughness will decrease with increasing angle of shoot, except for 2 minutes shot duration. It is very clear that the surface roughness decreased for longer shootings. The highest surface roughness (1.934 m) was obtained at 2 minutes of duration and 45^{0} of shooting angle. The surface smoother (1.278 m) was gained at a similarity angle with 6 minutes.

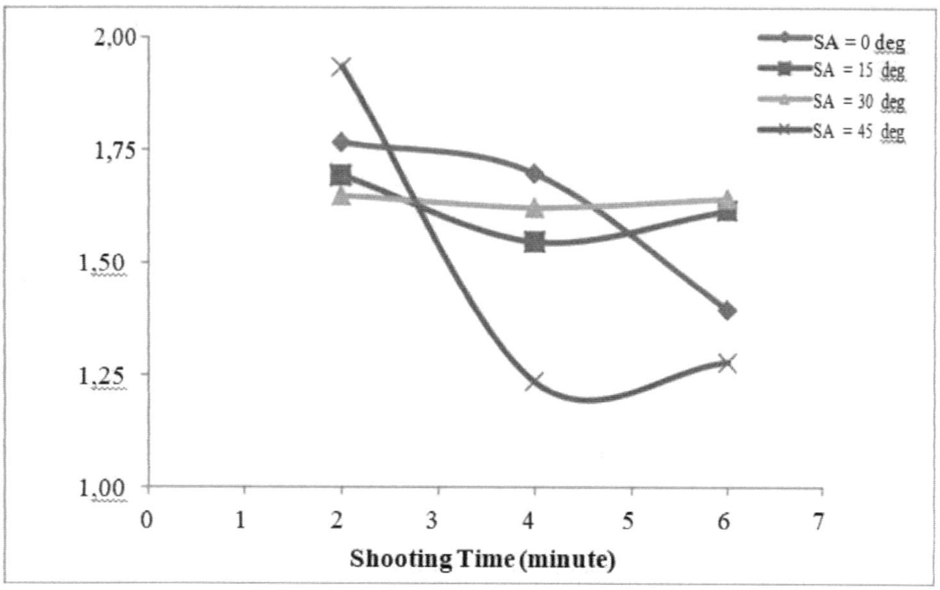

Figure 4. Data on surface roughness for different shooting angle (S_A)

Figure 4 presents that surface roughness is strongly affected by the duration of shot for both shooting time of 0° and 45°. It is very clear that the surface roughness decreased for longer shootings. But for the shooting angle of 15° and 30°, surface roughness has a similarity result for all durations.

These results had been proven by Hasan et al. in their investigation. It was found that increasing shot peening leads to increase residual stress, surface roughness for shot peening of austenite stainless steel 316L [9]. This can be attributed to the fact that as the pulsed current decreases, discharges strike the surface of the sample less intensely, and the resulting better erosion effect leads to a smoother surface. Furthermore, as the pulse-on duration decreases, the amount of heat energy transferred to the sample surface decreases, and so less material melts. The fact that the surface roughness decreases with decreasing discharge energy has been described in the literature [9,10].

5. Conclusion

Based on the experimental results through the shot peening process of stainless steel AISI 316L, it can be concluded that the shooting time (duration) has a significant effect on surface roughness. Meanwhile, the shooting angle of 0° and 45° will be making the surface roughness is higher. But for the shooting angle of 15° and 30°, the surface roughness has a similarity result.

References

[1] Nikmah A, Izak Rudyardjo D, Ady J and Taufiq A 2019 Studies on Density, Corrosion Rate and Hardness Characteristics of Stainless Steel Implanted by Nitrogen Ion *IOP Conf. Ser. Mater. Sci. Eng.* **515**

[2] Elias C N, Oshida Y, Lima J H C and Muller C A 2008 Relationship between surface properties (roughness, wettability and morphology) of titanium and dental implant removal torque *J. Mech. Behav. Biomed. Mater.* **1** 234–42

[3] Piattelli A, Scarano A, Piattelli M and Calabrese L 1996 Direct bone formation on sand-blasted titanium implants: an experimental study *Biomaterials* **17** 1015–8

[4] Aparicio C, Gil F J, Fonseca C, Barbosa M and Planell J A 2003 Corrosion behaviour of commercially pure titanium shot blasted with different materials and sizes of shot particles for dental implant applications *Biomaterials* **24** 263–73

[5] Rupp F, Scheideler L, Rehbein D, Axmann D and Geis-Gerstorfer J 2004 Roughness induced dynamic changes of wettability of acid etched titanium implant modifications *Biomaterials* **25** 1429–38

[6] Arifvianto B and Mahardika M 2011 Effect of sandblasting and surface mechanical attrition treatment on surface roughness, wettability, and microhardness distribution of AISI 316L *Key Engineering Materials* vol 462 (Trans Tech Publ) pp 738–43

[7] Arifvianto B, Mahardika M, Dewo P, Iswanto P T and Salim U A 2011 Effect of surface mechanical attrition treatment (SMAT) on microhardness, surface roughness and wettability of AISI 316L *Mater. Chem. Phys.* **125** 418–26

[8] Liu Y G, Li M Q and Liu H J 2017 Nanostructure and surface roughness in the processed surface layer of Ti-6Al-4V via shot peening *Mater. Charact.* **123** 83–90

[9] Jiang X P, Wang X Y, Li J X, Li D Y, Man C-S, Shepard M J and Zhai T 2006 Enhancement of fatigue and corrosion properties of pure Ti by sandblasting *Mater. Sci. Eng. A* **429** 30–5

[10] Hassan K S, Alwan A S and Abbass M K 2017 Effect of nitriding and shot peening on corrosion behavior and surface properties of austenite stainless steel 316L *Dubai UAE* **19** 395–401

IOP Conf. Series: Materials Science and Engineering **885** (2020) 012021 doi:10.1088/1757-899X/885/1/012021

The simulation buoy shape bullet of sea wave energy absorption based on diameter variations

Jufri[1,3], Daeng Paroka[2], Jalaluddin[1] and Rustan Tarakka[1]

[1]Department of Mechanical Engineering, Universitas Hasanuddin
[2]Department of Ocean Engineering, Universitas Hasanuddin
[3]Politeknik ATI Makassar

Email: jufri@atim.ac.id

Abstract. The use of fossil energy on a large scale to meet energy demand will be a threat to life in the form of air pollution, global warming and threats to the availability of fossil energy. Therefore, researchers feel it is important to conduct research with the aim of studying the potential of new and renewable energy from ocean wave energy absorption using buoys by varying the absorbent diameter of the buoyancy form. This potential is a great opportunity to meet the energy demand in coastal areas and small islands in Indonesia. This study focuses on the potential to absorb buoys carried out in transitional seas with a depth of 5.0 meters, variations in buoy diameter ranging from 1.0 meters to 10.0 meters. The wave force absorbed by the buoy is calculated using the Strip Theory Method including the kinetic force caused by the movement of the buoy. The highest efficiency of buoyancy absorption efficiency was 37.17% for floating buoys with diameter of 2.5 meters and meeting frequency of 1.379 rad/sec and the lowest 36.58% for diameter of 10.0 meters and meeting frequency of 1,463 rad / sec.

1. Introduction

More than 80% world energy demand are supplied by fossil energy [1,2]. The massive used of fossil energy to meet these energy needs will be a threat to human survival. The first is a threat to the environment, namely air pollution and global warming [3,4]. The second threat is the threat of the depletion of fossil energy, the world fossil fuel reserves data for OPEC has only reserves of up to 80 years and Non-OPEC has only 20 years of reserves which is a threat to the availability of energy sources for the lives of our generation in the future [3,5]. Therefore there is a need to maximize the use of existing new and renewable energy potentials to overcome the two threats mentioned above. The potential of renewable energy such as solar energy, photovoltaic, ocean energy, hydro energy, biomass, biodiesel and geothermal energy is quite large, but its use is still minimal [6–9]. Efforts to maximize the use of renewable energy continue to be carried out, as evidenced by the existence of various studies conducted in various uses of new and renewable energy in accordance with the potential and characteristics of each regions.

Indonesia as an archipelago geographically stretches from 60 North Latitudes to 110 South Latitudes and 920 to 1420 East Longitude, which consists of large and small islands of around 17,504 islands. Three quarters of this region is the sea (5.9 million km^2), with a coastline length of 95,161 km, the second longest after Canada [10,11]. Based on Indonesia's geographical conditions, the abundant potential of marine energy to be used includes tidal energy, ocean current energy, ocean wave energy, and ocean thermal energy. Of the four ocean energy potentials, the most developed and trend by researchers is the ocean wave energy potential. Therefore researchers feel it is important to conduct research with the aim of studying the potential of new and renewable energy from ocean wave energy by varying the diameter of the absorbent buoy and the renewable energy source. In this research an assumption is built that the variation in buoyancy diameter will affect the efficiency of wave power absorption by bouy. Energy absorption is obtained from the theoretical power of sea wave multiplication results with buoy diameter [12].

Some countries have used ocean wave energy converter technology, such as Archimedes Waveswing (AW) 0.25 MW in Portugal 2008, CETO in Australia 2008, Pelamis 0.75 MW in Portugal 2008, Wave Dragon 11 MW in Nissum Bredning Scotland 2003, Limpet 0 , 10 - 0, 50 MW on Islay Scotland 2000, Oyster on the Orkney Coast of Scotland, PowerBuoy 0.5 MW in Atlantic City 2005, and AquaBuOY 0.25 MW Washintong US 2004 [13]. Utilization of wave energy converter technology that uses ocean wave energy above provides many advantages namely this potential is abundant, renewable, available, and environmentally friendly [10]. However, all of the wave energy converter (WEC) designs cannot be directly applied in Indonesian sea areas. This is due to differences in the characteristics of sea waves from countries that have implemented the WEC design with the characteristics of Indonesian sea waves. The characteristics of the height and frequency of Indonesia's sea waves based on estimated data of average sea wave heights, significant heights of 0.2 - 2.0 meters and average maximum height of 0.4-2.5 meters. Wave frequencies above 3 meters are only 0% -5% [14,15].

Therefore, researchers feel it is important to conduct research with the aim of studying the potential of WEC based on the characteristics of sea waves in the Indonesian Archipelago. The main component that plays the biggest role in absorbing the wave energy of the WEC is the buoy, so this research is focused on researching the optimal absorption by the bullet shape buoy. The selection of bullet buoy forms used in this study is based on previous research which states that the bullet buoy shape has the greatest efficiency among other forms [16]. The assumption used is varying the diameter of the buoy will affect the efficiency of absorption of wave power by the buoy. The reason for this assumption is that buoy absorption is a product of theoretical wave power available multiplication buoy diameters. The things observed and analyzed in this study are the processes of utilizing hydrodynamic forces arising from ocean waves. Theoretical study of the power of wave hydrodynamics has considerable challenges, where the diffraction force is relatively complex because it consists of various excitation, radiation and attenuation forces [17,18]. The hydrodynamic excitation force of ocean waves absorbed by the oscillating buoy float is obtained using the Strip Theory Method. Results The Strip Theory Method is used as input to determine the generator power output, using relevant mathematical equations.

2. Modeling of the absorber float

2.1. Absorbent Floating Buoy

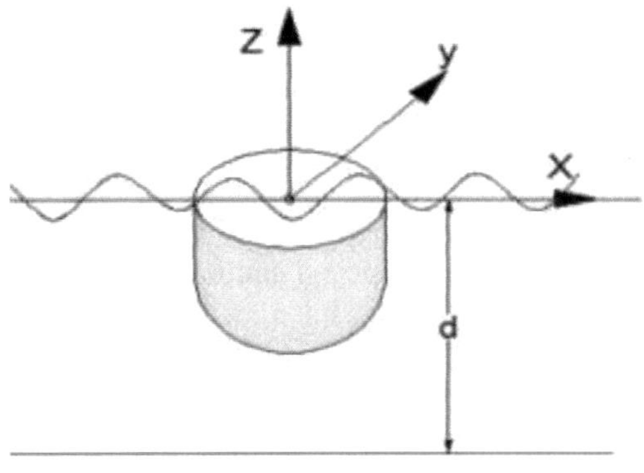

Figure 1. Schematic of floating buoy coordinates

The right hand coordinate system is used to define fluid action on the float. Cartesian coordinate system x, y, z) is assigned to the body. The positive x-direction leads to the bow, positive z-direction is directed upward, and the z = 0 (or xy plane) field coincides with the calm surface of the water

2.2. Mass-Spring-Damper Buoy System

To simplify the problem in the analysis of the floating buoy behavior as shown in Figure 1, it can be assumed that the oscillating behavior of the mechanical motion is the same so that it can be modeled as the spring behavior in Figure 2.

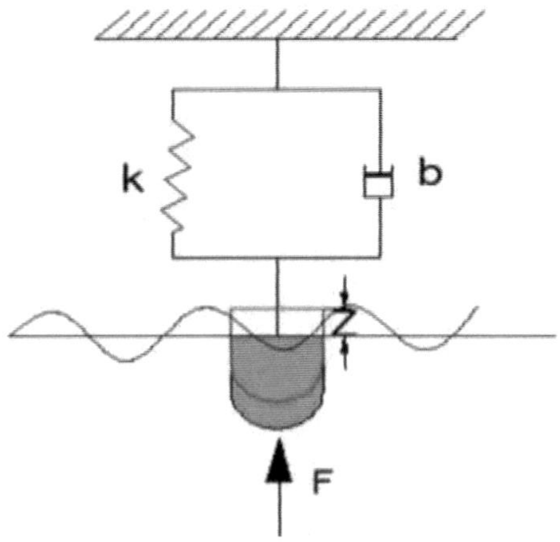

Figure 2. Spring-damped oscillating motion

3. Mathematical equation

3.1. Wave velocity

Wave Velocity be calculated based on category of grouping, deep sea, transition sea or shallow sea. The area of research carried out is limited to the transitional sea so that the equation can be written [19].

Individual Wave Velocity

$$C = \frac{L}{T} = \frac{\omega}{k}, L = \frac{2\pi}{k}, T = \frac{2\pi}{\omega} \quad \text{......................(1)}$$

Individual Wave Velocity

$$C_g = C.n \quad \text{..(2)}$$

The value of n can be calculated using the equation:

$$n = \frac{1}{2}\left(1 + \frac{2kh}{cosh(2kh)}\right) \quad \text{................................(3)}$$

3.2. Water Depth Condition

Wave propagation speed Velocity can be calculated based on grouping categories, deep sea, transitional sea or shallow sea. The area of research carried out is limited to transitional seas so that the equation can be written [19].

Wave Individual Velocity :

$$C = \frac{L}{T} = \frac{\omega}{k},\ L = \frac{2\pi}{k}\ ,\ T = \frac{2\pi}{\omega} \text{ ....(4)}$$

Wave Goup Velocity

$$C_g = C.n \text{ ....(5)}$$

The value of n can be calculated using the equation :

$$n = \frac{1}{2}\left(1 + \frac{2kh}{\cosh(2kh)}\right) \text{.................................(6)}$$

3.3. Buoy Motion

To simplify the case, the movement of the buoy is only observed from the rise and fall of the water by the excitation of the surface movement of the water waves. Approaches to equations that are buoys can be written:

$$Z_t = Z \times \sin(\omega t + \varepsilon)\ \text{[m] ....(7)}$$

The movement of the buoy can be solved by using a second-order differential equation, such as the forced spring motion, mass system and dampers. The buoyancy response force is calculated based on Newton's law. The forces involved as in Figure 2 above, consist of an external harmonic force acting on the system ($F\cos(\omega t)$), add mass force ($a\ddot{z}$), damping force ($b\dot{z}$) and recovery force (cz). So the equation can be formulated in a mathematical equation as follows:

$$a\ddot{z} + b\dot{z} + cz = F\cos(\omega t) \text{ ....(8)}$$

where: a, b, c are add mass mass, attenuation, and recovery constellation; F is a coercive function at frequency ω

Based on the equation (5) response force of the buoy harmonic motion, the magnitude of the heave (z direction direction) can be found by changing equation:

$$z = \frac{F}{\sqrt{(c - a\omega^2)^2 + b^2\omega^2}} \times \cos(\omega t) \text{ ....(9)}$$

3.4. Drag Force

On floating objects moving oscillations will experience drag depending on the speed opposite to the movement of the buoy. Inhibition force, the addition of drag force (Fdrag) due to the effect of sea water viscosity. The magnitude of the Fdrag for an oscillating cylinder in a wave, can be calculated using the Morison Equation [20].

$$F_{drag} = \frac{1}{2}\rho U^2 C_D.D \text{ .... (10)}$$

Where Cd drag coefficient, U vertical water particle velocity (m/s), D cylinder diameter (m), ρ volum of bullet buoy (m).

3.5. Force Froude-Krilov section
Floating objects experience a force called the Froude-Krilov section style whose magnitude can be calculated using the equation:

$$f_3 = \rho\xi gbe^{ikx} - e^{-kds}\left[\frac{N}{m}\right] \text{ ....(11)}$$

Where b is the total width of the buoy (m) that is in contact with the surface of the water.

The magnitude of the excitation force experienced by floating buoys depends on the distance of the heave vertical movement and the magnitude of the excitation force per unit length (8) as the equation below [12]:

$$F_e = f_3.z.\sin(\omega t + \alpha)\ [N] \text{ ....(12)}$$

3.6. Wave Energy
The energy that can be produced by each wave unit correlates with the wave height associated with the equation [19]:

$$\bar{E} = \frac{E}{L} = \frac{\rho g H^2}{8}\ [\text{J/m}^2] \text{ ....(13)}$$

here is where: $\rho\left[\frac{kg}{m^3}\right]$, $g\left[\frac{m}{s^2}\right]$, H [m] density, gravity acceleration, and significant wave height

3.7. Wave Power
The power of each unit of the wave per unit length can be calculated by using the product of the multiplication of the average wave energy with the wave group velocity as shown in the following equation [20]:

$$P_\omega = \bar{E}xC_g\ \ [Kw/m] \text{ ....(14)}$$

3.7.1. Wave Power Absorbtion by Bouy Float. Wave power can be absorbed by buoys, calculated using the following equation [12]:

$$P = P_\omega xd\ \ [Kw] \text{ ....(15)}$$

Where d [m] is a horizontal extension of the buoy that is passed by the sea wave.

3.7.2. Actual Wave Power absorbtion. Absorption of ocean waves by floating buoys is an excitation representation of power in the form of lift and absorption of power in the form of radiation needed to move up and down the floating buoy, so the equation can be written as [20]:

$$P_a = \frac{1}{2}x|F_e||\dot{z}|.cos\varphi \text{ ....(16)}$$

3.8. Efficiency of Power Conversion
The efficiency of power conversion by a buoy can be calculated by comparing equation (13) with equation (12) such as [12]:

$$\eta_a = \frac{P_a}{P}x100\% \text{ ....(17)}$$

IOP Conf. Series: Materials Science and Engineering **885** (2020) 012021 doi:10.1088/1757-899X/885/1/012021

4. Parameter design

4.1. Simulation buoy design

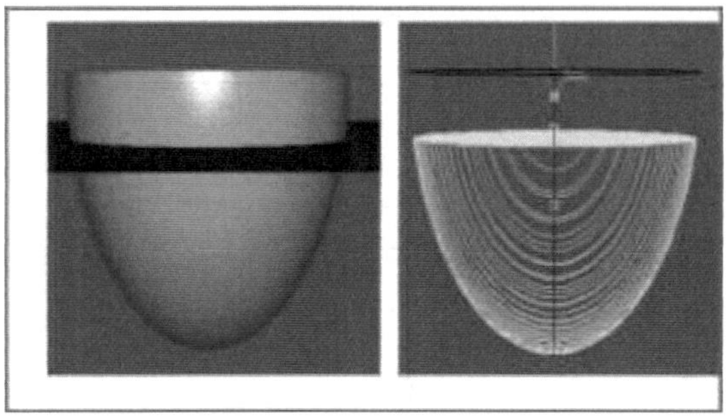

Figure 3. Shape of Simulation buoy

4.2. Diameter size variations

Table 1. Diameter size variations

Diameter	Unit
1	m
1.5	m
2	m
2.5	m
3	m
3.5	m
4	m
5	m
6	m
8	m
10	m

4.3. Setting Simulation Parameters

Table 2. Setting Simulation Parameters

Parameter	Quantity	Unit
Modal period	9.988	s
Characteristic wave height	1.000	m
Spectrum type	Jonswap	-
Wave heading	3.142	rad

IOP Conf. Series: Materials Science and Engineering **885** (2020) 012021 doi:10.1088/1757-899X/885/1/012021

5. Simulation results and discussion

5.1. Simulation Results

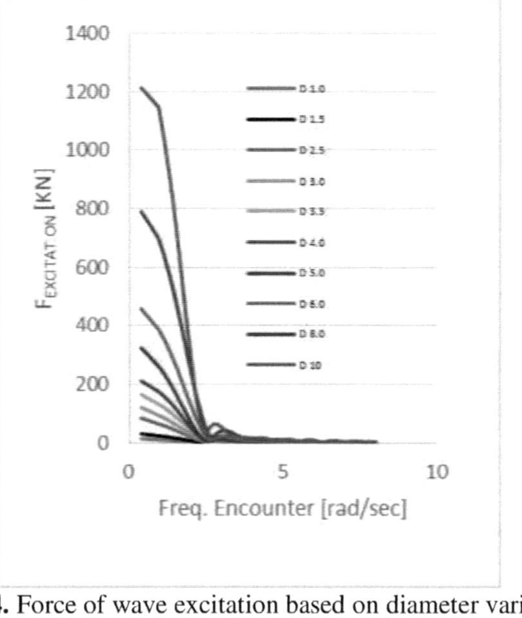

Figure 4. Force of wave excitation based on diameter variations

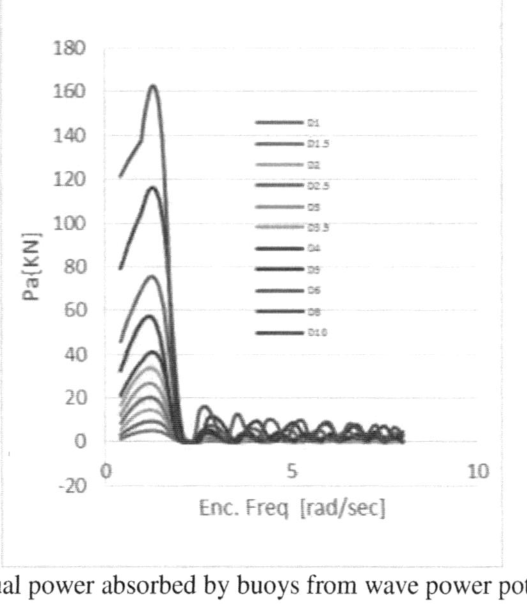

Figure 5. Actual power absorbed by buoys from wave power potential

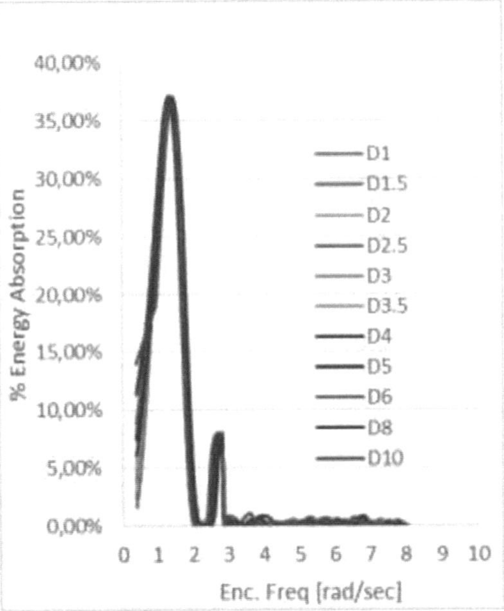

Figure 6. Absorbent absorption efficiency based on diameter variations

5.2. Validation of Results

This paper aims to examine the application of ocean wave energy by floating buoys based on variations in diameter, simulation data compared with experimental results in terms of the results of previous studies [14,16,18] to verify simulation results and calculations using mathematical equations. The buoys used in this simulation are buoys with a diameter of 1.0, 1.5, 2.0, 2.5, 3.0, 3.5, 4.0, 5.0, 6.0, 8.0, and 10 meters. Excitation force is calculated for each diameter of the test buoy. Simulation testing is carried out on regular waves with a wave height of 1 meter with transitional sea conditions. The simulation results obtained from images 3,4,5 and 6 obtained a good suitability of trend charts with experimental results

5.3. Absorption Power and Efficiency

In the calculation and analysis, the hydrodynamic parameters are obtained using Strip Theory Method. After calculations and simulations using relevant equations results are obtained as shown in Figures 3, 4 and 5. The diameter has an influence on the absorption force of the excitation force. On picture. 3 the excitation forces that can be absorbed from each float vary by diameter. The ability to absorb the excitation style of float floats directly in diameter. The magnitude of the diameter influences the absorption force of the excitation force because by extension the magnitude of the diameter will directly affect the width of the cross section of the float in contact with water (equation 9). Similarly, the ability of buoys to absorb power from wave power as shown in Figure 4 also has direct headlines with a diameter.

The diameter has an influence on the efficiency of weathering absorption. In addition, the diameter of the float also affects the radiation attenuation coefficient, b (ω) and the Fexc (ω) excitation force. Fexc and b (ω) are also directly proportional to the diameter of the buoy which will provide a drag to the heave rotating force like the results seen in Figure 5. From the variation in diameter observed, the diameter obtained gives optimal efficiency.

6. Conclusion

This research was conducted by varying the buoy diameter from 1.0 to 10 meters for determine the effect of the buoy diameter on the absorption efficiency of power in a regular wave with a height of 1 meter in the transition ocean area within 5 meters. Simulation results obtained from figures 3,4 and 5 obtained a good trend graph suitability with experimental results. Based on the research results obtained as shown in Figures 3, 4 and 5, some conclusions can be drawn:

1. The diameter and draft style of the float affect absorption. With increasing diameter, power absorption increases and is inversely proportional to the draft buoy force. For example, buoys with a 10 m diameter are absorbed 162.79 KN to 822% more than power is absorbed from buoys with a diameter of 3 m.
2. Based on the results of the efficiency obtained, the 2.5-meter diameter gives the maximum efficiency of 11 diameter variations observed at 37.17% and the smallest at 10-meter diameter buoys
3. Bullets buoys with diameter of 2.5 m can be considered in further research for use in absorbing wave energy converter around the Indonesian island area characters height wave seas

References

[1] Energy W 2016 World Energy Resources *World Energy Resour.*
[2] Schmid J Renewable Energy : Fraunhofer Institute for Wind Energy and Energy System Technology, Germany
[3] Perera F 2018 Pollution from fossil-fuel combustion is the leading environmental threat to global pediatric health and equity: Solutions exist *Int. J. Environ. Res. Public Health* **15** 16
[4] Shahzad U and August R 2015 Global warming: Causes, effects and solutions *Durreesamin J.* **1**
[5] Birol. D F 2017 Key World Energy Statistics IEA *IEA*
[6] Faizah S I and Husaeni U A 2018 Development of consumption and supplying energy in Indonesia's economy *Int. J. Energy Econ. Policy* **8** 313–21
[7] Moriarty P and Honnery D 2012 What is the global potential for renewable energy? *Renew. Sustain. Energy Rev.* **16** 244–52
[8] Saygin D, Rigter J, Caldecott B, Wagner N and Gielen D 2019 Power sector asset stranding effects of climate policies *Energy Sources, Part B Econ. Planning, Policy* **14** 99–124
[9] Firdaus A A, Yunardi R T, Agustin E I, Nahdliyah S D N and Nugroho T A 2020 An improved control for MPPT based on FL-PSo to minimize oscillation in photovoltaic system *Int. J. Power Electron. Drive Syst.* **11** 1082–7
[10] Lasabuda R 2013 Regional development in coastal and ocean in archipelago perspective of the Republic of Indonesia *J. Ilm. Platax* **1** 2
[11] Indonesia S 2017 *Statistics of marine and coastal resources 2016* (Statistics Indonesia)
[12] Binh P C, Tri N M, Dung D T, Ahn K K, Kim S-J and Koo W 2016 Analysis, design and experiment investigation of a novel wave energy converter *IET Gener. Transm. Distrib.* **10** 460–9
[13] Meisen P and Loiseau A 2009 Ocean energy technologies for renewable energy generation *Glob. Energy Netw. Inst.*
[14] Taryono 2016 Long And High Wave Analysis For KRI TNI-AL Operation Of The Navy In Indonesia Waters *J. Chart Datum* **1**
[15] Burman K and Walker A 2009 *Ocean energy technology overview*
[16] Kara F 2017 Control of Wave Energy Converters for Maximum Power Absorption with Time Domain Analysis *J. Fundam. Renew. Energy Appl.* **7**
[17] Khojasteh D and Kamali R 2016 Evaluation of wave energy absorption by heaving point absorbers at various hot spots in Iran seas *Energy* **109** 629–40
[18] Antonio F de O 2010 Wave energy utilization: A review of the technologies *Renew. Sustain. energy Rev.* **14** 899–918
[19] Kalofotias F 2016 *Study for the hull shape of a wave energy converter-point absorber: design optimization & modeling improvement* (University of Twente)
[20] Journee J M J and Pinkster J A 1997 *Offshore hydromechanics*

IOP Conf. Series: Materials Science and Engineering **885** (2020) 012022 doi:10.1088/1757-899X/885/1/012022

Computational studies of pressure distribution of flow through inline dimpled plate

M Setiawan Sukardin[1,2], Nasaruddin Salam[2], Rustan Tarakka[2], Jalaluddin Haddade[2], and Muhammad Ihsan[3]

[1]Politeknik ATI Makassar, Indonesia
[2]Department of Mechanical Engineering, Hasanuddin University, Gowa, Indonesia
[3]Sekolah Tinggi Teknik Baramuli, Pinrang, Indonesia

Email: setiawan_mkz@yahoo.co.id

Abstract. Dimple plates are widely used for construction and vehicles, especially in car bodies, trains, and aircraft wings. The dimples surface helps to overcome turbulent airflow around vehicles, thereby delaying the separation point and producing fewer vertices and drag. This study aims to predict the pressure distribution occurring over the plate surface with the formation of inline dimple rows on the upper back end. The test is carried out with Computational Fluid Dynamic (CFD) FLUENT program. The test model has a dimension of 300 mm in length, 100 mm in width, and dimples ratio (DR) of 0.5. Dimples are arranged inline as much as 1 to 6 rows. Upstream velocities through dimple plates range from 10 m/s to 20 m/s. The results of the study show that the minimum pressure coefficient occurs at the top front of the plate due to the flow separation that occurs at the front end of the plate. At x/L=0.25 x/L=0.5 and x/L=0.75 (before the dimples formation), a flow pattern returns to normal, this can be seen from the distribution of the relative pressure coefficient that does not change. At the measurement point x/L=0.95 (after dimples formation), there is a decrease in the pressure coefficient. This decrease occurs due to changes in flow characteristics. The pressure coefficient does not change significantly in dimple variations of 2 to 4 rows. Meanwhile, the pressure coefficient decreases with the variation of the 5 and 6-row dimples.

1. Introduction

Recent developments in the vehicle have heightened the need for aerodynamic drag reduction. There have been studies concerning the development of the method for surface friction drag reduction, and the efforts have progressed widely in two ways: methods for delaying the transition of the laminar-turbulent boundary layer and methods for changing or adjusting the structure of turbulent boundary layer [1]. Passive techniques that have been extensively investigated for turbulent drag reduction include dimples formation on the body surface. The dimpled surface on a golf ball is a well- known application of a non-smooth surface for drag reduction [2]. Several attempts have been made to investigate the reduction in the aerodynamic drag of a vehicle using dimple surface. Wang et al. have conducted numerical studies on generic vehicles to optimize aerodynamic drag using dimple arrays [3]. The dimples surface helps turbulent airflow around the vehicle, thereby delaying the separation point and producing smaller wake and lower form drag. Livya et al. (2015) have evaluated aircraft airfoil drag at 0° - 20° variation angle of attack, which applies the formation of one dimple [4]. Types of dimples that are applied include square, semi-spherical, compound (a combination of semi-spherical and square) each in the form of concave (inward), and convex (outward). The flow through the dimples forms a small separation bubble. Bubble formation accelerates flow between dimples on the airfoil surface and the boundary layer changes from laminar to turbulent. This transition flow delays the occurrence of flow separation as a cause of decreased drag. Other studies have applied dimple rows on the edge of the back of the Ahmed body with a surface inclination of 25°. The computational approach used is the k-epsilon model on ANSYS Fluent. Dimple ratio variation (DR) were 0.005, 0.2, 0.4 and 0.5 at a speed of 40 m/s where the greatest drag reduction of 1.95% were obtained in the DR 0.4 when compared to the model without dimples [5]. In another study, Zhou et al. (2016) have examined the flow of fluid across the surface of the dimples, with the distribution of surface pressure coefficients measured in a dimple with respective Reynolds numbers; Re=8.2K, Re=36.7K Re=50.5K [6]. Computationally and experimentally, airflow resistance through the dimpled plates has been reviewed in inline and zigzag formations with plates without dimples. The smallest Cd in the inline formation occurs in 2 lines of dimples, while the Cd in the zigzag formation is 1 row of dimples. Decrease in the percentage of the coefficient of resistance between the plates without dimples and the zigzag formation dimpled plates was 5.88%, while the inline formation dimpled plates were 8.65%. [7].

Large-scale asymmetrical vortex structures are formed inside the dimples. The source is located in the front wall of the dimples. Vortex structure tends to intersect the median dimples. In oscillating motion, large-scale asymmetric vortices are periodically released over the back wall at an angle that increases with increasing flow velocity [8]. Dimple, as a passive control, triggers instability that causes significant momentum transport. The shear layer is formed as the flow separation through the first two lines of dimples becomes unstable, and the coherent vortex collection. When vortex develops through plates or dimples, the dynamics of flow become very different due to changes in momentum transport across the boundary layer on staggered arrangement dimples formation 2 and 8 [9].

Fluid flowing on a curved flat plate causes flow separation and boundary layer changes. Separation is caused by the influence of normal force pressure, which is usually called form drag or pressure drag. This study is expected to predict the pressure distribution as a surface drag caused by variations in the number of inline dimple rows on the plate in the quest of an optimal number of dimple rows on objects that have a significant drag form.

2. Method

The test model is in the form of a plate with inline dimple formation. Each model was applied in a variety of dimples rows. Each plate has a dimple row, which is applied to the plate top, with a variation of 1 to 6 rows of dimple arrays. The x-axis is in line with the direction of flow velocity U_0. The distance between the center of the dimples on the x-axis and y-axis is 9 mm, respectively. Dimples ratio R/D=0.5. The dimensions of the plate are 300 mm long, 100 mm wide and 5 mm thick. Airflow velocities U_0 through the surface of the dimpled plate very successively 10 m/s, 12 m/s, 16 m/s, 18 m/s, and 20 m/s. The tapping position of the measurement of the pressure distribution in the workpiece model were at x/L=0.1 x/L=0.25 x/L=0.5 x/L=0.75 (before the dimples) and x/L = 0.95 (after the dimples).

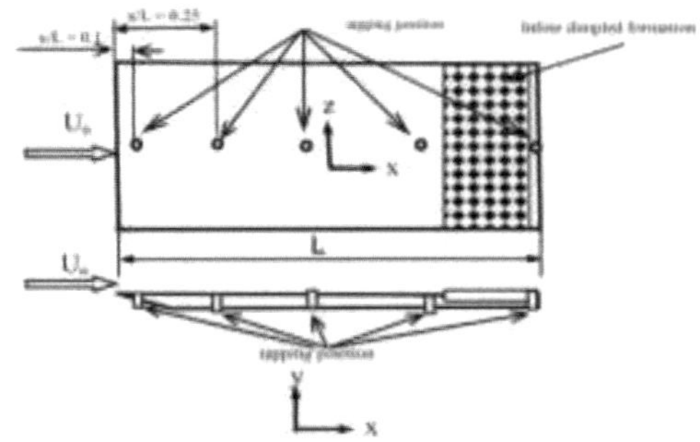

Figure 1. The position of dimples on the plate

Figure 2 illustrates the diameter of the dimples D relative to the distance between the center of the dimple point on the x (Lx) axis and the distance between the center of the dimples on the z-axis (Lz).

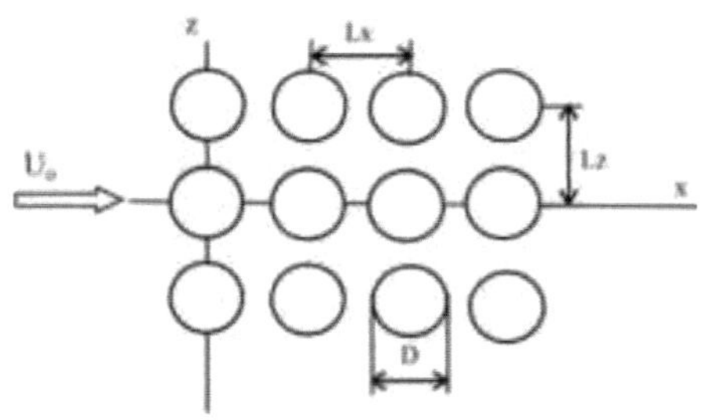

Figure 2. Dimension of inline dimple formation

The computational analysis uses CFD simulation based on ANSYS FLUENT 18.0. The simulation stage starts from drawing the workpiece model using Autodesk Inventor. The meshing process uses ANSYS ICEM CFD 18.0. The assembly of meshing was conducted using tetrahedrons. The number of elements is 1,249,265, and the number of nodes is 227,901. Figure 3 is an example of the result of typical meshing. Figure 4 is a detail of the results of the dimpled plate meshing. The next process is the setup process in ANSYS FLUENT 18.0, simulated in the condition of Re=303K, where air velocity U_0 in a row were 10 m/s, 12 m/s, 14 m/s, 16 m/s, 18 m/s, and 20 m/s on each number of rows dimples. The iteration is attempted to 200 times, followed by the calculation process. Computational results are in the form of pressure distributions at the nodes as measurement points, respectively at x/L=0.1, 0.25, 0.5, 0.75 before the dimples, and 0.95 after the dimples.

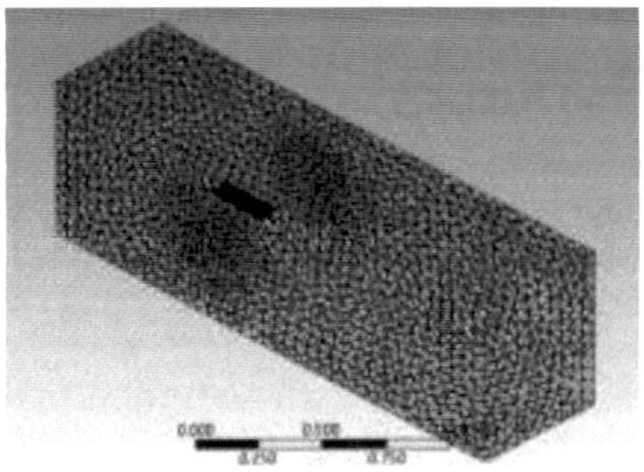

Figure 3. Meshing of test object with 6-row inline dimple formation

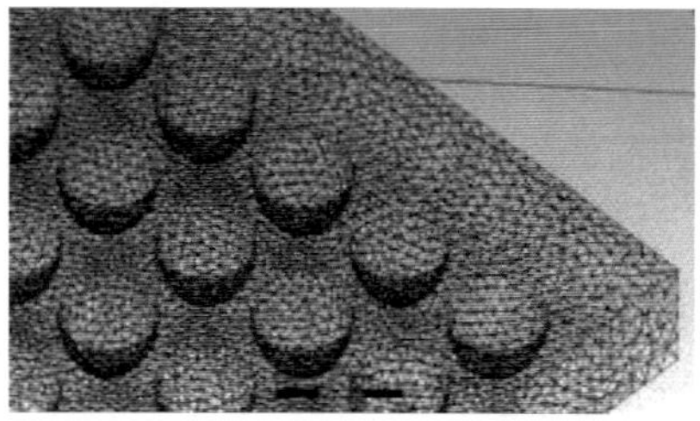

Figure 4. Detail of Meshing of the dimpled plate

Pressure drag is the most common form used to define drag on objects. Fluid drag occurs due to the flow, filling the space behind the object, causing a pressure difference between the upstream and downstream flow. Total pressure at the back is lower than that of the front of the object, causing backward suction. The pressure force on flat areas of the plate, which is perpendicular to the flow, causes an enormous drag effect compared to the pressure force on both sides of the flat plate inline to the direction of flow. To acquire the value of this pressure drag force, we need pressure distribution data along the surface of the model. The pressure drag value, a dimensionless parameter, is the coefficient of C_p pressure and commonly expressed as [8].

$$cp = \frac{P-Po}{\frac{1}{2}\rho Uo^2} \qquad (1)$$

p = pressure on the surface of the test model
(Pa) p_o = streamlined pressure or flow line (Pa)
ρ = fluid density
(kg/m^3) U_0 = upstream
speed (m/s)

3. Result and discussion

The present study consists of computational fluid dynamics for variations of the dimple rows on the upper backside of the plate. Upstream velocities range from 10 m/s to 20 m/s. Figures 5 to 9 show that at the measurement point x/L=0.1 (front edge of the plate), there is a decrease in pressure in each test model. It shows that the minimum pressure coefficient occurs on the top side of the front plate due to flow separation that occurs at the front end of the plate. At x/L=0.25, x/L=0.5, and x/L=0.75 (before the dimples formation), the flow pattern returns to normal, which can be seen from the distribution of the relative pressure coefficient unchanged.

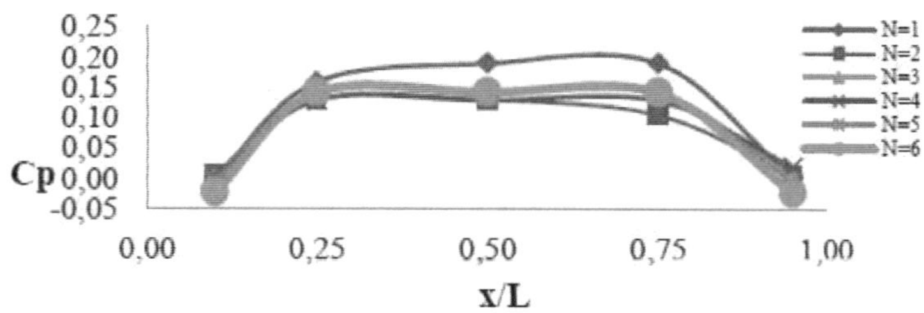

Figure 5. Pressure coefficient for various number of dimple rows of plate with the tapping position, upstream Uo = 10m/s

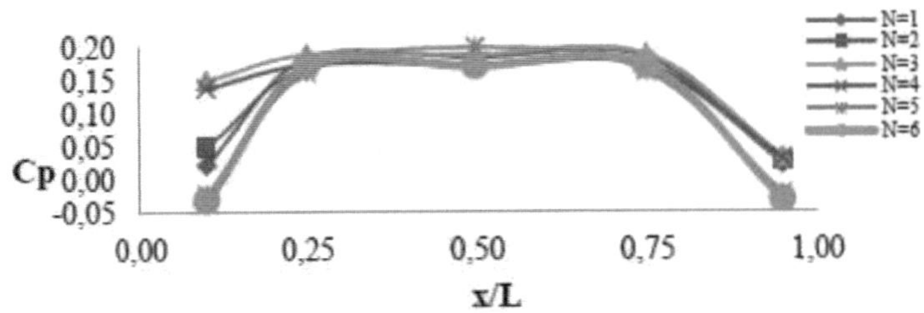

Figure 6. Pressure coefficient for various number of dimples rows of plate with the tapping position, upstream Uo = 12m/s

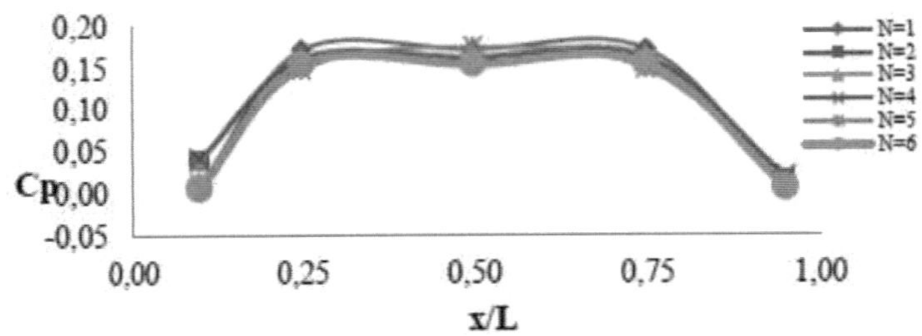

Figure 7. Pressure coefficient for various number of dimples rows of plate with the tapping position, upstream Uo = 14 m/s

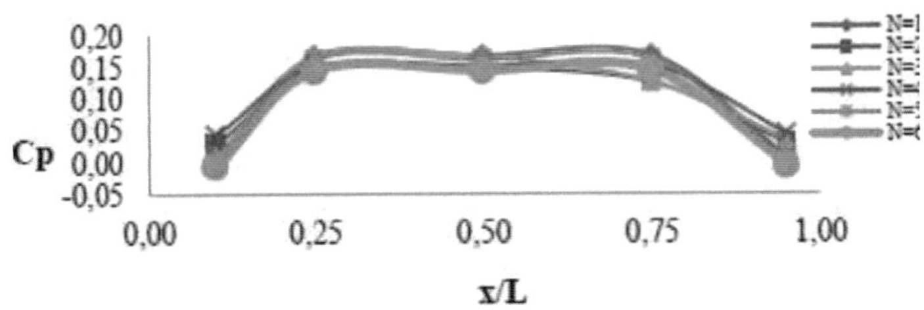

Figure 8. Pressure coefficient for various number of dimples rows of plate with the tapping position, upstream Uo = 16 m/s

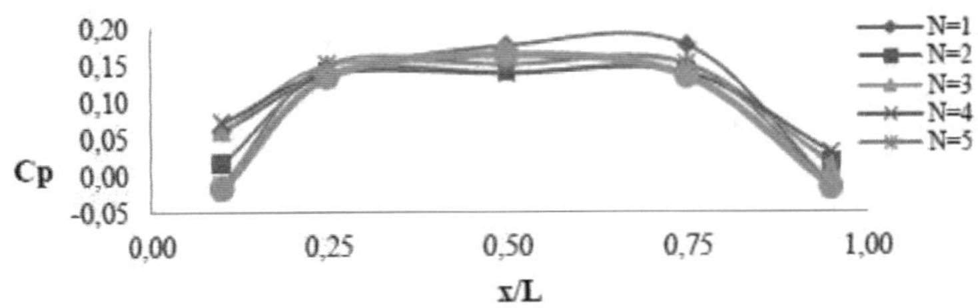

Figure 9. Pressure coefficient for various number of dimples rows of plate with the tapping position, upstream Uo = 18 m/s

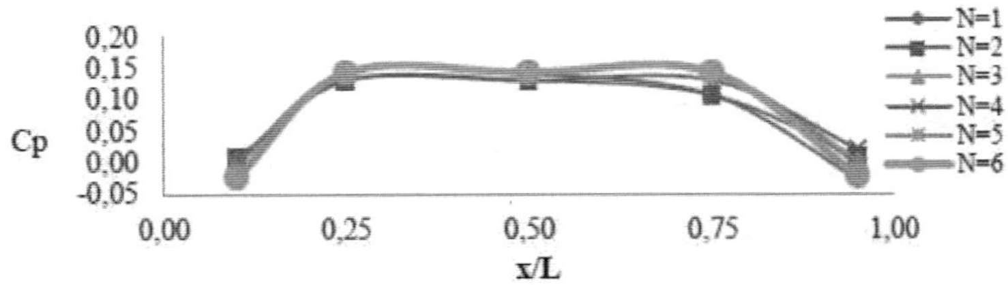

Figure 10. Pressure coefficient for various number of dimples rows of plate with the tapping position, upstream Uo = 20 m/s

Figures 5 to 10 show that at the measurement point x/L=0.1 (front edge of the plate), there is a decrease in pressure in each test model. It shows that the minimum pressure coefficient occurs on the top side of the front side of the plate due to flow separation on the location. At x/L=0.25, x/L=0.5, and x/L=0.75 (before the dimples formation), the flow pattern returns to normal, as indicated by the consistent distribution of the pressure coefficient.

Table 1. Pressure coefficient (C_p) at point after dimple array with x/L=0.95.

Upstream, U_0	Number of dimples rows N					
(m/s)	1	2	3	4	5	6
10	0.001	0.030	0.039	0.032	0.008	0.000
12	0.022	0.025	0.033	0.031	-0.022	-0.032
14	0.009	0.015	0.010	0.020	0.013	0.006
16	0.006	0.027	0.034	0.043	-0.002	-0.007
18	-0.014	0.018	0.013	0.031	-0.012	-0.017
20	-0.029	0.007	0.003	0.020	-0.005	-0.023

From table 1 above, a graph is made of the relationship between the pressure coefficient C_p and the number of dimples N lines in the upstream constant $U_{0,}$ as in the following figure.

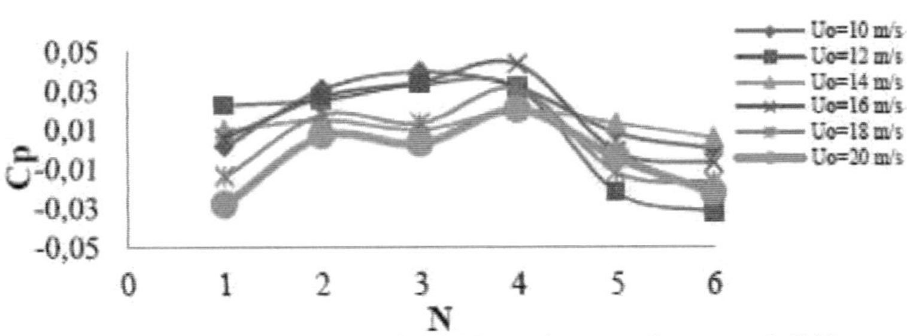

Figure 11. Pressure coefficient at point x/L = 0.95 N number or of rows and different upstream velocity Uo

Figure 11 shows the pressure coefficient at point x/L=0.95 of dimple rows variation at different upstream velocity. The figure also shows that regardless of the variations of the number of rows, no significant change in the pressure coefficient is generated for all upstream speeds. High coefficient of pressure C_p tends to occur in variations of 2 and 4 rows. This shows that the addition of dimple rows up to 4 rows does not contribute to significant friction at the plates. This phenomenon is in line with the smallest coefficient of resistance obtained in 2 rows of dimples [7]. Numbers of rows of 2 to 4 rows can be an alternative to passive flow control, which serves to delay the separation of the surface shape of the objects that change shape drastically, including the shape of the bluff body. Another finding of the research is the development of micro-sized Kelvin-Helmholtz vortex structures over the dimple indentation which confirms the results of Zhou et al. [6]

4. Conclusion

This study concerns to examine numerically the effect of dimple rows in a range of upstream velocity from 10 m/s to 20 m/s on the pressure distribution over a flat plate. From the simulation results, the following conclusions can be drawn. The minimum pressure coefficient occurs at the top front of the plate due to the flow separation that occurs at the front end of the plate. At x/L=0.25, x/L=0.5, and x/L=0.75 (before the dimples formation) a flow pattern returns to normal as indicated by the relatively unchanged distribution of the pressure coefficient. At the measurement point x/L=0.95 (after dimples formation), there is a decrease in the pressure coefficient. This decrease occurs due to changes in flow characteristics. Fluid flow through the surface of the dimple causes low-speed recirculating flow, which produces Kelvin-Helmholtz vortex structures. The pressure coefficient does not change significantly in dimple variations of 2 to 4 rows, whereas in dimple variation of 5 and 6 rows, the pressure coefficient decreases.

Acknowledgments

The authors gratefully acknowledge financial support from the Ministry of the Industry Republic of Indonesia under the 2019 F.Y. Grant.

References

[1] Viswanath P R 2002 Aircraft viscous drag reduction using riblets *Prog. Aerosp. Sci.* **38** 571–600
[2] Alam F, Steiner T, Chowdhury H, Moria H, Khan I, Aldawi F and Subic A 2011 A study of golf ball aerodynamic drag *Procedia Eng.* **13** 226–31
[3] Wang Y, Wu C, Tan G and Deng Y 2017 Reduction in the aerodynamic drag around a generic vehicle by using a non-smooth surface *Proc. Inst. Mech. Eng. Part D J. Automob. Eng.* **231** 130–44
[4] Livya E, Anitha G and Valli P 2015 Aerodynamic analysis of dimple effect on aircraft wing *Int. J. Mech. Aerospace, Ind. Mechatron. Manuf. Eng.* **9**
[5] Wong S F and Dol S S 2016 Simulation Study on Vehicle Drag Reduction by Surface Dimples *Int. J. Mech. Mechatronics Eng.* **10** 560–5
[6] Zhou W, Rao Y and Hu H 2016 An experimental investigation on the characteristics of turbulent boundary layer flows over a dimpled surface *J. Fluids Eng.* **138**
[7] Salam N, Tarakka R, Jalaluddin J and Muh S 2018 Fluid Flow Resistance Through Hemispherical Dimpled Plates in Parallel and Zigzag Configurations *Int Rev Mech Eng IREME* **12**
[8] Voskoboinick V A, Turick V N, Voskoboinyk O A, Voskoboinick A V and Tereshchenko I A 2018 Influence of the deep spherical dimple on the pressure field under the turbulent boundary layer *International Conference on Computer Science, Engineering and Education Applications* (Springer) pp 23–32
[9] Beratlis N, Balaras E and Squires K 2014 Effects of dimples on laminar boundary layers *J. Turbul.* **15** 611–27

IOP Conf. Series: Materials Science and Engineering **885** (2020) 012023 doi:10.1088/1757-899X/885/1/012023

The development of premarital mobile learning version 3.0 with jquery technology to be learning flexibility premarital course application

Herlinah[1], Baso habibi[2], and Abd Rochman[1]

[1]Technical Information STMIK Handayani Makassar
[2]Computer Systems STMIK Handayani Makassar

Email: linaherlinah@handayani.ac.id

Abstract. This research activity aims to develop premarital mobile learning applications v 2.0 to become premarital mobile learning v 3.0 based online with the app store. The focus of the application development is more on the flexibility of the application, namely the ease in the process of editing the application's animated video and e-book content, which will be adjusted to the applicable regulations in the Ministry of Religion. The software v 3.0 premarital mobile learning application that will be designed to build will also be added to the forum feature as a means of communication between the headman, as well as the process of question and answer between the premarital interviewees with the premarital course participants. The approach method used is object-oriented methodology, by implementing UML diagrams as functional modeling, using Javascript programming language, back server database language with php, and using jquery as an interface developer. The framework of application software engineering development with software Development Life Cycle (SDLC), namely the stages of planning, analysis, design, implementation, testing, and maintenance. The result of the research achieved are (1). Developing premarital mobile learning application software v 2.0 into premarital mobile learning application v.30 with all the ease of editing the material to be adjusted to the applicable provisions, material content stored in the html tag server (2). The development of version 3.0 with the flexibility of e-book learning, multimedia animation and forums as a means of communication between the premarital interviewees and the premarital course participants.

1. Introduction

The development of Information and Communication Technology in the current millennial era, which is called as Industry 4.0 Evolution, is already inevitable. The digital age cannot be avoided; therefore, it must be enjoyed and utilized. The development of information and communication technology has penetrated all fields of work, including the development of learning models with multimedia technology that aims to achieve quality education. The target of achieving quality education is marked by developing a multimedia-based learning model with e-learning, m-learning, distance learning, known as online learning, which is widely used in Indonesia today. Known as online learning, which is widely used in Indonesia today. Multimedia-based Pre-Marriage Education with Pre-Marriage Mobile Learning is a learning application software that was built and produced as a development model for the learning of pre-marital education activities in the Office of Religious Affairs. Provisions for the implementation of pre-marital education in accordance with the regulations of the Decree of the Minister of Religion (KMA) No. 477 of 2004 [1], about providing insight into marriage in the household to the bride and groom. Regulation of the Director-General of Islamic Community Guidance Number DJ.II / 542 of 2013[2], concerning Guidelines for Organizing Pre-Marriage Courses at the Office of Religious Affairs (KUA). Then in Regulation of the Minister of Religion (PMA) No. 19 of 2018 concerning Supervision instructs officials who have duties in the field of Islamic Community Guidance in the Ministry of Religion of the Regency/City to supervise the implementation of Marriage Guidance in KUA. As well as the Implementation of Marriage Guidance regulated in the Decree of the Director-General of Islamic Community Guidance Number 379 of 2018 concerning Guidelines for the Implementation of Marriage Marriage Guidance for Prospective Brides. Conceptually the bride and groom course / premarital education held at the Office of Religious Affairs (KUA) is conducted for 10 working days with a duration of 24 hours of study. There are some course materials that are presented during the teaching and learning process in accordance with the syllabus of pre-marital course materials published by the ministry of religion. These activities are intended to provide sufficient understanding for the bride and groom before the marriage is launched. But in reality, in the field, the contextual can not be fully carried out in accordance with applicable regulations, including the difficulty in managing the time for the bride and groom status as Employees; both civil servants, BUMN employees, private and military employees. Similar to the implementation of marriage guidance (big win) which is carried out by the Islamic Community Guidance of the ministry of religion in the regency/city also experiences several obstacles including improper budget allocation, obstacles for marriage guidance participants who have a residence far from the location of the marriage guidance held. The premarital mobile learning application software was first announced on September 22, 2015, at the Office of Religious Affairs in the Mandai District of the Ministry of Religion in Maros Regency. Then premarital learning application software is used as a learning model to accompany pre-marital education activities face-to-face between speakers and participants of pre-marital courses in several religious affairs offices within the scope of the Maros district religious ministry. Based on the aforementioned things that inspired the need for the development of ICT-based learning media by using Android-based pre-marital mobile learning application software produced. Product Pre-marital mobile learning application version 1.0, which was previously produced, using an object-oriented programming language (OOP) with E-Clips as software programming. The results of the implementation of the application and application system trials on pre-marital course participants, as well as based on input from the functional officials of the prince as a pre-marital education resource at the office of religious affairs, the development of a mobile application for pre-marital marriage based on Android by utilizing Unity 3D technology. Unity 3D technology is generally widely used to create android-based game applications, still experiencing development so that it can also be used for android-based application development [3]. Pre-marriage m-learning application software with version 1.0 and version 2.0 that has been produced and then socialized and implemented in pre-marital education activities in the office of religious affairs is very helpful for prenuptial course participants or brides in accessing premarital materials, but there are still some deficiencies in facilities and features and content material that is still in the form of images (jpg). Therefore, it will be designed to build application flexibility that can be done in the process of editing the material in accordance with the regulations for the implementation of premarital courses or marriage guidance that applies to the ministry of religion [4].

2. Extension-rule based theorem proving method

2.1. Policy on premarital education activities

The juridical basis on which the legal basis for conducting the bride and groom courses is:

a. Decree of the Minister of Religion (KMA) 477 in 2004 [1].
   Through the Decree of the Minister of Religion (KMA) No.477 of 2004 concerning providing insight into marriage and households to the bride and groom, the government mandates that before

the marriage takes place, each prospective bride and groom must first be given insight about the meaning of a household through a bride-to-be course (sustain).

b. Regulation of the Director-General of Islamic Community Guidance, No. DJ.II / 491 of 2009. Regulation of the Directorate General of Islamic Community Guidance, No. DJ.II / 491 of 2009. About the bride and groom course. As a basis for the application of the bride and groom course as a condition of marriage registration This regulation is intended to increase understanding and knowledge about the life of the household/family in realizing a sakinah, mawaddah and rahmah family and reducing the number of disputes, divorce, and domestic violence.
c. Circular Letter of the Director-General of Islamic Community Guidance. DJ.II / PW.01 / 1997/2009 With the issuance of the Circular Letter of the Director-General of Islamic Community Guidance Number DJ.II / PW.01 / 1997/2009 regarding the skinny bride and groom brides, which is a follow up to the Regulation of the Director-General of Islamic Community Guidance, No. DJ.II / 491 of 2009, making Suscatin's steps clearer. The birth of regulations regarding the bride and groom course is a form of government's real concern for the high divorce rate and domestic violence cases in Indonesia.
   By attending the bride and groom course, the bride and groom couples who want to pitch to the marriage level will be equipped with basic material knowledge and skills about married life. Thus it is hoped that the bride and groom will have insight and knowledge about domestic life, which in turn will be able to gradually minimize the number of divorces and domestic violence.
d. Regulation of the Director-General of Islamic Community Guidance No.II / 372 of 2011
   With the Regulation of the Director-General of Islamic Community Guidance No.II / 372 of 2011 concerning guidelines for organizing pre-marital curricula and syllabus courses, it can be more directed towards achieving pre-teaching targets.
e. Decree of the Director-General of Islamic Community Guidance No. 373 of 2017 concerning Technical Guidance for Marriage Guidance for Prospective Brides
f. Regulation of the Minister of Religion of the Republic of Indonesia Number 19 the year 2018 Concerning Marriage Registration for Supervision instructs officials who have duties in the field of Islamic Community Guidance in the Regency / City Ministry of Religion to supervise the implementation of Marriage Guidance in KUA. This Marriage Guidance activity within the Ministry of Religion work unit is funded through the National Budget and NR NRB, which runs in accordance with applicable laws and regulations.

2.2. Mobile learning

The use of information and communication technology in the world of education continues to develop in various strategies and patterns, which basically can be grouped into e-Learning systems as a form of learning that utilizes electronic devices and digital media, as well as mobile learning (m-learning) as a form of learning which specifically utilizes mobile communication devices and technology. The level of penetration of mobile devices is very high, the level of use is relatively easy, and the price of devices that are increasingly affordable, compared to PC devices, is a driving factor that is increasingly expanding opportunities for the use or application of mobile learning as a new trend in learning, which forms a paradigm that can increase efficiency and the effectiveness of the process and learning outcomes of students in Indonesia.

Mobile Learning is a learning model that is carried out between places or environments using technology that is easy to carry when the learner is in a mobile/cellphone condition. With its various potentials and strengths, Mobile Learning is expected to be able to be an alternative learning source that can improve the efficiency and effectiveness of the process and learning outcomes of students in Indonesia in the future. There are three functions of Mobile Learning in classroom learning activities (classroom instruction), namely as supplementary (optional) which is optional (optional), complementary (complimentary), or substitute (substitution) [5]:

1. Supplement (additional) Mobile Learning functions as a supplement (additional), namely: students have freedom of choice, whether to use Mobile Learning material or not. In this case, there is no obligation for students to access Mobile Learning material. Even if it is optional, students who use it will certainly have additional knowledge or insight.
2. Complement (supplementary) Mobile Learning functions as a complement (complimentary), namely: the material is programmed to complement the learning material received by students in the classroom. Here it means that Mobile Learning material is programmed to become reinforcement or remedial material for students in participating in conventional learning activities.
3. Substitution (substitute) Some universities in developed countries provide several alternative models of learning activities for students. The goal is that students can flexibly manage their lecture activities in accordance with the time and daily activities of students. There are three alternative models of learning activities that students can choose from, namely: fully face-to-face (conventional) 2) partially face-to-face and partly through the internet 3) entirely through the internet.

2.3. Mobile jquery technology
Jquery mobile is a framework created for designing cross-platform mobile device applications. Part of the application that can be directly dealing with the user (User Interface). JQuery Mobile is an HTML5-based user interface system designed to create responsive websites and applications that can be accessed on all smartphones, tablets, and desktop devices.

JQuery Mobile was developed as an alternative that answers the need for the development of mobile web applications that are easy and have a lot of potentials. JQuery Mobile is possible to create mobile web applications whose appearance and interactivity are consistent across all devices that support, as well as having sophisticated user interface (UI) capabilities. JQuery Mobile is an HTML5-based user interface system, and the development of the popular JQuery and jQuery UI libraries is currently popular. The advantages of jQuery are that the code is lightweight, adopts a design that can adapt to various types of mobile devices, and accentuates semantic markup. Therefore, the design is very flexible and has a theme that can be adjusted [3].

2.4. Parallel proving algorithm based on semi-extension rule
The method used is an object-oriented methodology, by implementing UML diagrams as functional modeling, using the Javascript programming language, the database back server language with php, and using Front-end mobile jquery as an interface developer. The framework of application software engineering development with software Development Life Cycle (SDLC), namely with the stages of planning, analysis, design, implementation, testing, and maintenance [6].

3. Experimental Result

3.1. Premarital m-learning technology
In the beginning, premarital mobile learning application software was given the name of multimedia-based premarital education, with the initial concept that premarital education activities could integrate the concept of face-to-face learning by utilizing information technology devices with presentation files. But in reality, the implementers of these activities at the office level of religious affairs do not all have ICT infrastructure in the form of OHP (Overhead Projector) and LCD (Liquid Crystal Display) media because these media can be an option in learning. Apart from the limitations of ICT equipment that is not owned, sometimes the compatibility of the material with the media to be used is not appropriate because each learning equipment media has advantages and disadvantages.

With the development of the use of mobile phone technology as a learning medium as in several research results such as the Mobile School Service developed by Zoran Vucetic, et al.; in 2010, where the mobile phone technology was utilized as a medium of learning media for students at the University of Novi Sad, Zrenjanin, Serbia. In addition, mobile phone technology is utilized in the world of education, such as research into the development of mobile Phone-Based Learning designs in SQL courses in the department of engineering education in informatics Undiksha.

Furthermore, premarital mobile learning application software began to be developed by utilizing mobile phone devices that are already owned by participants of the premarital course. The application software is designed to be built using an object-oriented programming language (OOP) with E-Clips as programming software. The prenuptial mobile learning application software version 1.0 produced can have an e-book feature that contains the content of premarital course material. The e-book material prepared in the application is adjusted to the syllabus determined by the religious material. Premarital e-learning Mobile eBook 1.0 can display all pre-marital course material; the appearance can be enlarged and reduced (touch). But in version 1.0 application software that was built was not yet equipped with audiovisual or multimedia to show the process of marriage flow to premarital participants.

To overcome the weakness of the application, then version 1.0 was developed into a premarital mobile learning application version 2.0, which was built using unity technology. Pre-marital mobile learning application features version 2.0 is equipped with an e-book and video animation of marriage flow diagrams and reproductive health animations. However, the e-book feature is still lacking because the material content contained therein is still in the form of jpg files that cannot be edited and added material in accordance with regulations that apply to the implementation of bride and groom courses, premarital courses or marriage guidance that applies to the Ministry of Religion of the Republic Indonesia.

3.2. Premarital marriage planning with mobile jquery technology

To overcome the shortcomings of premarital mobile learning application software that has been implemented on users, and found several weaknesses, version 3.0 will be developed. Premarital mobile learning technology using mobile jquery will provide solutions to any changes in the content of premarital course materials in accordance with regulations of the ministry of religion.

Pre-marital mobile learning application design with mobile jquery technology created can be a responsive and compatible application that will work on all popular smartphone, tablet, and desktop platforms.

Following is the appearance of the premarital mobile learning application software interface developed using jquery mobile technology:

1. Display Login Menu

This display is intended for participants of premarital courses, pre-marital education speakers and premarital mobile learning application administrators by granting username and password permissions.

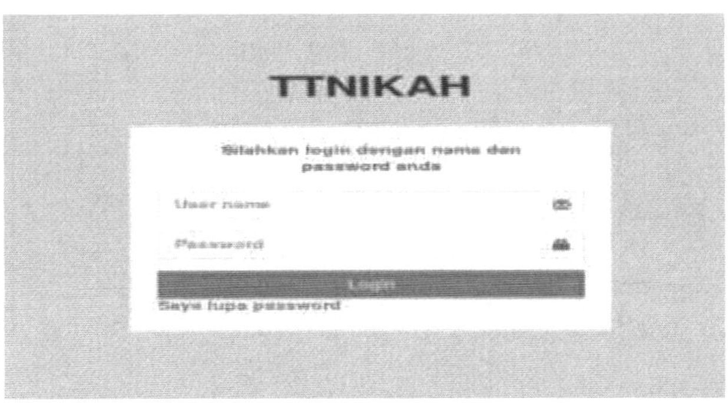

Figure 1. Login Menu

2. Display the Main Form Menu

In this main Menu Display, if the participant is a pre-marital course participant, a pre-marital course guide or tutorial will be shown containing all material in accordance with the syllabus or curriculum. In this view, it will be equipped with a video animation of a marriage flow chart and reproductive health animation, the material of which works with the health center or health institution. In this main menu, a futuristic digital consultation menu will be added between the general public and various household, marriage, and divorce matters that apply to religious affairs offices.

Figure 2. Main Menu Display

3. Display the Form Editor Menu

The appearance of this editor form menu contains an application mastering intended for the administration team to manage the contents of the material or video that will be displayed on the home menu.

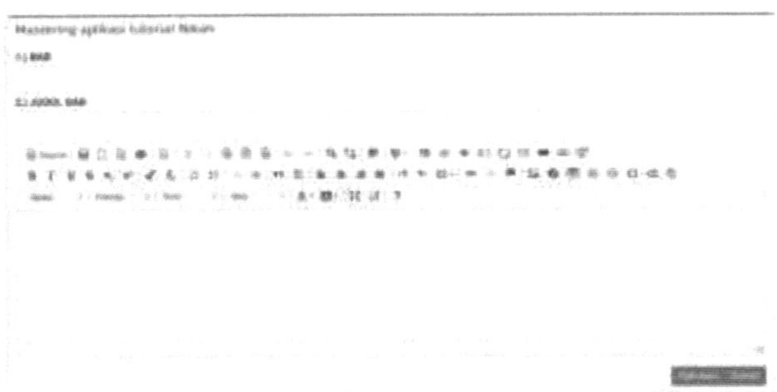

Figure 3. Display the Form Editor Menu

4. Display Form display mobile application tutorial premarital courses

On this menu, display contains an electronic book pre-marital course material that is tailored to the syllabus and curriculum that applies to the ministry of religion

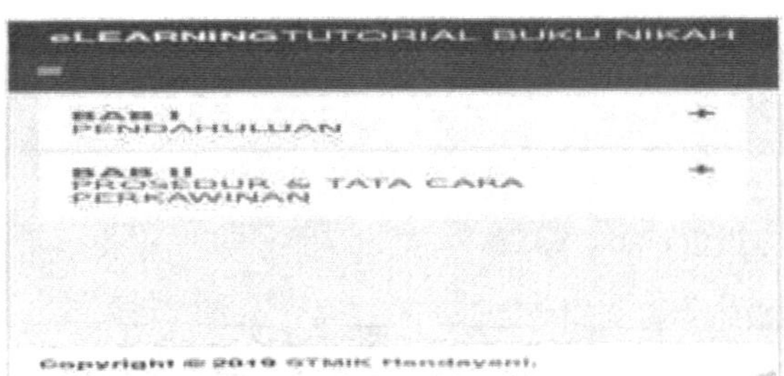

Figure 4. Display the Form Editor Menu

Based on the results of the premarital mobile learning application interface design with mobile jquery technology that has been developed, the futuristic application software will be more flexible with the ease of doing the editing process and adding material that applies in accordance with policies in the ministry of religion, repairing animated videos according to the material content and will add online consultation menu between pre-marital course participants (prospective brides) and resource persons for pre-marital education activities, as well as between the general public who need solutions to household problems, marital problems, and divorce in accordance with operational standards (SOP) for the upstream workgroup on religious Affairs office.

4. Conclusions

Based on the development of application design results that have been carried out, the following conclusions are concluded:

1. Developing premarital mobile learning applications with material content flexibility in accordance with the curriculum applicable to the ministry of religion, the resulting application features make it easy for any policy changes to also make changes to the contents of the application material.
2. Premarital mobile learning application software with mobile technology makes it easy to implement the interface design of the resulting menu form to adapt to various types of mobile devices used by application users.

References

[1] Anon 2004 Decree of the Minister of Religion No. 477, 2004 - Marriage Registration *scribd*
[2] Direktur Jenderal Bimbingan Masyarakat Islam 2013 *Peraturan Direktur Jenderal Bimbingan Masyarakat Islam, Pedoman Penyelenggaraan Kursus Pra Nikah* (Indonesia)
[3] Lutfi M and Nasrah A *Android-Based Mobile Learning Application Development Using Unity Technology*
[4] Habibi B 2017 Android-Based Mobile Learning Pranikah Application Design in BP4 Ministry of Religion, Maros Regency **9** 6
[5] Manib A 2012 Mobile Learning *J. UPI*
[6] Nugroho A 2010 *Object-Oriented Software Engineering with the USDP Method* (Andi Yogyakarta)

IOP Conf. Series: Materials Science and Engineering **885** (2020) 012024 doi:10.1088/1757-899X/885/1/012024

Automatic buoy system for position control based on global positioning system (GPS)

Z Akbar, Z B Hasanuddin and A E U Salam

Department of Electrical Engineering, Universitas Hassanuddin, Makassar, Indonesia

Email: zainal@atim.ac.id

Abstract. This study aims to design a buoy using an electric motor as a driver to maintain the position based on the coordinates of the Global Positioning System (GPS) that has been determined. This control system is designed to replace the existing buoy system, namely a buoy using a chain to maintain its position. The method used in the study is an electric motor control method using a microcontroller with GPS coordinates as the return point (home coordinates). The solar cell system that connected to the battery is used as a source of energy for the buoy control system. The result of this device is the distance between the return point and the stop point of less than 12 meters.

1. Introduction

The use of buoys is not new in Indonesia. Until now, the buoy system is still being developed for other needs. The development of this buoy has been utilized for various purposes, such as navigation, tsunami monitoring, research, fisheries, and energy [1]. According to Chandra, who developed the buoy for early warning of the pollution of the Pekalongan fishing port by using three sensors, namely temperature, salinity, and dissolved oxygen, as well as the telemetry system running well [1]. For other purposes, data collection on ocean currents and observations in Indonesian waters require data that can be directly identified in real-time by Purba et al. By examining GPS Drifter Combined (GERNED) with float tracking [2].

In other studies, Ranjith created an automatic limiting warning system for fishermen using Global Positioning Systems (GPS) and Global Systems for Mobile Communication (GSM) techniques [3]. In the conventional design, the buoy uses a chain and anchor to maintain its position so that the buoy is not carried away by ocean waves. This will be difficult if it will be used in deep water areas because it requires a long chain to place it. As did Joe et al. develop a buoy robot system that is always in position automatically without a retaining line. This prototype uses a wave glider platform to hold the buoy position without mooring and develops a power generation system using a self rectifying wave turbine (WTS) system [4].

IOP Conf. Series: Materials Science and Engineering **885** (2020) 012024 doi:10.1088/1757-899X/885/1/012024

From the description above, this research will discuss an automatic buoy system for position control using a global position system (GPS) and a driving motor to maintain the buoy position at a predetermined location.

2. Methods

This tool uses components that are easily available on the market. In designing this buoy, several steps are needed, including the following:

2.1. Planning System

The planning in this system can see in figure 1.

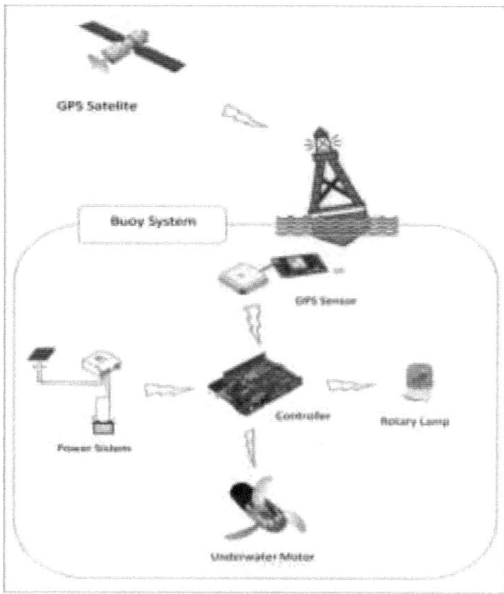

Figure 1. Planning of system

2.2. Design System

The design includes laying electric motors, control rooms, and solar panels. The design in this system can see in figure 2.

Figure 2. Design of system

2.3. Hardware Design

Hardware design consists of solar cells, global position system sensors (u-blox 6), compass sensors (HMC5883L), Arduino UNO, and electric motors (MKGP-G1100). The hardware design can be seen in Figure 3 below:

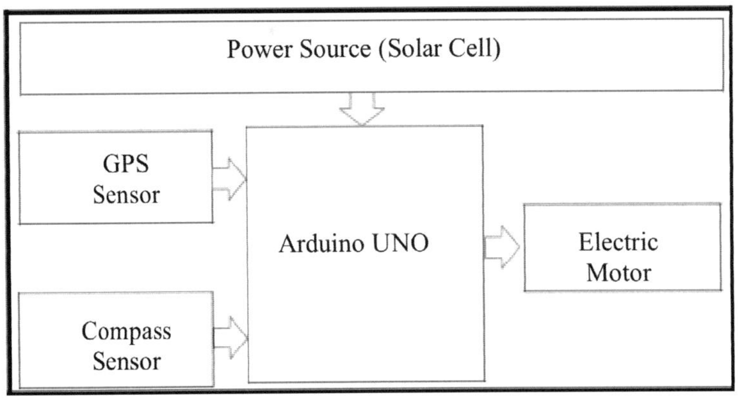

Figure 3. Hardware design

2.4. Selection of components uses, such as:

2.4.1. Global Positioning System (GPS) sensor. Is a navigation system or satellite-based positioning. The global positioning sensor system can see in figure 4.

Figure 4. Global Positioning System (GPS) sensor

*2.4.2. Compass Sensor. The s*ensor functions as a determinant or moves the buoy. This module requires a power of 5 volts with a current consumption of 15 mA. There are two ways to get compass data, from PWM or I2C connections. The compass sensor system can see in figure 5.

Figure 5. Compass Sensor

2.4.3. Microcontroller Arduino. A microcontroller on duty receives the coordinates provided by the GPS, processes them, and issues command to the motor. Arduino UNO can see in figure 6.

Figure 6. Arduino UNO

2.4.4. Motor Direct Current. It is a lifebuoy that has the ability to be underwater. Motor Direct Current can see in figure 7.

Figure 7. Underwater Motor

3. Result and Discussion

3.1. System Analysis
In reviewing its components, in-depth identification and evaluation of problems, constraints, and needs are needed so that they can be incorporated into improvements to this system.

3.2. Problem Analysis
Waves that are common in water areas are a major problem. With so many waves, the frequency makes the buoy uneasy.

3.3. Discussion results
Tools in research still need the additional design so that the movement is more efficient. This design is mainly the bottom of this buoy. With the improvement of the basic design of the buoy, it is expected that the buoy can move well.
In this study, a buoy can survive at a distance of 12 meters from the specified coordinates.

4. Conclusions
In designing the buoy, it still needs some improvements in the design and control of the motor. In the float test, the average coordinates of the buoy are 12 meters from the coordinates set.

IOP Conf. Series: Materials Science and Engineering **885** (2020) 012024 doi:10.1088/1757-899X/885/1/012024

Reference

[1] Chandra H, Sianturi D S A and Firdaus Y 2014 Uji Performansi Buoy Pluto Untuk Peringatan Dini Pencemaran Perairan *J. Kelaut. Nas.* **9** 151–60

[2] Purba N P, Harahap S A, Prihadi D J, Faizal I, Mulyani P G, Fitriadi C A, Pangestu I F, Atmoko P D, Alfath A and Sitio J T 2017 Pengembangan Instrumen Lagrangian Gps Drifter Combined (Gerned) Untuk Observasi Laut *J. Kelaut. Nas.* **12** 109–16

[3] Ranjith S, Shreyas K and Karthik R 2017 Automatic Border Alert System for Fishermen using GPS and GSM techniques *Indones. J. Electr. Eng. Comput. Sci.* **7** 84–9

[4] Joe H, Kim M, Wi S-M, Kwon H and Yu S-C 2014 Development of mooring-less robotic buoy system using wave powered renewable energy *2014 Oceans-St. John's* (IEEE) pp 1–6

Designing a road tool for blind people using the computer minimum system

AR Hakim[1] and F Metandi[2]
[1]Lecturer Computer Engineering, Samarinda State Polytechnic
[2]Lecturer Informatic Techniqu, Samarinda State Polytechnic

Email : rofiq_93@yahoo.com

Abstract. This research aims to design blindness devices namely sticks using a minimum computer system, expected to help blind people in their activities and interact with their surrounding environment. This research uses descriptive method, where in the process of conducting research through the stages of Information System development using a waterfall model which is one of the models in the SDLC (System Development Life Cycle). The stages carried out in this study are comprehensive analysis of development needs and System planning.The results of this study are in the form of blind road aids using the minimum computer system. Where after having this system design, it can proceed to the implementation stage

1. Introduction

The blind are part of the people who have mobility towards the environment in social life. The mobility released by the blind is not limited in terms of social aspects, such as community acceptance, but is also seen as facilitating facilities and infrastructure so as to facilitate mobility for the blind to carry out their activities.

Pedestrians who are blind must use special markings that are easily recognized by other road users. Signs for people with visual impairment may consist of assistive devices which are equipped with reflectors or sounds or red cloth [1]

A blind person is a blind person Stick is one of the blind facilities that is used every day in its mobility. The current blind stick is a conventional stick that is used on the blind where this stick responds to the compilation of the tip of the stick about the object it is attending, where the object is still in the stick. The stick cannot provide information to the user if the object is outside of the stick. Someone must help him help him.

The problem solved in this study is How to research aids for the Blind using the Minimum Computer System ". It is expected that the results of this research can be done by making tools for the Blind using a Computer Minimum System, so that it will really help the blind people with daily activities.

2. Reseasrch Methods

IOP Conf. Series: Materials Science and Engineering **885** (2020) 012025 doi:10.1088/1757-899X/885/1/012025

2.1 Research Scope

a. The scope of this research is for the Blind in Samarinda City
b. This research uses an Arduino minimum system and an Ultrasonic Sensor
c. The Information Technology Development Method used in this study is Waterfall model

2.2 Blind

The blind are individuals who have obstacles in vision. The blind can be classified into two groups, namely: total blindness (Blind) and low vision. Definition of the blind according to Somantri (2006) is an individual who has vision impairment or vision accuracy of less than 6/60 after being corrected or no longer has vision. People with visual impairments are individuals whose senses of vision (both) do not function as a channel for receiving information in daily activities like a sighted person (Somantri, 2006). Because the blind have a fairness in the sense of sight, the learning process emphasizes the other senses, the sense of touch and the sense of hearing. Therefore the principle that is considered in providing teaching to blind individuals is that the media used must be tactual and sound, for example the use of braille writing, embossed images, model objects and tangible objects. While the media that is voiced is a tape recorder and JAWS software.

2.3 The Blind Stick

Conventional blind people sticks are a long, straight walking aid that is the mobility aid most widely used for blind people. For most blind blind sticks in the form of a long stick that is still conventional, that is, the blind blind can be folded. The visually impaired stick is generally made from a tubular rod made of hollow aluminum with an outer radius of 6 mm (with a radius of 4 mm)

A good blind stick handle itself is a handle that is wrapped like a tennis racket with a thickness of about 200 mm from the top of the aluminum tube. At the bottom end of the stick, covered with a material made of plastic. The blind person is given a white and red color as a marker that indicates the person with disabilities. The placement of colors as markers is under the handle. The length of the stick is the height of the solar plexus of someone who uses it (measured from the handle to the tip of the stick). The general design of a conventional stick is depicted in Figure 1 at the far left, where each stick has a height that is relative to each user. The angle formed by the use of blind sticks generally ranges from 450.

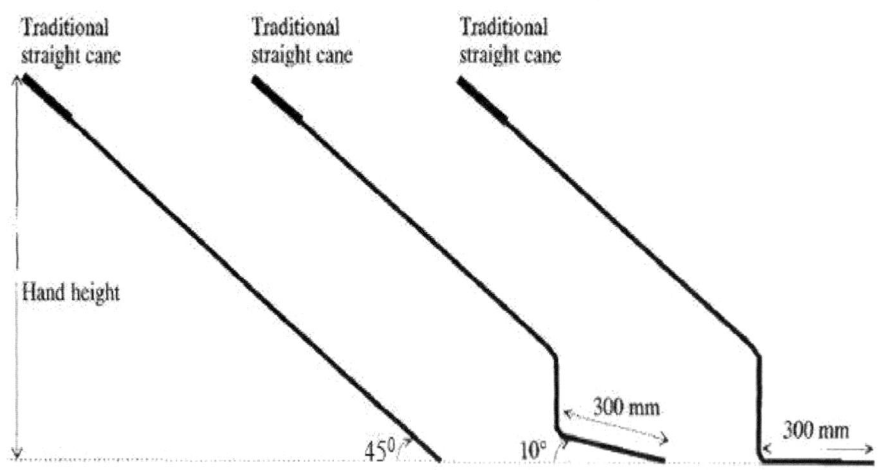

Figure 1. Model of a conventional blind person stick [2]

The types and types of blind people are very diverse. Type of blind stick as shown in figure 1 is a type of blind stick that has a safe distance from the object. The conventional stick in Figure 1 is used more because of its simple and foldable shape. A white stick is used by many people who are blind or blind, both as a means of mobility and as a courtesy to others. In addition to the above models there are at least five different varieties of these aids, each serving slightly different needs. The five stick varieties, namely:

a. The long stick type, a white stick (conventional), also known as the Hoover stick, after Dr. Richard Hoover as the designer made this tool primarily as a mobility tool used to detect objects in the user's path. The length of the stick type depends on the user's height, which extends from the floor to the user's arm. This type of stick is widely recommended.
b. Kiddie stick, a stick that works in the same way as an adult long stick, but is designed to be used on children.
c. Stick Identification, a stick that is used to remind others of the visual impairment experienced by the carrier. This stick is lighter and shorter than conventional sticks, and is not used as a mobility tool.
d. Supporting Sticks, are white support sticks designed to offer physical stability to users with visual impairments.
e. Mobility Sticks, are sticks made of aluminum, graphite-plastic or fiber-reinforced plastics, and are the simplest blind blind sticks.

3. Result and Discussion

3.1. System Analysis

A walker for the Blind uses the Minimum Computer System in the form of a stick that the system will work by: 1) People with visual impairment will use tools such as conventional blind people sticks, which are held and moved in front of the right side to the left and vice versa continuously 2) The front end emits ultrasonic signals or waves towards the front right and left 3) The ultrasonic sensor receives reflections from the ultrasonic wave if at a certain distance from the end of the stick there is a solid object blocking the ultrasonic waves from the tip of the walking aid 4) The results of ultrasonic wave reflection will be processed by a minimum computer system to determine the distance of the stick with objects in the surrounding area, especially the front right and left 5) At a certain distance from a solid object in front of the end of the stick, there will be a warning from the stick in the form of beep or tone from the speaker. The frequency and volume of the sound will get faster and higher if the distance to the object is getting closer

3.2 User Analysis

The user is a blind person who can already use conventional sticks for daily activities, not limited by age and sex. Domicile of users around the city of Samarinda

3.3 Data analysis

A road aid for the Blind uses the Minimum System The computer will send, receive and output data, i.e.

a. sending ping data in all directions in the form of ultrasonic signals or waves to the front right and left continuously during use.
b. receive data from reflected objects in the front area in the form of ultrasonic signals or waves
c. processing data in terms of analog and digital voltage in the minimum system
d. output data in the form of sound through the speaker

3.4 Technology Analysis

Technologies that will be used in this study include:

Hardware :

a. Arduino UNO minimum system
b. Ultrasonic Sensor
c. Sound Booster
d. Speaker
e. Power supply
f. Blind Stick
g. Computers The computer is used to create programs that will be embedded in the computer's minimum system memory.

Software :

a. Application to program minimum computer systems, such as Arduino IDE, Bascom and others
b. C Programming , Basic, Assembler

3.5. Design

3.5.1. Diagram Block. Based on the results of the analysis that has been done, a road diagram for the Blind uses a Computer Minimum System, can be seen in Figure 2

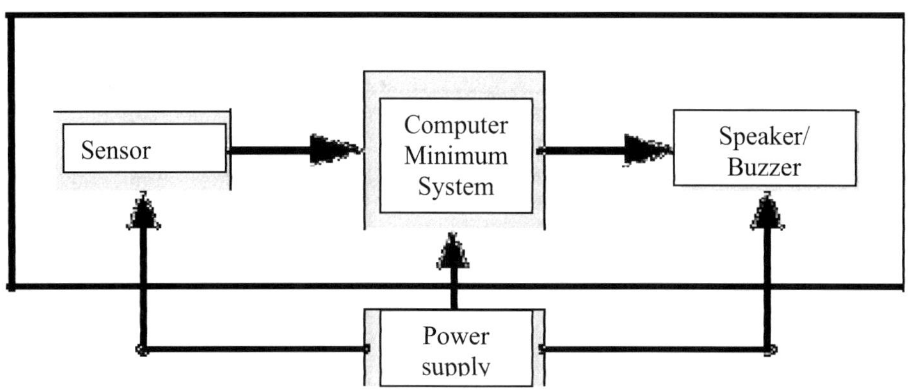

Figure 2. Block Diagram of a Walking Aid for the Blind using the Minimum System

3.5.1.1. Sensor. The sensor functions to receive and read ultrasonic waves which are reflections of objects in the front area of the blind blind stick. The ultrasonic signal is generated by a transmitter that is integrated with the sensor.

3.5.1.2. Computer Minimum System. The minimum computer system receives data from sensors in the form of analog or digital electrical signals, then processes the data in the form of a decision whether there is an object in front of the blind stick and the distance between the tip of the blind stick with the object in the front area. This information will be issued in the form of a voice signal with a certain frequency and volume in accordance with the distance recorded.

3.5.1.3. Speaker. Frequency and volume information from the minimum computer system will be received by the speaker to be converted into sound to be heard by blind stick users, so blind stick users can avoid objects in front of them

3.5.1.4. Power supply. All block diagrams require a DC mains supply to be provided by the power supply block. A simple power supply is needed but able to meet the needs of tool

3.5.2. Module Design

3.5.2.1. Ultrasonic Ping Sensor. Ultrasonic sensor is a sensor that converts physical quantities (sounds) into electrical quantities. In this sensor ultrasonic waves are generated through an object called piezoelectric. Ultrasonic waves are waves with a frequency above the frequency of the sound wave that is more than 20 KHz. As already mentioned, the ultrasonic sensor consists of an ultrasonic transmitter circuit called a transmitter and an ultrasonic receiver circuit called a receiver.

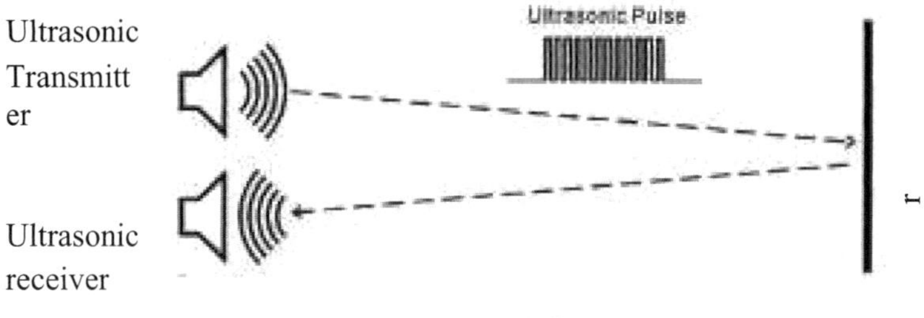

Figure 3. The working principle of ultrasonic sensors

Source: Neti Triyani (2006) Design and Realization of Microcontroller-Based Human Weight and Height Measuring Devices [3]

Figure 4. Ultrasonic Sensor

Ultrasonic signal generated will be emitted from the ultrasonic transmitter. When a signal hits a barrier, this signal is reflected, and is received by an ultrasonic receive. The signal received by the receiver circuit is sent to the microcontroller circuit for further processing to calculate the distance to the object in front of it (the reflecting plane). The working principle of the ultrasonic sensor can be shown in figure 3

IOP Conf. Series: Materials Science and Engineering **885** (2020) 012025 doi:10.1088/1757-899X/885/1/012025

The working principle of the ultrasonic sensor is:

a) The signal is emitted by an ultrasonic transmitter. The signal has a frequency above 20kHz, usually used to measure the distance of objects is 40kHz. The signal is generated by an ultrasonic transmitter circuit.
b) The emitted signal will then propagate as a sound signal / wave with a sound speed that ranges from 340 m / s. The signal will then be reflected and will be received again by the Ultrasonic receiver.
c) After the signal reaches the ultrasonic receiver, then the signal will be processed to calculate its distance.

3.5.2.2. Computer Minimum System. The Minimum System, commonly called Sismin, according to Hendawan (2009 H 5), is a circuit that is designed to use the minimum possible components to support the microcontroller as desired. Sismin has programmable input / output support and On-Chip RAM. Sismin can be made very flexible depending on the application to be made. In general, a microcontoler requires two elements (besides the power supply) to function: Crystal Oscillator (XTAL), and RESET Circuit, these 2 elements are the main conditions for the formation of Sismin (Minimum System). Analogy of the function of the Crystal Oscillator is the heart of the human body. The difference, the heart pumps blood and all of its contents, while XTAL pumps data. And the function of the RESET circuit is to make the microcontroller restart the reading of the program, it is needed when the microcontroller has a problem in executing the program.

Figure 5. Minimum System

3.5.2.3. Software design

For the operation of the hardware which is a walker tool for the Blind using the Minimum Computer System, software will be embedded in the minimum computer system. The software consists of programs made using the language C, Basic or Asembler will contain requests related to continuous sensor reading and data processing from the sensor to determine the distance between the sensor and objects in the area received. Then the results of this data processing will determine the output in the form of sound with a certain frequency and intensity via speaker or buzzer. Figure 6 shows the design of a software flowchart for aids for the Blind using the Computer Minimum System.

Design for the output of the process of detecting the distance of the object measured to exceed a certain limit of objects in the front area of 200 cm. If the distance of the object to the sensor is more than 200 cm then Buzzer Off and if the distance is <200 cm and if> 150 cm then Buzzer On with a sound interval of 500ms, an alarm will sound, if the distance is <150 cm and if> 100 cm then Buzzer On with a sound interval of 300ms, it will emit an alarm sound, then if the distance <100 cm and> 50 cm then Buzzer On with a sound interval of 200ms, then if the distance <50 cm and> 30 cm then Buzzer On with a sound interval of 100ms, it will issue alarm sound, and the next stage if the distance <30 cm and if> 2 cm then Buzzer On with a sound interval of 20ms, it will emit an alarm sound.

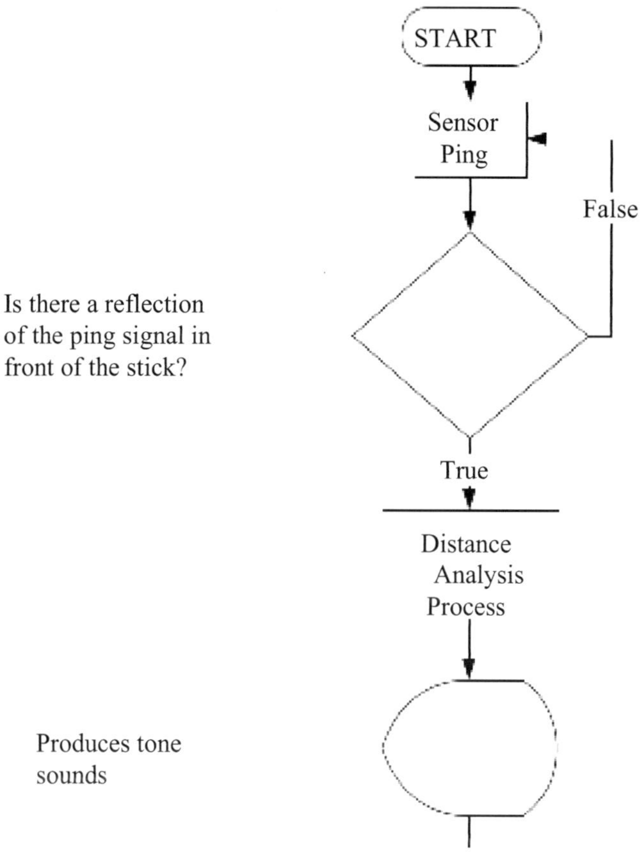

Figure 6. System Flowchart

4. Conclusion

The results of the design of blind walking aids with a minimum computer system using components namely ultrasonic sensors as input, a minimum computer system for processing data and speaker or buzzer outputs of multiple sounds. Blind hearing aids with a minimum computer system is designed simply with one kind of warning that is the tone of sound with different frequencies according to the distance between the sensor and objects in the front area.

References

[1] Pemerintah P, Indonesia R, Dan L, Pangan I, Indonesia P R and Umum K 1999 bahwa salah satu tujuan pengaturan, pembinaan, dan pengawasan pangan adalah terciptanya perdagangan pangan yang jujur dan bertanggung jawab;
[2] Schellingerhout R, Bongers R M, Van Grinsven R, Smitsman A W and Van Galen G P 2001 Improving obstacle detection by redesign of walking canes for blind persons *Ergonomics* **44** 513–26
[3] Wanti S 2010 Perancangan dan Realisasi Alat Pengukur Berat dan Tinggi Badan Manusia dengan

Keluaran Suara Berbasis Mikrokontroler AVR ATMega 8535 *Proy. Akhir pada Fak. Elektro dan Komun. IT Telkom Bandung tidak diterbitkan*

IOP Conf. Series: Materials Science and Engineering **885** (2020) 012026 doi:10.1088/1757-899X/885/1/012026

Simulation of generate value for development of biomass plants in Indonesia: a feasibility study

Safril, F Sumasto, M Wirandi and F Imansuri

Department of Automotive Industrial Engineering. STMI Polytechnic of Jakarta. Ministry of Industry Republic of Indonesia. DKI Jakarta. Indonesia

Email: safril@kemenperin.go.id

Abstract. The main purpose of this paper is to define the generating value of biomass plants in Indonesia by a feasibility study. It is important to measure the optimal location to build biomass plants in Indonesia. Besides that. the need of supply of electrical energy in Indonesia is always increase in recent years. Based on data. three cities in Indonesia that potential become location to build those biomass plant from total organic wastes to generate total electricity through conversion of methane gas that produced from waste. In this paper. proposed method of feasibility study performed to achieve optimal value in the construction of biomass power plants and the optimal location to implant. The analysis through the data of total electricity generated from gas that calculated on feasibility aspect values on NPV. IRR. and PI. The result show. in three cities of Indonesia produce a positive feasibility value based on value of IRR that greater than 20% tax so that the Government of Indonesia can implementing biomass plants on those three cities

1. Introduction

In many countries. organic waste treatment has been carried out with the aim of utilizing organic waste that is converted into electrical energy. The development of environmentally friendly technology is a top priority in each country as the main manifestation in reducing organic waste and helping to reduce the greenhouse effect caused by polluted air from waste that decomposes in reservoirs. Sustainable waste management is an important part of general efforts to reduce pollution and greenhouse gas emissions and to reduce global climate change [1]. Anaerobic digestion has emerged as a successful and promising technology for organic waste management as part of a growing effort in developing alternative energy sources. and strengthening environmental regulations on waste disposal and policies around the world from reducing greenhouse gas emissions [2].

In Indonesia. the development of anaerobic organic waste management technology is one of the top priorities. The Indonesian government has planned the development of renewable technology as a supplier of additional electrical energy to various regions. Based on politics and economics is very worrying over trends in the use of fuel from petro-chemicals that are not sustainable. This gives more attention to the use of biomass as a renewable energy to produce a decrease in fossil fuels [3]. CO_2 in biogas comes from organic waste released into the atmosphere on a relatively short time scale [4]. To prevent this. the biogas production process by anaerobic fermentation from organic matter is the most widely used technology and has the highest popularity [5]. Biogas not only contains methane gas and carbon dioxide. but also other gases such as hydrogen sulfide. siloxanes. halogenated hydrocarbons. or ammonia produced during the anaerobic digestion stage [6,7].

The application of anaerobic co-digestion technology for organic waste treatment has increased in recent years [8] and anaerobic waste treatment for solid waste co-digestion is considered a clean energy technology that is able to convert energy directly from organic waste by microorganisms [9,10]. Bio digester technology as an organic waste treatment is an effort to help reduce water pollution at the concentration of polluted organic waste. Utilization of bio digester technology refers to the process of mixing organic waste that is transformed by microorganisms in the absence of oxygen. so that it will produce methane gas which is expected to provide electrical energy through the conversion process.

2. Literature Review

In the application of biomass-based energy supply improvement activities. several countries have already done so. In Bolivia [11]. the development of the framework is prioritized as optimization in the transportation of biomass base materials to expand biomass applications in the energy in Bolivia. A critical review of the maturity status of the four most promising thermochemical conversion routes has been conducted in Malaysia [12] which has a very positive opportunity as a country in the tropics with abundant agricultural.

In Europe industries. the supply of biomass is the main focus because the production and consumption of biomass for the industry are quite balanced [13] where the opportunity to increase the supply of biomass production is the main point to maintain environmentally friendly sustainability. The application of biomass consumption in Indonesia is somewhat lacking due to the lack of biomass power plants in Indonesia. Related to power plants. it is necessary to have good planning in terms of location and implementation because it requires a considerable investment. For wide application in Indonesia. feasibility studies are needed. especially for big cities that can supply basic materials for biomass that will later be converted into electrical energy.

3. Method

In this research method. applying the mathematical model method to get the value of the net present value (NPV). Internal Rate Return (IRR) and Profitability Index (PI) with a tax value assumed to be 20%. In this paper. discuss and calculate the Pay Back Period (PP) and the Average Rate of Return (ARR). The data obtained is based on the organic waste treatment process through the digestive process. processed through fermentation to produce methane gas that is converted into electrical energy. Data processing uses input data based on the value of investment reduced by the value of expenses. the work process with an estimate of 3 working months and the results obtained are based on the benefits of processing waste into methane gas that is converted into electrical energy. The study made the cities of Jakarta. Bandung. and Surabaya in Indonesia as objects with total waste that can be seen in Table 1. Design of biomass reactor design by co-digestion treatment. with reactor design planning per unit of 1.050 m^3 and at each reactor construction issued costs 50 USD per m^3 [14].

Data is taken from the analysis process carried out from Farizal. et al. [15]. Where in it is known that the most total waste is generated in Jakarta City. ranked second is the most total waste generated in Bandung City and ranked third with the most total waste being in Surabaya City. The data obtained is based on the organic waste treatment process through the digestive process. processed through fermentation to produce methane gas that is converted into electrical energy. Data processing uses input data based on the value of investment reduced by the value of expenses. the work process with an estimate of 3 working months and the results obtained are based on the benefits of processing waste into methane gas that is converted into electrical energy. The total waste will affect the methane gas produced to be converted into electrical energy as the final results are shown in Table 1. In the digestive process of organic waste. methane is produced through the processing of organic waste in the reactor room. The total digester in the design of the reactor design is shown in Table 2.

Table 1. Data Collection of Organic Waste

City	Organic Waste	Methane	kWh
Jakarta	13.902.789	302.251.383	2.446.477.101
Bandung	5.105.438	106.749.742	864.051.628
Surabaya	4.787.838	104.696.554	847.432.750

Table 2. Reactor Design

City	Volume	Digester	Total Digester
Jakarta	10.427.092	1.050	9.931
Bandung	3.829.079	1.050	3.647
Surabaya	3.590.879	1.050	3.420

The design of the digester in Jakarta City with a volume of organic waste of 10.427.092 m3 required 9.931 ≈ 10 reactors with the planning of one reactor with dimensions of 1.050 m^3. Planning for reactor construction in Bandung City and Surabaya City with dimensions of 1.050 m^3 so that it needs the construction of a chancellor about 3.647 ≈ 4 reactors for Bandung City and 3.420 ≈ 4 reactors for Surabaya City.

4. Result and Discussion

The data is processed using a mathematical model by planning in advance to find the value of initial investment and revenue obtained by the company based on electricity sales and waste transportation services. Planning on the design of biomass reactors in three cities. namely Jakarta. Bandung and Surabaya. based on the calculation. obtained the value of fund management shown in Table 3.

Table 3. Fund Management

City	Investment (USD)	Operational Cost (USD)	Value of receiving funds (USD)
Jakarta	522.285.727	316.401.360	598.788.102
Bandung	191.775.680	138.977.904	223.565.806
Surabaya	179.823.267	144.401.904	217.523.487

The calculation of investment funds is carried out with the assumption of total expenditure in the construction of a biomass reactor of \$ 50 per m^3 and the average of Tax Object Sales Value assumption in each city:

(a) Tax Object Sales Value in Jakarta is 285.7 USD per m^2
(b) Tax Object Sales Value in Bandung is 39.3 USD per m^2
(c) Tax Object Sales Value in Surabaya is 46.4 USD per m^2

The revenue value is calculated based on the total sales of electricity to the State Electricity Company (PLN) with an average value taken from three cities in Indonesia around 0.07 USD per m^2.

Table 4. shows the value of EAT (Earning After Tax) calculated based on the value based on Table 3. The calculation is done with the assumption that the value of the tax or Degree of Freedom is 20%. Planning is calculated over the next 5 years with an increase in inflation estimated at 35% to 38% per year.

IOP Conf. Series: Materials Science and Engineering **885** (2020) 012026 doi:10.1088/1757-899X/885/1/012026

Table 4. Earning After Tax (EAT)

Years	Earning After Tax (USD)		
	Jakarta	Bandung	Surabaya
2020	225.909.393	67.670.321	58.497.266
2021	304.977.681	91.354.934	78.971.309
2022	414.769.646	124.242.710	107.400.981
2023	568.234.415	170.212.513	147.139.344
2024	784.163.493	234.893.268	203.052.294

Payback Period (PP) in the three cities. obtained positive results that are estimated to not more than 5 years have earned income. In this case. the planning for building a biomass co-digestion reactor is feasible (Table 5.). Average Rate Return (ARR) obtained based on the results of calculations from data processing that has been described and known. The ARR value obtained during the Payback term expiration period shows a very favorable and feasible value (Table 5.).

The Net Present Value (NPV) in Table 5 shows a positive NPV value. Based on these results. it can be concluded that construction carried out in three cities will have a good and positive impact on the company. Profitability Index (PI) is considered positive if it is greater than 1 (one) time of receipt. In this case. the revenue obtained by the company must not be less than 1 time the receipt. When the receipt of less than 1 time the receipt occurs. the company will be considered a loss. However. in the calculation of the feasibility study of three cities in Indonesia showed acceptance more than once (Table 5.). The PI calculation results show that the construction of a biomass reactor is feasible. Internal Rate Return (IRR) is shown in table 5. The results show the IRR obtained a positive value by exceeding the assumed Degree of Freedom value of 20%. The calculation is obtained based on the interpolation of negative NPV and positive NPV.

Table 5. The Results of Calculations on Feasibility

Value	Jakarta	Bandung	Surabaya
PP	4.7 years	4.7 years	3.11 years
ARR	101.114.290.50	24.708.994.08	19.691.435.43
NPV	91.241.119.74	18.543.072.10	32.547.994.86
PI	1.17	1.10	1.18
IRR	25.00	23.00	25.01

The results show that Jakarta has the potential to get the most profitable among the three cities. For the initial implementation. it is recommended to build a power plant in Surabaya. Due to Surabaya City has the fastest PP so that it can be a more precise depiction and pilot project.

5. Conclusion

The feasibility study in three major cities in Indonesia. namely Jakarta. Bandung. and Surabaya. produced several conclusions with different values in its application. Jakarta City has a payback period of 4.7 years and will experience a return on capital or after that generate a net profit. ARR value of 101.114.290 USD. NPV 91.241.119 USD and PI of 1.17 times in the receipt of profits from the construction of a biomass power plant reactor with an IRR value of 25% greater than the tax determination of 20%. Based on the value of the IRR. it can be concluded that the feasibility study in Jakarta City is considered positive and feasible to be implemented.

In Bandung City. it is known that the value of the payback period is 4.7 years smaller than 5 years. so it is considered feasible to apply. Furthermore. the **ARR** value of 24.708.994 USD. the **NPV** of 18.543.072 USD and the IRR value of 23% are greater than the tax determination of 20%. In the construction of biomass reactors with organic waste as raw material to produce methane gas to be converted into electrical energy with Profitability Index of 1.10 times acceptance. It can be concluded that the construction of biomass reactors in Bandung City is considered feasible to apply.

Based on data processing in Surabaya City. the result show obtained number of values that were considered positive. Payback Period in 3.11 years were considered positive because less than 5 years. Obtained ARR 19.691.435 USD was positive. NPV 32.547.994 USD was stated positively. PI that more than 1 times. 1.18 > 1 times in development the reactor is considered feasible and IRR is 25% > 20% that it is declared positive. Data processing in three cities in Indonesia concludes that a feasibility study on the construction of a biomass reactor with organic waste as raw material is considered feasible in implementation and the city suggested as the initial implementation is Surabaya city in terms of speed in the payback period.

Acknowledgments
This research is supported and funded by Directorate of Research and Community Engagement (P2M). STMI Polytechnic of Jakarta. Nomor: SP DIPA-019.10.2.579350/2019.

References

[1] Jyothilakshmi R and Prakash S V 2016 Design, fabrication and experimentation of a small scale anaerobic biodigester for domestic biodegradable solid waste with energy recovery and sizing calculations *Procedia Environ. Sci.* **35** 749–55

[2] Hreiz R, Adouani N, Jannot Y and Pons M-N 2017 Modeling and simulation of heat transfer phenomena in a semi-buried anaerobic digester *Chem. Eng. Res. Des.* **119** 101–16

[3] Howe D, Garcia-Perez M, Taasevigen D, Rainbolt J, Albrecht K, Li H, Wei L, McDonald A and Wolcott M 2019 Thermal pretreatment of a high lignin SSF digester residue to increase its softening point *J. Anal. Appl. Pyrolysis* **142** 103691

[4] Awe O W, Zhao Y, Nzihou A, Minh D P and Lyczko N 2017 A review of biogas utilisation, purification and upgrading technologies *Waste and Biomass Valorization* **8** 267–83

[5] Okonkwo U C, Onokpite E and Onokwai A O 2018 Comparative study of the optimal ratio of biogas production from various organic wastes and weeds for digester/restarted digester *J. King Saud Univ. Sci.* **30** 123–9

[6] Lichtfouse E, Schwarzbauer J and Robert D 2011 *Environmental chemistry for a sustainable world: volume 2: remediation of air and water pollution* vol 2 (Springer Science & Business Media)

[7] Oktavitri N I, Pratiwi W B, Purnamasari I, Hayati M, Rosita M, Fitrianingtyas and Hadinnata S 2019 Anaerobic Digestion of Slaughterhouse Wastewater : CO2 Capture of Biogas Using Chlorella vulgaris *Indones. J. Chem* **19** 1–8

[8] Usack J G and Angenent L T 2015 Comparing the inhibitory thresholds of dairy manure co-digesters after prolonged acclimation periods: Part 1–Performance and operating limits *Water Res.* **87** 446–57

[9] Ratanatamskul C and Manpetch P 2016 Comparative assessment of prototype digester configuration for biogas recovery from anaerobic co-digestion of food waste and rain tree leaf as feedstock *Int. Biodeterior. Biodegradation* **113** 367–74

[10] Geraldi A, Ni'matuzahroh, Wanguyun A P, Hariyanto S and Nur B 2019 Bioprospecting Of Cellulolytic And Biosurfactant- Producing Bacteria For Organic Waste Treatment *Pollut. Res.* **38** 114–7

[11] Morato T, Vaezi M and Kumar A 2019 Developing a framework to optimally locate biomass collection points to improve the biomass-based energy facilities locating procedure–A case study for Bolivia *Renew. Sustain. Energy Rev.* **107** 183–99

[12] Chan Y H, Cheah K W, How B S, Loy A C M, Shahbaz M, Singh H K G, Shuhaili A F A, Yusup S, Ghani W A W A K and Rambli J 2019 An overview of biomass thermochemical conversion technologies in Malaysia *Sci. Total Environ.* **680** 105–23

[13] Malico I, Pereira R N, Gonçalves A C and Sousa A M O 2019 Current status and future perspectives for energy production from solid biomass in the European industry *Renew. Sustain. Energy Rev.* **112** 960–77
[14] Cheng S, Li Z, Mang H-P, Huba E-M, Gao R and Wang X 2014 Development and application of prefabricated biogas digesters in developing countries *Renew. Sustain. energy Rev.* **34** 387–400
[15] Farizal F, Aji R, Rachman A, Nasruddin N and Mahlia T M I 2018 Indonesia's Municipal Solid Waste 3R and Waste to Energy Programs *Makara J. Technol.* **21** 153–9

Sustainable development impact of implementing electric taxis in Jakarta: A cost-benefit analysis

F Sumasto, F Imansuri, M Agus, Safril and M Wirandi

Department of Automotive Industrial Engineering, STMI Polytechnic of Jakarta, Ministry of Industry Republic of Indonesia, DKI Jakarta, Indonesia

Email: f-sumasto@kemenperin.go.id

Abstract. Due to issues in global warming in the world, the application of replacing fossil fuels to electricity is important to reduce air pollution, especially in the metropolitan area. In 2016, total emissions in Indonesia increased from 11th to 10th in the world and were the highest in ASEAN, so this became the focus in order to reduce emissions. What issues make efforts to change from conventional taxis to electric taxis an important measurement to achieve targets in energy-saving and reduce emissions because electric taxi has the potential to resource savings and to improve urban air quality. This study was conducted to diagnose the potential application of electric taxis by conducting a feasibility study, social impact, and environmental analysis in the implementation of the electric taxis to measure sustainability aspects. In the calculation process, the maintenance cost is estimated to be fixed. The results of this study from the cost-benefit analysis are expected to be able to contribute to the government and the private sectors for decision making towards increasing sustainability in Indonesia.

1. Introduction

The progress of a country can be seen from the factor of vehicle ownership and housing. Vehicles are strongly related to transportation. Transportation is one of the basic economic and social growth in a country. In Indonesia, the transportation sector creates many opportunities to employ people, for example, as drivers. The transportation sector, especially online transportation, can absorb a lot of labor, for example, such as motorbikes, catches, and online taxis. In addition, online transportation supports or provides services between goods and food, which means it has a positive impact on other businesses. Transportation also contributes to connecting people, which is a positive impact. In its negative impact, transportation certainly impacts air pollution, accidents, traffic jams, and noise.

Transportation is one source of GHG emissions and the main principle of air pollution, especially in urban areas. Transportation accounts for around 14% of global GHGs [1]. Increasing the amount of transportation is usually characterized by increasing urbanization, thereby increasing the demand for transportation itself, especially in developing countries like Indonesia. Based on the WHO, the GHG index in Indonesia accounts for around 9.2% of the world's GHGs. The GHG level of the transportation sector in Indonesia is high, especially in urban areas such as Jakarta, which reaches 24.71% [2]. To reduce GHG levels, one of them is changing the mode of transportation from conventional to electric transportation.

Under normal circumstances, electric vehicles do not produce GHG emissions or air pollution. Even if electrical energy sources are not obtained from combustion such as hydroelectric power, nuclear power plants or renewable sources, the use of electric vehicles will increasingly have a positive impact on the environment and energy transfer [3]. The transition from conventional vehicles to electric vehicles is very difficult to implement because investing in electric vehicles requires a large investment fund. The reason is the price of electric vehicles that are more expensive when compared directly with the prices of conventional vehicles and fossil fuels.

It is reflecting on the difficulty of making a transition where the government must take the lead in starting the transition. At present, the private sector in Indonesia has begun the transition to electric vehicles, especially in public transportation. The establishment of a more sustainable urban mobility in Jakarta City has begun to be implemented by the private sector. The private sector sells transportation services with a taxi fleet. The launch of the Electric Taxi fleet starts in October 2018 with 19 fueling stations. From the implementation of electric taxis, it is expected to become more competitive, technically, financially, and environmentally. The public or other private parties may not be fast enough to accept this change. In this case, this allows the government to collaborate with the private sector, or the government can take the initiative to make the transition from conventional vehicles to electric vehicles to official vehicles in government.

Taxis in urban areas generally have characteristics of longer daily trips and longer distances than private cars. Private cars are usually only in the parking lot most of the day, so it does not affect greenhouse gas emissions. In terms of the time to stop on the road, taxis tend to be far longer than other vehicles, which will lead to taxis that will be much greater, causing an increase in greenhouse gas emissions and air pollutants compared to other vehicles. The distance traveled by taxi per day is much longer than a private car that usually sits in a parking lot most of the day. Based on these things, the environmental impact arising from conventional taxis will be disproportionately higher when compared to comparable private vehicles [4].

Based on that information, this study was conducted to diagnose the potential application of electric taxis by conducting a feasibility study, social impact, and environmental analysis in the implementation of the electric taxis to measure sustainability aspects di Indonesia. The paper is organized as follows. Section 2 describes the literature review and previous researchers about electric taxis. Section 3 describes the method to collect data and assumptions. Section 4 describes the results and discussion about feasibility and cost-benefit. Section 5 discusses some are concluding marks.

2. Literature Review

The application or transition from conventional vehicles to electric vehicles is basically to implement sustainable pillars that are expected to have a positive impact on human life. From an economic perspective, there is research on life cycle costs and the optimal capacity of batteries that have been carried out in China [5] and the US [6]. Then there is the research about scheduling that has been done in Taiwan by mixing conventional taxis and electric taxis [7]. From an environmental perspective, there is research on solar electricity production and the conversion of conventional taxis into electric taxis in Chile [8] and research on the magnitude of the impact of carbon dioxide resulting from the transition from conventional taxis to electric taxis in Brazil [9]. Research that examined the eco-environment as a result of the shift towards electric taxis has been done in Beijing [10] and Seoul [4] Previous studies conducted on electric taxis mostly came from countries with high levels of greenhouse gas emissions except for Chile. When viewed regionally, these studies are in the American, European, and East Asian regions. China and the USA rank first and second for the world's greenhouse gas emissions index, while Indonesia ranks increased from 11th to 10th in the world and first ranks in the Asia Pacific [11].

Based on sustainability pillars, namely from the economic, social, and environmental aspects, which focus on renewable energy. Then seen from the aspect of developing countries and geographical regions of Indonesia where Indonesia really needs public transportation that is environmentally friendly and policy aspects where the ranking of GHG emissions in Indonesia continues to increase. Thus, an assessment of the feasibility study is needed to see as good an impact caused both economically, socially, and environmentally in Indonesia in making the transition to electric vehicles, especially in Jakarta City.

3. Method

Conventional taxis and electric taxis are the targets chosen in this study. This study uses face-to-face interviews with taxi drivers in gathering the required data. Conventional taxis that are used as objects are taxis with Toyota Vios models. The electric taxis used as the object are taxis with the BYD e6 A/T model. The data is obtained from face-to-face interviews and secondary data (Table 1.). The result from face-to-face interviews is mileage per day, trips per day, operational time per day, and maintenance cost per month. The secondary data are the initial investment is obtained for 30 cars, fuel capacity, price of fuel, and fuel used per km.

Table 1. Data obtained from interviews and secondary data.

Explanation	Toyota Vios	BYD e6 A/T
Investments for 30 units (IDR)	6,600,000,000	11,325,000,000
Fuel Capacity	42 liters	75 kWh
Mileage (km/day)	210	210
Trips per day	6	7
Operational Time (hour/day)	16	16
Maintenance Cost (monthly)	2,923,000/unit	20,000,000/unit
Average Tariff per Trip (IDR)	170,000	197,000
Fuel used per km	0.09 liter	0.3 kWh

In addition to primary data from face-to-face interviews and secondary data, there are several assumptions. It is assumed there is no Value Added Tax (VAT) on Luxury Goods for electric taxis because they are still in the pilot project category and are assumed to be free from import duties. Besides, income tax is assumed to be pessimistic with a value of 20%, and the differentiation for taxis is 10% per year.

For data processing, this study calculates earnings after tax (EAT) to be able to facilitate the calculation of the Payback Period (PP). The Payback Period value is obtained by interpolating from the difference in investment value and earnings after tax per year. The final point of the interpolation results will be a minimum period to reach the Payback Period. Average Rate Return (ARR) is used to see the value of returns using the average value of EAT and the average value of the investment. This transition project will be considered feasible to be implemented if the ARR results show a positive value.

4. Result and Discussion

Table 2. shows the results of the calculations obtained by EAT for five years with a degree of freedom of 20% for conventional taxis and electric taxis. The interpolation results show that the transition project for electric taxis will reach the Payback Period of 3.489 years (Figure 1.). Table 3. shows the Average Rate of Return, Net Present Value (NPV), Profitability Index (PI), and Internal Rate Return (IRR) for both conventional taxis and electric taxis. ARR for implementing electric taxis in Jakarta City is 4.12% or equal IDR 2,211,291,506. The NPV for implementing electric taxis in Jakarta City is IDR 4,441,776,499. The relationship between cost and benefit in the implementation of electric taxis showed positive results with a PI of 1.34. The results show the IRR obtained a positive value with a value of 30.01.

Basically, it needs an assessment of the impact of the transition of conventional vehicles to electric vehicles, which is done by assessing the feasibility study. There are three feasibility assessments in this paper, namely: economic impacts, social impacts, and environmental impacts. The economic impact can be seen from the assessment of the feasibility study by knowing the value of the Present Net Value (NPV) and sensitivity factors to the comparison between conventional taxis and electric taxis. The calculation of electric vehicles is assumed not to apply Value Added Tax (VAT) for Luxury Goods and import duties. The economic impact on conventional vehicles is still considered feasible if the Value Added Tax (VAT) on the cost of Luxury Goods and vehicle import duties are eliminated. Under the policy scenario, the impact of the transition to electric taxis has a positive economic impact.

Table 2. Earning after tax for conventional and electric taxis.

	Earning After Tax	
Year	Conventional Taxis	Electric Taxis
2020	4,828,992,000	5,804,352,000
2021	6,277,689,600	7,545,657,600
2022	8,223,773,376	9,884,811,456
2023	10,855,380,856	13,047,951,122
2024	14,437,656,539	17,353,774,992

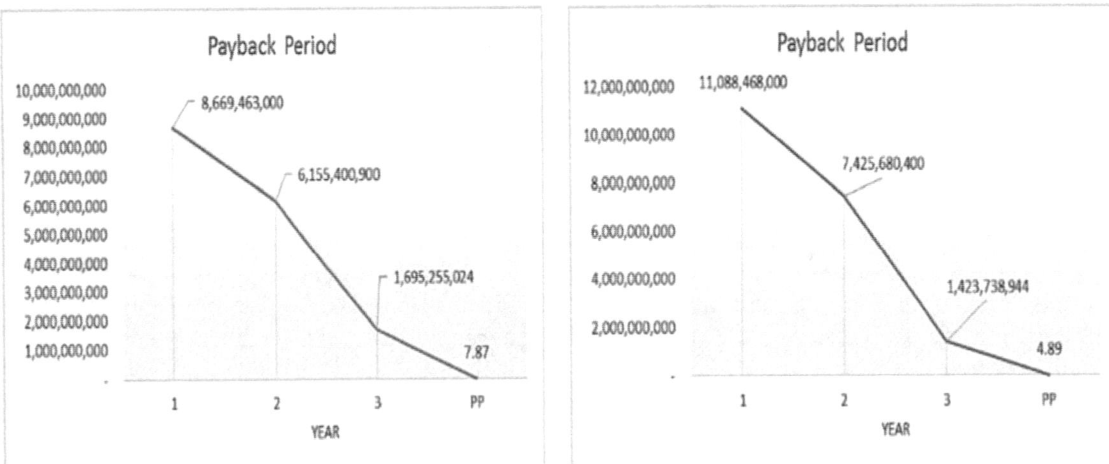

(a) (b)

Figure 1. Payback Period for (a) Conventional Taxis and (b) Electric Taxis by interpolating from the difference in investment value and EAT per year.

Table 3. The results of calculations on conventional and electric taxis feasibility.

Value	Conventional Taxis	Electric Taxis
ARR	2,045,497,027	2,211,291,506
NPV	3,189,628,464	4,441,776,499
PI	1.33	1.34
IRR	29.01	30.01

The sensitivity factor (Figure 2.) is obtained through an interpolation process based on the effect of increased processing costs due to inflation and the effect of investment on NPV. Sensitivity values are seen in positive and negative NPV, where a positive Degree of Freedom factor is found at 30%. The sensitivity factor on a positive NPV with the Degree of Freedom 30% will affect the company's cash value, which refers to the Internal Rate Return (IRR). The sensitivity value obtained is based on the zero points of net cash received by the company, the Degree of Freedom limit cannot be less than 20% if it exceeds the assumption then it is assumed that electric taxis in Indonesia are not feasible to be implemented. Whereas the sensitivity factor on negative NPV with a Degree of Freedom value of 31%, can be seen in Figure 2. This provides an explanation of the economic impact based on a Degree of Freedom sensitivity value of more than 20%, which is declared feasible in the implementation of electric taxis in Jakarta City.

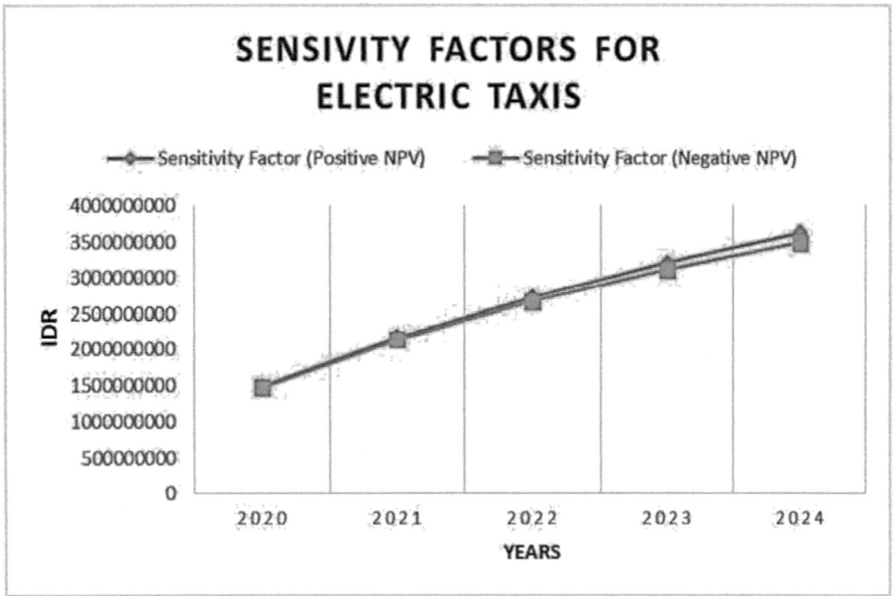

Figure 2. Sensitivity factors in positive and negative npv for electric taxis.

Social impacts have a significant effect if the government provides incentives for exemptions and import duties. However, the social impact on the community can also be seen from environmentally friendly electricity, and also reduces the costs of refueling. It is known that the current price of electricity is IDR 1,026 per kWh while the price of fossil fuels is IDR 6,450 per liter, conventional vehicles will cost IDR 580.50 per kilometer, and electric taxis will spend IDR 153.90 per kilometer. In the scenario of conventional taxis and electric taxis traveling 210 kilometers per day, the cost savings incurred by electric taxis are 3.77 times more than conventional taxis. Conventional taxis will cost IDR 121,905 at a distance of 210 kilometers, while electric taxis will only cost IDR 32,319. This will provide a good social impact on electric taxi users in Jakarta City, which can save up to 73.48% per day, which should be spent on conventional taxis.

It is assumed that in one day, 30 conventional vehicles are operating with a distance of 210 km per vehicle and assuming each vehicle consumes 1 liter of fuel every 11.11 kilometers. Table 4. shows the environmental impacts that occur based on the level of pollution generated by vehicle fumes per day from emission load estimation in Indonesia [12]. If electric vehicles operate in Indonesia with a total of 30 vehicles, it will reduce SO_2 and NOx air pollution by 2.47% per day.

Table 4. The environmental impact obtained from each conventional vehicle per day.

Category	CO (g/km)	HC (g/km)	NOx (g/km)	CO_2 (g/km)	SO_2 (g/km)
Motor Cycle	0.088	0.037	0.77	152.972	0.021
Car (Gasoline Fuel)	0.252	0.025	15.12	437.062	0.197
Car (Diesel Fuel)	0.018	0.001	1.85	30.517	0.233
Bus	0.069	0.008	24.74	119.890	1.933
Truck	0.053	0.011	28.10	91.552	1.302
Total	0.480	0.083	70.58	831.992	3.686

5. Conclusion

In a small effort to improve urban air quality in Indonesia, investigate the economic feasibility of the transition from conventional taxis to electric taxis in Jakarta City. This study uses an approach to replace current conventional taxis into electric taxis. The transition to the public transportation sector will be relatively easy if the government can provide incentives in the form of free import duty for the procurement of electric taxis by private sectors.

In the cost-benefit analysis for the transition to electric taxis, it gives a significant economic impact on the cost of purchasing a vehicle based on the given scenario. On the benefits side, the increase in social benefits is seen from the reduction or more cost savings incurred up to 73.48% in the use of electric taxis compared to conventional taxis (gasoline fuel). Improving air quality in Jakarta City is one of the benefits to the environment. From the environmental side, it is reducing air pollution by 2.47% per day (SO_2 and NOx) in the implementation of 30 electric taxis. The results show the implementation of electric taxis in Jakarta City has a positive impact on aspects of sustainability in terms of economic, social, and environmental. The policy that can be taken by the government is to collaborate with certain private sectors to be able to provide free import duty for the procurement of electric vehicles that are used as public transportation.

Acknowledgments

This research is supported and funded by the Directorate of Research and Community Engagement (P2M), STMI Polytechnic of Jakarta, Nomor: SP DIPA-019.10.2.579350/2019.

References

[1] Santos G 2017 Road transport and CO_2 emissions: What are the challenges? *Transport Policy* **59** 71–4

[2] Center for Data and Information Technology ESDM 2017 Study on the Use of Local Emission Factors (Tier 2) in the Energy Sector GHG Inventory (Jakarta: Center for Data and Information Technology ESDM)

[3] Aziz M, Huda M, Budiman B A, Sutanto E and Sambegoro P L 2018 Implementation of Electric Vehicle and Grid Integration *2018 5th International Conference on Electric Vehicular Technology (ICEVT)* (Surakarta) pp 9–13

[4] Kang SC and Lee H 2019 Economic appraisal of implementing electric vehicle taxis in Seoul *Research in Transportation Economics* **73** 45–52

[5] Deyang K, Dan M and Minmin W 2016 A Simulation Study of Upgrading Urban Gasoline Taxis to Electric Taxis *Energy Procedia* **104** 390–5

[6] Hu L, Dong J, Lin Z and Yang J 2018 Analyzing battery electric vehicle feasibility from taxi travel patterns: The case study of New York City, USA *Transportation Research Part C: Emerging Technologies* **87** 91–104

[7] Lu C-C, Yan S and Huang Y-W 2018 Optimal scheduling of a taxi fleet with mixed electric and gasoline vehicles to service advance reservations *Transportation Research Part C: Emerging Technologies* **93** 479–500

[8] Girard A, Roberts C, Simon F and Ordoñez J 2019 Solar electricity production and taxi electrical vehicle conversion in Chile *Journal of Cleaner Production* **210** 1261–9

[9] Teixeira A C R and Sodré J R 2016 Simulation of the impacts on carbon dioxide emissions from replacement of a conventional Brazilian taxi fleet by electric vehicles *Energy* **115** 1617–22

[10] Shi X, Wang X, Yang J and Sun Z 2016 Electric vehicle transformation in Beijing and the comparative eco-environmental impacts: A case study of electric and gasoline powered taxis *Journal of Cleaner Production* **137** 449–60

[11] Janssens-Maenhout G, Crippa M, Guizzardi D, Muntean M, Schaaf E, Olivier J G J, Peters J A H W, Schure K M 2017 *Fossil CO_2 and GHG emissions of all world countries* (Luxembourg: Publications Office of the European Union)

[12] Novianti S and Driejana 2010 The influence of factor emission characteristics in transport-induced nitrogen oxides (NOx) emission load estimation (Case study: karees area, bandung) *Jurnal Teknik Lingkungan* **16** 185–98

IOP Conf. Series: Materials Science and Engineering **885** (2020) 012028 doi:10.1088/1757-899X/885/1/012028

Production capacity planning using RCCP method with CPOF approach: a case study in an automotive Industry

Siti Aisyah[1], Humiras Hardi Purba[2] and R F Dewarani[1]

[1]Polytechnic STMI, Ministry of Industry Republic of Indonesia, Cempaka Putih, Jakarta

[2]Master of Industrial Engineering Program, Mercu Buana University, Jakarta, Indonesia

Email : lalita1712aisyah@gmail.com

Abstract: The consumer motorcycle soaring demand cause automotive manufacturing has an obstacle in fulfilling them. The motorcycle manufacturing realizes that the soaring demand has a constraint in terms of production load or meet. The aim of this study was to conduct an RCCP analysis with the CPOF approach in the XYZ Company assembling line. The problem that occurred in the assembling line consists of 3 working groups, namely sub assy, mainline RH, and mainline LH. The solution to the problem in the assembling line is done by the Rough Cut Capacity Planning-RCCP method of Capacity Planning Using Overall Factor Approach-CPOF. Based on the result using the CPOF approach, the production load of each workgroup was calculated as follows: sub assy of 115,690 minutes, mainline RH of 186,000 minutes and mainline LH of 218,701 minutes. This result showed an overload capacity in the assembly line and led to a production loss of 868 units in January, 836 units in February, and 867 units in March. It is possible to solve this underload problem through overtime and work weekends.

1. Introduction

The XYZ is an automotive manufacturing company that produces a motorcycle. In an effort to meet the needs of consumers properly and quickly, the company has an obstacle in terms of meeting a load of production in accordance with existing capacity. The imbalance between the capacity and load was due to many work elements of an operator. The problem of imbalanced production capacity occurred in the assembly line consisted of 3 workgroups, namely sub assy, mainline RH, and mainline. To fulfilling the production load or consumer demand for the capacity by the XYZ Company is very influential on the production schedule and can cause losses in terms of time and production. The problem in this assembling line can be solved by the Rough Cut Capacity Planning (RCCP) method with the Capacity Planning Using Overall Factor Approach (CPOF) approach. RCCP is a plan to determine the capacity to meet the Master Production Schedule. Good support from vendors in supplying automotive parts or components largely determines the assembling. One important in automotive manufacture is the supply of spare parts for customers, where customers must quickly get it [1]. The application of the CPOF approach required the input of the Master Production Schedule (MPS), the total production time, and the proportion of resource use. The Capacity calculation using this method is expected to provide improvement propose to the XYZ Company that the planning production schedule can run well.

Rough Cut Capacity Planning (RCCP) is a plan to determine the capacity to meet the Master Production Schedule (MPS). The RCCP validates the MPS; it is useful to determine specific sources that are expected to become a potential obstacle [2, 3].

Capacity Planning with Overall Factors (CPOF) approach are required three inputs, namely: MPS, the total time to produce a product, and the proportion time source are used. The step CPOF multiplied the total time of each family to the number of MPS to obtain the total time needed for the factory to search MPS. This total time is then divided into the usage time of each source by multiplying the total time by the proportion of resource use [2-4].

The normal time is obtained by multiplying the cycle time and the rating factors. The calculation rating factor is taken if the operator working at an unnatural speed, thus the results of the time calculation should be normalized first. The formula is Wn = Cycle Time x (1 + Rating Factor) [5]. Standard time is the time that needed naturally by a normal worker to complete a task carried out in a good work system. The formula is WS = Normal Time x (1 + Allowance) [5, 6].

2. Research Method

In this study, using *the Rough Cut Capacity Planning* (RCCP) method and *Capacity Planning with Overall Factors* (CPOF) approach purposed to figure out production capacity planning. To achieve the objectives of the research, the steps taken are as follows:

2.1. Data collection

The data required is data on the number of workdays in 3 months (January 20 days, February 19 days and March 19 days), actual working hours (Monday - Thursday is 485 minutes, and Friday is 445 minutes), production process, total production time (53.63 minutes), total demand for three months (January = 3,500 units, February = 3,000 units, and March = 3,200 units) and the production process efficiency set by the company (January = 75%, February = 72% and March = 73%)

2.2. Data Processing

- Calculate normal time: Wn = Wc x (1- Rating Factor) [5]
- Calculate standard time: Ws = Wn x (1 + Allowance) [5, 6]
- Calculate the working hours available [7]:
- Calculating Utilities [8]: (Number of actual requests: total hours worked) x 100%
- Calculate Available Capacity [8]: Available working hours x efficiency x utility

The operating system needs control and integration from the manufacturer, supplier, and logistics partner [9].

2.3. Analysis

The capacity analysis is carried out using the CPOF approach, and proposed capacity building is carried out using the CPOF approach.

3. Result and Analysis

3.1. Calculation of Available Work

3.1.1. Time. Measurement of available work hours was performed to obtain information about the work hour of the operator in the assembly line of PT XYZ. The calculation was done for January, February, and March using the following formula:

Work Hours on Day = Actual Work Hours x Number of Work Day

Calculation example for January:

Work Hour Monday – Thursday=
Actual Work Hours x Number of Work Day = 485 minutes x 16 days = 7,760 minutes
Work Hour on Friday = 445 minutes x 4 days = 1,780 minutes
Total Work Hour = 7,760 + 1,780 = 9,540 minutes

Table 1. Recapitulation of calculation of work hour/month.

Month	Work Hour (minute)
January	9.540
February	9.055
March	9.055

3.2. Calculation of Rough Capacity CPOF Technique
Data input in the CPOF technique is standard time, production quantity per month, and proportion. Calculation steps in CPOF technique as below:

Example task calculation bracket upper in January:
Total capacity requirements = total monthly production x total production time Total capacity requirements = 3.500 units x 53,63 minutes = 187.705 minutes

The proportion of resource use = the operating time of an activity / the total standard time The proportion of resource use = 1,53 minutes / 53,63 minutes = 0,02853

Time required by a task = total capacity needs x proportion of resource use The time required by a task = 187.705 x 0,02853 = 5.355

Table 2. Calculation of rough capacity.

Task	Proportion	Periode			Total Time (minute)
		Jan (minute)	Feb (minute)	Mar (minute)	
Sub Assy					
Bracket Upper	0.02853	5.355	4.590	4.896	14.841
Main Line RH					
Assembling of Swing Arm, Prop Stand	0.03431	6.440	5.520	5.888	17.848
Main Line RH					
Numbering Frame	0.04456	8.365	7.170	7.648	23.183

After the data obtained from the rough capacity calculation, then the analysis is carried out with the CPOF approach to determine the capacity to meet the load or work. As for the analysis of the CPOF approach can be seen in table 3:

Table 3. Analysis of the CPOF approach.

Description	Jan	Feb	Mar	Total
Bracket Upper				
Available time (minute)	9.540	9.055	9.055	27.650
Available capacity (minute)	4.007	3.325	3.569	10.901
Rough Capacity (minute)	5.355	4.590	4.896	14.841
Underload (minute)	-1.348	-1.265	-1.327	-3.940
Loss (multiplied by task)	881	827	867	2.575

Assembling of *Swing Arm, Prop stand*				
Available time (minute)	9.540	9.055	9.055	27.650
Available capacity (minute)	4.865	3.977	4.297	13.139
Rough Capacity (minute)	6.440	5.520	5.888	17.848
Underload (minute)	-1.575	-1.543	-1.591	-4.709
Loss (multiplied by task)	856	839	865	2.559
Numbering Frame				
Available time (minute)	9.540	9.055	9.055	27.650
Available capacity (minute)	6.296	5.150	5.553	16.999
Rough Capacity (minute)	8.365	7.170	7.648	23.183
Underload (minute)	-2.069	-2.020	-2.095	-6.184
Loss (multiplied by task)	866	845	877	2.587

Loss (multiplied by task) was obtained through the calculation of capacity under load (minute) divided by standard time. All tasks listed above represented the workgroup of sub assy, mainline RH and mainline LH that experienced capacity under load, thus improvement is required to meet the production schedule.

3.3. Capacity Improvement Proposal using the CPOF Approach

The existence of capacity underload insists the company balances production load with available capacity. One effort to increase the capacity is to schedule overtime or work weekends. Example: The work hours that cannot meet the company's production load target, thus the company should be doing work weekends on Saturday for 4 (four) weeks, with the actual work hour of 7 hours, or about 420 minutes. So it has 1,680 minute work weekends and added overtime for 2 hours or 120 minutes after actual work hours for 13 days on January, 2 hours or 120 minutes after actual work hours for 16 days on February, and 2 hours or 120 minutes after actual work hours for 15 days on March.

Table 4. Improvement proposal.

Description	Jan	Feb	Mar	Total
Bracket Upper				
Available time (minute)	12.780	12.655	12.535	37.970
Efficiency (%)	0.75	0.72	0.73	-
Utilization (%)	0.56	0.51	0.54	-
Available Capacity (minute)	5.368	4.647	4.941	14.956
Rough Capacity (minute)	5.355	4.590	4.896	14.841
Underload (minute)	+13	+57	+45	+115
Assembling of *Swing Arm, Prop stand*				
Available time (minute)	12.780	12.655	12.535	37.970
Efficiency (%)	0.75	0.72	0.73	-
Utilization (%)	0.68	0,61	0,65	-
Available Capacity (minute)	6.518	5.558	5.948	18.024
Rough Capacity (minute)	6.440	5.520	5.888	17.848
Underload (minute)	+78	+38	+60	+176
Numbering Frame				
Available time (minute)	12.780	12.655	12.535	37.970
Efficiency (%)	0.75	0.72	0.73	-
Utilization (%)	0.88	0.79	0.84	-
Available Capacity (minute)	8.435	7.198	7.686	23.319
Rough Capacity (minute)	8.365	7.170	7.648	23.183
Underload (minute)	+70	+28	+38	+136

4. Conclusion

Based on the result of data analysis performed, the conclusion drawn is as follows:

1. Results of analysis using the CRP method CPOF approach for a workgroup of sub assy, mainline RH, and mainline LH showed that all tasks conducted by the three workgroups in the assembly line had available production capacity that was lower than the production load. This problem led to a lack of quantity produced in January, February, and March of 868 units, 863 units, and 867 units, respectively.
2. Improvement proposed to balance the capacity and load is to schedule overtime and work weekends. In January, there should be work weekends on Saturday for four weeks with previous actual work time from 420 minutes to 1,680 minutes and overtime of 2 hours for 13 days. In February, there should be work weekends on Saturday for four weeks with previous actual work time from 420 minutes to 1.680 minutes and overtime of 2 hours for 16 days. In March, there should be work weekends on Saturday for four weeks with a previous actual work time of 420 minutes to 1.680 minutes and overtime of 2 hours for 15 days.

References

[1] Purba H H, Mukhlisin and Aisyah S 2018 Productivity Improvement Picking Order by Appropriate Method, Value Stream Mapping Analysis, and Storage Design: A Case Study in Automotive Part Center *Manag. Prod. Eng. Rev.* **9** 71-81

[2] Gaspersz V 2004 *Production Planning and Inventory Control (PPIC): Berdasarkan Pendekatan Sistem Terintegrasi MRP II dan JIT Menuju Manufakturing 21* (Jakarta: PT Gramedia Pustaka Utama)

[3] Fogarty D W, John H B and Hoffmann T R 1991 *Production & Inventory Management* (Ohio: South-Western Publishing Co.)

[4] Suryadhini P 2014 *Modul Praktikum Perencanaan dan Pengendalian Produksi* (Jakarta)

[5] Sutalaksana I Z A and Ruhana 2006 *Teknik Perancangan Sistem Kerja* (Bandung: ITB Bandung)

[6] Wignjosoebroto S 2006 *Pengantar Teknik dan Manajemen Industri* Edisi 2

[7] Handoko T H 1999 *Dasar-dasar Manajemen Produksi & Operasi* (Yogyakarta: BPFE)

[8] Kusuma H 2002 *Manajemen Produksi (Perancangan dan Pengendalian Produksi)* (Yogyakarta: Andi)

[9] Purba H H, Fitra A and Nindiani A 2019 Control and Integration of Milk-Run Operation in Japanese Automotive Company in Indonesia. *Manag. Prod. Eng. Rev.* **10** 79-88

IOP Conf. Series: Materials Science and Engineering **885** (2020) 012029 doi:10.1088/1757-899X/885/1/012029

Production capacity planning in motorcycle assembly line using CRP method at P T XYZ

Humiras Hardi Purba[1], Siti Aisyah[2] and R F Dewarani[2]

[1] Master of Industrial Engineering Program, Mercu Buana University, Jakarta, Indonesia

[2] Polytechnic STMI, Ministry of Industry Republic of Indonesia, Cempaka Putih, Jakarta

Email : hardipurbaipb@gmail.com

Abstract. PT XYZ is an automotive manufacturing company that produces a motorcycle. In an effort to meet the needs of consumers precisely and quickly, PT XYZ deals with problems in terms of fulfilling the capacity in accordance with consumer demand. The problem of imbalanced production capacity occurred in assembly line 3 that produced motorcycle of FU types, consisting of 3 workgroups, namely sub assy, mainline RH, and mainline LH. The imbalance between the capacity and load was due to many work elements of an operator. Thus the load on each work in the three workgroups was greater than the available capacity. This problem was solved using the method of Capacity Requirement Planning/CRP. Based on the result using CRP, the production load of each workgroup was calculated as follows: sub assy of 115,690 minutes, mainline RH of 186,000 minutes and mainline LH of 218,701 minutes. This result showed an overload capacity in the assembly line and led to a production loss of 868 units in January, 836 units in February, and 867 units in March. It is possible to solve this under loading problem through overtime and work weekends.

1. Introduction

PT XYZ is an automotive manufacturing company that produces a motorcycle. In an effort to meet the needs of consumers properly and quickly, PT XYZ deals with problems in terms of fulfilling the capacity in accordance with consumer demand. The problem of imbalanced production capacity occurred in assembly line 3 that produced motorcycle of FU types, consisting of 3 workgroups, namely sub assy, mainline RH, and mainline LH. The imbalance between the capacity and load was due to many work elements of an operator. Thus the load on each work in the three workgroups was greater than the available capacity.

The balance between production load or consumer demand and capacity to be fulfilled by PT XYZ greatly affects the consistency of master production schedule and generates a loss in term of time and production; hence the problem found in assembly line 3 (three) of PT XYZ should be solved using the method of Capacity Requirement Planning/CRP.

Capacity Requirement Planning (CRP) is a method applied to plan the production capacity requirement; thus production process in a company runs well and in accordance with the production plan that has been designed by the company [1]. CRP has the ability to adjust the level of order arrival to the available capacity by identifying the work center area that is in overload or underload situation.

Calculation of capacity using this method is expected to provide improvement proposals to PT XYZ in order to balance between consumer ordering and capability of production capacity in PT XYZ. Thus the planned production schedule is applied properly.

1.1. Problem statement

The imbalance between capacity and consumer demand or load leads to problems in capacity requirement planning. The problem formulation in this study includes:

1. How is the analysis result of Capacity Requirement Planning (CRP) applied in the workgroup of sub assy, mainline RH, and mainline LH in assembly line 3 (three) of PT XYZ ?
2. How is the improvement proposal to production capacity in the workgroup of sub assy, mainline RH, and mainline LH using the CRP method ?

1.2. Research Purpose

The purpose of this study is listed as follows:

1. To produce the analysis of Capacity Requirement Planning (CRP) applied in the workgroup of sub assy, mainline RH, and mainline LH in assembly line 3 (three) of PT XYZ
2. To determine the improvement proposal concerning the production capacity in the workgroup of sub assy, mainline RH, and mainline LH using the CRP method.

2. Theoretical background

Capacity Requirement Planning is a process to determine the workload of each work center based on the production schedule [2]. CRP is an important step in controlling manufacture. The capacity requirement planning CRP examines the assumption and identifies areas with capacity overload and underload. Thus planner will be able to take appropriate action. In Capacity Requirement Planning (CRP), the workload in each work station is determined. Three things are required To determine/plan the capacity requirement [2]: 1. Work Center, 2. Load, 3. Capacity.

Analysis of Capacity Requirement Planning (CRP) requires separated calculation related to the need for setup time and run time. Analysis of CRP is more detailed than that of RCCP since CRP analysis requires information of standard setup time and standard run-time per unit item to be made. The calculation of operation time per unit in CRP analysis is done using the formula as follows [3].

3. Result and Discussion

PT XYZ only was found to have one shift work in a day, starting from 07.30 WIB to 16.30 WIB. There were 20 workdays in January, 19 workdays in February, and 19 workdays in March. Actual work time from Monday-Thursday was 485 minutes, while it was 445 minutes on Friday. The data required for calculation of capacity using the CRP method included the production quantity in the assembly line of PT XYZ. The quantity of motorcycle production based on information obtained from PT XYZ reached 3,500 units in January, 3,000 units in February, and 3,200 units in March.

3.1. Calculation of normal time

The normal time for each task is calculated by multiplying cycle time obtained by rating factors. The normal time for each task in the workgroup of sub assy, mainline RH, and mainline LH was calculated using the following formula [4, 5]:

$$\text{Normal time (Wn)} = \text{Ws x (1+Rating Factor)}$$

Moreover, the result of normal time calculation for each task is presented in Table 1.

Table 1. Calculation of normal time.

Task	Ws (minute)	Rating Factor	Wn (minute)
Sub Assy (Bracket Upper)	1.20	1+0.09	1.31
Main Line RH (Assembling of Swing Arm, Prop stand)	1.44	1+0.09	1.57
Main Line LH (Numbering Frame)	1.84	1+0.11	2.04

3.2. Calculation of Standard Time

Standard time is measured by multiplying normal time by an allowance that has been determined previously. The normal time for each task in the workgroup of sub assy, mainline RH, and mainline LH was computed using the following formula [4,5]:

$$\text{Standard Time (WS)} = \text{Wn x } (1 + \text{Allowance})$$

The data of the calculation result of standard time are shown in Table 2 below.

Table 2. Calculation of standard time.

Task	Wn (minute)	Allowance	Wn (minute)
Sub Assy (Bracket Upper)	1.31	0.17	1.53
Main Line RH (Assembling of Swing Arm, Prop stand)	1.57	0.17	1.84
Main Line LH (Numbering Frame)	2.04	0.17	2.39

3.3. Calculation of Available Work Time

Measurement of available work time was performed to obtain information about the work hour of the operator in the assembly line of PT XYZ. The calculation was done for January, February, and March using the following formula [6]:

$$\text{Work Time on Day} = \text{Actual Work Time x Number of Work Day}$$

Calculation example for January:

- Work Time Monday – Thursday = Actual Work Time x Number of Work Day = 485 minutes x 16 days = 7,760 minutes
- Work Hour on Friday = 445 minutes x 4 days = 1,780 minutes
- Total Work Hour = 7,760 + 1,780 = 9,540 minutes

Table 3. Recapitulation of calculation of work time/month.

Month	Work Time (minute)
January	9,540
February	9,055
March	9,055

3.4. Efficiency and Utilization

Efficiency in the assembly line of PT XYZ in January, February, and March was measured to be 75%, 72%, and 73%, respectively. Utilization is the ratio of comparison between the required work time and available work time. The formula of utilization is as follows [7]:

$$\text{Utilization} = (\text{Actual work time for production / Available work time}) \text{ x } 100\%$$

IOP Conf. Series: Materials Science and Engineering **885** (2020) 012029 doi:10.1088/1757-899X/885/1/012029

Table 4. Calculation of utilization.

Task	Utilization (%)		
	January	February	March
Sub Assy (Bracket Upper)	0.56	0.51	0.54
Main Line RH (Assembling of Swing Arm, Prop stand)	0.68	0.61	0.65
Main Line LH (Numbering Frame)	0.88	0.79	0.84

3.5. Calculation of Available Capacity

Calculation of available capacity was performed for each task in the workgroup of sub assy, mainline RH and mainline LH. The formula used to calculate available capacity is as follows [7]:

Available Capacity = Available Work Time x Efficiency x Utilization

Table 5. Calculation of available capacity.

Description	January	February	March
Bracket Upper			
A (minute)	9,540	9,055	9,055
Efficiency (%)	0.75	0.72	0.73
Utilization (%)	0.56	0.51	0.54
Available Capacity (minute)	4,007	3,325	3,569
Assembling of *Swing Arm, Propstand*			
A (minute)	9,540	9,055	9,055
Efficiency (%)	0.75	0.72	0.73
Utilization (%)	0.68	0.61	0.65
Available Capacity (minute)	4,865	3,977	4,297
Numbering Frame			
A (minute)	9,540	9,055	9,055
Efficiency (%)	0.75	0.72	0.73
Utilization (%)	0.88	0.79	0.84
Available Capacity (minute)	6,296	5,150	5,553

Based on the result of calculation for available capacity, the quantity of product expected to produce in the assembly line of PT XYZ: In January, assembly line could produce 2,632 units, resulted from the calculation of total available capacity in a minute divided by total standard time. According to the schedule, the assembly line should have produced 3,500 units in January. Hence 868 units should be added. Moreover, available capacity in February reached 2,164 units, yet the schedule of PT XYZ targeted 3,000 units. Thus 836 units were still required. In March, the calculation showed that the assembly line only produced 2,333 units of the required that amounted to 3,200 units. Therefore, there was a lack of 867 units.

3.6. Calculation of Load using CRP Method

Before the result of actual requirement or termed load is obtained, total operation time should be determined. Total operation time is calculated using several formulas: The formula for calculating setup time per unit is as follows [8]:

Setup Time/Unit (Minute) = Setup Time/Lot (Minute) / Lot Size The formula for calculating operation, Time per unit is as follows:

Operation Time Per Unit = Run Time / Unit + Setup Time Unit The formula for calculating total operation time is as follows:

Total Operation Time = Lot Size x Operation Time Per Unit

After all those formulas were determined, Total Operation Time was calculated. Example of calculation is written below:

3.7. Bracket Upper

Lot Size in the assembly line of PT XYZ amounted to 60 units. Since information obtained from PT XYZ showed that the production preparation time required by the company reached 0.0543 minutes, setup time/unit calculated was 0.0009 minute. Moreover, the operation time/unit of 1.5309 was determined by adding Run Time (standard time) with setup time/unit. The last one, total operation time, was obtained by multiplying lot size by Operation time/unit, resulted in 91.85 minutes. After *total operation time was determined,* production load was further calculated using the following formula [7-9]:

Load = (Total Operation Time x Production Quantity per Month) / Lot Size

The result is presented in Table 6 as follows:

Table 6. Calculation of load.

Work Group	January	February	March
Bracket Upper (min)	5,358	4,593	4,899
Assembling of *Swing Arm, Prop stand* (min)	6,092	5,222	5,570
Numbering Frame (min)	8,367	7,172	7,650

4. Analysis and Discussion

4..1. Analysis of Capacity using CRP Method

Following data processing, CRP analysis was applied to determine whether capacity and load were in balance or not. Increased productivity is done by improving the working method of the picking process, and the operating system needs control and integration from the manufacturer, supplier, and logistics partner [10, 11]. The result is presented in Table 7 as follows:

Table 7. Analysis of CRP.

Description	January	February	March	Total
Bracket Upper				
Available Time (min)	9,540	9,055	9,055	27,650
Available Capacity (min)	4,007	3,325	3,569	10,901
Load (min)	5,358	4,593	4,899	14,840
Underload (min)	-1,351	-1,268	-1,330	-3,949
Loss (multiplied by task)	883	829	869	2,581
Assembling of Swing Arm, Prop stand				
Available Time (min)	9,540	9,055	9,055	27,650
Available Capacity (min)	4,865	3,977	4,297	13,139
Load (min)	6,442	5,522	5,890	17,854
Underload (min)	-1,577	-1,545	-1,593	-4,715
Loss (multiplied by task)	857	840	866	2,563
Numbering Frame				

Available Time (min)	9,540	9,055	9,055	27,650
Available Capacity (min)	6,296	5,150	5,553	16,999
Load (min)	8,367	7,172	7,650	23,189
Underload (min)	-2,071	-2,022	-2,097	-6,190
Loss (multiplied by task)	867	846	877	2,590

Loss (multiplied by task) was obtained through the calculation of capacity under load (minute) divided by standard time. All tasks listed above represented the workgroup of sub assy, mainline RH and mainline LH that experienced capacity under load, thus improvement is required to meet the production schedule.

4.2. Capacity Improvement Proposal using CRP Method

The existence of capacity underload insists the company balances production load with available capacity. One effort to increase the capacity is to schedule overtime or work weekends.

Table 8. Improvement proposal.

Description	January	February	March	Total
Bracket Upper				
Available Time (min)	12,780	12,655	12,535	37,970
Efficiency (%)	0.75	0.73	0.75	
Utilization (%)	0.56	0.51	0.54	
Available Capacity (min)	5,368	4,647	4,941	14,956
Rough Capacity (min)	5,358	4,593	4,899	14,850
Underload (min)	+10	+54	+42	+106
Assembling of Swing Arm, Prop stand				
Available Time (min)	12,780	12,655	12,535	37,970
Efficiency (%)	0.75	0.73	0.75	
Utilization (%)	0.68	0.61	0.65	
Available Capacity (min)	6,518	5,558	5,948	18,024
Rough Capacity (min)	6,442	5,522	5,890	17,854
Underload (min)	+76	+36	+58	+170
Numbering Frame				
Available Time (min)	12,780	12,655	12,535	37,970
Efficiency (%)	0.75	0.73	0.75	
Utilization (%)	0.88	0.79	0.84	
Available Capacity (min)	8,435	7,198	7,686	23,319
Rough Capacity (min)	8,367	7,172	7,650	23,189
Underload (min)	+68	+26	+36	+130

Table 8 shows that available production capacity will be able to meet the production load compared to the previous situation. Lack of production due to capacity under load, which reached 868 units in January, 836 units in February, and 867 units in March could be fulfilled.

5. Conclusion

Based on the result of data analysis performed, the conclusion drawn is as follows:

1. Results of analysis using the CRP method for a workgroup of sub assy, mainline RH, and mainline LH showed that all tasks conducted by the three workgroups in the assembly line had available production capacity that was lower than the production load. This problem led to a lack of quantity produced in January, February, and March of 868 units, 863 units, and 867 units, respectively.

2. Improvement proposed to balance the capacity and load is to schedule overtime and work weekends. In January, there should be work weekends on Saturday for four weeks with previous actual work time from 420 minutes to 1,680 minutes and overtime of 2 hours for 13 days. In February, there should be work weekends on Saturday for four weeks with previous actual work time from 420 minutes to 1.680 minutes and overtime of 2 hours for 16 days. In March, there should be work weekends on Saturday for four weeks with a previous actual work time of 420 minutes to 1.680 minutes and overtime of 2 hours for 15 days.

6. Recommendation

Moreover, the recommendation for this study is listed as follows:

1. The company should check production capacity underload and adjust it to overtime and work weekends in order to achieve production load and production capacity balance.
2. Capacity improvement proposal using the CRP method should be applied by the company on condition that the schedule proposal for overtime and work weekends is approved by the company.

References

[1] Fogarty D W, Blackstone J H and Hoffmann T R 1991 *Production & Inventory Management* (Ohio: South-Western Publishing Co)

[2] Indrajit E and Djokopranoto R 2006 *Perkembangan Integrasi Perencanaan, dari Material Requirement Planning (MRP) sampai ke Enterprise Resource Planning (ERP)*

[3] Gaspersz V 2004 *Production Planning and Inventory Control (PPIC): Berdasarkan Pendekatan Sistem Terintegrasi MRP II dan JIT Menuju Manufakturing 21* (Jakarta: PT Gramedia Pustaka Utama)

[4] Wignjosoebroto S 2006 *Pengantar Teknik dan Manajemen Industri* (Surabaya: Guna Widya)

[5] Sutalaksana I Z A and Ruhana 2006 *Teknik Perancangan Sistem Kerja* (Bandung: ITB Bandung)

[6] Handoko T H 1999 *Dasar-Dasar Manajemen Produksi dan Operasi* edisi pertama (Yogyakarta: BPFE)

[7] Suryadhini P 2014 *Modul Praktikum Perencanaan dan Pengendalian Produksi* (Jakarta)

[8] Kusuma H 2002 *Manajemen Produksi (Perancangan dan Pengendalian Produksi)* (Yogyakarta: Andi)

[9] Ishak A 2010 *Manajemen Operasi* (Yogyakarta: Graha Ilmu)

[10] Purba H H, Mukhlisin and Aisyah S 2018 Productivity Improvement Picking Order by Appropriate Method, Value Stream Mapping Analysis, and Storage Design: A Case Study in Automotive Part Center *Management and Production Engineering Review* **9** 71-81

[11] Purba H H, Fitra A and Nindiani A 2019 Control and Integration of Milk-Run Operation in Japanese Automotive Company in Indonesia *Management and Production Engineering Review* **10** 79-88

IOP Conf. Series: Materials Science and Engineering **885** (2020) 012030 doi:10.1088/1757-899X/885/1/012030

The implementation of preventive maintenance using machine damage analysis: a case study of power plant machine

Sri S Islam, Tika Lestari, Anisa Fitriani and Dilla Ayu Wardani

Supply Chain Management and Smart Manufacturing Research Unit, Faculty of Engineering and Technology, Sampoerna University, Jakarta Indonesia

Email: sri.islam@sampoernauniversity.ac.id

Abstract. Maintenance has an important role in supporting the manufacturing process. The irregular maintenance in the industry can cause downtime and discontinue the production process. The importance of maintenance can be seen in PT PLN Sektor Tello, one of the industries which supply the electricity in Makassar. Since the electricity contributes to the huge impact on the life's sector, there should be preventive maintenance to all of the machine components in order to ensure that the components can run well and reduce the impact. To analyze the PT PLN Sektor Tello machine's performance, the researchers use the Failure Mode and Effect Analysis (FMEA) to identify the type of failures and recommend the maintenance plan. The FMEA's result will be used to calculate the RPN (Risk Priority Number) for each failure and provide the recommendations that will be conducted at maintenance when the RPN's value is equal or more than 125. By using the FMEA, we found three types of machine failures, such as the component's leakage, the machine's knock, and the broken component.

1. Introduction

Maintenance is the process of correcting the failure and improve the performance of the system [1,2]. The maintenance is essential in the industry because of the needs of high-quality performance of automatization and mechanization to support the production process that has been increased through the years. If the machine does not run effectively, the production process will be discontinued immediately. Therefore, the implementation of maintenance of a machine before it is failure becomes a trend in the industry since it can reduce the cost needed for repairing and maintaining the quality of the final product [3].

According to [4], maintenance can be divided into four categories, which are reactive, preventive, predictive, and proactive maintenance. Reactive maintenance is the process of performing or correct the fault in the system after the failure exists. Different from reactive maintenance, preventive maintenance is the process to avoid the failure of the system before the failure exists by doing time-scheduled maintenance. Predictive maintenance is the process of showing the failure of a system before it stops production. Last but not least, proactive maintenance is the process where the maintenance expert does something to solve the problem. From those four types of maintenance, the preventive maintenance commonly applied in many processes since it can minimize the cost needed to

repair the system and expedite the production process by keeping the quality and effectiveness of the final product [5–7].

The analysis that will be used when applying the preventive maintenance in this study is Failure Mode and Effect Analysis (FMEA). FMEA is the method designed to identify the cause and effect of possible failure on the system for a given product or process [8,9]. The FMEA can be used to address the most and less serious concern about the failure and recommend users to do the corrective action appropriately. Thus, it also helps the manufacturers to define which type of maintenance action they will use to minimize the repair time along with its failure's occurrence.

Practically, the FMEA has been used in many aspects of life: healthiness, manufacturing, and safety. One of the applications in using FMEA in healthiness aspects is using the FMEA analysis to find the causes of the increasing number of NICU because of the Central line-associated bloodstream infections (CLABSIs) over two years periods [8–10]. In here, the FMEA methods are applied and resulted in decreasing the CLABSIs rate from 2.6 to 0.8 per 100-line days and increasing of quality improvement and safety in the NICU. Moreover, the FMEA also can be used in manufacturing to find out the possible failure of the Automated Storage/Retrieval Systems (AS/RS)'s the new design [8].

2. Research Method

This study utilizes preventive maintenance using Failure Mode and Effect Analysis (FMEA) to evaluate the maintenance action needed on PT.PLN Sektor Tello. PT. PLN Sektor Tello is a company that has an important role in supplying electricity that is equal to 197,61, in total. One of the machines that have an important role in producing electricity is a generator.

During the operation, the damage was found in several components of the diesel-electric power generation machine that resulted in considerable losses. Therefore, new preventive maintenance planning is needed to avoid further damage to the machine and to maximize the production process. In this study, the researcher applies quantitative methods—Company survey, observation, interview, and documentation analysis process. Data collection was conducted on (February 2019).

As shown in Figure 1, we use Failure Mode and Effect Analysis (FMEA) to identify the machine component that often experiences damage and also the causes. The outputs of these processes are about types of failures that potentially occur in the production process, the effect of the failure mode, and Risk Priority Number (RPN) in each failure. In FMEA conventional assessment, the RPN value was comprised of three elements, namely the severity of the failure affect (S), Probability of the occurrence of the failure mode (O), and the probability of failure detection (D) [11].

Risk Priority Number is calculated by the following formula [12]:

$$RPN = S \times O \times D \text{ ....(1)}$$

S = Total value agreement with the FMEA team about the effect of the unavailability of the spare part.
O = The frequency of the specific failure rate of production activity produces a form of waste.
D = Measurement of the ability to detect or control failures that might occur

Each element is evaluated by a scale of 1 to 10. Therefore, higher value is indicated the higher risk of the failure mode occurrence [11]. In this research, [12] the maximum standard to propose the maintenance action is on ≥ 125, thus, determined based on the multiplication of the mean value of each criterion (Severity, occurrence, and detection) and in accordance with the agreement of the relevant company.

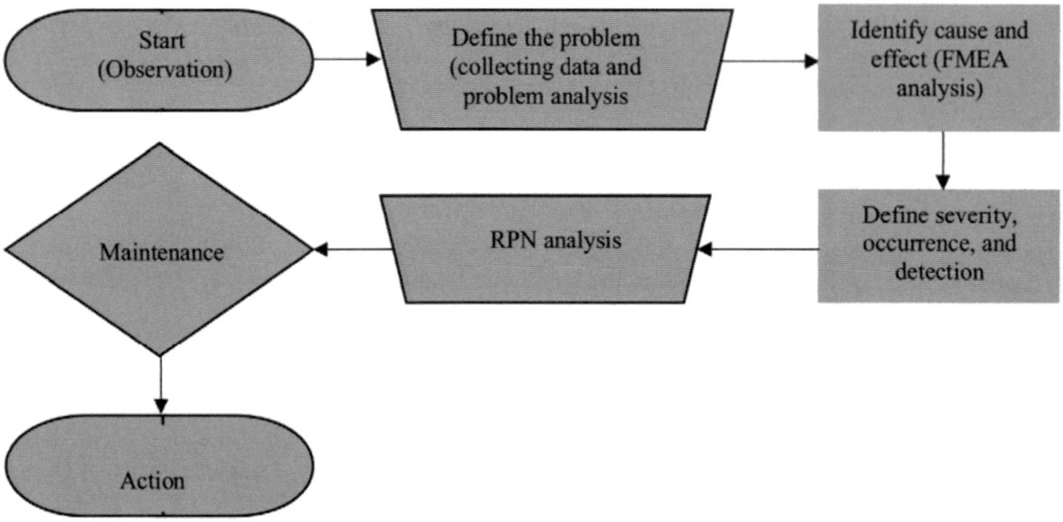

Figure 1. Research methodology.

3. Result and Discussion

Based on a case study we conducted at PT PLN Tello Sector, we found that some machines in the PLTD experienced a failure during operation and needed preventive maintenance planning to prevent those failures. Before planning corrective actions, an analysis of the types of failures that often occur along with the effects of failures is carried out using FMEA, as shown in Table 1.

Table 1. FMEA analysis.

Item	Failure Mode	Effect Analysis	Cause of Failure	Observation	S	O	D	RPN	Proposed Action
SW Diesel	A fan JCW radiator element is leaking	Stop operating	The clutch has worn	Visual observation by the operator	7	6	3	126	Radiator element should be replaced
	Hard knock onto the cylinder	Stop operating	Failure in the lubrication process	Visual observation by the operator	8	6	3	144	Adjust the Clearance inhaust and exhaust valve
	The CVS lubricant pipe on the outlet of the engine is leaking	Stop operating	Leakage on CVS lubricant pipe packaging	Visual observation by the operator	8	6	3	144	Replace the O-ring
	Cooling water jacket pump coupler is broken	The process is diverted into another pump	Cooling water jacket pump coupler is broken	Visual observation by the operator	6	5	4	120	No change
	Leakage on quick lane lubrication	Stop operating	O-ring for CVS valve in broken	Visual observation by the operator	8	6	3	144	Replace the O-ring

	The fuel oil booster pump does not operate.	Stop operating	Fuel filter does not work properly	Visual observation by the operator	8	5	3	120	No change
	Steam leaking from manhole boiler	The operation of MFO is interrupted due to slow steam production	Low pressure	Visual observation by the operator	8	3	3	72	No change
Mitsubishi	Rocker arm lubricant oil pressure is low	Stop operating	Filtered element is dirty	Parameter control	9	4	4	144	Clean the oil filter
	Intercooling radiator fan belt strap comes off	Still operating	Intercooling radiator fan belt strap comes off	Visual observation by the operator	6	4	4	96	No change

Based on the FMEA table above, we can conclude that there are five engine components that must be repaired. The five elements are the leakage of fan JCW radiator, the hard knock of the cylinder, the leakage of the CVS lubricant pipe on the outlet of the engine, Leakage on quick lane lubrication, and the low pressure of rocker arm lubricant oil. All of those components have RPN is more than 126 and need a component's replacement.

4. Conclusion

The objectives of this research are to identify the failures mode and effects of the machine and to recommend an effective maintenance action for PT.PLN Sektor Tello, specifically on SW Diesel and Mitsubishi generator. Based on the above findings, the writers conclude that there is five engine component that must be repaired. Thus components are the leakage of fan JCW radiator, the hard knock of the cylinder, the leakage of the CVS lubricant pipe on the outlet of the engine, Leakage on quick lane lubrication, and the low pressure of rocker arm lubricant oil. As a result of our findings, we recommend PT PLN Sektor Tello to applied 125 work hour (P1) of the weekly maintenance schedule for the hard knock of the cylinder and Leakage on quick lane lubrication, 250 work hours (P2) of the mid-month maintenance schedule for the leakage of fan JCW radiator and CVS lubricant pipe packing, then 1500 work hours (P4) of maintenance for rocket arm lubricant oil pressure, or quarterly maintenance schedule. Acknowledgment

5. Acknowledgment

We are thanking PT PLN Sektor Tello, who has given us a chance to observe the company's machine condition. Moreover, we also grateful to our heads of faculty for giving us the funding to finish our research.

Reference

[1] Institute of Electrical and Electronics Engineers 1990 *IEEE Standard Glossary of Software Engineering Terminology* vol 121990 (New York: Institute of Electrical and Electronics Engineers)

[2] Zahrotul H, Retno L N D and Lianny N 2019 The correlation between Goat Maintenance Management to the incidence of Gastrointestinal Parasite infections *Indian J. Public Heal. Res. Dev.* **10**

[3] Ben-Daya M and Duffuaa S O 1995 Maintenance and quality: the missing link *J. Qual. Maint. Eng.* **1** 20–6

[4] Ollila A and Malmipuro M 1999 Maintenance has a role in quality *TQM Mag.* **11** 17–21

[5] Basri E I, Razak I H A, Ab-Samat H and Kamaruddin S 2017 Preventive maintenance (PM) planning: a review *J. Qual. Maint. Eng.* **23** 114–43
[6] Ahmad R, Kamaruddin S, Azid I and Almanar I 2011 Maintenance management decision model for preventive maintenance strategy on production equipment *J. Ind. Eng. Int.* **7** 22–34
[7] Kay E 1976 The effectiveness of preventive maintenance *Int J Prod Res* **14** 329–44
[8] Chandonnet C J, Kahlon P S, Rachh P, DeGrazia M, DeWitt E C, Flaherty K A, Spigel N, Packard S, Casey D and Rachwal C 2013 Health care failure mode and effect analysis to reduce NICU line–associated bloodstream infections *Pediatrics* **131** e1961–9
[9] Mangeli M, Shahraki A and Saljooghi F H 2019 Improvement of risk assessment in the FMEA using nonlinear model, revised fuzzy TOPSIS, and support vector machine *Int J Ind Erg.* **69** 209–16
[10] Sutrisno A, Gunawan I and Tangkuman S 2015 Modified Failure Mode and Effect Analysis (FMEA) Model for Accessing the Risk of Maintenance Waste *Procedia Manuf* **4** 23–9
[11] Lo H-W, Liou J J H, Huang C-N and Chuang Y-C 2019 A novel failure mode and effect analysis model for machine tool risk analysis *Reliab Eng Syst Saf* **183** 173–83
[12] Thurnes C M, Zeihsel F, Visnepolschi S and Hallfell F 2015 Using TRIZ to Invent Failures – Concept and Application to go Beyond Traditional FMEA *Procedia Eng* **131** 426–50

Decision support system for selection of tourism in Tana Toraja using technique for order preference method by similarity to ideal solution (topsis)

Anton Topadang and Agus Triyono

Information Technology Department, Politeknik Negeri Samarinda

Email: antontpd@gmail.com

Abstract. Tana Toraja is one of the favorite tourist destinations in South Sulawesi. In Tana Toraja, there are many tourist sites that offer various advantages each. This can make tourists confused in determining the right tourist location to visit. Therefore we need a system that can provide the right information in determining the exact location of a tourist destination. This study aims to make it easy for prospective visitors to choose tourism objects in Tana Toraja that they wish. The method used in this study is the TOPSIS and fuzzy methods that can provide an alternative ranking of tourist attractions. The result of this Decision Support System is to produce recommendations for tourist attractions that can be visited by tourists.

1. Introduction

Tana Toraja is one of the districts in South Sulawesi, which is geographically located to the north of the island of Sulawesi. Tana Toraja is a tourist destination that is in high demand by tourists, both domestic and foreign. The tourist attractions offered are increasingly diverse such as natural landscape tours, traditional cemeteries, traditional houses, custom signs 'tuka, solo signs', traditional markets, culinary, waterfalls, and so on. To be able to assist tourists in determining the desired tourist location in accordance with their preferences and priorities, a decision support system will be able to assist in the decision-making stage. Decision support system (SPK) is one of the tools that can be used in decision making for decision-makers [1]. Decisions offered by decision support systems tend to be fast and quantitatively the best choices based on the level of importance/weight of the criteria given by management as the decision-maker [2]. With the help of a decision support system, decision making that is quite complex can be shortened. Tana Toraja is one of the districts in South Sulawesi, which is geographically located to the north of the island of Sulawesi. Tana Toraja is a tourist destination that is in high demand by tourists, both domestic and foreign.

The tourist attractions offered are increasingly diverse such as natural landscape tours, traditional cemeteries, traditional houses, custom signs 'tuka, solo signs', traditional markets, culinary, waterfalls, and so on. To be able to assist tourists in determining the desired tourist location in accordance with their preferences and priorities, a decision support system will be able to assist in the decision-making stage. Decision support system (SPK) is one of the tools that can be used in decision making for decision-makers. Decisions offered by decision support systems tend to be fast and quantitatively the best choices based on the level of importance/weight of the criteria given by management as the decision-maker. With the help of a decision support system, decision making that is quite complex can be shortened. The determination of tourist sites has been done in several studies, both using decision support systems and user-profiles. They are building a decision support system for website-based tourism for the Nigerian region using artificial intelligence. The advantage of this SPK is that it helps tourists in determining their planned visit in a shorter time and reduces travel costs [3].

SPK for tourism Mahamud, Masron, and Mohamed (2013) [4], which were built in combination using Geographic Information Systems (GIS), where SPK can display maps of tourist areas. This type of SPK can display data of the closest locations, public places around, and the route to a certain location. In this SPK, there are no input criteria or weights according to tourist conditions. The use of SPK, which received weighting criteria, was carried out by Taluay et al. (2015) [5], using the Analytic Hierarchy Process (AHP) for tourism in Talaud Island. SPK, which is built in this system, does not manage vehicle/attraction data provided by tourist sites. The basis of a decision support system is the characterization carried out by an interactive computer system that helps decision making using the help of data and models in overcoming unstructured problems. This system is designed to assist in making decisions, starting with identifying problems, taking relevant data, approaching decision making, and evaluating choices interactively [6].

2. Technique For Order Preference By Similarity To Ideal Solution (Topsis)

Technique For Order Preference by Similarity to Ideal Solution (TOPSIS) is a method that was first introduced in 1981. This method provides a hypothesis related to two artificial alternatives, namely alternatives for optimal solutions and alternatives for non-optimal solutions. The basis of this rationale comes from the concept of compromise, where alternative solutions have the shortest distance to the optimal solution and have the longest distance from the non-optimal solution Lokare and Jadhav (2016) [7]. The steps used in finding solutions with the TOPSIS method in general done by passing the following stages :

a. Determine the criteria used as points that determine whether an option is the best solution in a case. Criteria can be any factors that influence decision making.
b. Weighting criteria. Where at this stage, the criteria that influence decision making are given a weighting value based on the presentation of these criteria in determining decision making.
c. Building a normalized decision matrix. Elements of this normalization are shown in Equation (1) below:

$$r_{ij} = \frac{x_{ij}}{\sqrt{\sum_{i=1}^{m} x^2_{ij}}} \quad (1)$$

Where : rij = the result of normalization of decision matrix R; i = 1,2,3, ..., m; j = 1,2,3, ... m;

d. Construct the weighted normalized decision matrix shown in Equation (2) below:

$$V = \begin{bmatrix} w_1 r_{11} & \cdots & w_n r_{1n} \\ \vdots & \ddots & \vdots \\ w_1 r_{m1} & \cdots & w_n r_{mn} \end{bmatrix} \quad (2)$$

e. Determine the negative ideal solution and the positive ideal solution shown in Equation (3) and Equation (4) below:

+= {(max Vij)(min Vij| j '),
i = 1,2,3,…,m} = { 1+, 2+,…,+} (3)

−= {(max Vij)(min Vij| j '),
i = 1,2,3,…,m} = { 1−, 2−,…,−} (4)

f. Calculate the separation.

It is a way to calculate or measure the distance between alternative solutions with positive ideal solutions and negative ideal solutions shown in Equation (5) and equation (6) below:

$$S_i^+ = \sqrt{\sum_{j=1}^{n} (v_{ij} - v_j^+)^2} \quad (5)$$

$$S_i^- = \sqrt{\sum_{j=1}^{n} (v_{ij} - v_j^-)^2} \quad (6)$$

With i = 1,2,3,….,n;

g. Calculating the proximity relative to the optimal solution shown in Equation (7) below:

$$C_i^+ = \frac{S_i-}{S_i^- + S_i^+} \quad (7)$$

With 0<C<1 dan i=1,2,3,…,n;

h. Provides alternative ranking that can be sorted based on the value of Ci + from the largest to the smallest.

3. Research Methodology

The stages of this research are initial research, determination of criteria and alternatives, determination of matrices and weight calculations, normalization of decision matrices, the weighting of normalized decision matrices, determination of positive ideal solutions and negative ideal solutions, calculation of distances to positive ideal solutions and negative ideal solutions, weight calculation preferences for each alternative, and determine the ranking and recommendations of the best attractions to visit.

4. Result And Discussion

4.1. Analysis and identification of system requirements

The steps of analyzing and identifying system requirements are:

4.1.1. Determine alternatives. There are 15 tourist attractions as an alternative in this study are Arum Jeram, Bebo 'Tumbang Datu, Buntu Burake, Buntu Kalando, Buntu Tondon, Kambira, Lemo, Makula', Pango-Pango, Sarambu Assing, Sassa ', Suaya, Tampangallo, Tilangan' and Tumakke.

4.1.2. Determine the criteria. There are four criteria used to determine tourist destinations, namely: C1: Cost, C2: Distance, C3: Transportation, and C4: Facilities.

4.1.3. Table of preference. The following is a table of preferences or importance of each criterion as in table 1 as follows:

Table 1. Preference or importance of each criterion.

Price	Distance	Transportation	Facilities	Level of Importance
< 50,000	<5	Very less	Very less	1
51,000 – 100,000	5-10	Less	Less	2
110,000 – 150,000	11-15	Enough	Enough	3
151,000 – 200,000	16-20	Well	Well	4
>200,000	>21	Very good	Very good	5

4.1.4. Criteria weights. Decision making gives weight preferences based on the importance of the following criteria: W = (20, 30, 30.20) as can be seen in Table 2 below:

Table 2. Criteria of weights.

Criteria	Type	Weight (w)	Weight(w)/100
Price	cost	20	0.2
Distance	cost	30	0.3
Transportation	benefit	30	0.3
Amenities	benefit	20	0.2

4.2. Formation of preference weights and decision matrices

Based on alternative match rating data obtained from the initial selection results as in table 3.

Table 3. Alternative data.

Name of Tourist Object	Id	Price	Distance	Transportation	Facilities
Arum jeram	A1	450,000	20	less	enough
Bebo' tumbang datu	A2	70,000	8	good	good
Buntu burake	A3	65,000	2	good	good
Buntu kalando	A4	70,000	7	good	good
Buntu tondon	A5	20,000	1	good	good
Kambira	A6	75,000	11	good	good
Lemo	A7	50,000	9	good	good
Makula'	A8	60,000	10	good	good
Pango-pango	A9	110,000	7	enough	very good
Sarambu assing	A10	130,000	35	good	good
Sassa'	A11	70,000	9	good	good
Suaya	A12	75,000	12	good	good
Tampangallo	A13	75,000	9	good	good
Tilanga'	A14	60,000	13	good	good
Tumakke	A15	65,000	12	enough	good

From table 1 and table 3, a matching rating can be obtained from each alternative to each criterion outlined in table 4 below.

Table 4. Match Rating Table of each alternative for each criterion.

Id	Price	Distance	Transportation	Facilities
A1	5	4	3	3
A2	2	2	4	4
A3	2	1	4	4
A4	2	2	4	4
A5	4	1	4	4
A6	2	3	4	4
A7	2	2	4	4
A8	2	2	4	4
A9	3	2	3	5
A10	3	5	4	4
A11	2	2	4	4
A12	2	3	4	4
A13	2	2	4	4
A14	2	3	4	4
A15	2	3	3	4

4.3. *Completion using the TOPSIS method*

4.3.1. Make a normalized decision matrix. Calculate the normalized decision matrix weighted by Cost, Distance, Transportation, and Facilities using the formula in the following equation 2: The results of the calculation of matrix normalization can be seen in the following table 5.

Table 5. Normalized matrix.

Alternative	C1	C2	C3	C4
Normalization	R1	R2	R3	R4
A1	0.4927	0.3867	0.2027	0.1928
A2	0.1971	0.1933	0.2703	0.2571
A3	0.1971	0.0967	0.2703	0.2571
A4	0.1971	0.1933	0.2703	0.2571
A5	0.3941	0.0967	0.2703	0.2571
A6	0.1971	0.2900	0.2703	0.2571
A7	0.1971	0.1933	0.2703	0.2571
A8	0.1971	0.1933	0.2703	0.2571
A9	0.2956	0.1933	0.2027	0.3214
A10	0.2956	0.4834	0.2703	0.2571
A11	0.1971	0.1933	0.2703	0.2571
A12	0.1971	0.2900	0.2703	0.2571
A13	0.1971	0.1933	0.2703	0.2571
A14	0.1971	0.2900	0.2703	0.2571
A15	0.1971	0.2900	0.2027	0.2571

4.3.2. Make the weighted normalized decision matrix. After making the Normalized Matrix, we make the weighted normalized decision matrix. Criteria weights at this stage come from the weighting carried out by the TOPSIS method, as in the following settlement:

Table 6. Weighted normalization data.

Alternative	C1	C2	C3	C4
Weighted Criteria	Y1	Y2	Y3	Y4
A1	0.0985	0.1160	0.0608	0.0579
A2	0.0394	0.0580	0.0811	0.0771
A3	0.0394	0.0290	0.0811	0.0771
A4	0.0394	0.0580	0.0811	0.0771
A5	0.0788	0.0290	0.0811	0.0771
A6	0.0394	0.0870	0.0811	0.0771
A7	0.0394	0.0580	0.0811	0.0771
A8	0.0394	0.0580	0.0811	0.0771
A9	0.0591	0.0580	0.0608	0.0964
A10	0.0591	0.1450	0.0811	0.0771
A11	0.0394	0.0580	0.0811	0.0771
A12	0.0394	0.0870	0.0811	0.0771
A13	0.0394	0.0580	0.0811	0.0771
A14	0.0394	0.0870	0.0811	0.0771
A15	0.0394	0.0870	0.0608	0.0771

4.3.3. Determine the positive ideal solution matrix (A +) and the negative ideal solution matrix (A-)
Determine the positive and negative ideal solution matrix that is by finding the value of each of these criteria included in the cost or profit category, to calculate the positive ideal solution matrix (+):

a. The Positive Ideal Solution Matrix (A^+)

$y1+=min$ {0,1696;0,0264;0,0245;0,0264;0,0075;0,0283;0,0188;0,02260,0415;0,0490;0,0264;0,02 83;0,0283;0,0226;0,024} = 0,0394

$y2+=min$ {0,1152;0,0461;0,0115;0,0403;0,0058;0,0634;0,0518;0,05760,0403;0,2016;0,0518;0,06 91;0,0518;0,0749;0,0691} = 0,0290

$y3+=max$ {0,0410;0,0820;0,0820;0,0820;0,0820;0,0820;0,0820;0,08200,0615;0,0820;0,0820;0,08 20;0,0820;0,0820;0,0615} = 0,0811

$y4+=max$ {0,0579;0,0771;0,0771;0,0771;0,0771;0,0771;0,0771;0,02260,0964;0,0771;0,0771;0,07 71;0,0771;0,0771;0,0771} = 0,0964

b. The Negative Ideal Solution Matrix (A^-)

$y1-=max$ {0,1696;0,0264;0,0245;0,0264;0,0075;0,0283;0,0188;0,02260,0415;0,0490;0,026 4;0,0283;0,0283;0,0226;0,024} = 0,0985

$y2-=max$ {0,1152;0,0461;0,0115;0,0403;0,0058;0,0634;0,0518;0,05760,0403;0,2016;0,0518;0,06 91;0,0518;0,0749;0,0691} = 0,1450

$y3-=min$ {0,0410;0,0820;0,0820;0,0820;0,0820;0,0820;0,0820;0,08200,0615;0,0820;0,0820;0,08 20;0,0820;0,0820;0,0615} = 0,0608

$y4{-}{=}min\{0{,}0579;0{,}0771;0{,}0771;0{,}0771;0{,}0771;0{,}0771;0{,}0771;0{,}02260{,}0964;0{,}0771;0{,}0771;0{,}07$
7 1;0,0771;0,0771;0,0771} = 0,0579

4.3.4. Find the distance between the weighted values of each alternative to the positive and negative ideal solutions. Determination of the distance value of a positive ideal solution is obtained from the calculation of the search for the distance value between the weighted values, by using equation three above, so the results can be seen in the following table 7.

Table 7. Matrix of positive and negative ideal solutions.

Alternative distance	Positive (+)	Negative (-)	(D+) + (D-)
A1	0.1139	0.0290	0.1429
A2	0.0348	0.1088	0.1437
A3	0.0193	0.1332	0.1525
A4	0.0348	0.1088	0.1437
A5	0.0439	0.1210	0.1648
A6	0.0611	0.0874	0.1485
A7	0.0348	0.1088	0.1437
A8	0.0348	0.1088	0.1437
A9	0.0405	0.1030	0..1435
A10	0.1192	0.0483	0.1676
A11	0.0348	0.1088	0.1437
A12	0.0611	0.0874	0.1485
A13	0.0347	0.1088	0.1437
A14	0.0611	0.0874	0.1485
A15	0.0644	0.0850	0.1494

4.3.5. It is finding the preference value for each alternative. The preference value for each alternative is obtained from the results of the process of calculating the proximity of each alternative to a positive ideal solution and a negative ideal solution so that it is found that the alternative that has the greatest value is the best alternative.

$$V1 = \frac{0{,}0864}{0{,}0864 + 0{,}2035} = 0{,}2980$$
$$V2 = \frac{0{,}2162}{0{,}2162 + 0{,}0485} = 0{,}8168$$

....

$$V15 = \frac{0{,}1985}{0{,}1985 + 0{,}0714} = 0{,}7355$$

Furthermore, it can be seen in the following table 8.

Table 8. Table Preference values for each alternative.

Name of Tourist Object	Alternative	V
Arum jeram	A1	0.2030
Bebo' tumbang datu	A2	0.7576
Buntu burake	A3	0.8735

Buntu kalando	A4	0.7576
Buntu tondon	A5	0.7338
Kambira	A6	0.5885
Lemo	A7	0.7576
Makula’	A8	0.7576
Pango-pango	A9	0.7178
Sarambu assing	A10	0.2884
Sassa’	A11	0.7576
Suaya	A12	0.5885
Tampangallo	A13	0.7576
Tilanga’	A14	0.5885
Tumakke	A15	0.5691

They are then sorted according to Ranking from the largest to the smallest, as in the following table 9.

Table 9. Ranking results with TOPSIS.

Name of Tourist Object	Final	Ranking
Buntu burake	0.8735	1
Bebo’ tumbang datu	0.7576	2
Buntu kalando	0.7576	3
Lemo	0.7576	4
Makula’	0.7576	5
Sassa’	0.7576	6
Tampangallo	0.7576	7
Buntu tondon	0.7338	8
Pango-pango	0.7178	9
Kambira	0.5885	10
Suaya	0.5885	11
Tilanga’	0.5885	12
Tumakke	0.5691	13
Sarambu assing	0.2884	14
Arum jeram	0.203	15

5. Conclusion

Based on the results of the research that has been done, the conclusion can be drawn that is:

1) The decision support system for the selection of tourist attractions in Tana Toraja is built through several stages, namely: a study of literature, determine alternatives and criteria, determine matrices and weights, Normalize Decision Matrices, Weighting Normalized Decision Matrices, Determination of Positive Ideal Solutions and Negative Ideal Solutions, Distance Calculation Towards Positive Ideal Solution and Negative Ideal Solution, Calculation Weight Preference for Each Alternative and Determine Ranking of Tourist Attraction.
2) Testing with the TOPSIS method was successfully carried out by giving results in the form of ranking tourist sites that can be recommendations for potential tourists to make a decision to visit.
3) Analysis of calculations using the TOPSIS method can be used as a reference in developing into an application that can be accessed via the internet.

References

[1] Effendy F 2020 Mobile Based Decision Support System of Supplier Evaluation *Int. J. Interact. Mob. Technol.* **14**

[2] Retrialisca F, Effendi Y A and Nuzulita N 2019 Decision Support System and Recommendation on SBMPTN Try-Out with Analytic Hierarchy Process (AHP) *International Conference on Computer Science, Information Technology, and Electri* (Jember)

[3] Asafe Y N, Enaholo A E, Bolaji A and Olubukola O 2013 Web-based Expert Decision Support System for Tourism Destination Management in Nigeria *Int. J. Adv. Res. Artif. Intell* **2** 59-6

[4] Mahamud S, Masron T and Mohamed B 2013 Graphical User Interface For Tourism Decision Support System (TDSS) *Proceedings of International Conference on Tourism Development* pp 267- 277

[5] Taluay H R, Seminar K B and Monintja D R O 2015 Development of Web-Based Tourism Decision Support System in Talaud Island Regency *Int. J. Inf. Technol. Bus. Manag.* **39** 37-45

[6] Ginting G, Fadlina M, Siahaan A P U and Rahim R 2017 Technical Approach of TOPSIS in Decision Making International Journal of Recent Trends in Engineering and Research **3** 58–64

[7] Lokare V T and Jadhav P M 2016 Using the AHP and TOPSIS methods for decision making in best course selection after HSC 2016 *International Conference on Computer Communication and Informatics ICCCI 2016*

IOP Conf. Series: Materials Science and Engineering **885** (2020) 012032 doi:10.1088/1757-899X/885/1/012032

The Link Model of ERP-Usage, Absorptive-Capacity, and Task-Technology-Fit, to Task-Innovation in ERP User

A Velahyati Baharuddin

Department of Agroindustrial Engineering, Polytechnic ATI of Makassar, Makassar 90211, Indonesia

Email: Andi.velahyati@atim.ac.id

Abstract. An ERP system contains millions of information that is useful for its users. The use of an ERP system in an organization is usually mandatory, and users are required to use and understand the workings of an ERP system in completing their routine tasks. As soon as employees complete training classes and ERP systems are used routinely and intensely in their daily work (Go-Live), employees are required to obtain new knowledge or skills from the ERP system that has just been installed and complete their work differently by innovate their way in doing tasks. However, there have been many cases of the ERP system that led to the failure of users to improve their work habits. The purpose of this research is to develop a model that can explain the influence of absorptive knowledge capacity, the 'fit' of the technology with the work, and the decision-making process to the task innovation of ERP users. The model was built by adapting the model Task-Technology Fit and Absorptive Capacity of previous studies related to MIS. The collection of the data by a survey conducted on ten organizations ERP system users in Java and Sulawesi. Four hundred questionnaires distributed to 10 organizations ERP users engaged in mining, food processing, banking, public services, telecommunications, and government offices. Three hundred forty-five questionnaires were returned, with 342 questionnaires able to be processed. Data were analyzed using multivariate statistical techniques Structural Equation Modelling (SEM)-PLS. This research has developed a potential model in the role of studying the effects of ERP Usage, Absorptive Capacity, and Task-Technology Fit to Task Innovation in the context of enterprise applications at the level of individual analysis. Reasons and contributing factors were described in the final section of this study.

1. Introduction

Enterprise Resource Planning (ERP) is an integrated business solution that integrates all business functions of an organization into one functional system. An ERP system contains millions of information that is potentially useful to a user. The use of ERP systems in an organization is usually mandatory, where workers are required to use and understand the workings of ERP systems in completing their routine tasks. As soon as employees complete the training class and the ERP system starts to be used routinely and intensely in their daily work (Go-Live), employees are required to be able to take new knowledge or skills from the ERP system that has just been installed and do their work in a way different and better than before.

Research-research on MIS has used the use of technology and individual performance as a measure of the success of the impact of information system technology. The right measure in measuring it is to find out how much technology affects individual users [1].

Task-Technology Fit (TTF) is used to measure how far an information system technology fits into work and work processes so that it can help the work of its users. TTF was first used in the Task-to-Productivity Chain model by Goodhue and Thompson (1995) [2]. Goodhue and Thompson (1995) examined individual users of various types of technology to measure technology use and individual performance [2]. The influence of 'Fit' itself has been investigated before as in DeLone and McLean (1992), Benbasat et al. (1986) and Cooper and Zmud (1990), but new research by Goodhue and Thompson (1995) underlines the importance of compatibility between technology and work [3-5]. Jain and Kanungo (2006) adapted TTF to predict IS Use and its effect on individual performance, they examined the impact of office applications at the individual level [6].

The use of IT in supporting decision-making systems has been widely studied before, as in the research of Torkzadeh and Doll (1998), who examined how far technology is used in an organization for Decision Support, Work Integration, and Customer Service. Furthermore, in the research of Torkzadeh and Doll (1999), they then suggested four factors in measuring how far the impact of information system technology is Task Productivity, Task Innovation, Customer Satisfaction, and Management Control [7].

ERP system is seen as one of the company's knowledge providers, and to realize the existence of knowledge contained in the ERP system is influenced by the ability of the absorption of knowledge / Absorptive Capacity of the ERP users themselves. Absorptive Capacity in information systems research and management has been trusted as one of the important factors that influence the performance of individuals in understanding information system technology. Researches that use Absorptive Capacity in the context of information systems technology adapt from studies on Absorptive Capacity in the context of organizational management. The earliest research on Absorptive Capacity was research by Cohen and Levinthal (1990) [8]. Zahra and George (2002) then conceptualized the Absorptive Capacity into four dimensions, namely, the ability to recognize, understand, change, and use knowledge and information to achieve organizational goals [9]. Kositanurit et al. (2006), Park et al. (2007), Xu and Ma (2008), in their research, involved Absorptive Capacity in predicting the work productivity of an information system technology user [10-12]. Deng et al. (2008) used Absorptive Capacity in measuring IT Usage, Task Innovation, and Task Productivity of individuals using Computer-Aided Manufacturing technology [13].

This research seeks to combine Task-Technology Fit and Absorptive Capacity in measuring the use of ERP (ERP Usage) systems for decision making and their effect on innovation in working users (Task Innovation).

2. Methods

2.1. Questionnaire Design

The measuring instrument for the research model used in this study was a closed questionnaire. The target respondents of this study were employees of ERP system users who use ERP systems in their routine work. The questionnaire in this study used a seven-point Likert scale and adopted the questionnaire instrument in the research of Torkzadeh and Doll (1998), Jain and Kanungo (2006), Deng et al. (2008), Goodhue and Thompson (1995), and Zahra and George (2002) [6, 7, 9,13].

The questionnaire was distributed in two ways, namely by visiting the company directly and through an online survey using the Google Docs application and via email. The respondents of this study came from five ERP user companies. The ERP user company consists of five companies engaged in mining, food processing, banking, telecommunications, public services, and five government offices. The selection of the company where the research is based on the period of use of the ERP system in the company. The company determined as the object of this research is a company that studies are workers who use ERP systems in their daily routine and have used ERP systems for at least one year. Employees, as respondents in this study, are not limited to certain occupational levels or departments/divisions.

IOP Conf. Series: Materials Science and Engineering **885** (2020) 012032 doi:10.1088/1757-899X/885/1/012032

2.2. Sampling

This study uses a purposive sampling technique because it is adjusted to the research objectives of ERP system users like respondents. Sekaran (2003) states that the determination of sample sizes according to the rules of thumb is more than 30 and less than 500 [14]. Hair et al. (1998) recommend a sample size between 100 to 400 [15]. According to Wijanto (2008), to use the Weighted Least Square (WLS) estimate, it is determined that the rule of thumb of the sample is at least ten times the number of variables observed [16]. In this study, the number of variables observed was 29 items, so if using the WLS estimation method, the number of samples needed was 290 data. Demographic data of the respondents showed 69.3% were male, 32% of respondents in this study were 21-30 years old, and 30.7% were 31-40 years old. As many as 35.7% of respondents have used ERP systems for 2-5 years, and 42.7% of respondents have used ERP for more than five years.

3. Results

Data is processed using path modeling with partial least square (PLS) approach with the help of SmartPLS software. The measurement model was tested by testing the AVE, Cronbach's Alpha, and Discriminant Validity values. Table 1 shows the reliability test values of this research model.

Table 1. Results of reliability and discriminant validity.

	AVE	Cronbach's Alpha	Composite Reliability	ACAP	ERP Use	TI	TTF
ACAP	0.50	0.87	0.90	0.71	-	-	-
ERP Use	0.68	0.81	0.90	0.50	0.83	-	-
TI	0.82	0.79	0.90	0.52	0.44	0.91	-
TTF	0.52	0.82	0.88	0.63	0.46	0.39	0.72

The AVE value of all variables shows a value> 0.50, which means the variable can explain some of the average variances of the indicator. Cronbach's alpha values for all variables show more than the minimum value of 0.70. In addition, the value of composite reliability obtained from data processing shows that all latent variables have composite reliability values> 0.6. This shows that the indicators in this study are reliable in measuring each construct.

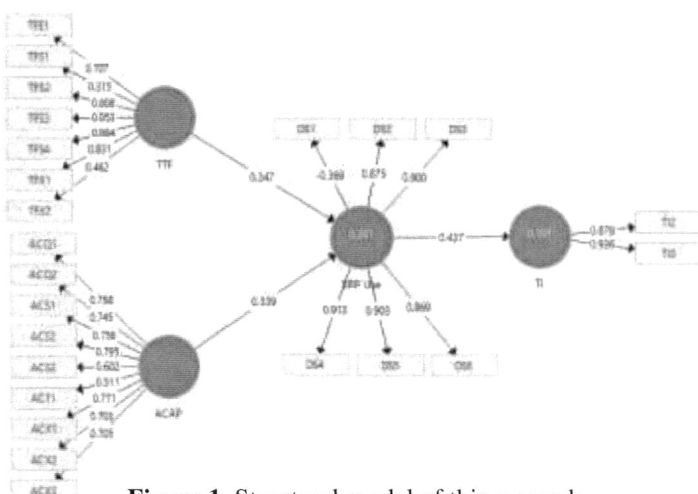

Figure 1. Structural model of this research.

The structural model evaluation is done by evaluating the goodness of fit model. The goodness of Fit value is obtained by calculating the square root value from the multiplication of the R2 value with the AVE value [17]. The structural goodness of fit value of this model is 0.40, which is greater than 0.36. This value is considered to meet the requirements for social and behavioral sciences studies [18]. It is performed by the bootstrap procedure to get the path coefficient value from the structural model. The results of the estimated parameters are in Figure 1.

Of the two variables that affect ERP use, the Absorptive capacity variable is more influential than task technology fit. While the influence of ERP use on task innovation amounted to 0.437, which means ERP use contributed 43.7% of the variance of task innovation.

4. Discussion

By evaluating the structural model, it can be interpreted that: 1) by increasing the value of task technology fit by one unit, it will increase ERP use by 0.247. This means, the more appropriate a job is to the technology used by a worker, it will also increase the level of use of the employee's ERP system in completing his work. 2) by increasing the value of absorptive capacity by one unit, it will increase ERP use by 0.3397. This means, the higher the ability of an employee to obtain knowledge from the ERP system he uses, the higher the level of ERP system usage in doing his daily work. 3) by increasing the value of ERP used by one unit, it will increase task innovation by 0.437. This means, the more often a worker uses an ERP system to complete their work, the more they will use their ability to innovate to complete their tasks.

5. Conclusion

This research has been able to prove the effect of task technology fit and absorptive capacity together influencing ERP use and its effect on the task innovation of an ERP system user. However, this study still has some shortcomings, including the lack of latent variables included in the model that can predict ERP use. In some studies, other variables have been investigated that might affect ERP use and task innovation, even in different technological contexts. Due to the limited time and cost in this study, these variables were not examined. Future studies are expected to be able to design a more comprehensive model in learning the task of innovation.

References

[1] Chang H, Chou H, Yin C and Lin C 2011 *ERP* Post-Implementation Learning, ERP Usage and Individual Performance Impact *Pacific Asia Conference on Information Systems (PACIS)* (Brisbane: AIS Electronic Library (AISeL))
[2] Goodhue D and Thompson R 1995 Task-Technology Fit and Individual Performance *MIS Quarterly* **19** 213-236
[3] DeLone W and McLean E 2003 The DeLone and McLean Model of Information Systems Success: A Ten Year Update *Journal of Management Information Systems* **19** 9-30
[4] Benbasat I, Dexter A and Todd P 1986 An Experimental Program Investigating Color-Enhanced and Graphical Information Presentation: An Integration of the Findings *Communications of the ACM* **29** 1094-1105
[5] Cooper R and Zmud R 1990 Information Technology Implementation Research: A Technological Diffusion Approach *Management Science* **36** 123-139
[6] Jain, V and Kanungo S 2006 *Performance Impacts of IS at The Individual Level: An Integrated Model. European Conference on Information Systems* (Gothenburg: AIS Electronic Library)
[7] Torkzadeh G and Doll W 1999 The Development of a Tool for Measuring the Perceived Impact of Information Technology on Work *Omega Int. J. Manag. Sci*, **27** 327-339
[8] Cohen, W and Levinthal D 1990 Absorptive Capacity: A New Perspective on Learning and

Innovation *Administrative Science Quarterly* **35** 128-152
[9] Zahra S and George G 2002 Absorptive *Capacity: A Review, Reconceptualization and Extension The* Academy of Management Review **27** 185-203
[10] Kositanurit B, Ngwenyama O and Osei-Bryson K 2006 An Exploration of Factors that Impact Individual Performance in an ERP Environment: An Analysis Using Multiple Analytical Techniques *Eur. J. Inf. Syst.* **15** 556-568
[11] Park J, Suh H and Yang H 2007 Perceived Absorptive Capacity of Individual Users in Performance of ERP Usage: The Case for Korean Firms *Information & Management* **44** 300-312
[12] Xu Q and Ma Q 2008 Determinants of ERP Implementation Knowledge Transfer *Information & Management* **45** 528-539
[13] Deng X, Doll, W and Cao M 2008 Exploring the Absorptive Capacity to Innovation/Productivity Link for Individual Engineers Engaged in IT Enabled Work *Information & Management* **45** 75-87
[14] Sekaran U 2003 Research Method for Business 4th ed (New York: John Wiley & Sons, Inc)
[15] Hair J, Anderson R, Tatham R and Black W 1998 *Multivariate Data Analysis* (New Jersey, USA: Prentice-Hall)
[16] Wijanto S 2008 *Structural Equation Modeling dengan LISREL 8.8: Konsep dan Tutorial* (Yogyakarta: Graha Ilmu)
[17] Ringle M C, Silva D, Bido D 2014 Structural Equation Modeling With The Smart PLS *Revista Brasileira de Marketing* **13** 56 -73
[18] Wetzels M, Odekerken-Schröder G and Oppen C V 2009 Using PLS path modeling for assessing hierarchical construct models: guidelines and empirical illustration *MIS Quarterly* **33** 177-195

Ergonomic risk analysis of lecture chairs at the engineering faculty, Hasanuddin University

Mulyadi[1], Andi Nurwahidah[2], Dedi Nismar Satria[1]

[1]Industrial Engineering, Hasanuddin University
[2]Industrial Engineering, Polytechnic ATI Makassar

Email: Mulyadihambali@ymail.com

Abstract. Uncomfortable college chairs will cause health problems such as low back pain, etc. In addition to health problems will also affect student productivity and enthusiasm for learning. This study will identify the risk of ergonomics from lecture chairs at the Faculty of Engineering UNHAS and also identify and provide recommendations for examples of lecture chairs that should be used. This research will use several ergonomics methods such as REBA and nordic body map. Nordic body maps are used to identify parts of the body that feel pain when sitting in a college chair, while the REBA method is used to identify ergonomic risks when sitting in a college chair. The results of identification using the Nordic Body Map method found that several parts of the student's body were aching like the neck, back, waist, buttocks. The next research method is the Rapid Upper Limb Assessment, the identification results obtained all students get a RULA 7 value which indicates changes in body posture or work facilities as soon as possible. The end of this study is the design of ergonomic lecture chairs in accordance with the wishes of students

1. Introduction

Ergonomics is the science that focuses on utilizing information about human capabilities and capabilities in supporting a working system that makes people able to live and also works on a better system, namely by achieving the goals needed through effective, efficient, safe and also comfortable. Ergonomics is not only important for company work, but the student's sitting position also needs to be identified so that the student is comfortable in the classroom. Complaints about workers' body postures are very common, this is due to the lack of knowledge of workers about the dangers of working in a work environment that is not ergonomic. Musculoskeletal disorders (MSDs) or musculoskeletal disorders are disorders that cause symptoms of pain in various body locations such as the neck, shoulders, wrists, hips, knees, and heels [1]. WHO (World Health Organization) states that this disorder is caused by several risk factors such as individual factors, employment factors and psychosocial factors [1]. The job factor in question is awkward posture, static and repetitive movements, temperature and vibration. Psychosocial factors have a monotonous work performance, little social interaction, lack of work control. Individual factors are more to gender, age, level of education, lifestyle, the habit of exercise and tenure [1] besides the work postures of the workers. the posture of a student sitting while attending a lecture can also have a large impact in the future. this research will focus on researching the ergonomics risk of college chairs. The condition of the chair of the university's engineering faculty class currently still uses the Chitose lecture chair that discusses chairs and tables, so students are less comfortable to sit for long periods of time. Uncomfortable college chairs will cause health problems such as lowback pain, etc. In addition, it will increase student motivation and learning enthusiasm. This research will discuss the ergonomics problem of lecture chairs at the Faculty of Engineering UNHAS and also the approval and giving examples of lecture chairs that must be used.

This research has been conducted by several people although with slightly different research objectives such as research conducted by Hermawan Yuni [2] who conducted research on the development and analysis of engine operator chair ergonomics with reserve engineering and RULA methods, research by hermanto et al [3] on the proposed design of sizes on tables and chairs Fold ergonomic learning for tenements, research by Pamungkas Gina et al [4,5] on the relationship between chair dimensions and lower back pain complaints using numerical and statistical scale measurement methods, research by Octavia et al [5] on ergonomic design of chairs and laptop desks in Catholic University of Severyangan using the method of NBM and REBA, and many other studies. while this study wants to identify risks when using lecture chairs and design ergonomic lecture chairs in accordance with the wishes of students where the research methods used are NBM, RULA, VoC, and HoQ

2. Research Methods

2.1. Nordic Body Map
Nordic body map is a research method that is often used to identify painful limbs when using a facility or doing a job. This assessment categorizes complaints into 4 types such as not sick, a little sick, sick and very sick.

2.2. RULA (Rapid Upper Limb Assessment)
This RULA method has often been used to identify ergonomic risks from a facility or work activity. RULA identifies the upper part of the body while the REBA method is for the whole body.

2.3. Voice Of Customer
Voice of customer method is used to identify consumer desires for a product. the purpose of using this method as a basis for designing ergonomic lecture chairs and in accordance with the wishes of students

2.4. House Of Quality / Quality Function Deployment
The house of quality method or quality function deployment is used to design lecture chairs in accordance with the results of the voice of the customer and answer all the desires of consumers by incorporating all student suggestions into the product design to be made

3. Result and Discussion
Lecture chairs that are currently used in lectures at the engineering faculty of Hasanuddin University still use folding chairs that blend into the table. lecture chairs that exist today are still not ergonomic. chairs cannot be adjusted in height, tables are very small in size, and hard seat is often complained of by students. students use the facility for hours so they often complain of pain when using the facility. in this study, several stages of research will be conducted. Study respondents were 18 students who were randomly selected and had different body shapes and gender. This data collection is done by photographing the position of students when sitting in a lecture chair and doing activities, as usual, they do each lecture. The first method used is a Nordic body map.

IOP Conf. Series: Materials Science and Engineering **885** (2020) 012033 doi:10.1088/1757-899X/885/1/012033

Table 1. Nordic Body Map

No	Part of Body	Scoring				Total
		1	2	3	4	
0	Neck		3	15		18
1	Nape of Neck		3	15		18
2	Left Shoulder		16	2		18
3	Right Shoulder	16	2			18
4	Left upper arm	15	3			18
5	Back			17	1	18
6	Right Upper Arm		15	3		18
7	Waist		2	16		18
8	Hip			18		18
9	Butt			18		18
10	Left Elbow	15	3			18
11	Right Elbow	16	2			18
12	Left Forearm	8	10			18
13	Right Forearm	16	2			18
14	Left Wrist	18				18
15	Right Wrist	18				18
16	Left Hand	18				18
17	Right Hand	18				18
18	Left Thigh	5	13			18
19	Right Thigh	5	13			18
20	Left Knee	18				18
21	Right Knee	18				18
22	Left Calf		18			18
23	Right Calf		18			18
24	Left Ankle	18				18
25	Right Ankle	18				18
26	Left Foot		18			18
27	Right Foot		18			18

From the data table above it is known that there are some parts of the body that hurt when they use college chairs for a long time. Parts of the body that often feel pain are the upper body parts due to the uncomfortable shape of the chair. Parts of the body such as:

1. Neck
2. Nape of Neck
3. Back
4. Waist
5. Hip
6. Butt

For body parts other than above students do not complain too much pain in those body parts. The cause of pain in some parts of the body is caused by the shape of the chair that is not ergonomic, can not be adjusted to the user's body.

Table 2. NBM explanation table

No	Parts of Body	Cause
1	Neck	chairs that are connected to a desk thus limiting student movements and causing students to bend down because the height of chairs and tables cannot be adjusted
2	Nape of Neck	chairs that are connected to a desk thus limiting student movements and causing students to bend down because the height of chairs and tables cannot be adjusted
3	Back	the back of a chair that is hard and cannot support your back properly causes the backs of students to ache
4	Waist	the seat less soft and the shape of the chair that is not ergonomic causes pain in this body part
5	Hip	the seat less soft and the shape of the chair that is not ergonomic causes pain in this body part
6	Butt	the seat less soft and the shape of the chair that is not ergonomic causes pain in this body part

After using the Nordic body map method, this study continued with RULA to analyze the ergonomic risk of student posture when sitting in a chair and doing activities

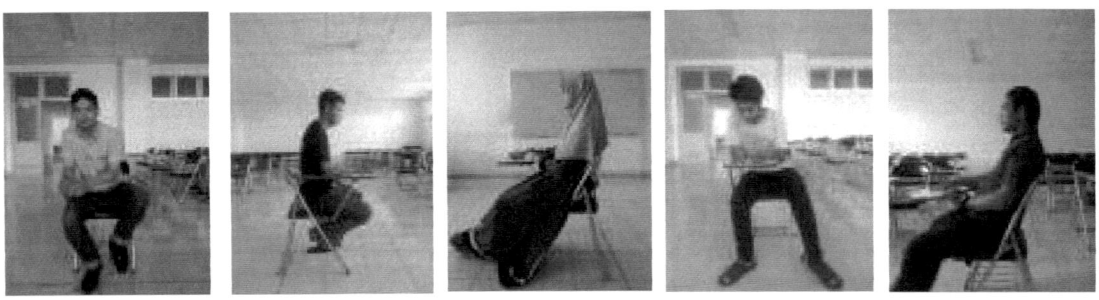

Figure 1. Student Posture

Table 3. Table A RULA

Tabel A Wrist posture score									
Upper Arm	Lower Arm	1		2		3		4	
		Wrist twist							
		1	2	1	2	1	2	1	2
1	1	1	2	[illegible]	2	2	3	3	3
	2	2	2	2	2	3	3	3	3
	3	2	3	3	3	3	3	4	4
2	1	2	3	3	3	3	4	4	4
	2	3	3	3	3	3	4	4	4
	3	3	4	4	4	4	4	5	5
3	1	3	3	[illegible]	4	4	4	5	5
	2	[illegible]	[illegible]	4	4	4	4	5	5
	3	4	4	4	4	4	5	5	5

4	1	4	4	4	4	4	5	5	5
	2	4	4	4	4	4	5	5	5
	3	4	4	4	5	5	5	6	6
5	1	5	5	5	5	5	6	6	7
	2	5	6	6	6	6	7	7	7
	3	6	6	6	7	7	7	7	8
6	1	7	7	7	7	7	8	8	9
	2	8	8	8	8	8	9	9	9
	3	9	9	9	9	9	9	9	9

Posture score A		Muscle use score		Force/load score		Wrist & arm score
4	+	1	+	0	=	5

Table 4. Table B RULA

Neck posture score	Tabel B Trunk Posture Score											
	1 Legs		2 Legs		3 Legs		4 Legs		5 Legs		6 Legs	
	1	2	1	2	1	2	1	2	1	2	1	2
1	1	3	2	3	3	4	5	5	6	6	7	7
2	2	3	2	3	4	5	5	5	6	7	7	7
3	3	3	3	4	4	5	5	6	6	7	7	7
4	5	5	5	6	6	7	7	7	7	7	8	8
5	[illegible]	[illegible]	[illegible]	[illegible]	[illegible]	[illegible]	8	8	8	8	8	8
6	8	8	8	8	8	8	8	9	9	9	9	9

Posture score B		Muscle use score		Force/load score		Wrist & arm score
8	+	1	+	0	=	9

Table 5. Table C RULA

Table C		**Neck, Trunk. Leg Score**						
		1	**2**	**3**	**4**	**5**	**6**	**7+**
Wrist/Arm Score	**1**	**1**	**2**	**3**	**3**	**4**	**5**	[illegible]
	2	**2**	**2**	**3**	**4**	**4**	**5**	**5**
	3	**3**	**3**	**3**	**4**	**4**	**5**	**6**
	4	**3**	**3**	**3**	**4**	**5**	**6**	[illegible]
	5	**4**	**4**	**4**	**5**	**6**	**7**	**7**
	6	**4**	**4**	**5**	**6**	**6**	**7**	**7**
	7	**5**	**5**	**6**	**6**	**7**	**7**	**7**
	8	**5**	**5**	**6**	**7**	**7**	**7**	**7**

From the results of the assessment using the RULA method, of the 18 respondents who were the object of the study, all received a score of 7 which indicates that immediate posture should be corrected because it will cause health problems in the future. This value is obtained from the student sitting position. Student sitting position is due to the height of the chair that cannot be adjusted, the height of the table that cannot be the desk, the table which is very small and is located on the right, the seat and back of the chair are not

IOP Conf. Series: Materials Science and Engineering **885** (2020) 012033 doi:10.1088/1757-899X/885/1/012033

soft and the curve of the chair is not ergonomic. The next step is to identify the desires of consumers, in this case, our students, from the results of interviews with several students, they complained:

1. Hard seat covers
2. Hard backrest
3. Chair height that cannot be adjusted
4. Lecture desk that cannot be folded
5. Width of the pedestal base that is not ergonomic
6. A left hand without support
7. The desk is too small

after listening to complaints and desires of students, continued with the house of quality method. by answering the desires of students into the design of chairs

IOP Conf. Series: Materials Science and Engineering **885** (2020) 012033 doi:10.1088/1757-899X/885/1/012033

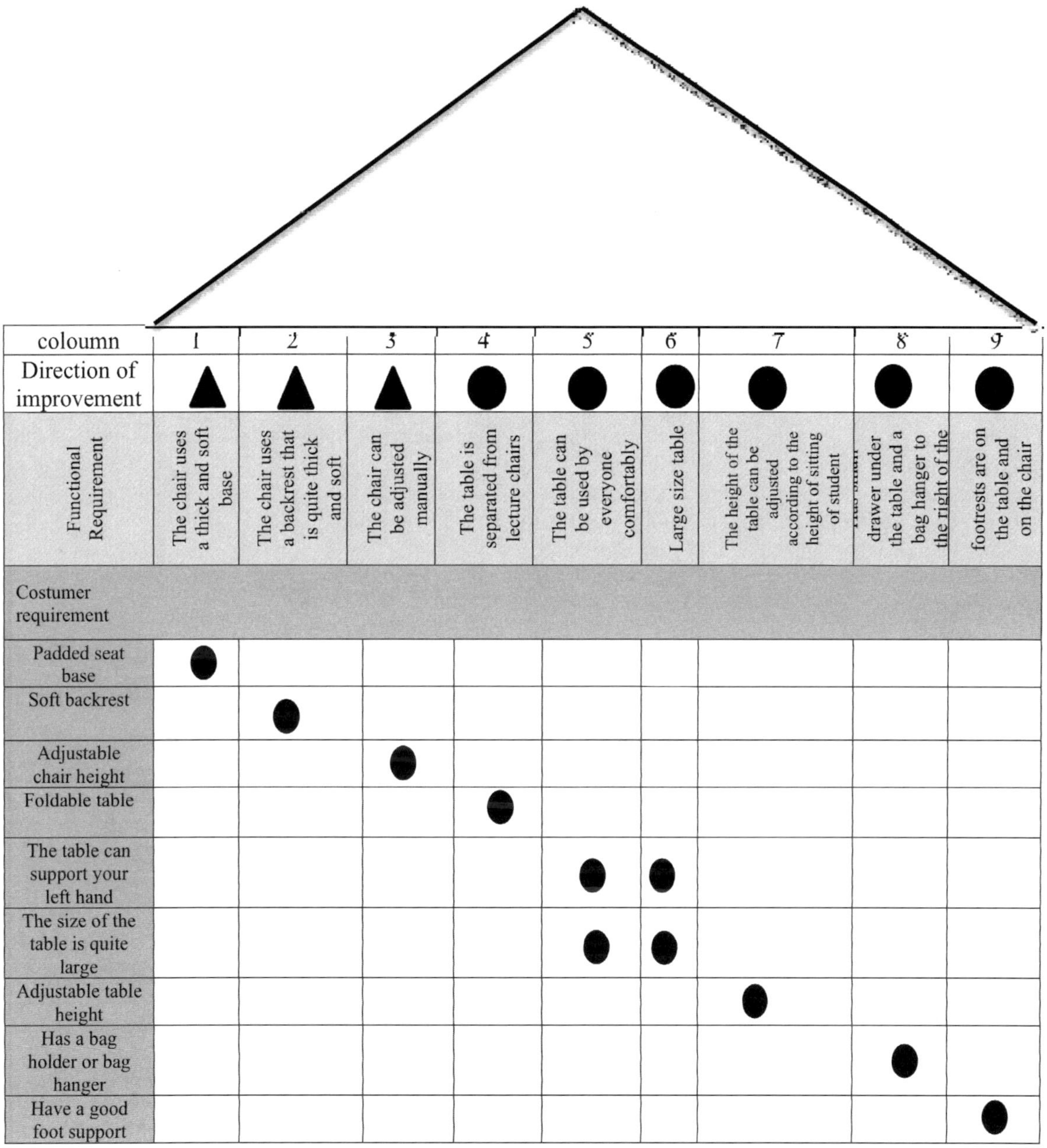

Figure 2. House of Quality

After identifying the desires of consumers which are translated into product specifications, a chair design will be made in accordance with the wishes of the students as shown below

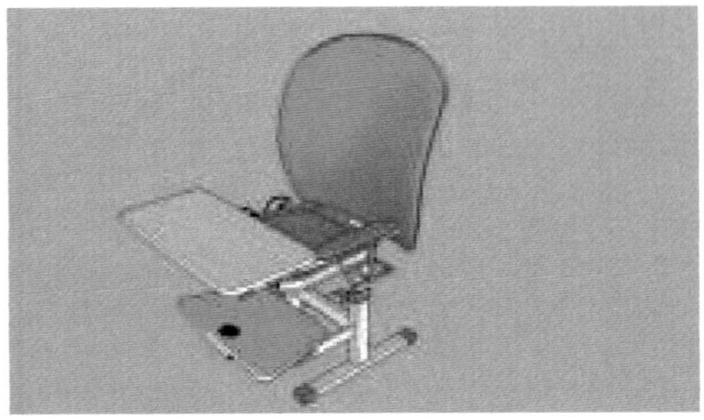

Figure 3. college chair design

4. Conclusion

The results of this study indicate that facilities in this case lecture chairs and student posture when sitting in a chair are very risky by getting a RULA score of 7, which means that facilities must be changed and changes in student posture when sitting. students also complained of aches/pains in several parts of the body such as the waist, back, neck, arms, etc. the next step is to identify the desires of students regarding the lecture chairs that they think are comfortable and good to use. 7 entries were translated into product specifications using the house of quality or quality function deployment

Reference

[1] Palupi, Dian 2017 *Metode REBA* Untuk Pencegahan Musculoskeletal Disorder Tenaga Kerja, *Jurnal Teknik Industri*, **18** 19-28

[2] Hermawan, Yuni, 2011 Pengembangan dan Analisis Ergonomi Kursi Operator Mesin Vulkanisir Ban dengan Metode Reserve Engineering. *Jurnal ROTOR*, **4**

[3] Hermanto 2017. Usulan Rancangan Ukuran pada Meja dan Kursi Lipat Belajar yang Ergonomis untuk Rumah Petak di Jakarta. ikraith-teknologi **1**

[4] Pamungkas, Gina Silvia, 2016 Hubungan Antara Dimensi Kursi dan Keluhan Nyeri Punggung Bawah Mahasiswa FK Undip. *Jurnal Kedokteran Diponegoro* **5**

[5] Octavia Hariandja, Johanna Renny 2013. Perancangan Kursi dan Meja Laptop yang Ergonomis di Universitas Katolik Parahyangan. *Jurnal Rekayasa Sistem Industri* **2**

Knowledge management as a source of innovation and business process development to improve the competitiveness of small and medium enterprises (SMES) in the madurese handicraft souvenir sector (case study: XYZ souvenir)

D G Ramadhan[1] and Sofiyanurriyanti[2]

[1]Politeknik ATI Makassar
[2]Universitas Teuku Umar

Email: della.ginza@atim.ac.id

Abstract. XYZ Souvenir is one of the UKM engaged in the creative and Madura souvenir craft made of flannelette. The problem faced today is how to maintain the sustainability of SMEs if they have different backgrounds and knowledge that make them have to share knowledge to maintain the sustainability of SMEs. Product innovation, optimization of raw materials, customer needs, target markets, marketing methods, are examples of knowledge that are not all known by each of the internal actors of SMEs. By implementing a knowledge management framework based on four KM conceptual models by Stankosky and mapping the Cimosa business process, which is intended to explore the tacit and explicit knowledge of internal actors related to the sustainability of SMEs. Qualitative methods are used in this study and the data collection model is in the form of interviews. With the implementation of the framework, there are nine critical success factors identified in the craft industry, especially those operating in the Madura region. Determinants of success are Price, Promotion, Packaging, Product Innovation, Online Stores, Product Quality, Finance, Human Resources, Collaborative Sources. In a four-action framework analysis, there are no success determinants that are eliminated because no success determinants are eliminated at an early stage so there is no reduction in the development of SME business processes in the reduction phase. The critical success factor development is carried out on the nine main criteria in the craft industry and creates two additional critical success factors, namely unique and trademark. In mapping, Cimosa's business processes, the process management phase consists of improving vision and mission, improving objective strategies, improving SOPs, and improving, evaluating, and monitoring processes. Support processes include human resources and IT process, production, and financial process, while the Core Process includes research and development, marketing process, and logistic process.

1. Introduction

The Indonesian Ministry of Industry targets Small and Medium Industries (SMEs) to grow above 10% by 2018. This target is above the growth of SMEs throughout 2017, which is projected to be around 9%. For this reason, a strong SME is needed that can compete in the international world. Indonesian SME products must go global. Entering the global market, Indonesian SMEs must be able to make a breakthrough. Breakthroughs include comparative studies abroad, exchanging information such as by improving the design, variety, and touch of product technology. Government policies, among others, make it easier to get raw materials. This is for the competitiveness of national SMEs to be more competitive on the global scene [1]. SMEs day-to-day business operations specifically require close attention [2–4]. This frequently leads to situations where insufficient time is available for strategic issues. This, in conjunction with lack of financial resources and expertise, very often results in most knowledge being kept in the minds of the owner and some key employees rather than recorded or shared through substitution arrangements [5,6]. Knowledge management is a surprising mix of strategies, tools, and techniques, some of which are nothing new under the sun. Storytelling, peer-to peer mentoring, and learning from mistakes, for example, all have precedents in education, training, and artificial intelligence practices. Knowledge management makes use of a mixture of techniques from knowledge-based system design, such as structured knowledge acquisition strategies from subject matter experts and educational technology (e.g., task and job analysis to design and develop task support systems [7].

XYZ Souvenir determines to have the right business model in accordance with the goals and objectives set by the XYZ Souvenir management. The business model will describe the rationale of how an organization creates, delivers and captures value [8–10]. Business Model is a hypothesis about how a company makes money in the long run: what the company will sell, and to whom, how the company will collect revenue, what technology will be used, when the company will rely on its business partners and about the cost management [8]. CIMOSA is a generic method that can be used regardless of the organizational model used. CIMOSA is divided into several levels depending on how detailed the company's business processes are to be mapped. CIMOSA not only provides integration of infrastructure concepts but also relates to two other concepts, namely life cycle (related to the development and maintenance of companies and product models) then frameworks (company development models and product models). CIMOSA is an architecture that directs and integrates hardware and industrial application funds, among other things, company activities. The need to integrate business dating from fast market evolution that forces corporate adaptation [7]. Referring to the problems faced by XYZ Souvenir, the purpose of this study is to identify the business models that are applied using the knowledge management approach and to formulate alternative business development strategies with the CIMOSA Business Model.

2. Methods

This research was conducted at XYZ Souvenir, located in Bangkalan, Madura, from June 2019 to August 2019. XYZ Souvenir was chosen deliberately on the basis that the company is currently developing its business. The data used in this study are primary and secondary data. Primary data were obtained from observations, focus group discussions, and direct interviews with internal and external respondents. Secondary data information obtained from company documents related to the problem together with supporting data from the literature, the Internet, scientific journals and the results of previous studies. This research is a descriptive case study that aims to make a more detailed, in-depth and comprehensive study of a particular object, which is usually relatively small in a certain period of time. This research uses the Business Model. Knowledge Management is used to identify the critical success factors of SMEs and the CIMOSA Business Model to formulate strategies for each XYZ Souvenir business development. Several interviews were also conducted to collect data. Respondents in this study were classified into internal and external respondents. Internal respondents were chosen based on their knowledge of the conditions of the XYZ Souvenir specifically and their full authority in determining the policies in the XYZ Souvenir. The research framework can be seen in figure 1.

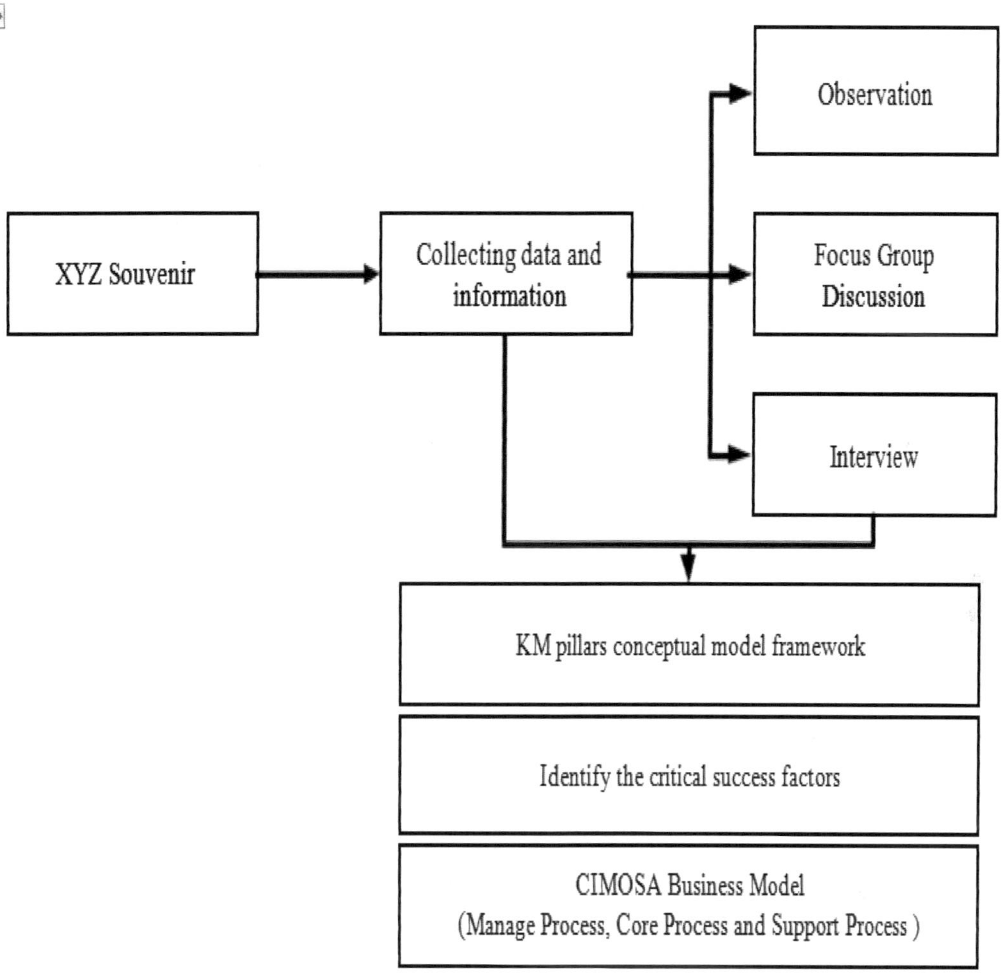

Figure 1. Reasearch framework

3. Result And Discussion

3.1. Business Model and Business Type of XYZ Souvenir
To find out the business processes of XYZ Souvenir SMEs, an analysis was conducted using the four KM pillars conceptual model framework by Stankosy (1999).

3.1.1. Leadership. The leadership applied to the XYZ Souvenir is based on the principle of kinship, although there has been a division of positions between members of the organization but applying leadership is still based on deliberation, this is what makes a conducive work climate for employees. Cultural aspects greatly affect the business process of XYZ Souvenir because the making of design and product innovation is based on Madura culture. One of them is the "carok" model, which is a one-on-one tradition of fighting using sickles. The business strategy adopted by XYZ Souvenir is based on the vision and mission that has been designed so that it can be developed and aligned with the ideals to be achieved by the company. The growth of XYZ Souvenirs from year to year increases with the segmentation of local people and foreign tourists. The leadership communication applied by XYZ Souvenir, both verbal and non-verbal communication, has applied a very good ethical concept because it is based on kinship.

3.1.2. Organization. The production processes and procedures applied by XYZ Souvenir have implemented SOPs in the manufacturing of various kinds of souvenirs produced. However, the application of product inspection must still get more attention, and this is due to differences in employee expertise so that the results given can also be different. The product workflow system that has been used is also quite good because there is no buildup of goods in the production process flow. Organizational education that is applied between employees and superiors can be established properly, as well as communication between employees.

3.1.3. Technology. The marketing process of XYZ Souvenir is carried out with websites and social media, and cooperation is also carried out with several shipping service companies so that the distribution of goods can be guaranteed and controlled. The payment process for online consumers is done through the Bank using a transfer facility between accounts so that the flow of funds can be clearly known. However, the application of technology in the XYZ Souvenir business process is still lacking, the production process is done manually and there is still little application of the technology. The use of communication technology is very good, both for internal and external communication of SMEs with the application of electronic media and the application of communication technology.

3.1.4. Learning

The use of instincts in the learning process of XYZ Souvenir is not widely used because it is more focused on the use of data, information and real facts. The application of innovations and inventions needs to be done regularly so that they can compete with other competitors. XYZ Souvenir carries out the learning process by providing supplies for employees through seminars, training, as well as SME forums and communities. Discussion forums are also conducted through internal and external organizations to share knowledge.

3.1.5. Multiple disciplines. XYZ Souvenir has not conducted regular evaluations of the application of SOPs and has not been able to optimally implement the automation system in the production process because the production process is still done manually. Organizational capacity building to handle internal and external functions and relationships is done regularly through internal meetings and forum discussions. The development of organizational behavior is carried out by attending seminars and sharing experiences between employees.

3.1.6. Environmental Instrument

XYZ Souvenir is very concerned about social groups and lifestyles in society. The groups and lifestyles that develop in this society, certainly determine the sustainability of the business, as a process that can never be separated from its environment. In addition, XYZ Souvenir also considers political risks and their effects on the organization. This is worth considering because changes in political actions and policies in a country can have a large impact on the country's financial sector and economy. Economic conditions also have a strong impact on the performance of each business because it can affect the income or expense of the business. The demographic environment of business processes can also affect market segmentation of the business processes that are carried out.

3.2. Identification Critical Success Factor of XYZ Souvenir

The following are critical success factors used in research:

3.2.1. Price. Price competition in handicraft SMEs is very tight, this is because pricing is the most important thing in business processes.

3.2.2. Promotion. Product marketing is one important factor in determining business success. With good marketing, management can introduce our products to consumers so that our products can be known and get a place in the hearts of consumers. Promotions and attractive offers can also increase consumer interest in buying our products so they can increase market share.

3.2.3. Packaging. Attractive packaging is also an important aspect of business processes. With a unique packaging can make the difference between each product with the same type.

3.2.4. Innovation Product. In the handicraft industry, aspects of innovation are very important to be developed. The development of time, technology, and the desire of consumers demand the creativity of the craft industry SMEs in fulfilling consumer satisfaction.

3.2.5. Online Shop. The rapid development of information technology requires handicraft industry SMEs to do marketing online so that it can make it easier to get product information and product purchases for consumers.

3.2.6. Quality Product. Product quality is very important to increase consumer confidence. By producing quality products that can foster a high sense of trust for SMEs and consumers.

3.2.7. Financial. The financial situation of SMEs is also an important aspect in the development of business processes. With a stable financial situation, SMEs can develop both in terms of business processes and human resource development.

3.2.8. Human Resource. Human resources are an important aspect of SME businesses, especially craft. Competent resources in their field can have an impact on product variety and SME product development.

3.2.9. Source Cooperate. The cooperative relationship between suppliers of raw materials and SMEs, as well as the relationship between SMEs and product distributors, is certainly an important aspect in the success of an SME business.

3.3. Blue Ocean Strategy.
Analysis of the addition of a critical success factor to XYZ Souvenir, so that it can move from the red ocean strategy to the blue ocean strategy can be seen in figure 2.

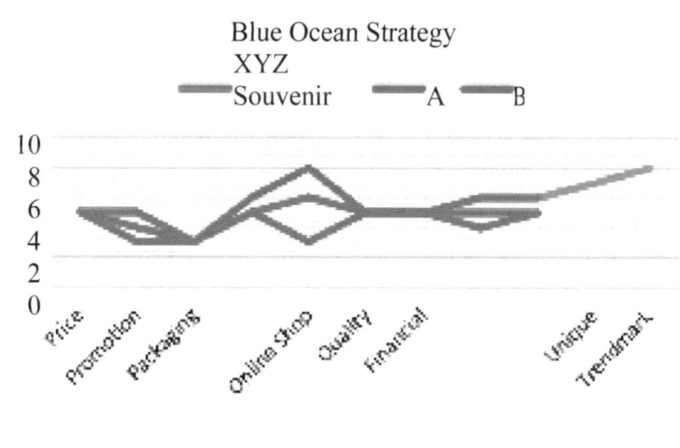

Figure 2. Blue ocean strategy

The four action frameworks for enhancing the development of XYZ Souvenir business processes are as follows:

3.3.1. Eliminate. At this stage, the elimination of critical success factors is carried out that are considered not important in the development of business processes. There is no critical success factor eliminated in XYZ Souvenir.

3.3.2. Reduce. At this stage, a critical success factor reduction is carried out, which is the impact of elimination because there is no critical success factor that is eliminated in the initial stages, so there is no reduction in the development of XYZ Souvenir business processes.

3.3.3. Raise. At this stage the critical success factor development is carried out which is considered important in developing business processes. For XYZ Souvenir, the development of critical success factor is carried out on ten main criteria in the craft industry including, price, promotion, packaging, product innovation, product quality, customer relations, financial, human resources, and corporate resources. This is done because there is a close relationship. Between critical success factors, so when an increase is made to one critical success factor, it will have an impact on other critical success factors.

3.3.4. Create. At this stage, the development of the idea of adding a critical success factor is developed to make XYZ Souvenir get out of the red ocean strategy competition. There are two additional critical success factors that are carried out at this stage, namely unique and trademark.

3.3.4.1. Unique. What is meant by unique criteria at this stage is the different types of raw materials used. Batik raw materials in making souvenirs can be created by forming their own motives that are not owned by other competitors.

3.3.4.2. Trademark. The creation of a trademark can have a very good impact on business development. By having its own trademark XYZ Souvenir can have its own market share that is not owned by other competitors. Trademark, which is meant by this criterion, is the creation of Madura special characters. This character can also be developed in collaboration with the Indonesian animation industry so that it can market the character and become an icon of Madura Island.

3.4. Cimosa Business Process
An overview of the XYZ Souvenir business process mapping can be seen in figure 3.

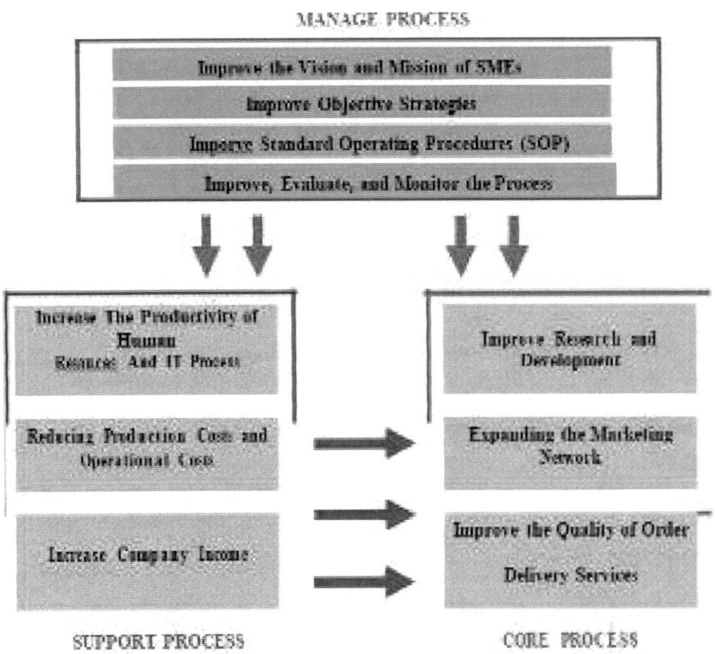

Figure 3. XYZ Souvenir CIMOSA business model

An overview of the XYZ Souvenir business process mapping based on the predetermined critical success factor is as follows:

3.4.1. In managing the process, the critical success factor is quality, so that the design of the strategy that will be used is improving the vision and mission of SMEs, improving objective strategies, improving standard operating procedures (SOPs), and also improving, evaluating, and monitoring the process.

3.4.2. In the supporting process, the critical success factors are price, packaging, financial, and company sources. The design strategies to be used are improving the productivity of human resources and it processes, reducing production costs and operating costs, and increasing company revenues.

3.4.3. In the core process, critical success factors are promotion, product innovation, online stores, uniqueness and trademarks. The design strategies to be used are improving research and development, expanding marketing networks and improving the quality of order delivery services.

4. Conclusion

The conclusion from the results of this study, there are nine determinants of the success of XYZ Souvenir that can be identified, namely price, promotion, packaging, product innovation, online stores, product quality, finance, human resources, and collaborative resources. The strategies that emerge in the business processes mentioned above in the first process based on critical success factors in the Manage Process, Core Process, and Support Process are:

4.1. Manage Process

Improving the vision and mission of SMEs, improving objective strategies, improving standard operating procedures (SOPs), and also improving, evaluating, and monitoring the process.

4.2. Core Process

Improving the productivity of human resources and it processes, reducing production costs and operating costs, and increasing company revenues.

4.3. Support Process

Improving research and development, expanding marketing networks and improving the quality of order delivery services.

References

[1] Wagu 2018 *IKM Indonesia Go Global* (Jakarta, Indonesia: Majalah Gema Industri Kecil)

[2] Hofer C W and Charan R 1984 The transition to professional management: Mission impossible? *Am. J. Small Bus.* **9** 1–11

[3] Tanti H 2017 The Role of Relationship Marketing in Small and Medium Enterprises (SMEs) *J. Pengur.* **48** 137–48

[4] Utama A A G S U, Millati I, Kurniawansyah D, Astuti P P D and Setyowati Y 2019 Increasing the existence of SMEs in Banyuwangi digital commerce through designing decision support system (DSS) for Banyuwangi mall site *Int. J. Civ. Eng. Technol.* · **10** 1572–1481

[5] Bridge S and O'Neill K 2012 *Understanding enterprise: Entrepreneurship and small business* (Macmillan International Higher Education)

[6] Wong K Y and Aspinwall E 2004 Characterizing knowledge management in the small business environment *J. Knowl. Manag.*

[7] Kaplan R S, Davenport T H, Robert N P D K S, Kaplan R S and Norton D P 2001 *The strategy-focused organization: How balanced scorecard companies thrive in the new business environment* (Harvard Business Press)

[8] Eisenmann T 2001 *Internet Business Models: Text and Cases* (New York: McGraw-Hill/Irwin)

[9] Odeniyi O A, Lawal N T, Kareem A E and Adebowale A 2015 An Appraisal Of Groupon E-

Business Model *Int. J. Sci. Technol. Res.* **4** 291–7
[10] Osterwalder and Pigneur 2015 *Business Model Generation* (Jakarta: . PT Elex Media Komputindo)

IOP Conf. Series: Materials Science and Engineering **885** (2020) 012035 doi:10.1088/1757-899X/885/1/012035

Application of line balancing using the heuristic method to equalize the production line at PT.Bogatama Marinusa Makassar

Ahmad Sawal and Ainun Jariah Hamzah

Agro Industrial Engineering, Polytechnic of ATI Makassar, Makassar, 90211, Indonesia

Email: sawal@atim.ac.id

Abstract. PT. Bogatama Marinusa is one of the fishery industry sectors located in South Sulawesi that moves in shrimp processing and frosting as pioneer of fishery industry in southern Sulawesi. PT. Bogatama Marinusa certainly wants the planned production target to be achieved as expected, where every day the company processes raw materials with the capacity of approximately 575 kg for *vannamei* shrimp and 425 kg for *black tiger* shrimp. At deheading work station (shrimp head separation) and sortation area (sorting) materials buildup occurs in the production flow (bottle neck) which results an idle time to other work stations. With the existence of these problems, it is necessary to do production line balancing in order to minimize idle time and be able to balance the production lines so that the works carried out effectively and efficiently. The method used for overcoming the production line imbalance is by applying line balancing, namely the heuristic method, which is calculation using Regional Approach (RA) and Ranked Positional Weight (RPW) methods. Line efficiency (LE) increased by 83.33%, smoothing index (SI) decreased by 160.90, balance delay (BD) decreased by 16.66%, and idle time (IT) decreased by 264 minutes, and production capacity could be increase to 1,022 kg means that the results are far better than the current conditions.

1. Introduction

1.1. Background

The current development of the food industry demanding companies to keep survive and develop. Companies that are able to survive and develop well can certainly increase competitive advantage in the industrial world. Therefore, companies must have effective management operations in determining the number of workers and the balance of work seen from the efficiency of production process time so wasting of time and costs can be avoided which can harm the company in achieving expected level of production [1].

PT. Bogatama Marinusa is one of the fishery industry sectors located in South Sulawesi that moves in shrimp processing and frosting as pioneer of fishery industry in southern Sulawesi. The raw materials used are vannamei shrimp (white shrimp) and black tiger shrimp (tiger shrimp).

PT. Bogatama Marinusa aim to achieve every planned production target as expected, every day the company processes raw materials with a capacity of approximately 575 kg for vannamei shrimp and 425 kg for black tiger shrimp to keep up the demand. However, the processed production capacity reaches 1 ton per day resulting a buildup on the production flow (bottle neck) spesifically in some work stations, one of which is deheading area and sortation area while the other work stations are idle, this is a big challenge for the company to create more efficient and effective production process. The balance of production line is important to note because it will affect the smoothness of the production process including balancing the assignment of several work stations and minimizing the total waiting time (idle time) to obtain a smooth production flow in order to produce targeted output, labor and equipment in reducing production cycle times which will provide substantial savings to the production costs [2]. Production lines in the deheading department (cut shrimp head) and sorting areashave obstacles caused by the production process which is done manually with minimal use of the machine. Human capability limitation in sorting process become one cause of the build-up factor and long waiting time, to overcome these obstacles it is necessary to do the research to the work station especially on deheading area and sortation area to minimize buildup and waiting time.

1.2. Matter Formulation
Based on the background described above, the formulation of the problem in this study is how to balance the production line using the heuristic method in the form of calculations using the Regional Approach (RA) and Ranked Positional Weight (RPW) methods?

1.3. Research Purposes
Based on the matter formulation, the purpose of this study is to determine the balance of the production line using the heuristic method in the form of calculations using the Regional Approach (RA) and Ranked Positional Weight (RPW) methods.

2. Methods
Data obtained from the results of the study are processed and analyzed by applying Line Balancing with calculations using the Regional Approach (RA) method and Ranked Positional Weight (RPW) which is a method to solve the problem of determining the optimal number of work stations. To solve the problem of production line, the steps that must be done are as follows: [3,4]

2.1. Data Adequacy Test
Testing the adequacy of data is needed to ensure that the data collected is objective enough , below are the formulation:

$$N' = \left[\frac{k/s\sqrt{N\Sigma x2-(\Sigma x)2}}{\Sigma x}\right] \qquad (2.1)$$

2.2. Data uniformity test
To ensure that the data gathered comes from the same system, it is very necessary to test the data uniformity. Therefore we need to test the uniformity of data to separate data which has different characteristics.

$$\sigma = \sqrt{\frac{\Sigma(Xi-x)^2}{N-1}} \qquad (2.2)$$

$$BKA = \overline{\mathrm{X}} + \mathrm{K}.\ \sigma \qquad (2.3)$$

$$BKB = \overline{X} + K.\sigma \qquad (2.4)$$

2.3. Creating Precedence Diagram
Precedence diagram which is a graphical description of the sequence of work operations, as well as dependence on other work operations that aim to facilitate the control and planning of related activities.

2.4. Determine the work station (work station)
Work station (work station) is a place where track carried out. After determining the cycle time, then determine the minimum number of work stations that will be formed using the following equation

$$K = \frac{\sum ti}{CT} \quad (2.5)$$

2.5. Calculating Cycle Time (CT)

Cycle time Is time needed to make one product unit in a work station. Determination of cycle time done by taking prime factorization from the total time element of the company's work. In designing the production line balance for a certain number of production, the cycle time must be equal or greater than the largest operating time which is the cause of bottle neck.

$$STmax \leq CT \leq \sum ti \quad (2.6)$$

2.6. Calculating production capacity

Calculating production capacity is needed to find out how much production capacity in one production line for one production cycle.

$$Capacity = \frac{Daily\ Production\ Time}{Cycle\ Time} \quad (2.7)$$

2.7. Calculate Line Efficiency (LE)

Line efficiency is the ratio of the total work station time divided by the cycle time multiplied by the work station. A good production path has a high track efficiency value which indicates that all work stations have a closer time to the specified cycle time. So it can be said that the higher the value of track efficiency, the track will become better.

$$L\,E = \frac{\sum_{i=1}^{k} STi}{(K)(CT)} \quad (2.8)$$

2.8. Calculate Smoothing Index (SI)

Smoothing index is an index that has a relative smoothness of the balancing from production line. A good production path has a smoothing index value that is close to zero. In other words the smaller the smoothing index value the better

$$SI = \sqrt{\sum_{i=1}^{K}(STmax - STi)} \quad (2.9)$$

2.9. Calculate balance delay (BD)

Balance delay is a measure of trajectory inefficiency resulting from actual idle time due to less than perfect allocation between work stations. A good production line has a balance delay value close to zero, which means that the idle time is small for all work stations. The smaller the balance delay, the better a production line.

$$BD = \frac{(K\ X\ CT) - \sum_{i}^{n} ti}{(K)(CT)} \quad (2.10)$$

2.10. Idle time (IT)

Idle time is the difference between cycle time (CT) and station time (ST). The idle time occurs in each work station. Idle time occurs when the processing time on a work station is smaller than the cycle time.

$$IT = K\,(CT) - \sum ST \quad (2.11)$$

IOP Conf. Series: Materials Science and Engineering **885** (2020) 012035 doi:10.1088/1757-899X/885/1/012035

3. Results and Discussion

3.1. Data Collection Results

Based on the results of direct observation in the production section for the initial process of PT. Bogatama Marinusa Makassar, namely in department receiving, deheading, sortation area, and chillroom, the data for operating elements of work time were obtained. Data collection carried out is the process time data from each work element with 18 observations by averaging the time to get the accuracy results. Then the data processing time on the track can be seen in Table 1 below.

Table 1. Description of Work Station Operations

Operation	Description	Predecessor Event	Station Time (Minutes)
A	Unloading of *vannamei shrimp*	-	50
B	Unloading of *black tiger shrimp*	-	48
C	*Pre-washing* of *vannamei shrimp*	A	52
D	*Pre-washing* of *black tiger shrimp*	B	55
E	Weighing 1 (*head on*) *vannamei*	C	34
F	Weighing 1 (*head on*) *black tiger*	D	28
G	Deheading (shrimp head cutting)	E, F	230
H	*Washing* 1	G	64
I	Weighing 2 (*head less*) *vannamei*	H	37
J	Weighing 2 (*head less*) *black tiger*	H	39
K	Sortation of *vannamei shrimp*	I	258
L	Sortation of *black tiger shrimp*	J	260
M	*Washing* 2	K, L	53
N	Weighing 3 (*head less*) *vannamei*	M	48
O	Weighing 3 (*head less*) *black tiger*	M	38
P	Storagement in *chill room* (Temporary Storage)	N, O	26
	Total		1.320

(Source: Data processing, 2019)

3.2. Data Processing

3.2.1. Data Adequacy Test. Based on data that has been observed at PT. Bogatama Marinusa, can be seen in table 1, then the adequacy of the observation data will be tested, there are two factors that are used to determine the test of data adequacy, namely the level of confidence and degree of accuracy. In this study used a confidence level of 95% = 2 and a degree of accuracy of 10% = 0.1.

Table 2. Recapitulation of the data adequacy test

No	Data Adequency Test			
	Operation	N	N'	Information
1	A	18	13,77	Adequate
2	B	18	11,38	Adequate
3	C	18	7.95	Adequate
4	D	18	3,65	Adequate
5	E	18	3,42	Adequate
6	F	18	15,58	Adequate
7	G	18	4,04	Adequate
8	H	18	17,66	Adequate
9	I	18	4,51	Adequate
10	J	18	15,39	Adequate
11	K	18	2,27	Adequate
12	L	18	2,02	Adequate
13	M	18	0,88	Adequate

14	N	18	10,68	Adequate
15	O	18	17,94	Adequate
16	P	18	16,69	Adequate

(Source: Data processing, 2019)

3.2.2. Data Uniformity Test. Based on data that has been observed at PT. Bogatama Marinusa, can be seen in table 1, then the uniformity of the observation data will be tested, this uniformity test is done with the aim of identifying the existence of data that deviates significantly from the actual average due to data which has too large or too small amount. From the data tested, the control limit will be obtained, namely the Upper Control Limit (BKA) and Lower Control Limit (BKB). In this study using 95% of confidence level = 2 and uniformity of data being tested using Minitab software.

Table 3. Recapitulation of data uniformity tests

No	Operation	x	Deviation Standard	K	BKA	BKB	Information
1	A	50	9.54	2	69.08	30.92	Even
2	B	48	8.33	2	64.66	31.34	Even
3	C	52	7.54	2	67.08	36.92	Even
4	D	55	5.37	2	65.74	44.26	Even
5	E	34	3.23	2	40.46	27.54	Even
6	F	28	5.68	2	39.36	16.64	Even
7	G	230	23.81	2	277.62	182.38	Even
8	H	64	13.91	2	91.82	36.18	Even
9	I	37	4.04	2	45.08	28.92	Even
10	J	39	7.87	2	54.74	23.26	Even
11	K	258	20.01	2	298.02	217.98	Even
12	L	260	19.02	2	298.04	221.96	Even
13	M	53	2.56	2	58.12	47.88	Even
14	N	48	8.07	2	64.14	31.86	Even
15	O	38	8.28	2	54.56	21.44	Even
16	P	26	5.46	2	36.92	15.08	Even

(Source: Data processing, 2019)

3.2.3. Describes the Presedence Diagram. Based on table 1, precedence diagram will be made, the following is a precedence diagram (Figure 1) based on a description of the department's operational receiving area (receipt of raw materials), deheading area (cut shrimp head), sorting area, and chilling room (temporary storage) .

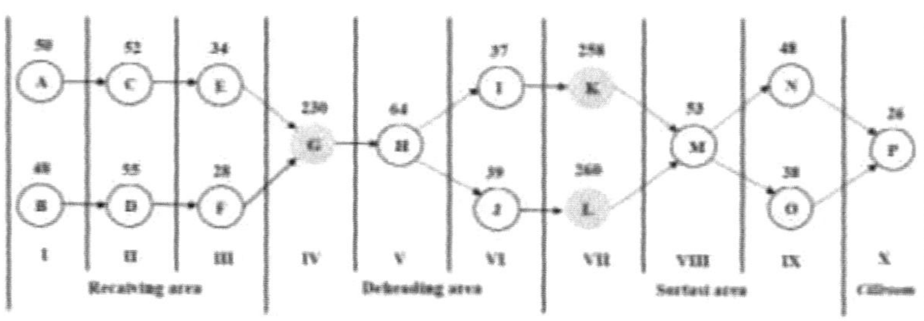

Figure 1. Precedence diagram

3.2.4. Line Balancing Analysis in Actual (Initial) Conditions. Based on the picture 1 precedence diagram above, the working elements of the initial conditions can be seen in table 4 where there are 10 work stations, can be seen the work load at stations 4 and 7 is the maximum load that is on the current production trajectory as for operating time can be seen in table 4 as follows :

Table 4. Work station initial conditions

Working Station	Element	Station Time (ST) (Minutes)
I	A, B	98
II	C, D	107
III	E, F	62
IV	G	230
V	H	64
VI	I, J	76
VII	K, L	518
VIII	M	53
IX	N, O	86
X	P	26
Total		1.320
Line efisiensi (LE)		25,48%
Smoothing Index (SI)		1.297,14
Balance Delay (BD)		74,51%
Idle Time (IT)		3.860 minutes
Production capacity		0,521 Ton (521,23 kg)

(Source: Data processing, 2019)

3.2.5. Analysis for Proposed Conditions Using The Regional Approach (RA) Method

1. Calculating CT determination (cycle time)

Determining the cycle time is obtained from the prime number of the total work station time, the cycle time must be greater or equal to the maximum time of the work station. Prime numbers for total work station time 1,320 minutes are 2, 3, 4, 5, 6, 8, 10, 11, 12, 15, 20, 22, 24, 30, 33, 40, 44, 55, 60, 66, 88, 110, 120, 132, 165, 220, 264, 330, 440, 660, 1,320 then the possible alternative time for cycles (CT) are 264, 330, 440, 660, 1,320 because the value is out of 260 minutes which is the biggest work station time. The value of the cycle time (CT) chosen is factorization 264. Where the provisions can be seen in equation (2.6) as for the determination of the cycle time (CT) can be seen as follows. Where:

STmax: 260 minutes
CT: 264 minutes
$\sum$ti: 1,320 minutes
Settlement:
$STmax \leq CT \leq \sum ti$
$260 \leq 264 \leq 1,320$
Value of CT = 264 is obtained

2. Alternative Improvement of Track Balance using The Regional Approach (RA) Method

Based on the picture 1 precedence diagram above, the merging of work elements of repair conditions using the regional approach (RA) method for alternative can be seen in table 5, among others as follows:

Table 5. Alternative experiments

Station	Element	ST	CT - STi	(STmax - STi)2
I	A,B,C,D,E	239	25	441
II	F, G	258	6	4
III	L	260	4	0
IV	K	258	6	4
V	H,J,I,N,O	226	38	1.156
VI	M, P	79	185	32.761
Total		1.320	264	34.366

Line Efficiency (LE)	83,33 %
Smoothing Index (SI)	170,75
Balance Delay (BD)	16,66 %
Idle Time (IT)	264 minutes

3.2.6 Line Balancing Calculations for Proposed Conditions using Ranked Positional Weight (RPW) Method

1. Calculating CT Determination (cycle time)
Determining the cycle time is obtained from the prime number of the total work station time, the cycle time must be greater or equal to the maximum time of the work station. Prime numbers for total work station time 1,320 minutes are 2, 3, 4, 5, 6, 8, 10, 11, 12, 15, 20, 22, 24, 30, 33, 40, 44, 55, 60, 66, 88, 110, 120, 132, 165, 220, 264, 330, 440, 660, 1,320 then the possible alternative time for cycles (CT) are 264, 330, 440, 660, 1,320 because the value is out of 260 minutes which is the biggest work station time. So that the value of the cycle time (CT) chosen is factorization 264. Where the provisions can be seen in equation (2.6) as for the determination of the cycle time (CT) can be seen as follows.
Where :
STmax: 260 minutes
CT: 264 minutes
$\sum$ti: 1,320 minutes
Settlement:
$STmax \leq CT \leq \sum ti$
$260 \leq 264 \leq 1{,}320$
Value of CT = 264 is obtained

2. Alternative Repair of Track Balance by Using The Ranked Positional Weight (RPW) Method
Based on the figure 1. precedence diagram above, the merging of the working elements of the repair condition by using the ranked positional weight (RPW) method for alternative 2 can be seen in table 4.14, among others as follows:

Table 6. Alternative experiments

Station	Element	ST	CT - STi	(STmax - STi)2
I	A,B,C,D,F	233	31	961
II	E, G	264	0	0
III	H,J,I	140	124	15.376
IV	L	260	4	16
V	K	258	6	36
VI	M,N,O,P	165	99	9.801
Total	1.320		264	26.190
Line Efficiency (LE)				83,33 %
Smoothing Index (SI)				161,83
Balance Delay (BD)				16,66 %
Idle Time (IT)				264 minutes

(Source: Data processing, 2019)

3.3. Proposed Improvements
Based on the previous discussion, it is known that the proposed improvement using the alternative positional weight (RPW) method calculation 3 has the best results so that the researchers propose to improve the production trajectory based on the results as many as 6 work stations. The following is a picture of the proposed improvement in the production line at PT. Bogatama Marinusa can be seen in figure 2 below:

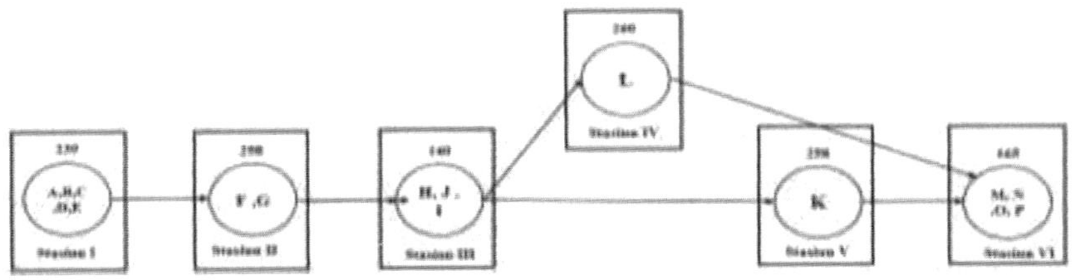

Figure 2. Proposed production lines

From the results of the proposed improvement by balancing the production line based on the Ranked Positional Weight (RPW) method carried out by combining work elements so the company can move or assign workers to a process that lacks workforce, namely at the work station deheading area (cutting of shrimp head) and station work sortation area (sorting). If after the transfer of labor has been carried out and it turns out that there is an excess of labor in each process and there is no process or station that has a shortage, labor elimination or reduction of this can occur due to employees in the deheading area (cutting of shrimp head) is a non-daily employee, but of course this can provide cost savings to the company.

4. Conclusion

Based on observations made at PT. Bogatama Marinusa Makassar is known for the current condition of having 10 work stations divided into 16 work elements by having a line efficiency level (LE) of 25.48%, smoothing index (SI) with a number of 1,297.14, balance delay (BD) with a number 74 , 51%, idle itime (IT) 3,860 minutes and only able to process raw materials as much as 521.23 kg.

5. Suggestions

Based on the conclusions from the calculation results, the company needs to apply the Ranked Positional Weight (RPW) method to determine the ideal work station in order to minimize idle time and eliminate obstacles then provide smooth production processes in the receiving area, deheading area (cutting head shrimp), sorting area, and cillroom (temporary storage warehouse).

References

[1] Gaspersz V 2004 *Production planning and inventory control* (Jakarta: PT Gramedia Pustaka Umum)
[2] Lamatinulu 2011 *Sistem Operasi* (Makassar: Cahaya Baliputra)
[3] Wignjosoebroto S 2003 *Pengantar teknik dan manajemen industri* (Surabaya: Guna Widya)
[4] Suryadi 1996 *Pengantar Teknik Industri* (Jakarta: Publisher Gunadarma)

IOP Conf. Series: Materials Science and Engineering **885** (2020) 012036 doi:10.1088/1757-899X/885/1/012036

Working posture analysis on musculoskeletal disorder to workers in the sack warehouse PT Makassar Tene Makassar

Arminas, M Basri and I RAulia

Industrial Engineering, Polytechnic ATI Makassar

Email: Arminas.atim@yahoo.com

Abstract. PT Makassar Tene is a factory that produces refined sugar from raw sugar. But in the sorting of auxiliary materials such as sacks is still done manually. The process of sorting sacks at PT Makassar Tene uses female workers over the age of 40 years. The conditions of work attitude carried out by some female workers are still many that are not ergonomic. The wok attitude is sitting, standing, hunchback and lifting weights more than a maximum burden that is tolerated by a woman's body which is 10 kg if done continuously. Pain complaints have been felt by workers due to inappropriate work postures, in the form of pain in the hips, waist, knees, shoulders and upper arms. The existing problems need to be evaluated on physical factors in sack sorting activities by identifying and analyzing the overall work posture. The methods used are OWAS and REBA. Both of these methods are able to define the movement of all parts of the worker's body so that it can improve overall body movement. Therefore, this study carried out observations to determine the level of risk caused by work posture when carrying out sack sorting activities by comparing the OWAS and REBA methods in female workers PT Makassar Tene. Based on the results of the study, it is known that there are several identified sorting activities of sacks at risk for musculoskeletal disorder (MSDs) in female workers. The activity are the process of lifting sacks to a pallet with a level of risk from slightly dangerous to very dangerous in the OWAS method and from medium to high in the REBA method. And the process of arranging sacks with a level of risk is very dangerous in the OWAS method and medium to high in the REBA method. So that it is necessary to improve the posture of the worker as soon as possible, especially in the activities of lifting and arranging sacks.

1. Introduction

According to Apep and Syafei (2002). Manual material handling activities are at risk of musculoskeletal disorders (MSDs), namely an injury to muscles, nerves, tendons, bones, joints, cartilage caused by work activities [1]. Furthermore, Lee and Han (2013) report that odd work postures are a major cause of musculoskeletal disorders [2]. Some examples of odd work postures include reaching back, turning, working with height above the head, wrists bending, kneeling, bending forward and backward bending, even squatting. Therefore, it is necessary to design a working system by taking into account the ergonomic rules so that workers can carry out their work safely and comfortably. Sack sorting process at PT Makassar Tene uses female workers aged over 40 years. The sorting process is Carried out sequentially by taking sacks from unsorted sacks locations with a weighing of 308-312 kg per pallet, bringing the sacks to the sorting location by using a hand truck as far as 8 meters, carrying out sacking, and lifting and stacking per 100 sacks have been sorted. There are still many conditions of working

IOP Conf. Series: Materials Science and Engineering 885 (2020) 012036 doi:10.1088/1757-899X/885/1/012036

posture Carried out by some female workers that are not ergonomic. The working attitude is sitting, standing, hunching, and lifting weights more than the maximum load tolerated by a woman's body, which is 10 kg if done continuously. Complaints of pain in parts of the body have been felt by workers due to Inappropriate posture, in the form of pain in the hips, waist, knees, shoulders and upper arms. In this study, work posture will be analyzed using OWAS and REBA methods. Both Reviews These methods can define the movement of all parts of the body to workers, especially among female workers who sort sacks so that they can improve overall body movement.

2. Research methods

In the data analysis stage, REBA (Rapid Entire Body Assessment) and OWAS (Ovako Working Posture Analysis System) methods are used by analyzing the work postures of female workers in carrying out the sack sorting process roommates done repeatedly at PT Makassar Tene.

3. Result and discussion

Nordic Body Map questionnaire is used to find out the complaints felt by workers when doing work activities. Based on Table 1, the data Recapitulation from the Nordic Body Map collected from all labor in sorting section sack, there are several parts of the body that were Complained of experiencing pain. The most common complaints felt by production workers during work

Table 1. Nordic Body Map

No	Part of Body	Score				Total	Percentage
		Ida	Hj Malaniah	Raudiah	Hasnah		
0	Neck	1	2	2	2	7	43.75
1	Nape of Neck	1	1	1	1	4	25
2	Left Shoulder	1	3	3	4	11	68.75
3	Right Shoulder	2	3	3	4	12	75
4	Left upper arm	3	2	2	3	10	62.5
5	Back	3	4	3	4	14	87.5
6	Right Upper Arm	3	3	2	3	11	68.75
7	Waist	4	4	4	4	16	100
8	Hip	4	4	4	4	16	100
9	Butt	1	1	1	1	4	25
10	Left Elbow	1	1	1	2	5	31.25
11	Right Elbow	1	1	1	2	5	31.25
12	Left Forearm	1	1	1	1	5	31.25
13	Right Forearm	2	1	1	1	6	37.5
14	Left Wrist	1	1	1	2	5	31.25
15	Right Wrist	1	1	1	2	5	31.25
16	Left Hand	1	1	1	1	4	25
17	Right Hand	1	1	1	1	4	25
18	Left Thigh	1	1	1	1	5	31.25
19	Right Thigh	1	1	1	1	5	31.25
20	Left Knee	3	2	4	3	12	75
21	Right Knee	3	2	4	3	12	75
22	Left Calf	2	2	2	2	8	50
23	Right Calf	2	2	2	2	8	50
24	Left Ankle	1	1	1	1	4	25

25	Right Ankle	1	1	1	1	4	25
26	Left Foot	1	1	1	1	4	25
27	Right Foot	1	1	1	1	4	25

Research of working posture aims to Determine Whether work posture when carrying out the work is already classified as a good method and not risky for the workers. Activities arrangement based on working hours in the sack warehouse are:

Table 2. The division of activities based on hours worked

No.	Time	Activities
1	08:00 to 10:00 a.m.	Sacks being sort in a standing position
2	10:00 to 12:00	Sacks being a lift and being sort on the Pallet
3	12:00 to 13:00	Break
4	13:00 to 15:00	Sacks being sort in a sit position
5	15:00 to 16:00	Sacks being a lift and being sort on the Pallet

Source: PT. Makassar Tene. 2018

Below are the results of the work posture analysis using the OWAS and REBA methods for each activity during the sack sorting.

Figure 1. Worker posture when working

IOP Conf. Series: Materials Science and Engineering **885** (2020) 012036 doi:10.1088/1757-899X/885/1/012036

After the measurement angle of REBA done, as shown in Table 3, Scoring posture being performed on the sorting process items, namely:

Table 3. REBA Score for sorting activities

body Posture	Score	Description	Final Score
Neck	2	42o Forward	2
trunk	3	39o Forward	3
legs	1	Legs Stand Steady	1

Table A		Neck											
		1				2				3			
	Legs	1	2	3	4	1	2	3	4	1	2	3	4
Trunk Posture Score	1	1	2	3	4	1	2	3	4	3	3	5	6
	2	2	3	4	5	3	4	5	6	4	5	6	7
	3	2	4	5	6	4	5	6	7	5	6	7	8
	4	3	5	6	7	5	6	7	8	6	7	8	9
	5	4	6	7	8	6	7	8	9	7	8	9	9

Table B		Lower Arm					
		1			2		
	Wrist	1	2	3	1	2	3
Upper Arm Score	1	1	2	2	1	2	3
	2	1	2	3	2	3	4
	3	3	4	5	4	5	5
	4	4	5	5	5	6	7
	5	6	7	8	7	8	8
	6	7	8	8	8	9	9

	Table C											
Score A	Score B											
	1	2	3	4	5	6	7	8	9	10	11	12
1	1	1	1	2	3	3	4	5	6	7	7	7
2	1	2	2	3	4	4	5	6	6	7	7	8
3	2	3	3	3	4	5	6	7	7	8	8	8
4	3	4	4	4	5	6	7	8	8	9	9	9
5	4	4	4	5	6	7	8	8	9	9	9	9
6	6	6	6	7	8	8	9	9	10	10	10	10
7	7	7	7	8	9	9	9	10	10	11	11	11
8	8	8	8	9	10	10	10	10	10	11	11	11
9	9	9	9	10	10	10	11	11	11	12	12	12
10	10	10	10	11	11	11	11	12	12	12	12	12
11	11	11	11	11	12	12	12	12	12	12	12	12
12	12	12	12	12	12	12	12	12	12	12	12	12

Figure 2 . REBA Score for sorting activities

After submitting the final score of table A and table B into Table C, the resulting a final score of 4. In sack sporting activity, there is a short repetition of movements in checking and flipping one by one sack that has been checked the caused +1 score for this activity. When there are 20 bags of sacks, one sack workers will put in between the sacks to be sorted as the caused separators +1 score for this activity. Total of REBA Activity Score Table C and the resulting score of 6. The score indicates Action Level 2 roommates mean moderate risk level and work posture improvement is required.

IOP Conf. Series: Materials Science and Engineering **885** (2020) 012036 doi:10.1088/1757-899X/885/1/012036

Based on the identification work posture with OWAS and REBA method in the sorting process in the warehouse Sacks sack PT Makassar Tene, the obtained results:

Table 4 . Working Posture OWAS recapitulation results and REBA

		Owas		Reba	
Time	activity	Category	Risk	Score	Risk
08:00 to 10:00 a.m.	Sorting (Standing)	2	slightly hazardous	6	Moderate risk level
10:00 to 12:00 15:00 to 16:00	lifting sacks	4	Very dangerous	11	Very high-risk level work
10:00 to 12:00 15:00 to 16:00	Develop sack	4	Very dangerous	6	Moderate risk level
13:00 to 15:00	sorting sack	2	slightly hazardous	5	Moderate risk level

Source: Data processed. 2019

From the recapitulation in Table 4:31, OWAS method of sorting activities in a standing position has slightly dangerous risks with a score of 2. But the REBA method sacks sorting activities have a moderate risk with a score of 6 so that the necessary improvement work posture. This is due to workers who are bent posture when sorting sacks. Event raised sack on OWAS methods have dangerous risks with a score of 4. But the REBA method has a high risk with a score of 11. This is due to improper posture when lifting sacks and load the lift is 15-16 kg per 100 sacks so that the necessary improvement in posture at this time also taking into account the weight of the load is lifted. Arrange activities sack on OWAS method has a very dangerous risk with a score of 4 while the REBA method has a moderate risk with a score of 6. This is because the body posture when preparing to sack workers bent and tilted and knees bent so that the necessary improvement work posture at the time this. Sorting activities sacks in a sitting position have little risk of danger with a score of 2, while the REBA method has a moderate risk with a score of 5. This is because the posture of some workers at the time of sorting sacks still bent so that the results are relatively similar but slightly different.

4. Conclusion

Based on research that has been done, it can be concluded that there are Several sack sorting roommates activities identified as a risk in occurring musculoskeletal disorders (MSDs), the resulting OWAS Also very dangerous risk and high risk on REBA. Lifting Sacks onto the pallet and arranging sacks are roommates need to improve activities related to the posture of the worker's body as soon as possible / at this time, especially in the activity of lifting and arranging sacks.

Reference

[1] Rachmat A and Syafei M Y 2002 Analisis Musculoskeletal Disorders Dalam Perbaikan Tempat Kerja *Prosiding Seminar Nasional Teknik Industri III. Surakarta*

[2] Lee T-H and Han C-S 2013 Analysis of working postures at a construction site using the OWAS method *Int. J. Occup. Saf. Ergon.* **19** 245–50

IOP Conf. Series: Materials Science and Engineering **885** (2020) 012037 doi:10.1088/1757-899X/885/1/012037

Flood handling with priority scale determination in Samarinda Seberang from East Kalimantan Province Indonesia

SSN.Banjarsanti[1], Kukuh P.[1], and Theta Margaritifera[2]

[1]Civil Engineering Study Program, Samarinda State Polytechnic
[2]Civil Engineering Post-graduate Program, Gunadarma University, Depok

Email: ssnbanjarsanti@polnes.ac.id

Abstract. Handling of flooding caused by changes in land use often causes changes and problems in the magnitude of runoff coefficient, discharge, water level, and drainage channel dimensions, as happened in Samarinda Seberang. The purpose of this study was to plan a drainage channel according to the priority scale using the Rational Modification Method as a Hydrological Analysis and Manning as a Hydraulics Analysis. Handling method is carried out in the form of structuring and planning the drainage system according to priority scale based on inundation parameters, economic studies, social and transportation disturbances, loss of housing areas, and property rights. The higher the criteria for handling the area is a priority. The results showed that the highest priority scale (610) was obtained on Cipto Mangunkusumo street, and the lowest priority scale (220) was at Bung Tomo street and Hasanuddin street. Cipto Mangunkusumo street as the highest priority scale will be planned for the first priority drainage channel with a design flood discharge in the 10 year return period (Q) 7.79 m3 / sec, the drainage channel dimension consists of the basic width of the channel (b) 0.77 m ~ 0.80 m, channel peak width (T) 3.30 m, channel depth (h) 1.925 m, watch height (f) 0.6 m, total depth 2.52 m, and slope of channel wall (m) 1: 1).

1. Introduction

Changes in the development of land use often cause new problems, one of which is the functioning of existing drainage channels. Changes in land use will affect the magnitude of runoff coefficient (C), the runoff coefficient influences the amount of discharge (Q), water level (H), and dimensions of the drainage channel. One of them is in the Samarinda Seberang area, which is an urban area that is actively developing. Given the limited budget owned by the regional government, it is necessary to handle floods by planning a drainage channel according to the priority scale that meets technical standards, effectively and efficiently.

From the description above, the existing problems can be identified, namely, big is the design flood discharge for the return period 2, 5, 10 and 25 years, and how many dimensions of the cross-section of the drainage channel in Samarinda Seberang area which covers Jalan Bung Tomo, Cipto Mangunkusumo, Pattimura, Hasanudin and Mangkupalas and determining handling priority scale.

2. Research Methods

In this study, the Rational Modification Method was used as a Hydrological Analysis and Manning Method as a Hydraulics Analysis, based on previous studies and Journals/Proceedings on the

Effectiveness of Polder Doors as Flood Control Alternatives in the Sempaja Watershed of Samarinda City, East Kalimantan by SSN. Banjarsanti, 2015, which was published in the Proceedings of PIT HATHI / XXXII / 2015.

The analysis carried out includes:

a. Analysis of rainfall design with Gumbel Method and Person Type III Log
b. Analysis of the frequency of rainfall data
c. Test the frequency analysis of Smirnov-Kolmogorov and Chi-Square
d. Analysis of the design flood discharge with the Rational Modification Method

The truth of the conclusions made from the analysis of hydrological data is actually not absolutely correct, because the conclusion of the hydrological analysis is generally made based on sample data from the population. Therefore the application of opportunity theory is very much needed in the hydrological analysis. One of them is the Pearson Type III Log Distribution Application, with the Skill Coefficient Value

$$CS = \frac{n \cdot \sum\left(\log X - \overline{\log X}\right)^3}{(n-1)\cdot(n-2)\cdot\left(\overline{S\cdot\log X}\right)^3} \quad ; \log X = \frac{\sum \log X}{n}$$

$$\overline{S\cdot\log X} = \sqrt{\frac{\sum\left(\log X - \overline{\log X}\right)^2}{n-1}} \quad ; \log X = \overline{\log X} + k\cdot\left(\overline{S\cdot\log X}\right)$$

By:
Y : average value of Y
$\bar{Y}$: average value of Y
S: standard deviation from Y
k: characteristics of the Pearson type III log distribution
Source: (Asdak, Chay. 2010 [1]; Soemarto, CD, 1986 [2]; Soewarno, 1991 [3])

2.1. Debit Flood Plan (Q plan)
Flood discharge plan (Q plan) is the volume of water flowing through a cross-section of a channel. The most commonly used method in calculating the planned discharge due to rainwater is the Rational Modification Method, Q = 0.278. Cs. C. I. A

By: Q = maximum flood discharge (m3 / dt)
Cs = Retention coefficient = 2tc / (2tc + td)
tc = time of concentration (minutes)
td = channel duration in minutes (minutes)
C = flowing coefficient
I = intensity of average rainfall during arrival time from flood (mm / th)
A = area of drainage area (km2)
Source: (Imam Subarkah. 1980) [4]

2.2 Network Planning
Urban drainage design criteria have specificity because, for urban areas, there are additional design variables such as linkages with land use and city drainage master plans. Hydraulic Analysis as a limitation used in Channel or River Capacity Planning. The channel cross-section is planned to be trapezoidal and calculated based on the following formula [5,6].

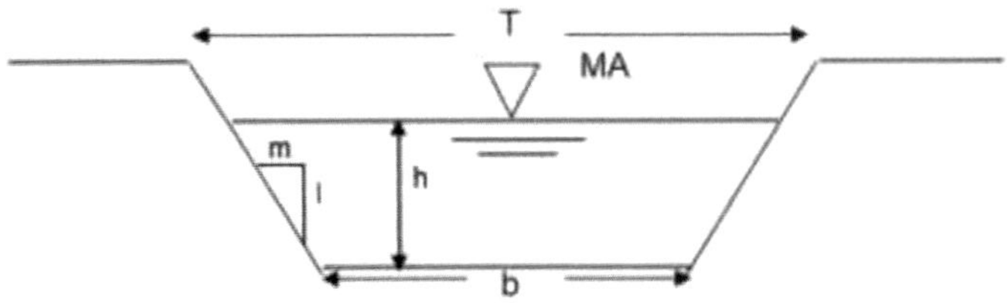

Figure 1. Cross-section of Trapezoidal Channel

$$Q = A\,V \; ; A = (b + mh)\,h \; ; \; P = b + 2h\sqrt{m^2+1} \; ; V = \frac{1}{n} . R^{\frac{2}{3}} . S^{\frac{1}{2}} \; ; \; R = \frac{A}{P}$$

By: Q = Plan debit (m3 / sec)
A = Area of wet cross section (m)
P = Wet channel circumference (m)
m = Slope of channel duct
R = Hydraulic finger (m)
b = Basic width Channel (m)

Table 1. Criteria for Inundation Handling Priority Determination

1. Inundation parameters	1. High inundation 2. Area of inundation 3. Duration of inundation 4. Frequency of inundation
2. Economic losses	
3. Social disturbances and government facilities	
4. Transportation losses and distributions	
5. Losses in residential areas	
6. Loss of personal property rights	

3. Results and Discussion

Location of inundation/flooding that occurred in Samarinda Seberang area, covering Jalan Bung Tomo, Cipto Mangunkusumo, Pattimura, Hasanudin, and Mangkupalas, as shown in the following figure 2. The processing of the data is as follows:

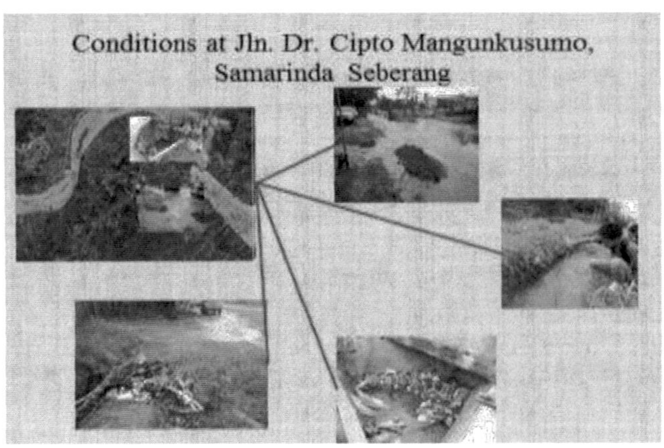

Figure 2. Location map
Source: Measurement Results (Survey)

3.1. Processing of Rainfall Data Into Rain Intensity

Hydrological analysis is needed to calculate the amount of design discharge that will be used in calculating drainage channel capacity. In this plan, rainfall data from the Temindung Airport Rainfall Recording Station, which are located 2 km from the study location, recorded (31 years) are presented in the following table:

Table 2. Maximum Daily Rainfall (mm) Per Year from Samarinda Temindung Airport Station

No.	Maximum daily rainfall number (mm)
1	115.80
2	105.60
3	85.70
4	80.50
5	108.90
6	97.30
7	89.40
8	105.30
9	94.30
10	90.00
11	141.80
12	82.00
13	79.10
14	94.60
15	85.00
16	117.10
17	83.80
18	101.60
19	66.30
20	39.00
21	118.20
22	108.00
23	132.10
24	94.40
25	73.00
26	60.20
27	86.50
28	105.50
29	98.90
30	96.00
31	102.50

Source: Temindung Airport Samarinda Rainfall Record Station

Rainfall calculation results are designed by the Person Type III Log Method

Table 3. Calculation of Rainfall Design with Log Pearson Type III Method

Return Periode (Year)	Probability (%)	Log X (mm)	X (mm)
2	50	1,987	78,885
5	20	2,050	101,175
10	10	2,073	118,430
25	4	2,090	128,948

Source: Calculation results

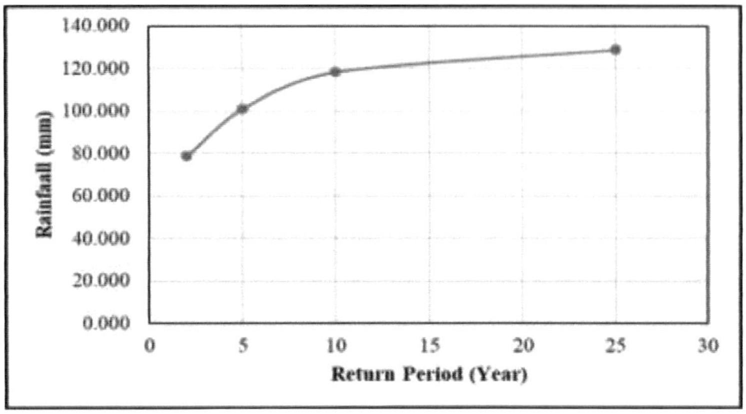

Figure 3. Graphic Relationship of Rainfall and Return Period

3.2. Frequency Conformity Test

Smirnov Test - Kolmogorov, this test is carried out to test the deviation in the horizontal direction, in order to determine the largest horizontal deviation of the theoretical distribution and empirical distribution (the difference or Δ). The testing stages are as follows:

- Number of data (n) = 31; Significant level α = 5%
- With n = 31 and α = 5%, then from the table, the relationship between n and α can find the critical price (Δcr) for the Smirnov Kolmogorov Test, which is (Δcr) = 0.29

$$\Delta cr = \left[\frac{(31-31)}{(31-31)}, (0{,}29 - 0{,}29)\right] + 0{,}29 = 0 + 0{,}29 = 0{,}29$$

The meaning of α = 5% (significance level of 5% means): approximately five out of 100 conclusions that will reject the hypothesis y that should be accepted or about 95% believe that we have made a correct conclusion.

$$P = \frac{10 \cdot m}{n+1} (\%) = \frac{100 \cdot 1}{31+1} = 3.125\%$$

Seen from the graph the relationship between percent chance of sceuronce (pr) or opportunity with XT so that the theoretical Dist. P. obtained al, for example, PT = 4.60

$$Kolom\ (7)'\Delta = |Po\ - \mathrm{PT}|(\%) = |4{,}76 - 4{,}60| = 0{,}16$$

Conclusion:

ΔP max = 13.71% = 0.1371 ≈ 0.14%; Significant degree (α) = 5%

Trust = 95%; Number of data (n) = 31; Δcritical = 0.29%; Because the value of maks P max <Δcr; Namely: 0.14 <0.29 - hypothesis accepted.

The conclusion is the hypothesis that the Type III Log Person Method is acceptable and meets the requirements.

Calculation of City Drainage Channels with Rational Modification Method

In the planning of the drainage channel, here is used the Rational Modification Method. The main drainage channel of the settlement is planned with T = 10 years, including:

1. Jl. Bung Tomo (Reel Road, Jl. Solid Karya, Jl. H. Jahrah)
2. Jl. Cipto Mangunkusumo

3. Jl. Sultan Hasanuddin
4. (Partly Jl. MAS Penghulu, Jl. Mangkupalas), partially Jl. Ampera.

Example Calculation of drainage channel discharge plan Jl. Bung Tomo, and the results of the debit calculation and the results of the subsequent channel dimensions are reflected in the table below. The calculation example is given to one of the Jalan Bung Tomo segments as follows:

1. Jl. Bung Tomo (Runoff Coefficient Value (C)
   Bung Tomo 1 0.38 Km2; Bung Tomo 2 0.10 Km2
   Bung Tomo 3 0.28 Km2; Bung Tomo 4 0.07 Km2

   Bung Tomo 2
   Value of runoff coefficient (C) land use in this area are:
   Settlements with values C = 0.83, Channel length, left and right = 741 m

1. Time of concentration

$$t1 = 0.0192 \left[\frac{L}{\sqrt{S}}\right] \times 0.77 \;\rightarrow\; S = 0.0025$$

$$= 0.0192 \left[\frac{200}{\sqrt{0.0025}}\right] \times 0.77 = 0.103 \text{ hours}$$

and t2 = v for concrete questions = 1.5 m/dt (see table)
= 0.23 hours; tc = t1 + t2 = (0.103 + 8.23) hours = 8,336 hours

2. Rain intensity

$$It_1 = \frac{R24}{24}\left[\frac{24}{tc}\right]^{2}/_{3} - for\ X10 \;= 118.43\ mm/hour$$

$$I_{10(1)} = \left[\frac{118.43}{29}\right]\left[\frac{24}{8.336}\right]^{2}/_{3} \;= 9{,}986\ mm/hour$$

$$Cs \;= \frac{2Tc}{2Tc+Ts}$$

$$\text{Where: Ts} = T_2 = \frac{2(8.836)}{2(8.836)+8.233} = 0.669$$

$$Q_1 = \frac{1}{3.6} \cdot Cs \cdot C \cdot I \cdot A = \frac{1}{3.6}(0.669)(0.083)(9.986) = 0.154\ m^3/dt$$

The following is given the Table of Debit Channel Calculation Result Plan in Samarinda Seberang District

Table 4. Debit Calculation Results Jl. Bung Tomo

Segment	1/3.6	Cs	C	I (mm/hour)	A (km^2)	Q (m^3/sec)	Qtotal (m^3/sec)
Bung Tomo 2	0.278	0.669	0.83	9.986	0.10	0.154	1.2
Bung Tomo 1	0.278	0.672	0.83	9.833	0.38	0.579	
Bung Tomo 3	0.278	0.670	0.83	9.872	0.28	0.154	
Bung Tomo 4	0.278	0.672	0.81	11.562	0.07	0.118	

Table 5. Recapitulation of Channel Dimension Calculation Results in Samarinda Seberang Area

Segment	T (m)	b (m)	H (m)	m	Material
Jl. Bung Tomo	1.70	0.50	1.40	1 : 1	Beton
Jl. Sultan Hasanudin	1.30	0.50	0.70	1 : 1	Pasangan Batu
Jl. Pattimura	1.70	0.50	1.40	1 : 1	Pasangan Batu
Jl. Cipto Mangunkusumo	3.30	0.80	2.52	1 : 1	Pasangan Batu

Table 6. Criteria for Determining Priority Handling Scale Scale in Samarinda Seberang Criteria for

Priority of Inundation Handling in Samarinda Seberang Subdistrict		Determining			
		Cipto M	Pattimura	Bung Tomo	Hasanuddin
1. Inundation parameters	a. Inundation height	100	100	30	25
	b. Area of inundation	0	25	0	25
	c. Duration of inundation	50	50	25	25
	d. Inundation frequency	100	100	100	25
2. Economic losses		65	30	30	30
3. Social disturbances and government facilities		65	30	65	30
4. Transportation losses and disruptions		100	30	65	30
5. Losses in residential areas		65	100	65	30
6. Loss of personal property rights		65	65	0	0
Total		610	530	400	220

Source: Calculation Results

Based on the assessment of the determinants of the priority scale of inundation handling with assessment analysis, the higher the value of the priority criteria for inundation handling, the regions will be prioritized to be dealt with based on inundation parameters, economic studies, social disturbances, transportation losses and disruptions, and losses in the regions housing and property rights, with the results of the priority areas for inundation and flooding in Samarinda Seberang District are Jalan Cipto Mangunkusumo with values (610), Pattimura with grades (530), Bung Tomo (400), and Hasanuddin (220).

4. Conclusions

Based on the analysis of priority areas for inundation and flooding, the highest to lowest priority scale values were obtained in Samarinda Seberang Subdistrict, namely the highest value of Jalan Cipto Mangunkusumo, namely (610), Pattimura with grades (530), Bung Tomo (400), and Hasanuddin (220). The results of the calculation of drainage channel planning that got the first priority scale, namely Jalan Cipto Mangunkusumo the priority scale value is 610, with the result of the largest design flood (Q) with a 10 year return period of 7.79 m3 / dt, with the results of the drainage channel planning dimension channel base width (b) 0.77 m ~ 0.80 m, channel peak width (T) 3.30 m, channel depth (h) 1.925 m, watch height (f) 0.6 m, total depth of channel (H) 2, 52 m, with the slope of the channel wall (m) 1: 1.

Recommendation

Given the budget constraints of the regional government, where flood control in the area must still meet technical standards, be effective and efficient, it should be done by determining the priority scale of handling inundation or flood areas so that the location of the area that has a higher inundation or flood influence is determined to become a top priority for immediate technical handling.

Acknowledgments
Thank you to the Directorate of Research and Community Service, the Ministry of Research, Technology and Higher Education which has funded this research in accordance with the Higher Education Leading Research Contract No: 862a / PL7 / LK / 2018, with Title Flood Control Based on Handling Priority Scale Samarinda Seberang District Samarinda City, East Kalimantan.

References
[1] Asdak C 2010 *Hydrology and management watershed* (Yogyakarta: Gajah Mada University Press)
[2] Soemarto C 1986 *Hydrological Engineering* (Surabaya: National Business)
[3] Soewarno 1991 *Measurement and Processing of River Flow Data (Hydrometry)* (Bandung: Nova Publisher)
[4] Subarkah I 1980 *Hydrology for Water Building Planning* (Bandung: Idea of Dharma)
[5] Te Chow V 2010 *Applied hydrology* (Tata McGraw-Hill Education)
[6] Raju K G R 1981 *Flow through open channels* (Jakarta: Erlangga)

Analysis of the impact of convection industrial waste

Nilda[1], Mulyadi[1], and Andi Nurwahidah[2]

[1]Industrial Engineering, Hasanuddin University
[2]Industrial Engineering, Polytechnic ATI Makassar

Email: Nilda.syamsul@yahoo.com

Abstract. Waste from the production of an industry is one of the problems that often arise. The statement about whether the waste will be thrown away or will be produced again to obtain products with added value to increase the economic value of residual waste production and reduce environmental impacts. This study will identify the impacts and causes of residual production waste in the convection industry and develop products from residual production waste so that we can reduce the environmental impact in the form of waste and also increase the economic value of products and of course become income for the industry. Initial identification is done by identifying waste from the production process in the convection industry. The list of wastes will be analyzed for their impacts and causes using the cause consequence analysis method. The initial method used was the fault tree analysis method to identify the cause of the waste that occurred and it was found that the average waste that occurred was caused by human error and the standard of clothing that was not good so it could not be continued to be produced. The next method is an event tree analysis that is used to identify the consequences if the waste is not handled properly. Solid waste from production if not managed properly will accumulate and can cause negative effects such as flooding. Processing waste into finished products must be supported for the sustainability of this earth, by providing financial assistance to small industries or households to develop products.

1. Introduction

The development of the manufacturing industry in Indonesia is one of the major contributors to the development of the Indonesian economy [1,2]. This can be seen from the increasing number of manufacturing industries in Indonesia, both large and medium industries. Some of the products from Indonesian industries have also gained control of the world market share. Based on data released by the United Nations Statistics Division in 2016, Indonesia ranks fourth in the world from 15 countries where the manufacturing industry contributed 22 percent to Gross Domestic Product. The development of the manufacturing industry at this time is certainly proof that Indonesian products have been accepted by many groups, but with such intense competition now, it is not impossible if one day the Indonesian industry will be unable to compete with industries from other countries. It is this highly competitive competition that demands the industry to increase the competitive advantage to win the competition.

The residual production waste from industry is one of the problems that often arise. A statement about whether the waste will be thrown away or will be produced again to get products with added value thereby increasing the economic value of residual waste and reducing the environmental impact. This research will identify the impacts and causes of residual waste produced in the convection industry and develop products from residual waste products so that we can reduce the environmental impact of waste and also add to the economic value of the product and of course, be an income for the industry.

Previous research on convection industry wastes was conducted by Idaman (2001) concerning the treatment of textile industry small-scale wastewater by the biofilter process [3]. Other research by Komarawidjaja (2016) concerning the distribution of textile industry liquid waste and its impact in several sub-district villages in Bandung [4]. A study by Mastur (2012), on the impact of the operation of the textile industry in the downstream watershed on the quality of well water and water supply in the city of Semarang [5]. Previous studies discuss more the impact of convection industry wastewater, Mazedul et al. (2014) raised research on the Environmental Sustainability Evaluation of Apparel Product: A Case Study on Knitted T-Shirts and concluded The unsustainable practices trends of poorly designed, raw materials selection, processing, manufacturing, and disposing of apparel products major contributing factor this problem [6]. The purpose of this study was to evaluate the environmental sustainability of various branded T-shirts and find out various sustainable issues for greener apparel products. While this study will discuss the causes and impacts of convection industry solid waste. This study uses the cause consequence diagram method to identify the impacts and causes of solid waste and provide an overview of the treatment of solid waste

2. Research Methods

This research focuses on identifying the impacts and effects of convection industry waste. The research method used is caused consequence diagram method. This method is a combination of two methods, namely the method of fault tree analysis and event tree analysis. This event tree analysis method identifies the effects of existing waste while the fault tree analysis method identifies the cause of the waste

3. Result and Discussion

The initial stage of this research is to identify the types of solid waste from the production of the convection industry. At this initial stage, it was identified that there were 6 types of solids, most of which were fabric waste from the material cutting process. The following are the types of solid waste produced:

1. fabric leftover from the fabric cutting process
2. defective clothes due to the wrong sewing process
3. defective clothes due to the wrong cutting process
4. Deformed clothes because of poor fabric quality
5. liquid waste from the screen printing process
6. bottles of screen printing paint

From the results of identification in each production process in the convection industry, it was found that the most waste occurred was the remnants of discarded cloth and clothing that was defective and could not be sold.

Figure 1. Types of convection industrial waste

The 6 types of waste that have been identified are then identified using the fault tree analysis method to determine the potential causes of the waste. Below is the identification of the causes and impacts of one of the convection industry wastes

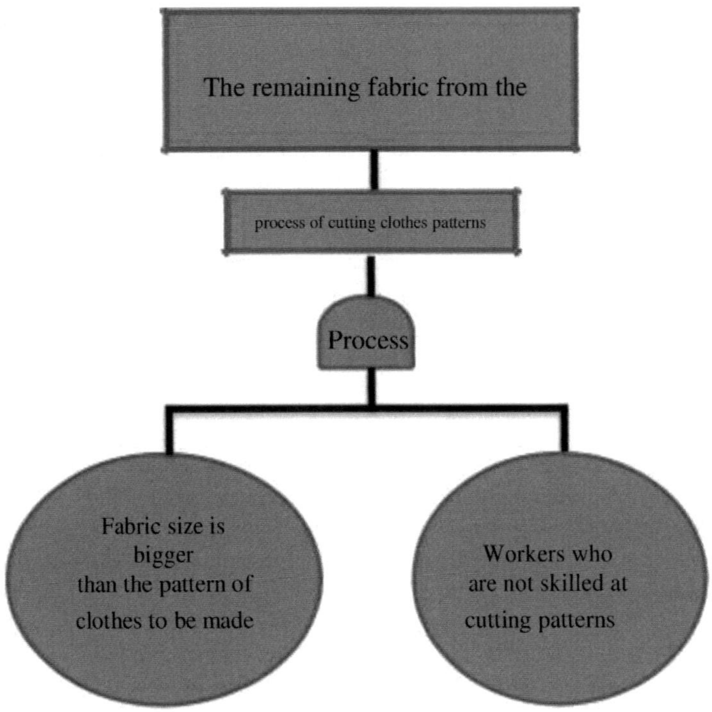

Figure 2. Fault Tree Analysis for remaining fabric from the process of cutting clothes pattern

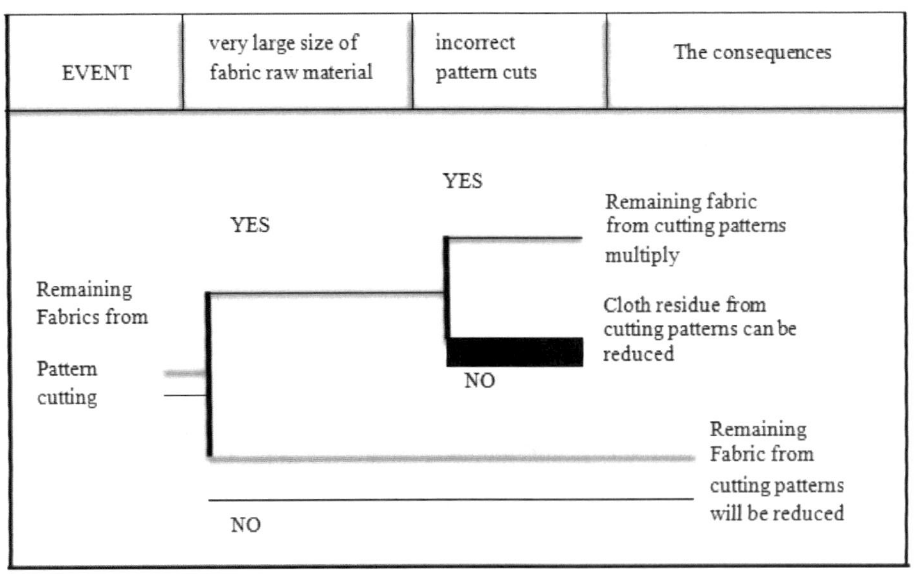

Figure 3. Event Tree Analysis for Remaining fabrics from pattern cutting

The results of the identification of several types of waste were obtained. The identified waste is a type of waste originating from the main raw materials from the convection industry, namely fabrics such as the results of pieces of cloth that are not used/wasted, clothes that are sewing incorrectly and cannot be sold anymore, clothes that are damaged due to wrong pieces, fabrics that are not used because incorrectly cut patterns and fabrics that cannot be used because of fabric defects that are congenital. The results of identification using the FTA method show that the waste from the fabric resulting from pattern cutting is a type of waste that cannot be avoided because at the time of cutting the fabric that follows the pattern will leave a cloth that will definitely be discarded. The second type of waste is the wrong sewing clothes, this type of defect, if it cannot be repaired, will become waste, this type of waste is caused by human error because workers are not trained so many clothes are sewing incorrectly and do not fit the desired specifications. Other types of waste are clothes that are defective and cannot continue to be produced due to poor quality raw materials such as tearing and others. The identification results of the event tree analysis method are known that all types of defects that result from the production process in the convection industry if not managed properly will cause major environmental problems whether it's in the form of solid waste from the production process or liquid waste from the clothes printing process. Solid waste in the convection industry, if not managed, will cause flooding because it is disposed of and makes garbage accumulate, which we know our country is still very poor in its solid waste management. For liquid waste, itself must also have good treatment from the beginning because it will endanger the environment with chemicals contained therein.

4. Conclusion

The initial identification process to determine the type of waste from the production process of the convection industry is found there are 4 types of waste namely cloth leftover from cutting fabric, clothes that are defective due to wrong sewing, clothes that are damaged because of scissors process, clothes that are defective because of poor fabric quality. After identification with cause consequence diagram methods (fault tree analysis and event tree analysis), it is known that the impact of convection industry waste, in addition to being able to pile up into the waste, can also cause material losses to the company. The results of the identification of the causes of waste are largely due to human error, and workers are less professional, and also the type of raw material is not good. In the development of waste management of the convection industry at this time still, by selling the waste to collectors, the convection industry is advised to manage its own waste to become a product so can have added value to the company and not contribute to the environment

Reference

[1] Sugiharti L, Purwono R, Primanthi M and esquivias miguel angel 2019 Indonesia Industrial Productivity Growth: Evidence of Re-industrialization or De-industrialization? *Period. Polytech.* **27** 108–18

[2] Tadjoeddin M Z, Auwalin I and Chowdhury A 2017 Revitalising Indonesia's Manufacturing *Eur. J. East Asian Stud.* **16** 124–53

[3] Said N I 2001 Pengolahan Air Limbah Industri Kecil Tekstil Dengan Proses Biofilter Anaerob-Aerob Tercelup Menggunakan Media Plastik Sarang Tawon *J. Teknol. Lingkung.* **2**

[4] Komarawidjaja W 2016 Sebaran limbah cair industri tekstil dan dampaknya di beberapa Desa Kecamatan Rancaekek Kabupaten Bandung *J. Teknol. Lingkung.* **17** 118–25

[5] Mastur Z and Rokhman F 2012 Dampak Pengoperasian Industri Tekstil di Das Garang Hilir Terhadap Kualitas Air Sumur dan Air Pasokan PDAM Kota Semarang *Sainteknol J. Sains dan Teknol.* **10**

[6] Islam M, Khan M and Rahman M 2014 Environmental sustainability evaluation of apparel product: a case study on knitted T-shirt *J. Text.* **2014**

IOP Conf. Series: Materials Science and Engineering **885** (2020) 012039 doi:10.1088/1757-899X/885/1/012039

Mixed matrix membrane of alginate zeolit crosslinked by PVA for separation of CO_2/CH_4

Syardah Ugra Al Adawiyah

Politeknik ATI Makassar

E-mail: syardah.ugra.su@gmail.com

Abstract. Mixed Matrix Membrane Alginate Zeolite Crosslinked by PVA for Separation CO_2/CH_4 was studied. Alginate/Zeolite/PVA mixed matrix membrane were prepared by solution casting. The research started by synthesis of Alginate/Zeolite/PVA membrane. The PVA 0.1% crosslinking alginate zeolite membrane (1:0.2) wa successfully synthesized. The membrane were characterized by Fourier Transform Infrared (FT-IR) and Barret Joyner Hallenda (BJH). The membrane permeability and selectivity of CO_2 and CH_4 were investigated. The effect of zeolite on the performance and the thickness of membranes of gas separation were analyzed. The result of permeability and selectivity show that membrane can flow CO_2 and resist CH_4. The effect of zeolite mass increasing on permeability of CO_2. The thickness of membrane influenced the flux of CO_2 gas. CO_2 permeability increased simultaneously with a decrease of a membrane thickness. According BJH result, pore diameter of membrane is 3.54 A, while kinetic diameter of CO_2 is 3.3 A and kinetic diameter of CH_4 is 3,8 A. This was explained performance of membrane was molecular sieving.

1. Introduction

Global warming is one of the major problems that face humanity. Global warming is caused by the emission of greenhouse gases, of which the main component is carbon dioxide (CO_2). CO_2 emissions have dramatically increased in the last 50 years, and are still continually increasing each year [1]. The technology of membrane has begun to be applied for the separation of CO_2 from natural gas. The advantages of this technology is a simple process, convenient, environmentally friendly, low energy consumption and operating costs [2]. His technology is commercially available for gas purification in 1980. Over the past years several studies are conducted in searching optimum membrane materials, membrane casting techniques and membrane module configurations for gas separation. Eficiency of membrane decreases with the passage of time which is the biggest challenge for industries [3].

In this method, a few components of the gas transported through a thin layer of membrane. Transport of each component is controlled by the partial pressure difference on the membrane. The membranes that can be used for gas separation have 2 types, i.e. porous membrane and non-porous membrane [4]. The transport mechanisms that can occur in gas permeation along the porous membrane is a bulk flow, molecular diffution and molecular sieveing [5].

The usefulness of alginate based on three main properties, the ability to: 1). soluble in water 2) increasing the viscosity of the solution to form a gel. 3). forming film and fiber [5]. Sodium alginate (NaAlg) is a cheap, stable and easily available polymer containing carboxyl groups [6]. NaAlg has very high hydrophilicity resulting from its carboxyl and hydroxyl groups [7], which is beneficial for the solubility of CO_2 in the membrane since water is needed for the preferential CO_2 permeation. Therefore, NaAlg should be a good candidate for CO_2 separation membrane materials [8].

Zeolite minerals, also known as natural sedimentary or natural occurring zeolites, are mainly composed of aluminosilicates with a three-dimensional framework structure bearing AlO_4 and SiO_4 tetrahedral. Natural zeolite mineral is suitable for use as the major component in the fabrication of mineral based membrane partly because it does not swell in water and easily forms a suspension for coating membrane on porous support [9]. Zeolite pore structure makes zeolites are often used as a tool for separating a smaller or larger substance from other substances. In use as a separator, zeolite is often synthesized then modified structure that can filter and separate the molecules of substances [10].

Polyvinyl alcohol (PVA) is an inexpensive polymer with an excellent film-forming ability used in membrane separation , as a fuel-cell electrolyte and in biological applications . PVA has a repeating hydroxyl group that makes it hydrophilic and offers good compatibility in aqueous solutions PVA induces extensive swelling in water, which decreases its mechanical stability. One way to improve membrane stability and mechanical properties in aqueous solutions is to incorporate an inorganic filler into the polymer matrix [11].

2. Experimental Section

2.1. Materials
Alginate, PVA, natural zeolite, CO_2 gas, CO_4 gas aquades, HNO_3, HCl, filter pape**r.**

2.2. Instruments
Glassware, weigher, magnetic stirrer, petri dish, oven, and set of gas permeability test equipment, Infrared Spectrophotometer, and Surface Area Analyser such as BET and BJH.

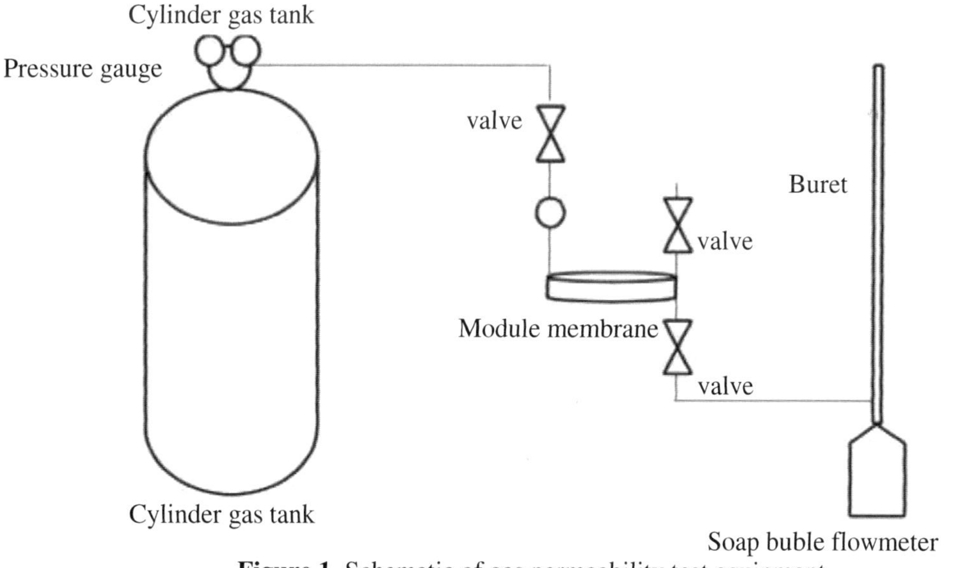

Figure 1. Schematic of gas permeability test equipment

2.3. Procedures

2.3.1. Zeolite activation. Natural zeolite crushed and sieved to 100 mesh screening passes. 20 g siftings washed with aquabidest. Then, the zeolite was dried at 110°C for 4 hours. Zeolite demineralized with a solution of 50 mL of 1 M HCl for 1 day. Furthermore, zeolite dealuminated with 50 mL of 8 M HNO_3 for 1 day. Zeolite oven at 110 °C for 4 hours and calcined at 600 °C for 3 hours.

2.3.2. Preparation of alginate-zeolite membrane an'd zeolite/PVA/alginate membrane. Materials such as alginate and zeolite dissolved in distilled water according to the composition specified (Table 1). Then stirred using a magnetic stirrer until homogeneous for 24 hours. Homogeneous mixture has been poured into a petri dish and then allowed to stand at room temperature to dry and peel. Membranes permeability tested using a permeability test equipment.

Table 1. Compotition of Membrane

PVA	Alginat: Zeolite Rasio
0.1%	1:0.1
0.1%	1:0.2
0.1%	1:0.3
0.1%	1:0.4
0.1%	1:0.5
0.1%	1:0.6

2.3.3. Permeability and selectivity. Single gas permeability of CO_2 and CH_4 were measured through all membranes by constant volume and variable pressure technique. Volumetric gas permeation rates were measured with soap bubble flow meter. The pressure difference of membranes were controlled at several pressure. The permeability test was repeated to the three different membrane samples and an average data was sermined. Pressure-normalized flux through the membrane is usually expressed as

$$\frac{P}{L} = \frac{Q_{stp}}{A_m(P_l - P_o)} \quad (1)$$

$\frac{P}{L}$ = permeability (GPU, 1 GPU=10^{-6} $cm^3_{(STP)}$ cm^{-2} s^{-1} $cmHg^{-1}$)
Q_{stp} = flow rate at standard conditions ($cm^3(STP)s^{-1}$)
A_m = membrane area (cm^2)
P_l = inlet pressure (cmHg)
P_o = out pressure (cmHg)

Selectivity of membrane can be sermined by relative permeability of components CO_2 and CH_4

$$\propto \frac{CO_2}{CH_4} = \frac{P/L(CO_2)}{P/L(CH_4)} \quad (2)$$

3. Result And Discussion

3.1. Characterization of the membranes

FTIR characterization that was carried out on the alginate/pva/zeolite membrane showed at (figure 2).

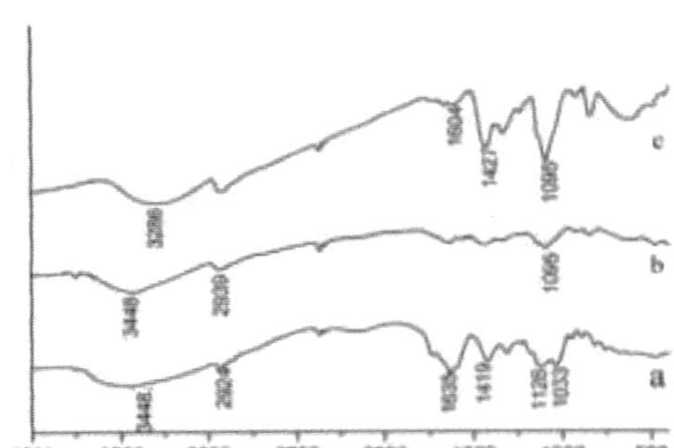

Figure 2. Spectrum of alginate/PVA/zeolite

Based on the analysis using the SEM (figure 3), the morfology of zeolite/PVA/alginate membranes is soft and dense. This figure shows a uniform distribution of zeolite inside the alginate matrix. This zeolite particles play a role in passing the gas through the membrane. In addition, the homogenity of the zeolite can affect permeability and selectivity of gas in separation process.

Identification of surface area is done through fisisorpsi N_2 gas at 77 K using microanalyzer NOVA 1200 (Quantachrome) using BET method (Brunaur Emmet Teller).

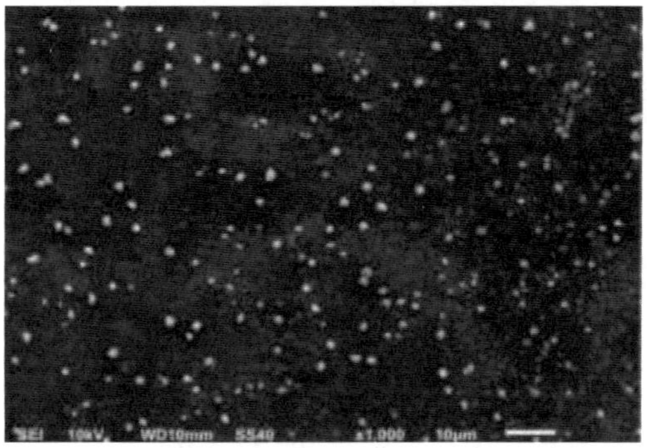

Figure 3. Analysis using the SEM

Based on this analysis, zeolite/PVA/alginate membrane has a specific surface area of 23.421 m^2/g and can produce the good permeability and selectivity of gas. Identification pore diameter is done through fisisorpsi N_2 gas at 77 K using microanalyzer NOVA 1200 (Quantachrome) using BJH method (Barret Joyner Hallenda). Based on this analysis, zeolite/PVA/ alginate membrane have a pore diameter of 3.54 Å, can produce good permeability and selectivity of gas.

3.2. Membrane permeability and selectivity

Table 2. Effect of zeolite loading for the CO_2 permeability using alginate-PVA-zeolite membrane

Pressure (Psi)	**CO_2 Permeability (GPU)**			
	Membrane 1: 0,1	Membrane 1:0,2	Membrane 1:0,3	Membrane 1:0,3
5	-	-	-	-
10	-	-	-	$3,15 \times 10^{-2}$
15	$4,6 \times 10^{-2}$	$1,92 \times 10^{-2}$	$2,4 \times 10^{-2}$	$6,28 \times 10^{-2}$
20	$5,77 \times 10^{-2}$	$55,7 \times 10^{-2}$	243×10^{-2}	278×10^{-2}
25	$45,7 \times 10^{-2}$	11500×10^{-2}		

Table 3. Effect of membrane thickness for the CO_2 permeability using alginate-PVA-zeolite membrane

Pressure (Psi)	**CO_2 Permeability (GPU)**		
	Membrane 50 ml (0,1589 mm)	Membrane 40 ml (0,1267 mm)	Membrane 30 ml (0,0899 mm)
5	-	-	-

10	-	-	-
15	9,38x10-2	1,92x10-2	3,18x10-2
20	11,8x10-2	55,7x10-2	55,5x10-2
25	4740x10-2	11500x10-2	

The permeability of CO_2 through this membrane increase with the increasing of applied zeolite mass and increasing of applied pressure . Discontinuities in permeability test occures at applied pressure of 25, 20 and 20 for membrane respectively. At applied pressure of 25 Psi, the CO_2 permeability of membrane 1:0,2 is much highest than was reported for other membrane. From the four types of membrane composition used and the permeability value of CO_2 gas, it can be concluded that the membrane composition 1:0,2 has the best quality compared to other membranes. The average permeability of the membrane and the high pressure can be applied to membrane used as an indicator of the membrane strength during permeability testing.

Using the composition 1:0,2, varying thickness used to see the effect of the thickness on the membrane permeability. Three variation of membrane thickness (0.0899 mm, 0.1267 mm and 0.1589 mm) are tested. In the permeability test using all membrane, the CO_2 gas can pass through membrane when the pressure of 15 Psi, while the CH_4 gas could not penetrate the membrane on some variation of the pressure until the membrane 0.0899 mm leak at a pressure of 25 Psi and 0.1267 mm, 0.1589 mm leak at pressure of 30 Psi. As shown at Table 4, the permeability of CO_2 through this membrane increase with the increasing of applied pressure in all membrane and membrane with thickness 0.1267 mm showed the best permeability compare to other membranes.

4. Conclusion

CO_2 permeability increased simultaneously with an increase of zeolite content in the polymer matrix. The thickness of membrane influenced the flux of CO_2 gas. CO_2 Permeability increased simultaneously with a decrease of membrane thickness. The PVA 0,1% crosslinking alginate zeolite membrane (1:0,2) shows the best permeability.

References

[1] Powell C E and Qiao G G 2006 Polymeric CO2/N2 gas separation membranes for the capture of carbon dioxide from power plant flue gases *J. Memb. Sci.* **279** 1–49

[2] Mulder M 2012 *Basic principles of membrane technology* (Springer Science & Business Media)

[3] Drioli E and Barbieri G 2011 *Membrane Engineering for the Treatment of Gases: Gas-separation problems combined with membrane reactors* vol 2 (Royal Society of Chemistry)

[4] Pabby A K, Rizvi S S H and Requena A M S 2008 *Handbook of membrane separations: chemical, pharmaceutical, food, and biotechnological applications* (CRC press)

[5] McHugh D J 2003 A guide to the seaweed industry FAO Fisheries Technical Paper 441 *Food Agric. Organ. United Nations, Rome*

[6] Wafiroh S, Pudjiastuti P and Sari I I 2016 The production of sulfonated chitosan-sodium alginate found in brown algae (Sargassum sp.) composite membrane as proton exchange membrane fuel cell (PEMFC) *AIP Conference Proceedings* vol 1718 (AIP Publishing LLC) p 50008

[7] Purwanti T, Syamsuar M, Hariyadi D M and Erawati T 2019 Characteristics And Release Of Gentamicin Sulphate From Sodium Alginate Microspheres Entrapped In Emulgel *Int. J. Drug Deliv. Technol.* **9** 704–10

[8] Yaqun Z H U, Zhi W, ZHANG C, Jixiao W and Shichang W 2013 A Novel Membrane Prepared from Sodium Alginate Cross-linked with Sodium Tartrate for CO2 Capture *Chinese J. Chem. Eng.* **21** 1098–105

[9] Scott J, Guang D, Naeramitmarnsuk K, Thabuot M and Amal R 2002 Zeolite synthesis from coal fly ash for the removal of lead ions from aqueous solution *J. Chem. Technol. Biotechnol. Int. Res. Process. Environ. Clean Technol.* **77** 63–9

[10] Howie R A, Zussman J and Deer W 1992 *An introduction to the rock-forming minerals* (Longman)

[11] Lue S J, Wang W-T, Mahesh K P O and Yang C-C 2010 Enhanced performance of a direct methanol alkaline fuel cell (DMAFC) using a polyvinyl alcohol/fumed silica/KOH electrolyte *J. Power Sources* **195** 7991–9

Worker fatigue in silo tower due to hot climate

A Nurwahidah[1], I Bakri[2], Mulyadi[2], N Ester Kelah[2]

[1] Industrial Engineering, Polytechnic ATI Makassar
[2]Industrial Engineering, Hasanuddin University

Email: Nurwahidah.andi@atim.ac.id

Abstract. The Tower Area at PT Eastern Pearl Flour Mills, which is located on the jetty, causes workers who are directly exposed to the reflection of sunlight. A hot work climate increases the physical fatigue of workers which can be measured by the strength of the handle and the stimulation reaction time. This research was conducted to identify the working climate conditions in the Tower area of PT. EPFM, conditions of work fatigue of workers and the influence of a hot work climate with work fatigue of workers. The linear regression method is used to determine the relationship between hot work climate and work fatigue. The hot working climate including ISBB and humidity respectively become independent variables or predictor variables while handheld strength and reaction time are dependent variables or response variables. Statistical test results with linear regression obtained ISBB and humidity significantly influence the decrease in the strength of the dominant hand grip (Sig. 0,000). ISBB and humidity significantly influence the decrease in the non-dominant grip strength (Sig. 0,000). ISBB and humidity significantly influence the increase in reaction time with regular stimulus (Sig. 0,000). ISBB and humidity significantly influence the increase in reaction time with random stimulation (Sig. 0,000). Recommendations that can be suggested to companies are the provision of mineral water and isotonic water and rest areas to support the recovery process, as well as provide information to workers about the effects of working in a hot environment.

1. Introduction

The design of a work system that is safe and comfortable as well as effective and efficient for its workers, requires many integrated factors that must be considered. These factors include humans, machines / technology, work methods, and work environment. Human interaction with machines implies that workers in carrying out work, in addition to dealing with machinery and equipment, another thing affecting the work of operators, is the work environment. The work environment influences operator performance and can even cause human error. Humans will be able to carry out activities if supported by a good work environment to create effective, comfortable, safe, healthy and efficient working conditions. Safe and healthy working environment conditions can make workers work normally both physically and mentally. Safe and healthy working conditions can also maximize productivity improvements. At the level of careless management of the quality of the work environment, opportunities to achieve targets in work productivity automatically become small. The low quality of the work environment in terms of physical and mental will cause pressure on workers so that many events disrupt work activities in the form of accidents and illness caused by work and the impact will endanger workers individually, in groups and even at the company level. PT Eastern Pearl Flour Mills uses sophisticated equipment in the loading and unloading process, but workers remain the main component that determines the success of the business in increasing productivity and work efficiency. Activities at the open dock exposed to direct sunlight have high heat pressure. High heat pressure produces a high work climate too. A high work climate will stimulate the body to sweat. So that over time the body experiences a lack of fluids. This is part of the general signs of fatigue.

Fatigue caused by a high work climate will have an impact on the workforce in doing their work If there are workers experiencing fatigue, workers will experience decreased concentration and increased fatigue in doing work that can trigger work accidents. Previous studies have been carried out by Sawant et al [1]. On the effects of temperature, relative humidity and water content on the germination of the percentage of wheat stored in different storage structures. Yildizel et al [2] raised the theme of the study of the impact of weather conditions on the health and performance of workers on construction sites. research by Balaganarajan et al [3] on climate change and its impact on employee productivity. Lundgren et al [4]. Conducted a study of the effects of heat stress on working population facing climate change. while this study aims to identify the relationship of work climate with the level of productivity of workers.

2. Research Methods

This study involved 13 workers in the Tower section of PT Eastern Pearl Flour Mills.The data taken in this study are as follows:

2.1. Work Climate Data

Data to find out the working climate condition is obtained by measuring ISBB and Humidity using the Heat Index WGBT Meter

2.2. Data Characteristics of Research Subjects

Data characteristics of research subjects are data on age, weight, and height.

2.3. Data on Work Fatigue

2.3.1. Handheld Strength Data, Handheld strength data is obtained through the measurement of handheld strength with an Electronic Hand Dynamometer.

2.3.2. Reaction Time Data, Reaction time data is obtained through measuring reaction time using the Reaction Timer (ERGO QUIP) tool.

Dependent sample t-test, often termed Paired Sample t-Test, is a type of statistical test that aims to compare the average of two groups that are in pairs. Paired samples can be interpreted as a sample with the same subject but undergo 2 different treatments or measurements.

Data processing and analysis techniques were performed using a regression test using the SPSS computer program Version 22.0, with the interpretation of the results as follows:

1) If P-value ≤ 0.01, the test results are declared very significant.
2) If P value$>$ 0.01 but ≤ 0.05, the test results are declared significant.
3) If P value$>$ 0.05 then the test results are declared not significant

3. Result and discussion

Maintenance activities are carried out when there are no ships docked at the dock (there are no loading/unloading activities). Maintenance activities are carried out by the same worker as the loading/unloading worker/operator and are carried out in just one work shift each day, from 8 AM to 4 PM. Maintenance activities are carried out in the form of machine maintenance and cleaning of the tower area. Cleaning the tower area also includes the area around the conveyor belt. Cleaning is done by pouring water through the hoses on each side and9 rubbing the mast to avoid corrosion because there is a lot of dust in the form of wheat and wind-borne pellets attached to the iron towers.

The average changes in the Temperature of the Wet Bulb Temperature (outdoor and full sun) every hour around the Tower of PT. Eastern Pearl Flour Mills can be seen in figure 1 below

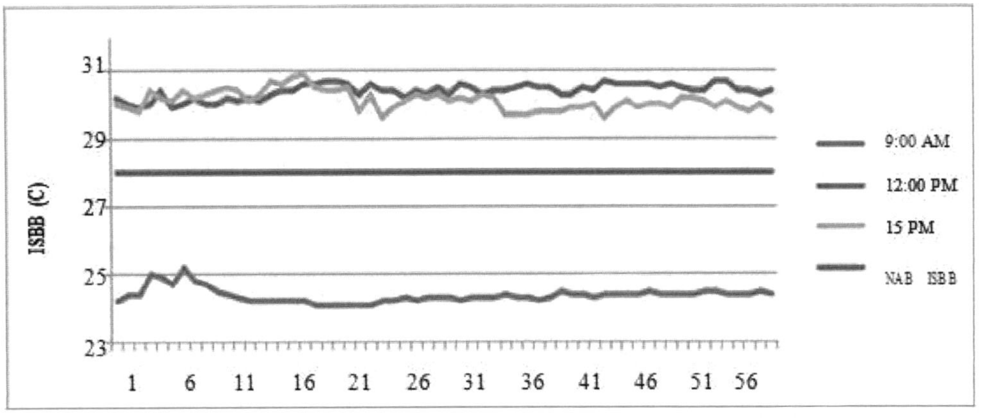

Figure 1. ISBB average in the tower area

Average hourly humidity changes around PT. Eastern Pearl Flour Mills can be seen in figure 2 below

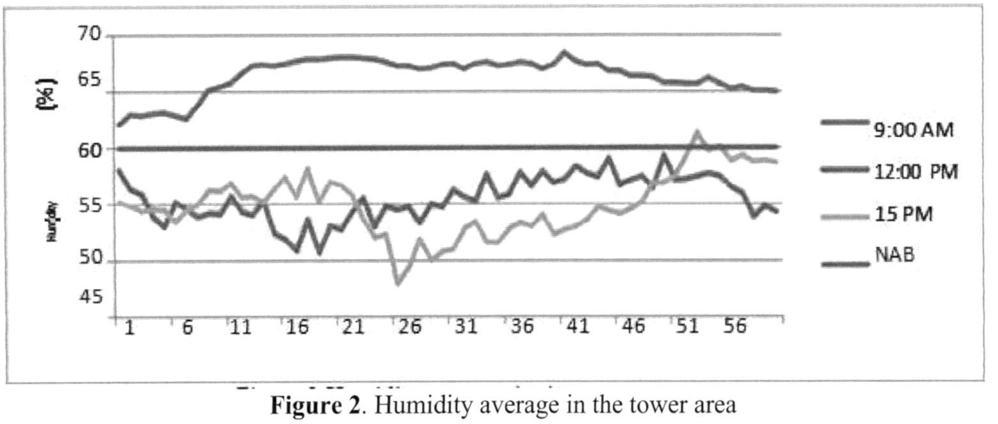

Figure 2. Humidity average in the tower area

Before measuring psychomotor abilities (reaction time) and physical abilities (handgrip strength), the researchers first took data on the age (years), height (cm) and weight (kg) of workers. Table 1 below shows data on the characteristics of the worker's body in the Loading / Unloading section at PT Eastern Pearl Flour Mills Tower.

IOP Conf. Series: Materials Science and Engineering **885** (2020) 012040 doi:10.1088/1757-899X/885/1/012040

Table 1. data on worker characteristics

Data	Min	Max	Average	Standard Deviation
Age (Years)	25.00	46.00	37.00	6.97
Weight (kg)	55.00	78.00	67.00	7.73
Height (cm)	150.00	168.00	163.00	5.56
Body Mass Index	21.48	30.22	25.20	3.12

3.1. Physical Ability Data

In this study, physical ability data was measured through the handgrip strength of workers by using an Electronic Hand Dynamometer. This measurement is carried out before starting work and after work to find out whether there is fatigue experienced by workers. Table 2 is show the average data of strength of workers at 09 AM, 12 PM and 03 PM using dominant and nondominant hands.

Table 2. Handgrip strength measurement data of workers

Name	09 AM		12 PM		03 PM	
	Dominant	Non Dominant	Dominant	Non Dominant	Dominant	Non Dominant
Worker 1	47,97	45,90	37,90	35,03	42,63	42,37
Worker 2	48,67	44,50	40,23	38,37	40,33	39,57
Worker 3	45,40	44,00	41,73	38,67	43,27	41,23
Worker 4	50,63	51,07	47,10	43,17	48,13	45,67
Worker 5	40,70	39,57	32,93	33,10	36,60	33,63
Worker 6	43,80	43,90	33,70	32,00	36,47	33,33
Worker 7	45,23	41,63	37,13	33,23	39,30	36,87
Worker 8	40,67	39,63	37,77	33,70	38,77	36,50
Worker 9	44,20	42,90	41,53	36,10	39,83	38,93
Worker 10	42,27	39,77	37,30	35,10	38,67	37,57
Worker 11	37,37	32,80	38,00	33,93	38,30	32,87
Worker 12	43,70	38,83	31,33	31,97	37,30	37,63
Worker 13	44,40	42,27	36,37	35,33	40,13	38,20
Average	44,23	42,06	37,93	35,36	39,98	38,03
Max	50,63	51,07	47,10	43,17	48,13	45,67
Min	37,37	32,80	31,33	31,97	36,47	32,87
Standard Deviation	3,58	4,31	4,15	3,15	3,19	3,68

3.2. Psychomotor Ability Data

In this study, the measurement of psychomotor abilities is done by measuring the reaction time, which is the time between giving a stimulus to someone to the reaction. Stimulation is carried out regularly and randomly using the Reaction Time Meter tool. In Table 3 the following data will show the average results of measuring reaction time with regular and random stimuli at 09 AM, 12 PM and 03 PM.

Table 3. Worker reaction time measurement data

Name	9 AM		12 PM		03 PM	
	Regular	Irregular	Regular	Irregular	Regular	Irregular
Worker 1	1,54	1,74	1,89	3,05	1,58	2,94
Worker 2	1,15	2,15	2,65	3,73	1,59	3,05
Worker 3	0,94	2,05	1,72	3,91	1,18	2,21
Worker 4	1,23	1,48	1,73	3,08	1,33	2,11
Worker 5	1,28	2,42	2,44	3,74	1,47	1,59
Worker 6	1,22	2,31	2,16	3,72	1,53	2,46
Worker 7	1,42	1,85	2,59	2,89	1,32	2,74
Worker 8	1,03	1,83	2,13	3,19	1,52	2,39
Worker 9	1,24	1,89	1,97	3,12	1,35	2,18
Worker 10	1,28	1,88	1,68	3,16	1,42	1,72
Worker 11	1,26	2,11	1,87	4,37	1,39	3,04
Worker 12	1,34	1,94	1,83	2,84	1,44	2,09
Worker 13	1,45	1,74	1,96	2,53	1,98	2,43
Average	1,26	1,95	2,05	3,33	1,47	2,38
Max	1,54	2,42	2,65	4,37	1,98	3,05
Min	0,94	1,48	1,68	2,53	1,18	1,59
Standard Deviation	0,16	0,25	0,33	0,52	0,19	0,47

To see how much the relationship of work climate with the physical ability (handgrip strength) and psychomotor responses (reaction time), Table 4 presents data on the r-value of the comparison of each independent variable and the dependent variable.

Table 4. R value data of the dependent variable and the independent variable (ISBB, humidity)

Dependent variable	Independent variable (ISBB, Humidity)	
	R	R^2
Handgrip strength (Dominant)	0,598	0,358
Handgrip strength (Non dominant)	0,608	0,369
Reaction Time (Regular)	0,824	0,678
Reaction Time (Irregular)	0,814	0,663

Table 5 can be seen the significance value between the dependent variable and the independent variable (ISBB, Humidity)

Table 5. significance of dependent and independent variables (ISBB, Humidity)

Independent Variable	Dependent Variable							
	Handgrip strength (dominant)		Handgrip strength (non dominant)		Reaction Time (Regular)		Reaction Time (Irregular)	
ISBB	0.209		0.105		0.000		0.000	
Humidity	0.302	0.000	0.161	0.000	0.000	0.000	0.000	0.000

The average work climate measurement results in the tower area for ISBB at 09:00 amounted to 24.50C, at 12:00 amounted to 30.60C and at 15:00 amounted to 30.00C. The measurement results of humidity at 09:00 amounted to 66.3%, at 12:00 amounted to 56.4% and at 15:00 amounted to 56.8%. The work climate measurement results for ISBB at 09:00 show conditions in accordance with the standards but at 12:00 and 15:00 show conditions not in accordance with the specified standards or NAV, while for humidity at 09:00 indicate conditions are not appropriate with standards but at 12:00 and 15:00 shows conditions according to standards.

The results of measurements of the physical abilities of workers in the tower area, based on the strength of the handgrip (Handgrip Strength) shows the symptoms of fatigue indicated by a decrease in the strength of the handgrip on the dominant hand from 44.23 ± 3.58 kg at 09:00 to 37.93 ± 4 , 15 kg at 12:00 but there was an increase in the strength of the dominant hand grip at 15:00 to 39.98 ± 3.19 kg. Non-dominant hands also experienced a decrease in handgrip strength from 42.06 ± 4.31 kg at 09:00 to 35.36 ± 3.15 kg at 12:00 but an increase in non-dominant handgrip strength at 15: 00 to 38.03 ± 3.68 kg. The measurement results of the psychomotor ability of tower area workers, based on the reaction time (Reaction Time) showed that fatigue symptoms were indicated by an increase in reaction time on regular stimulus from 1.26 ± 0.16 seconds at 09:00 to 2.05 ± 0, 33 seconds at 12:00 but the reaction time decreased at 15:00 to 1.47 ± 0.19 seconds. At the time of the random stimulus reaction, there was also an increase in reaction time from 1.95 ± 0.25 seconds at 09:00 to 3.33 ± 0.52 seconds at 12:00 but there was a decrease in reaction time at 15:00 to 2 .38 ± 0.47 seconds.

4. Conclusion

There is a moderate/sufficient influence / linear relationship (r = 0.598) between ISBB and Humidity to the decrease in dominant Handgrip strength in PT Eastern Pearl Flour Mills tower area workers. There is a strong influence / linear relationship (0.608) between ISBB and Humidity to the decrease of non-dominant Handgrip strength in the tower area workers at PT Eastern Pearl Flour Mills. There is a very strong influence / linear relationship (0.824) between ISBB and Humidity towards an increase in regular stimulus reaction time on the tower area workers at PT Eastern Pearl Flour Mills. There is a strong influence / linear relationship (0.814) between ISBB and Humidity to the increase in the random stimulus reaction time on the tower area workers at PT Eastern Pearl Flour Mills.

Reference

[1] Sawant A A, Patil S C, Kalse S B and Thakor N J 2012 Effect of temperature, relative humidity and moisture content on germination percentage of wheat stored in different storage structures *Agric. Eng. Int. CIGR J.* **14** 110–8

[2] Yildizel S A, Kaplan G, Arslan Y, Yildirim M S and Ozturk A U 2015 A study on the effects of weather conditions on the worker health and performance in a construction site *J. Eng. Res. Appl. Sci.* **4** 291–5

[3] Balanagarajan K and Gajapathy V 2018 Climate changes and its impact on employee productivity *Int. J. Appl. Eng. Res.* **13** 27–9

[4] Lundgren K, Kuklane K, Gao C and Holmer I 2013 Effects of heat stress on working populations when facing climate change *Ind. Health* **51** 3–15

IOP Conf. Series: Materials Science and Engineering **885** (2020) 012041 doi:10.1088/1757-899X/885/1/012041

Determination of lead time arrival of raw materials optimal vannamei shrimp at PT. XYZ Makassar

Puadi Haming and Rizqa Awaliyah

Industrial Engineering, Politeknik ATI Makassar

Email: adilajoa@gmail.com, awaliyahrizqa@gmail.com

Abstract. Inventory control is the efforts made by a company so that the production process can be optimally fulfilled with the smallest possible risk. The problem of raw material requirements and costs incurred for the production process is very important, because with the existence of raw materials the production process can run smoothly. This study aims to determine the amount of economical Vannamei shrimp raw material inventory and to know when the order should be done with PT. XYZ Makassar. Where the need for raw materials in 2017 was 1,855,558 kg. Therefore, companies must control the supply of raw materials that are economical and efficient in order to meet the production process for each specified period, determine the optimal waiting time for the arrival of raw materials and determine the quantity of orders that are economical. Based on the results of research determining the optimal waiting time for arrival of raw materials at PT XYZ Makassar using the Economic Order Quantity (EOQ) method, obtained an economical raw material inventory of 20,672 kg per message with a total inventory cost of Rp.136,586,577 each time message. With the Stock Out Cost (SOC) and Extra Carrying Cost (ECC) method, the optimal waiting time for arrival of raw materials is 3 days with a minimum cost of Rp. 1,836,979,830 and with the Reorder Point (ROP) method, the results of the order are made again at the level inventory in warehouse 18,555 kg.

1. Introduction

1.1. Background

PT. XYZ Makassar is a company engaged in shrimp processing and freezing. One of the raw materials used is Vannamei shrimp obtained from suppliers in Barru, Pangkajenne and the Islands, Makassar, Bulukumba, Kendari, Palu and Luwu which have been partners for so long. The company's finished products are only exported to Japan and not distributed domestically, so the quality of the seafood products is prioritized. To produce quality products, synergy between humans or workers is needed, with the material, technology and the environment.

According to Agus Ahyari (2003), The continuity of the production process within a company will be influenced by various factors including capital, technology, methods, inventory and labor [1]. One important part in the production process is inventory.Inventory control is a very important problem because the amount of inventory will determine or influence the smooth production process as well as the effectiveness and efficiency of each company. This inventory needs to be controlled regularly and periodically, starting from raw materials, semi-finished materials, to finished goods. Raw material inventory must be able to meet production needs, because if raw material inventory cannot be met, it will hamper the production process. Sometimes raw materials are still quite a lot but purchases have been made so that it results in the accumulation of raw materials in warehouses. This can reduce the quality of ingredients and will increase storage costs. Conversely, delays in ordering raw materials have an impact on not meeting the needs of products ordered by consumers. So it is very important to control inventory well.

PT. XYZ Makassar processes frozen raw shrimp which will be exported to Japan using the main raw material, Vannamei shrimp. Where the availability of main raw materials is to support the continuity of the production process within the company. From the results of the study there are obstacles in the company that is no consideration of determining the optimal waiting time for the arrival of raw materials, which results in companies often experiencing excess and lack of raw materials so that it impedes the company's production process.

Based on data from the amount of Vannamei shrimp raw material needs in 2017 reached 1,855,558 kg, while in 2016 the needs of Vannamei shrimp raw materials only reached 1,790,728 kg or an increase of 3.62%. To ensure the smooth and sustainable production, the company needs to control inventory by determining the optimal amount of raw material inventory using the Economic Order Quantity (EOQ) method. Next, According to Gitosudarmo Indriyo (2002) make an optimal waiting time for the arrival of raw materials by calculating re-order points using the calculation of the cost of Stock Out Cost (SOC) and Extra Carrying Cost (ECC) [2]. Stock Out Cost (SOC), which is the costs that have to be incurred due to the delay in the arrival of raw materials and Extra Carrying Cost (ECC), the costs that have to be incurred because the raw materials come too early. By using this method it is expected to be able to provide the best solution for the company where it can be set for an economical number of orders and when the company has to reorder.

1.2. Formulation of the problem
Based on the background that has been described previously, the formulation of the problem in this study is what is the waiting time (lead time) for the optimal arrival of Vannamei shrimp raw material at PT. XYZ Makassar

1.3. Research purposes
Based on the formulation of the problem, the purpose of this research is to determine the lead time for the optimal arrival of Vannamei shrimp raw material at PT. XYZ Makassar

2. Research Methodology
The data obtained is then observed and processed to obtain the desired results, and the data processing analysis stages are:

2.1. EOQ Method
a. According to T. Hani Handoko (1995), Calculate the number of orders in a single message by using the EOQ method [3].

$$EOQ = \sqrt{\frac{2.D.S}{H}}$$

Where,

EOQ : Optimum number of units per-order
S : Orderong cost for each order
D : Annual demand in unit for the inventory item
H : Holding or carrying cost per unit per year

b. According to Heizer and Render (2005), Calculates the total cost (TIC) of inventory based on the EOQ method [4].

$$TIC = S \times \frac{D}{Q} + H \times \frac{Q}{2}$$

Where,
TIC : Total inventory costs
Q : Number of unit per- order
S : Orderong cost for each order
D : Annual demand in unit for the inventory item
H : Holding or carrying cost per unit per year

c. According to Gitosudarmo Indriyo (2002), Calculate the ratio of Extra Carrying Cost (ECC) and Stock Out Cost (SOC) [2].

Extra Carrying Cost can be searched using the formula:

$$ECC = \frac{D}{\text{number of days / year x storage costs}}$$

Stock Out Cost can be searched using the formula:

SOC = Need for raw materials per day x Stock out cost

d. According to Sofyan Assauri (1998), Calculate the value of ROP (Reorder Point) [5].

Reorder Point = (LD x AU) + SS

Where,
LD : Lead time or waiting time
AU : Average unit or average usage during the unit waiting time
SS : Safety stock

3. Results and Discussion

3.1. Data Collection Results

As explained in the previous chapter, that to discuss problems in the company is used the calculation of Extra Carrying Cost (ECC) and Stock Out Cost (SOC).

PT. XYZ Makassar procures Vannamei shrimp raw materials which are closely related to the production process by ordering from suppliers who have become partners. The purchase of raw materials in 2017 amounted to 1,855,558 kg per month on the average is 154,630 kg at a price of Rp. 66,000 per Kg.

Table 1. Cost of Vannamei Shrimp Raw Materials per Kg in 2017

Name of goods	Unit	Order Amount (Kg)	Price/Kg
Vannamei Shrimp	Kg	1.855.558	Rp 66.000

Source: PT. XYZ Makassar, 2017

The cost of raw material inventory that must be borne by PT. XYZ Makassar includes:

3.1.1. Ordering Cost. The details of the costs incurred by the company for each order of raw materials are as follows:

Table 2. Ordering Costs

No.	Cost Type	Cost incurred
1.	Transportation costs	IDR 540,000 per Ordering
2.	Cost of Telephone	IDR 40,000 per ordering

3.	Cost of Listing	IDR 50,000 per ordering
4.	Cost of Unloading Raw Materials	IDR 130,000 per ordering
	Total	IDR 760,000 per ordering

Source: PT. XYZ Makassar, 2017

In table 2, it can be seen that the cost that must be incurred by the company in making each order during 2017 is IDR 760,000. These costs are the accumulation of costs incurred by the company in making each order.

3.1.2. Storage Cost. The cost of storing raw materials at PT XYZ Makassar is calculated in the form of a percentage that is the percentage of inventory value. The amount of raw material storage costs set by the company is 10% of the inventory price per kilogram per year and is a calculation of some storage costs such as packing costs, electricity usage, ice usage, warehouse administration costs, and reserve costs for possible damage to goods in inventory.

Table 3. Storage costs per kg

Cost	Amount
Price per Kg, IDR 66.000 x 10%	IDR 6,600/kg
Total	IDR 6,600/kg/day

Source : PT XYZ Makassar Tahun 2017

In the table 3, it can be seen that the storage costs incurred by the company is IDR.6,600 per kg per day.

Table 4. List of Raw Material Needs at PT XYZ Makassarperiod for 2017

No.	Month	Amount (Kg)
1	January	123,949
2	February	105,284
3	March	160,929
4	April	181,534
5	May	140,236
6	June	128,236
7	July	180,236
8	August	192,295
9	September	150,372
10	October	162,109
11	November	190,075
12	December	140,304
	Total	1,855,558

Source : PT XYZ Makassar Tahun 2017

Based on table 4 it can be seen that the total company's raw material needs in 2017 were 1,855,588 kg with an average number of orders of 154,630 kg, the cost of ordering each time ordering was Rp. 760,000 and storage costs required are Rp. 6,600.- per kg / day with a storage time of 3 days. Then from the data above the total inventory cost (TIC) of the company can be seen as follows:

IOP Conf. Series: Materials Science and Engineering **885** (2020) 012041 doi:10.1088/1757-899X/885/1/012041

$$TIC = \frac{D}{Q}S + \frac{Q}{2}H$$

Where,

TIC : Total inventory costs
Q : The number of items each time an order = 154,630 kg
S : Order cost for each order = Rp. 760,000
D : Annual demand for supplies = 1,855,588 kg
H : Per-unit storage cost = Rp. 6,600.

$$TIC \;= \frac{1.855.588\ \text{kg}}{154.630\ \text{kg}} Rp.760.000 + \frac{154.630\ \text{kg}}{2} Rp.6.600$$

$$TIC \;= Rp\ 519.399.138$$

3.2. Data processing

3.2.1. Calculation of inventory control using the EOQ method. Inventory analysis of the EOQ method uses QM For Windows V5 software for more precise accuracy and completeness of the calculation results including frequency and total cost based on the EOQ results in table 5 below

Table 5. Inventory Results EOQ method

Parameter	Value	Parameter	Value
Demand rate (D)	1855558	Optimal order quantity (Q*)	20672.22
Setup/ordering cost	760000	Maximum Inventory level (Imax)	20672.22
Holding/ carrying cost	6600	Average Inventory	10336.11
Unit cost	66000	Order per period (year)	89.76
		Annual Setup Cost	68218320
		Annual Holding Cost	68218320
		Total Inventory (holding + setup cost)	136436600
		Unit cost	122466800000
		Total cost	122603300000

Source : QM Data Processing for Windows V5

From table 5 above we get the EOQ value of 20,672.22 kg which means that the number of economic orders that must be ordered is 20,672 kg / one-time order and the results of orders per period (year) are 89.76 times means that the order frequency in one period is 90 times order.

3.2.2. Determination of TIC (Total Inventory Cost). The TIC formula is as follows:

$$TIC = S \times \frac{D}{Q} + H \times \frac{Q}{2}$$

Where,

TIC : Total inventory costs
Q : The number of items each time an order = 20,672 kg
S : Order fee for each order = Rp. 760,000
D : Annual demand for supplies = 1,855,588 kg
H : Perunit storage cost = Rp. 6,600

IOP Conf. Series: Materials Science and Engineering **885** (2020) 012041 doi:10.1088/1757-899X/885/1/012041

3.2.3. Comparison of the company's total inventory costs (TIC) with the total inventory costs (TIC) using the EOQ method

Table 6. Comparison of company TIC with EOQ method

Comparative Variable	Company	EOQ
Number of orders one time order	154,630 kg	20,672 kg
The frequency of ordering	12 times	90 times
TIC	IDR 519,399,138	IDR 136,586,577

Source: Data processed 2019

3.2.4. Comparison of Stock Out Cost (SOC) and Extra Carrying Cost (ECC). To calculate the re-order point, it must carefully calculate the timeliness because if the order regresses somewhat later than that time it will add to the purchase of basic materials or so-called Stock Out Cost (SOC), and if too early the extra carry ing cost (ECC) cost is calculated. Determination of Lead Time with minimum costs. From 10 historical data about Lead Time in 2017 showing the probability is:

Table 7. Lead time probability

Lead Time	Frequency	Probability
2 days	2 times	20%
3 days	5 times	50%
4 days	3 times	30%
	10 times	

Calculation of Extra Carrying Cost (costs incurred due to arrival of raw material orders too early)
Demand (D) : 1,855,588 kg
Storage Cost : Rp 6,600/kg

$$\text{ECC} = \frac{D}{\text{number of days / year x storage costs}}$$

$$ECC = \frac{1{,}855{,}588 \text{ kg} \times \text{Rp } 6{,}600}{300 \text{ hari}} = \text{Rp } 40{,}822{,}936/\text{day}$$

Calculation of Stock Out Costs (costs incurred for the purchase of substitute / substitution basic materials will come later than the coming order).
The need for raw goods per day:

$$\frac{1{,}855{,}588 \text{ kg}}{300\ days} = 6{,}185\ kg/day$$

Stock Out Cost = Rp 6,600/kg

$$\text{SOC} = \text{Need for Raw Goods per day x Stock out cost}$$

$$\text{SOC} = 6{,}185\ kg\ x\ 6{,}600 = \text{Rp } 40{,}821{,}100/\text{day}$$

3.2.5. Calculation Reorder Point (ROP)
Needs per day = (1,855,588)/300=6,185 kg
Needs during lead time = 3 x 6,185 kg = 18,555 kg

ROP = Need during lead time + safety stock
= 18,555 kg + 0 = 18,555 kg

3.3 Discussion
From the results of ECC and SOC calculations can be compared as follows:

Table 8. ECC and SOC calculation results

Lead Time	ECC		SOC		Total (Rp)
	Every order (Rp)	For one year (90 x messages) (Rp)	Each Order (Rp)	For one year (90 x messages) (Rp)	
2 day	0	-	44,903,100	4,041,279,000	4,041,279,000
3 day	8,164,587	734,812,830	12,246,300	1,102,167,000	1,836,979,830
4 day	36 ,740,642	3,306,657,780	-	-	3,306,657,780

Source: Data processed

From the comparison of ECC and SOC calculation results it is known that the Lead Time with the most minimal cost is 3 days with a total cost of Rp. 1,836,979,830, thus it can be determined 3 days before the raw materials run out the company must re-order raw materials.Order back (ROP) is done when the inventory in the warehouse is at the level of 18,555 kg.

4. Conclusion
Based on the calculation of raw material inventory at PT. Bogatama Marinusa Makassar can be known to the company in 2017 ordering raw materials with the number of raw materials each time ordering 154,630 kg with a total inventory cost of Rp. 519,399,138. While the optimal purchase quantity using the Economic Order Quantity (EOQ) method is 20,672 kg per message with a total inventory cost of Rp. 136,586,577 every time you order.

From the results of ECC and SOC calculations it can be concluded that the optimal waiting time or Lead Time is 3 days with a minimum total cost of Rp. 1,836,979,830. Thus it can be determined, 3 days before the raw materials run out the company must re-order with the inventory level at the warehouse at 18,555 kg.

5. Suggestion
Based on the conclusions from the calculation results, the company needs to apply the Economic Order Quantity (EOQ) calculation and determine the optimal waiting time by using the Stock Out Cost (SOC) and Extra Carrying Cost (ECC) cost calculations to determine the economic order quantity and when the company must make reorder (Reorder) for the sake of smooth production at the company.

References
[1] Ahyari A 2003 *Manajemen Bahan-Bahan: Efisiensi Persediaan Bahan* (Yogyakarta: BPFE Yogyakarta)
[2] Gitosudarmo I 2002 *Manajemen Keuangan Edisi 4* (Yogyakarta: BPFE)
[3] Handoko and Hani T 1995 *Manajemen Produksi dan Operasi* (Jakarta: Grasindo)
[4] Jay H and Barry R 2005 *Operations Management Edisi Ketujuh* (Jakarta: Salemba Empat)
[5] Assauri S 1998 *Manajemen Produksi dan Operasi* ed BPFE–UGM (Yogyakarta)

IOP Conf. Series: Materials Science and Engineering **885** (2020) 012042 doi:10.1088/1757-899X/885/1/012042

The effect of top roll defects on dyh5000 toyoda hara finisher draw frame on sliver unevenness at pt xyz

W A Subakdo
Akademi Komunitas Industri Tekstil dan Produk Tekstil Surakarta, Indonesia

Email: ardi_wawan@yahoo.com

Abstract. PT. XYZ is one of the spinning companies that produce cotton carded and Rayon yarn. One of the machines used in the yarn spinning process is the finisher draw frame. In the finisher draw frame process, slivers was produced. To maintain stability of the resulted sliver remain standard, it is necessary to check the sliver number (Ne1) and unevenness of the sliver (u%) with the uster uneveness tester tool. On the finisher draw frame, problems were often found, one of these was unevenness value (u%) that exceed the standard. The purpose of this study was to see the effect of top roll defects on unevenness (u%) and how to overcome these defects. After observing the sliver and testing it on the finisher draw frame, the slivers was tested using an uster unevenness tester 4. The results were not in accordance with company standards, which are 2.50% to 2.90%. The causes of the defect include: lapping material, use of equipment during lapping cleaning. To overcome the problem of defects, repairs were made by grinding the top roll or replacing it with a new top roll. After repairs or replaced with a non-defective top roll, the result is 2.82% and 2.88%

1. Introduction

PT XYZ is a national private textile company in the form of a limited liability company (PT) with venture capital originating from domestic investment (PMDN). PT. XYZ produces cotton carded and rayon yarn. The production process starts from the cotton raw material that is unraveled, mixed by the Blowing machine so that the fiber becomes more homogeneous. After the Blowing machine, cotton fiber is fed by the Chute feed to the Carding machine, the sliver from the carding machine is processed on the Drawing machine. The sliver of the drawing machine is then processed on the simplex machine, the result of the Simplex machine goes to the Ring Spinning machine, then is forwarded to the Winding machine, the result of the Winding machine is thread in the form of cones that enters the ultra violet chamber before entering the Packing section.

The observations were made on the cotton carded process on the finisher draw frame. The draw frame that was used in the company consists of two parts namely the breaker draw frame and the finisher draw frame. In principle, the draw frame does the process of stretching the sliver so that there is a reduction in material, alignment of the fiber, doubling and mixing of the material.

On the Finisher draw frame PT. XYZ occurs a problem that results in defects in the sliver produced, namely unevenness (U%) generated outside the predetermined standard. After testing the unevenness (U%) using the uster tester 4, a problem was found in the stretching area. After checking on the machine it was found the top roll.

Kuang (2015) [1], discusses the comparison of theoretical calculations of yarn unevenness with previous research using different threads. The research shows that the theoretical unevenness of the yarn is the effect of fiber length and fiber quality. Yan (2008) [2], presents a semi-empirical method for predicting yarn unevenness. This study explains the variation in the amount of fiber can be used as an index of fiber straightening in assessing yarn unevenness in terms of changes in the amount of fiber in the transverse section.

Jabbar (2013)[3],examines the impact of increasing card production rates on yarn unevenness, using the Box-Behnken experimental design method. Increase in card cylinder speed significantly decreases the yarn imperfections, without significantly affecting any other yarn parameter. An increase in card production rate results in a significant increase in yarn imperfections as well as yarn unevenness. The number of draw frame doublings not only significantly affects the yarn tenacity and elongation, but also yarn hairiness. However, the effect of draw frame doubling is not linear. By increasing the number of doubling up to a certain level, the yarn tenacity, The result is that yarn unevenness increased by increasing card production rate and total yarn imperfections increase by decreasing card cylinder speed and increasing card production rate on Ring Spun yarn.
Yan (2010) [2] describes the basis of Martindale's limit yarn unevenness, could not be used as a fully effective yardstick for evaluating the yarn unevenness caused by a specific spinning process. He constructs a new model with binomial probability distribution that describes the status of the fiber arrays in a yarn. The analysis of the model is done with the simulation of the parameters and the calculation of cotton ring-spun yarns is conducted. The new model can serve as a semi-empirical method for forecasting the yarn unevenness.

From some of the reasons above, the formulation of problems related to top roll is how the effect of top roll defects on the quality of unevenness (U%) and how to overcome top roll defects?

2. Methods

Data was obtained by observing the three factors that may be the cause of unevenness, such as: machines, humans, and the environment. Then the sliver is tested using Uster Tester 4. Uster Tester 4 can be used to measure u%, coefficient of variance (cv%) and average. These observations were made between July 3 and August 31, 2018

3. Results and discussions

The process on the finisher draw frame is an improvement process after the breaker draw frame, to produce better evenness. This process has the function so that the evenness of the machine gets better, the fibers in the sliver become more parallel with the sliver axis, the raw material derived from similar or unequal material is mixed and becomes more homogeneous, removes dust from the previous process. The finisher draw frame parts through which the material is as follows:

3.1. Feeder zone

The feeder zone consists of creel in which the sliver is fed. This section consists of sliver delivery to manage sliver movement, a feed roll to pull the sliver to be fed, sensors to detect broken slivers in the feed area and sliver conveying spoons to deliver sliver running to be more organized. In this area, efforts must be made as best as possible to prevent a draft, so that the surface velocity of all parts delivering slivers must be adjusted to the same. Before the sliver enters the drafting zone, the sliver is passed through the sliver guide which must be arranged so that slivers can be close together but do not overlap so that a sliver drawing is produced with good level. To ensure that the sliver is fed in a fixed number of duplicates, stop motion is installed in the creel area. If a break occurs, it will be easy to detect and a connection will be made by the operator.

3.2. Drafting zone

The drafting zone is the heart of the drawing process. In the drafting zone, a pair of scanning rollers was installed which serves to detect the number or size of the sliver fed to the stretching rollers. Stretching rollers consist of upper and lower rollers. Lower rollers are usually made of steel whose surface was smooth and grooved to avoid slippage of the fibers. The frontmost bottom rollers are fixed where they cannot be moved and the middle rollers and rear rollers can be shifted to get the desired setting distance. In the drafting or stretching area, the multiple slivers were stretched (draft) by pairs of stretch rollers which have different surface speeds. The speed of the stretch rollers which were in the front sequence is always faster than the speed of the rear surface, the higher the speed ratio between the rollers, the greater the value of stretching and vice versa.

3.3. Delivery zone

The speed of delivery depends on the type of material being processed and the production plan. In this area the sliver was collected from the drafting zone. After exiting the calender roll, the sliver is inserted into the coiler through the trumpet and the coiler tube. Sliver from coiler is collected in a can that was on a rotating can table.

The cause of the unevenness (U%) was observed in the machine, human, and environmental factors. Machine factor was include the defects in the top roll. Human factors that was occured because standard operational procedures are ignored. Then the observed environmental factors were temperature, humidity, and cleanliness of the machine area. During observation, temperature and humidity is in accordance the standard.

Then the following corrective actions are carried out: checking spare parts of machine and replacing defective top roll, ensuring that the room temperature is up to standard and maintaining the cleanliness of the machine area, and the operator is required to carry out standard operational procedures for machine maintenance. On the finisher draw frame, the causes of top roll defects include:

3.3.1. Lapping. Lapping is the situation of material which is rolled on the bottom or top roll. Lapping that is arises will affect the quality of the top roll.

3.3.2. Use of non-standard equipment when cleaning the lapping. The use of a cutter or knife is not recommended because it could cause top roll defects.

The results of sliver testing on the TOYODA - HARA finisher draw type DYH5000 in the year of manufacture 1990 at PT XYZ by using a uster tester 4 with a defective top roll spare part obtained a unevenness (u%) by 3.11% and 3.09%. These results are poor results and do not comply with company standards, which are valued at 2.50% - 2.90%. The test results can be seen in figure 1

USTER TESTER 4 – SX R 1.8 Tue 24.07.18 07:18 Operator Page 1

Style	18/CD	Sample ID	00075	Nom. Count	Nec 0.13	
Tests	2 / 1	v= 25 m/min	t= 1 min	Meas. Slot	1	Short staple

Daily Report Sliver

Article	COTTON	Material class	Sliver	Mach. Nr	COTTON
Uster Statistics					
Fiber	COTTON 4.1Micr 38mm 100%		1.2den 38mm 0%		1.2den 38mm 0%

Nr	U% %	CVm %	CVm 1m %	CVm 3m %	CVm 10m %	Ref Crit 1 %
1	3.11	3.91	1.47	1.14		-0.3
2	3.09	3.85	1.17	0.80		0.3
Mean	3.10	3.88	1.32	0.97		0.0
CV						
s						
Q95						
Max	3.11	3.91	1.47	1.14		0.3
Min	3.09	3.85	1.17	0.80		-0.3

Figure 1. Uster tester report before correction

Then improvement was made to the top roll, such as, by grinding, if not possible, then replace the defective top roll with the non-defective one. The results of sliver testing after correction were 2.82% and 2.88%. These resuts still comply to company standards. Test results can be seen in figure 2.

USTER TESTER 4 - SX R 1.8³ Tue 24.07.18 07:24 Operator Page 1

Style 18/CD Sample ID 00076 Nom. Count Nec 0.13
Tests 2 / 1 v= 25 m/min t= 2 min Meas. Slot 1 Short staple

Daily Report Sliver

Article COTTON Material class Sliver Mach. Nr COTTON
Uster Statistics
Fiber COTTON 4.1Micr 38mm 100% 1.2den 38mm 0% 1.2den 38mm 0%

Nr	U% %	CVm %	CVm 1m %	CVm 3m %	CVm 10m %	Ref crit ± %
1	2.82	3.55	1.25	0.93		-0.5
2	2.88	3.59	0.70	0.40		0.4
Mean	2.85	3.57	0.97	0.66		-0.05
CV						
s						
Q95						
Max	2.88	3.59	1.25	0.93		0.4
Min	2.82	3.55	0.70	0.40		-0.5

Figure 2. Uster tester report after correction

4. Conclusions

On the finisher draw frame, a decrease in the quality of the sliver unevenness (u%) occurs was mainly due to the defective top roll. To overcome the defective top roll is done by replacing the top roll, so that the sliver unevenness falls away according to the company standard.

Acknowledgement

The authors would like to thank Mr. Syaifuddin Yossi Yiko, for providing opportunity and support to prepare yarn samples for this study and all member of Textile Community College of Surakarta (Akademi Komunitas Industri Tekstil dan Produk Tekstil Surakarta)

References

[1] Kuang X, Hu Y and Yu C 2016 The theoretical yarn unevenness of cotton considering the joint influence of fiber length distribution and fiber fineness *Text. Res. J.* **86** 138–44
[2] Yan G, Zhu J and Yu C 2010 A new approach to theoretical yarn unevenness: a binomial distribution model *J. Text. Inst.* **101** 753–7
[3] Jabbar A, Hussain T and Moqeet A 2013 Impact of Carding Parameters and Draw Frame Doubling on the Properties of Ring Spun Yarn *J. Eng. Fiber. Fabr.* **8** 155892501300800220

Effect of government regulation on labor competitiveness: Study of nine major Indonesian companies

A A Aris and S Asmal

Department of Industrial Engineering, Universitas Hasanuddin, Makassar, Indonesia

E-mail: ir.affandy_agusmanaris@yahoo.com

Abstract. Free trade that drives competition in the industrial world drives the need for welding power by 45,000 against a background of specifications. Indonesia has a professional welding force of no more than 15,000 people spread across various countries triggering a deficit in domestic labor that should handle infrastructure development, sea tolls, power plants, and railways. The problems currently being faced by the industry in the context of increasing the competence and competitiveness of the welding industry workforce. The study purposes to analyze the effect of government regulation to labor competitiveness. This research was conducted in November 2016 - March 2017. Data obtained through interviews with 117 respondents distributed in 12 government companies with purposive sampling approach. Data statistical analysis technique is using Structural Equation Modeling. Government Regulation significantly and positively influences the competitiveness of workers with structural coefficient is 0.450 and p-value < 0.05. Indonesian Government can provide added value to the competitiveness of Indonesian workers through the use of authority to draw up regulations in the field of welding, set standard operational production and assist the recruitment process of workers.

1.Introduction

Free trade that drives competition in the industrial world drives the need for welding power by 45,000 against a background of specifications [1]. Indonesia has a professional welding force of no more than 15,000 people spread across various countries triggering a deficit in domestic labor that should handle infrastructure development, sea tolls, power plants, and railways [2]. This gap encourages training needs and the availability of technology becomes increasingly important to support work productivity [3].

The problems currently being faced by the industry in the context of increasing the competence and competitiveness of the welding industry workforce are as follows; the still limited number and low quality of competent certified human resources in the field of welding, the limited number of skilled operators (operators) in the field of welding, the intense competition in the welding sector of goods, and the lack of equipment and machinery to support the smooth running of products.

World trade agreements, including ASEAN Economic Community (AEC), ASEAN Free Trade Agreement (AFTA) and World Trade Organization (WTO), have determined that welding workers must have the ability to improve products so that competitiveness is a key requirement. The development of manufacturing competencies in particular welding to support the improvement of the quality of production is a major need. The welding industry, universities or related research institutions, welding equipment providers, welding service providers and expertise certification providers must work together for advanced welding workers and technology and be able to keep abreast of the latest technological developments [4].

2. Materials and Methods

2.1. Study location

This research was conducted in November 2016 - March 2017 in six regions; Jakarta, Bandung, Makassar, East Luwu, and Tangerang. The study population is spread over eight government institution: Ministry of Industry, Ministry of Manpower, Vocational Training Center, Central for Material and Technical Products, National Professional Certification Board, Professional Welding Certification Institute, Indonesian Welding Association, PT. Indonesian Ship Industry, PT. Vale, Eastern Pearl Flour Mills, PT. Dirgantara Indonesia, PT. Garuda Maintenance Facility, and PT. Pindad.

2.2. Data Collection and Analysis

Data were collected by interview using a structured questionnaire of 117 respondents drawn by purposive sampling. Measurement of data validity using confirmatory factor analysis of indicators of each latent variable (government regulations and labor competitiveness) (Figure 1).

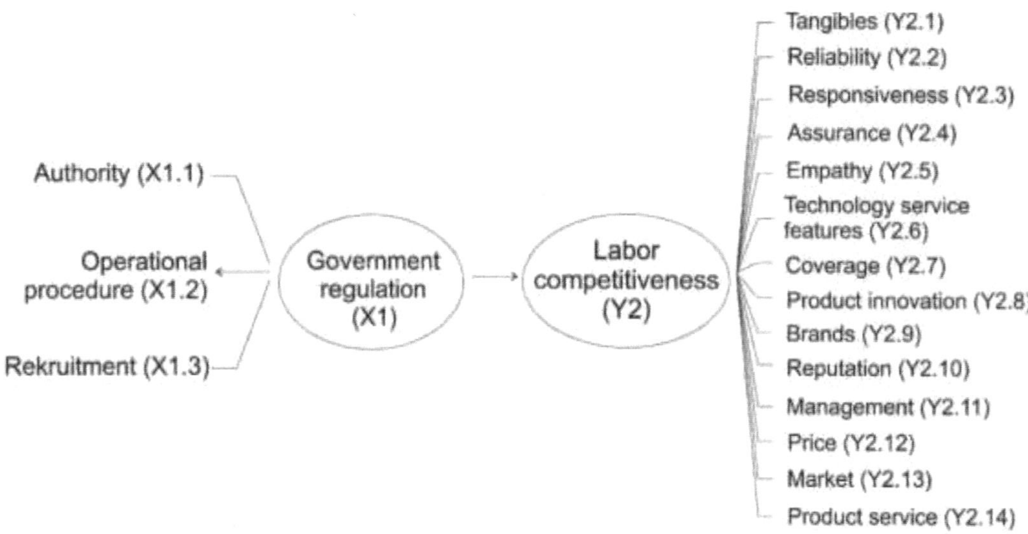

Figure 1. Hypothesis Testing Result using GeSCA

2.3. Statistical Analysis

Data were analyzed using GeSCA as a component-based Structural Equation Modeling (SEM) method [5][6]. The steps for modelling GeSCA structural equations through the following stages; designing structural models, designing measurement models, constructing path diagrams, converting path diagrams to equation systems, estimating and evaluating the goodness of fit, and testing hypotheses [7][8].

3. Results

3.1. Validity test of research instruments

Before the questionnaire was used as a data collection tool, testing the validity and reliability of the instrument was carried out. Testing the validity of the Pearson correlation, the instrument is declared valid if the correlation value > 0.3 (Table 1).

Table 1. Hypothesis testing results in structural model

	Indicators	Item	Correlation	Valid/Not Valid
Government Regulation (X1)	Authority (X1.1)	X1_01	0,390277778	Valid
		X1_02	0,388194444	Valid
		X1_03	0,397222222	Valid
	Operational Standards (X1.2)	X1_04	0,386805556	Valid
		X1_05	0,384722222	Valid
		X1_06	0,389583333	Valid
	Recruitment (X1.3)	X1_07	0,396527778	Valid
		X1_08	0,389583333	Valid
		X1_09	0,394444444	Valid
Labor Competitiveness (Y2)	Tangibles (Y2.1)	Y2_01	0,288888889	Valid
		Y2_02	0,319444444	Valid
		Y2_03	0,306250000	Valid
		Y2_04	0,284027778	Valid
	Reliability (Y2.1)	Y2_05	0,311805556	Valid
		Y2_06	0,302777778	Valid
		Y2_07	0,272916667	Valid
	Responsiveness (Y2.3)	Y2_08	0,333333333	Valid
		Y2_09	0,317361111	Valid
		Y2_10	0,272916667	Valid
	Guarantee (Y2.4)	Y2_11	0,342361111	Valid
		Y2_12	0,295833333	Valid
		Y2_13	0,322222222	Valid
	Empathy (Y2.5)	Y2_14	0,299305556	Valid
		Y2_15	0,301388889	Valid
	Technology Service Features (Y2.6)	Y2_16	0,340972222	Valid
		Y2_17	0,315277778	Valid
		Y2_18	0,310416667	Valid
		Y2_19	0,318055556	Valid
	Coverage Area (Y2.7)	Y2_20	0,275000000	Valid
		Y2_21	0,289583333	Valid
	Product Innovation Activities (Y2.8)	Y2_22	0,274305556	Valid
		Y2_23	0,336805556	Valid
		Y2_24	0,323611111	Valid
		Y2_25	0,313888889	Valid
	Brand Awareness (Y2.9)	Y2_26	0,297222222	Valid
		Y2_27	0,318750000	Valid
	Company Reputation (Y2.10)	Y2_28	0,314583333	Valid
		Y2_29	0,334722222	Valid
		Y2_30	0,293750000	Valid
	Management Capability (Y2.11)	Y2_31	0,284722222	Valid
		Y2_32	0,290972222	Valid
		Y2_33	0,306250000	Valid
	Competitive Prices (Y2.12)	Y2_34	0,289583333	Valid
		Y2_35	0,343750000	Valid
		Y2_36	0,320138889	Valid
	Market Coverage (Y2.13)	Y2_37	0,290277778	Valid
		Y2_38	0,349305556	Valid
	Product Delivery and Service	Y2_39	0,307638889	Valid
		Y2_40	0,282638889	Valid

(Y2.14)	Y2_41	0,320138889	Valid
	Y2_42	0,336805556	Valid

Based on Table 1, all indicators on each variable have a correlation value greater than 0.30 therefore the research instrument is declared valid. Meanwhile, the Cronbach's alpha value for all variables is greater than 0.60 thus it can be said that the research instrument is also reliable.

3.2. Structural model

Testing the structural model essentially tests the hypothesis in the study. Hypothesis testing is done by T-test (T-statistic) on each path of direct influence partially. The results of the complete analysis contained in the results of SEM analysis. Table 2 presents the results of testing the direct influence hypothesis.

Table 2. Hypothesis testing results in structural model

Relationship	Path Coefficient	p-value	Information
Government regulation (X1) Labor competitiveness (Y2)	0.450	4.32	0.000

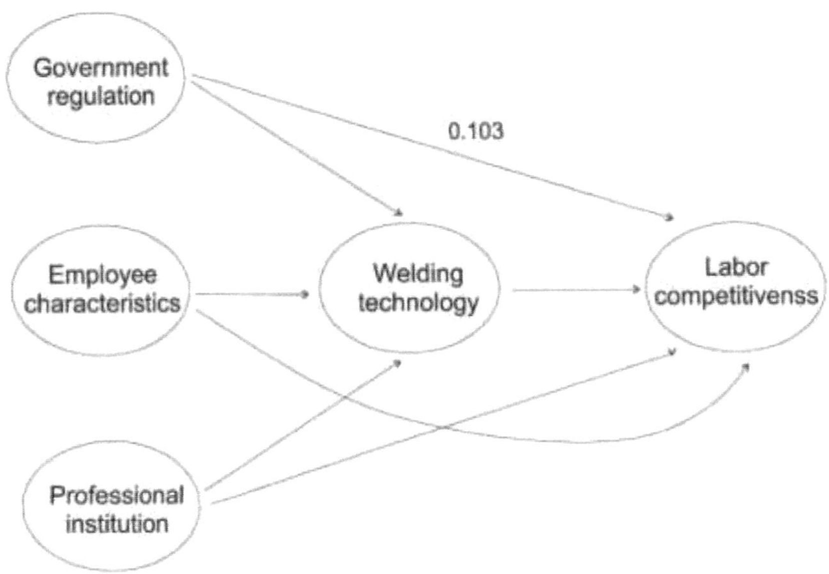

Figure 2. Hypothesis testing result in WarpPLS inner model

The Effect of Government Regulation (X1) on Labor Competitiveness (Y2), obtained a structural coefficient of 0.450, and a P-value of 0.000. Because the P-value <0.05, indicates that there is a significant and positive influence between Government Regulation (X1) on Labor Competitiveness (Y2). The higher the value of Government Regulation can affect the high Workforce Competitiveness (Y2). Thus, hypothesis 2 of this study was accepted.

4. Discussion

Government regulation is generally defined as a law that is imposed by the government on individuals and companies in the industrial sector to regulate and modify economic behavior. Conflicts can occur between public services and commercial procedures (e.g. maximizing profits), the interests of people who use these services (see market failures), and also the interests of those not directly involved in transactions (externalities) [9]. Therefore, most governments have some form of control or regulation to manage this possible conflict. The ideal goal of economic regulation is to ensure the safe and appropriate delivery of services, while not hampering the functioning and effective business development [10].

Labor competitiveness occurs when workers have entered the highly competitive labor market, where the level of wages is determined in the welding industry, not by each company, each company is a wage taker [10]. This means that actual equilibrium wages will be set in the market, and the supply of labor to each company is very elastic at the market level [11].

The results of this study complement various theoretical studies that reveal the relationship between government regulations and the competitiveness of Indonesian Workers. The results of data analysis and testing between variables found that government regulations which gave a significant and positive caregiving towards the competitiveness of Indonesian workers. The implication of labor competitiveness in this research is to get a formula that increases the competitiveness of workers is done by increasing government regulations. The theoretical implications of the study found that government regulation can improve the competitiveness of the workforce, both directly and indirectly to the use of technology. This indicates that the importance of government regulations in terms of authority, operational standards, and the recruitment process.

Welding technology management, worker characteristics and involvement of professional institutions encourage labor competitiveness. Without the management of welding technology, there is no visible effect between the characteristics of workers and professional institutions on the competitiveness of the workforce. The welding industry in Indonesia must apply government policy-based welding technology measurements (Decree of the Minister of Manpower and Transmigration of the Republic of Indonesia Number: KEP.342/MEN/X/2007 concerning the Establishment of Indonesian National Work Competency Standards) to ensure the high competitiveness of welding workers [12].

5. Conclusions

This study found that government regulations have a significant and positive effect on the competitiveness of Indonesian workers. The Indonesian government should develop and implement regulations that can provide added value to the competitiveness of Indonesian workers. The government can utilize the authority in determining regulations in the field of welding, setting standardized and clear operational standards, and the recruitment process (government officials) that can accommodate the competitiveness of welding workers in Indonesia.

References

[1] Helveston J P, Wang Y, Karplus V J and Fuchs E R H 2019 Institutional complementarities: The origins of experimentation in China's plug-in electric vehicle industry *Res. Policy* **48** 206–22

[2] ---------- 2019 Indonesia Welding Consumables Market: Positive Outlook of Power and Oil and Gas Industries will Propel Market to US$293 million by 2023 *Res. Transpar. Mark.*

[3] Onyango J W and Wanyoike D M 2014 Effects of training on employee performance: a survey

of health workers in Siaya County, Kenya *Eur. J. Mater. Sci.* **1** 11–5
[4] Gyasi E A, Kah P and Martikainen J 2014 Welding management as a tool for innovative, competitive and sustainable manufacturing: Case study–West Africa *Int. J. Dev. Sustain.* **3** 1782–93
[5] Chin W W 1998 Commentary: Issues and opinion on structural equation modeling *MIS Q.* **22** vii–xvi
[6] Fan X, Thompson B and Wang L 1999 Effects of sample size, estimation methods, and model specification on structural equation modeling fit indexes *Struct. Equ. Model. a Multidiscip. J.* **6** 56–83
[7] Tenenhaus M 2008 Component-based structural equation modelling *Total Qual. Manag.* **19** 871–86
[8] Oke A E, Ogunsami D R and Ogunlana S 2012 Establishing a common ground for the use of structural equation modelling for construction related research studies *Constr. Econ. Build.* **12** 89–94
[9] Armstrong A J and Jan H 2019 Government's Role in The Electricity Sector *The Renewable Energy Policy Manual* (U.S. Export Council for Renewable Energy)
[10] Besley T 2015 Law, regulation, and the business climate: The nature and influence of the World Bank Doing Business project *J. Econ. Perspect.* **29** 99–120
[11] ---------- Competitiveness Definition in Economics
[12] KEMENAKER-RI 2018 JDIH

The manufacture of automatic chicken feed making machines as an economical solution for chicken farmers in Indonesia

Amrin Rapi[1], Abdul Nasser Arifin[1], Asiyanthi Tabran Lando[2]

[1]Agro-Industrial Manufacture Engineering Department, ATI Makassar Polytechnic, Makassar, Indonesia
[2]Environmental Engineering Department, Hasanuddin University, Makassar, Indonesia

Email: abdulnasser@atim.ac.id

Abstract. Currently laying hens have a good chance to be developed in Indonesia, this breeding business experiences rapid development every year. The chicken farming industry has very interesting potential to be developed but there are still challenges in its development. One of the biggest challenges faced by chicken farmers in Indonesia today is the high cost of feed that they have to spend around 70% - 80% of the annual production cost. On the other hand, the key to success in the chicken farming industry is strongly influenced by meeting the need for adequate feed, both in quality and in quantity, this is because the feeding that is in accordance with the nutritional needs of chickens affects the amount and quality of produced. Based on these problems, we assume that the independent supply of poultry feed by breeders by providing livestock feed making machines using equipment and components that are widely available on the market today is an economic solution that can reduce production costs in a chicken farming industry in Indonesia. The manufacture of automatic chicken feed making machines 2.02 kg / hr for laying hens using 1400 rpm electric motor and automatic control system is expected to be an economical and practical solution in improving the quality of production of the chicken farm industry in Indonesia.

1. Introduction

According to handbook series of agribusiness manuals prepared by the FAO Investment Centre Division (ICD), the world egg production increases has tripled since 1970 when global output was about 20 million tons to 60 million tons in 2007 [1]. This tren showes like in Indonesia too, currently. The poultry industry in Indonesia follows the development of the global poultry industry which is experiencing rapid development from year to year. This industry is the main sector in supporting national economic growth in Indonesia where 10% (around 12 million) of the national workforce are currently absorbed in the poultry industry sector. With an economic value of USD 35 billion (Rp. 494 trillion) making this industrial sector very potential and interesting to develop in Indonesia now [2].

At present, the poultry industry in Indonesia has major obstacles in the availability of raw materials for feed production, where the supply of feed raw materials, which are still largely imported, such as corn 40-50 percent, soybean meal 95 percent; fish meal 90-92 percent, bone meal and vitamin / feed additives almost 100 percent imported [3]. Based on this problem, several studies have been carried out to find alternative substitutes for corn and soybean meal as the main raw material for making feed. One alternative that has been carried out by several breeders in Indonesia is to utilize food waste in the form of dry rice, rice bran, CGM, cassava flour, rijek noodles, sorting bread, rijek rice as an energy source to replace corn. As for substituting soybean meal commonly used tempe tofu making industry waste in the form of soybean epidermis as a source of protein.

The use of alternative raw materials in making poultry feed requires a machine that can process these raw materials into ready-to-use feed by poultry farmers to be very important. At present the widely available feed making machines on the market are feed pellet making machines with a price range of Rp. 1.7 million - 15 million per unit. This price does not include shredder (Hammer Mild) and mixing machine (Mixer) raw materials, which costs Rp. 10 million - Rp. 50 million per unit. This condition is certainly very burdensome for farmers in meeting their livestock feed needs.

To overcome these obstacles at this time, the design of poultry feed manufacturing has been carried out by several researchers, such as research on the design of a mixer and pellet maker using a PLC control system [4], then the corn sheller machine which is used as poultry feed [5], Furthermore, the feed making machine is carried out by PT. Popular Farm located in the city of Semarang, Indonesia. This feed making machine made by PT Poluper Farm uses grinder and vertical mixing methods to produce feed for the company's poultry feed needs.

Of the three feed making machines that have been made today, some of the fundamental problems faced by chicken farmers in Indonesia have not been resolved directly and clearly. Such as the use of raw material for feed substitutes for corn and soybeans as the main source of energy and protein for poultry, the integrated feed making machine has not been integrated in a whole machine series consisting of grinder-mixer-feed pellet makers and the unavailability of machines that use automatic control systems in one the working circuit of the machine.

Based on the conditions above, it is currently very necessary for a practical feed making machine and velcsibel in the use of alternative raw materials to substitute corn and soybeans in the manufacture of feed, a feed making machine that has a whole working circuit system in the process of making poultry feed consisting of a grinder process - mixing and printing pellets. The choice of pellets as a form of feed is because the pellets have more durable properties in terms of storage and practicality in their use. In addition, to improve the efficiency of the feed making machine, we need a machine that can operate automatically in each work series.

Therefore, we are trying to design a practical machine for making semi-automatic and flexible poultry feeds so that they can utilize alternative raw materials for poultry feed manufacturing. In addition, this machine will work in a whole work series of the process of making feed consisting of a grinder - mixer - pellet printer. The existence of this machine is expected to reduce the dependence on imported raw materials for making feed and increase the production of the poultry industry in Indonesia now.

2. Methodology

The design and assembly of the equipment was carried out at the ATI Makassar Polytechnic workshop. The design of the tool is done by using the AutoDesk Inventor 2017 software, using the Parametic Solid Modeling function in which this application can design and edit in the form of a solid model with data that has been stored in a data base so as to facilitate the design editing process. Furthermore, the Adaptive function of this software has the ability to analyze the friction of an animation of a tool and can adjust itself [6].

To test the success in the design of this poultry feed making machine, several mechanical testing parameters and tool function testing were established. In the mechanical testing parameters, testing is carried out on each feed manufacturing process consisting of mechanical testing of the working system of the grinding machine, mechanical testing of the working system of the mixing machine and mechanical testing of the working system of the printing machine. And constant loading at various turns, then from this process, the function test is carried out in parallel by measuring the results of grinding, mixing and printing results in the form of pellets produced, both in quality and quantity.

IOP Conf. Series: Materials Science and Engineering **885** (2020) 012044 doi:10.1088/1757-899X/885/1/012044

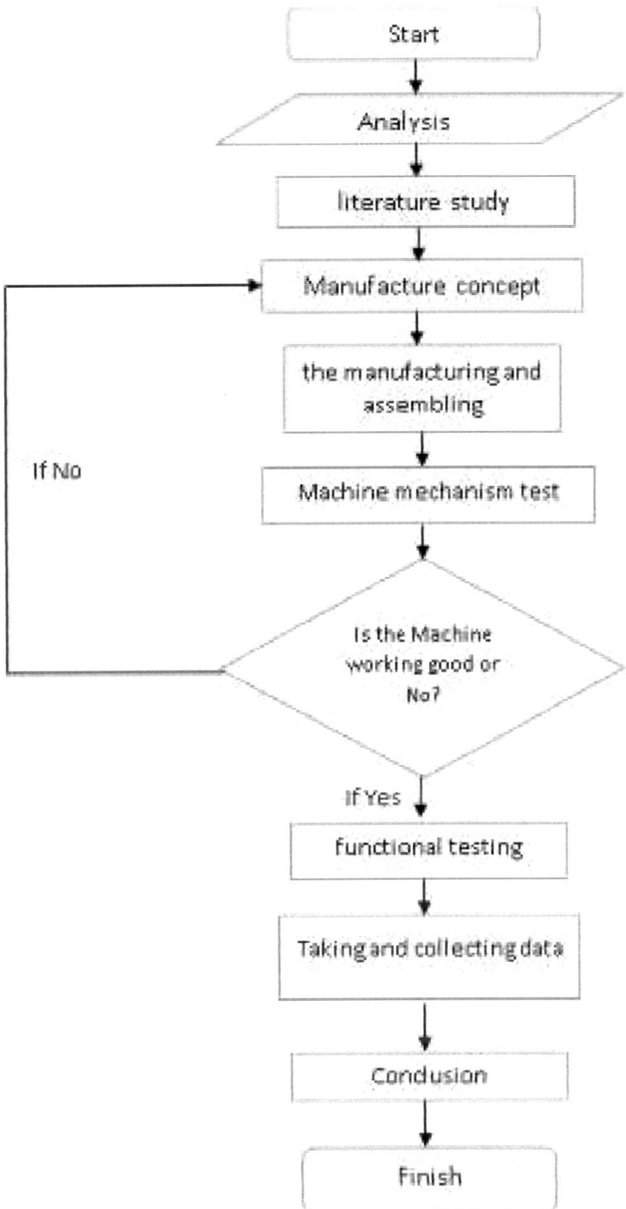

Figure 1. Flow chart research

In the main drive system of the engine using the main driving motor used is an AC electric motor with power (P_m) = 1 HP = 745.7 Watt, voltage (V_m) = 220 volts and rotation (n_m) = 1400 rpm. The results of this round are transmitted with pulleys using V-belts with motor pulley size (D_m) = 80 mm, diameter of the pulley grinder 1 (D_{p1})= 180 mm and grinder 2 (D_{p2}) = 80 mm, pulley mixer diameter (D_{mix}) = 155 mm and the diameter of the round reduser (D_r) =75 mm. To calculate the balance in each series, the equation is used:

$$n_2 = \frac{n_{1\,x}\, d_1}{d_2} \quad \ldots\ldots\ldots\ldots\ldots\ldots\ldots\ldots\ldots\ldots \quad 1$$

where: n_1 = pulley rotation speed of drive (rpm)
D_1 = diameter of drive pulley (mm)
n_2 = pulley rotational speed moved (rpm)
D_2 = diameter of pulley actuated (mm)

To calculate the length of the belt (L) be used the equation:

$$L = 2C + \frac{\pi}{2}\left(D_p + d_p\right) + \frac{1}{4C}\left(D_p - d_p\right)^2 \quad \ldots\ldots\ldots\ldots\ldots \quad 2$$

where: L = belt length (mm)
C = the distance of the pulley center point to the driven pulley (mm)
D_p = diameter of drive pulley (mm)
d_p = diameter of pulley actuated (mm)

Next to calculate the torque (T) and angular velocity (ω_p) drive for each work circuit using the equation:

$$T = \frac{P}{\omega_p} \quad \ldots\ldots\ldots\ldots\ldots\ldots\ldots \quad 3$$

$$\omega_p = \frac{2\pi \cdot n_p}{60} \quad \ldots\ldots\ldots\ldots\ldots\ldots\ldots \quad 4$$

where: T = torque (Nm)
P = Power shaft (Watt)
ω_p = angular shaft velocity (rad/s)

As for the safety factor was used 1.25-1.5 for load-controlled conditions and the working voltage can be determined in advance.

The loading on the function test is carried out with three different loading, each of which consists of 1 kg, 2 kg, and 3 kg of feed material at a constant rotation of 1400 rpm which will be processed into poultry feed. The raw material for making feed with the composition of each used consists of 50% dry rice as a source of carbohydrates, 20% rice bran, fish meal and starch 20% and 10% respectively. Furthermore, water is added as much as 0.5 liters of solvent.

3. Result and Discussion

From the results of the design carried out using Autodesk Inventor 2017 software, a design of the device was obtained, the researcher took a benchmark from the scale of planning that has been made either 2 dimensions or 3 dimensions. Below this is a 3-dimensional planning drawing created by the researcher.

The image description:

1. Chassis as a holder for all engine parts
2. Crusher Chamber as a home of a hammer mill
3. Cover as a crusher and hammer mill cover
4. Exit funnel as a conduit or a place for the release of expert materials that have been milled in cattle

5. Entrance funnel as a place to enter new poultry feed ingredients will be ground.
6. Knife crusher as a beater for poultry feed which is put into the crusher room
7. Crusher as a place to install the blade holder shaft, and also counter balance hammer mill
8. Blade spacers as a knife edge so that they do not come into contact with each other.
9. Knife holder shaft as a blade that will hit poultry feed ingredients
10. Material funnel as a place to collect feed ingredients
11. Mixer blades as a mixer for poultry feed dough consisting of 3 blades with a length of 180 mm wide by 20 mm in the shape of an anchor
12. Mixer container as a place for mixing poultry feed dough
13. Pellet maker as a dough maker which is driven by a screw conveyor with a hole size of 5 mm each
14. Speed modifier with a ratio of 1:30 from the motor speed
15. Electric motor; As the driving force of the poultry feed making machine.

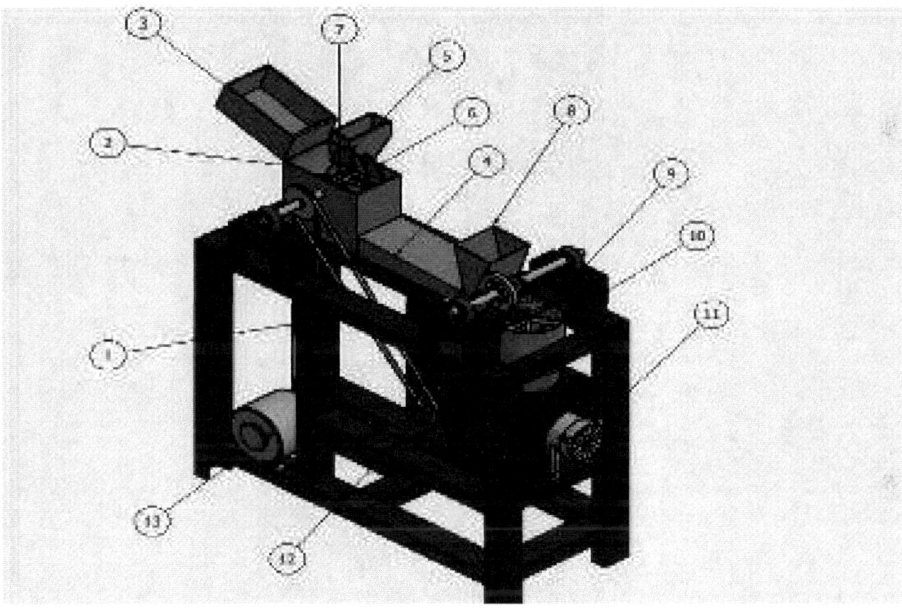

Figure 2. The results of the design of the feed making machine in 3 dimensions

The results of testing in the machine testing process are carried out in 3 times they are:

1. The first experiment was tested by entering 1 kg of poultry feed ingredients consisting of 500 grams of corn, 200 grams of fish meal, 200 grams of bran and 100 grams of starch with a constant rotation speed.
2. The second experiment was tested by entering 2 kg of poultry feed ingredients consisting of 1 kg of corn, 400 grams of fish meal, and 400 grams of bran and 200 grams of starch with a constant rotation speed.
3. The third experiment was tested by inserting 3 kg of poultry feed ingredients consisting of 1.5 kg of corn, 600 grams of fish meal, 600 grams of bran and 300 grams of starch with a constant rotation speed.

Figure 3. material funnel as a place to collect feed ingredients before mixing well

Figure 4. Mixing process

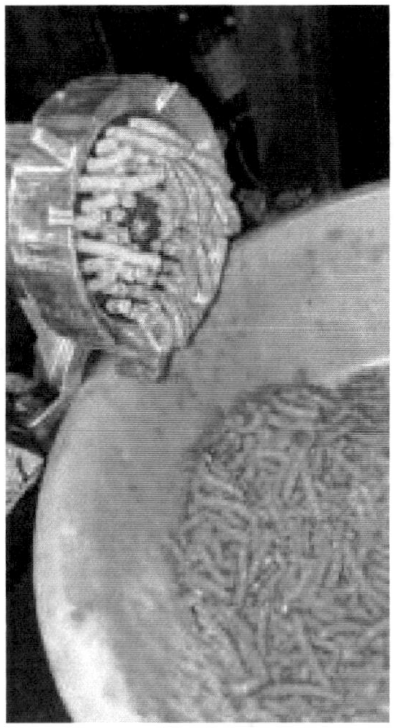

Figure 5. pellet making process as chicken feed

Table 1 Comparison of feed ingredient composition in three experiments

No.	Raw material	Persentasi (%)	Mass of raw materials (kg)		
			1	2	3
1	Dry rice	50	0.5	1	1.5
2	rice bran	20	0.2	0.4	0.6
3	fish meal	20	0.2	0.4	0.6
4	starch	10	0.1	0.2	0.3

From tests conducted using the comparison of ingredients above, the following results are obtained (see table 2)

Tabel 2. Test results

No.	Mass of raw materials (kg)	Time (menit)	Rotation (rpm)	Production capasity (kg/jam)
1	1	29.003	1400	2.06
2	2	59.316	1400	2.02
3	3	88.100	1400	2.04

4. Conclusion

From the results of the design this feed making machine concluded, Making feed using alternative raw materials is very possible using this machine, so that the content and composition of nutrients needed by chickens can be adjusted to the needs of farmers. Poultry feed ingredients that use alternative raw materials in the form of dry rice, rice bran can produce poultry feed in the form of pellets with an average production capacity of 2.04 kg / hour.

References

[1] FAO 2010 *Agribusiness handbook Poultry Meat and eggs* (Paris: FAO investment Center Division)
[2] CIPS *Reformasi kebijakan dalam Industri Unggas di Indonesia* (Jakarta: ResearchGate)
[3] Sayuti R and Noekman K M 2016 Industri perunggasan: memadukan pertumbuhan dan pemerataan *Forum Penelitian Agro Ekonomi* vol 20 pp 50–64
[4] Rosandy, Jimmy, Permana, Fandi S and Harianto 2014 Alat Pembuat Pakan Ternak Ayam Jenis Pellet *J. Control Netw. Syst.* 51–8
[5] Sugiyanto S and Ratih S Y 2017 Rekayasa Mesin Untuk Industri Kecil Pakan Ternak Unggas Di Klaten *Pros. SNATIF* 857–61
[6] Waguespack C 2010 *Mastering Autodesk Inventor 2010* (John Wiley & Sons)

IOP Conf. Series: Materials Science and Engineering **885** (2020) 012045 doi:10.1088/1757-899X/885/1/012045

Effect of types of control materials in a modified sedimentation equipment batch laboratory scale

Mustafa[1], Masing [2] and M Irwan [3]
[1,3]Applied S1 Study Program Applied Industrial Chemical Technology, Chemical Engineering Department Samarinda State Polytechnic
[2]D3 Electrical Engineering Study Program, Electrical Engineering Department Samarinda State Polytechnic

Abstract. The purpose of this research is to be able to recognize laboratory scale sedimentation devices in the process of deposition through batch system experiments in a cylindrical tub, to explain how the relationship between time, height of deposition and concentration of various solids with sedimentation rates, can compare the precipitating materials used in the sedimentation process both manually and automatically, and can determine the performance of tools and sedimentation rates by using a variety of sedimentation and flocculant materials used in the sedimentation process both without the addition of flocculants or with the addition of flocculants. In supporting this research, the method used is experimentation and analysis. The sample is a mixture of media that is water - depositional material (solids) where the depositional material has previously been sieved and cleaned of impurities and the variables analyzed were deposition time, deposition height, suspension concentration and concentration of flocculant addition. the settling material used varies, among others: 50 grams, 100 grams, 150 grams, 200 grams, and 250 grams, then put into a settling tube. While the results can be obtained from visualization data that is based on more accurate observations and data processing conditions (measurements) obtained in the form of the influence of time on settling rate, the influence of sedimentation column height on settling rate and the effect of solid concentration on settling rate. From the results obtained from various settling materials, the settling speed for CaCO3 precipitates without flocculants and with flocculants is 4.43 cm / minute and 3.86 cm / minute. For the settling speed for tapioca flour depositor without flocculant and with flocculant is 23.64 cm / minute and 23.47 cm / minute. And for settling speed for dry soil sedimentation material without flocculant and with flocculant is 18.60 cm / minute and 15.10 cm / minute and for settling speed of active carbon precipitation material without flocculant and with flocculant is 7.76 cm / minute and 7 .05 cm / minute.

1. Indroduction

In everyday life sedimentation is used to reduce superimposed materials such as turbidity in water and also has a role to reduce the content of certain pathogenic organisms that are in water. In addition, this sedimentation operation is also used on an industrial scale to reduce pollution from industrial waste. In industry, sedimentation is carried out in a continuous (often) manner which is often called a thickener. And in the chemical industry it is also often necessary to separate the components of a mixture so that it is separated into their respective fractions. A component has different fractions in terms of particle size, phase or chemical composition. Sedimentation itself is a process of separating solids from liquids (solid-liquid) by using the force of gravity to settle suspension particles. This separation process is based on the movement of solid particles through the fluid due to the gavitation force [1].

Sedimentation speed can increase with the presence of flocculants. The overall effect of flocculation is to create a fusion of fine particles into larger particles so that they can easily be deposited.

Merging between particles that can occur if there is contact between these particles. In its application, sedimentation is one of the most economical ways to separate solids from suspension, pulp or slurry. Sedimentation is also a process of separating solids in liquids which is often used in industry / factories and at the laboratory scale. Sedimentation can take place in batches or continuously, where in sedimentation in batches which are usually used at the laboratory scale, this is because at the laboratory scale experiments are carried out in stages (batch) not continuously (continuously) as in the industry /1 factory. One concrete example that requires sedimentation equipment is at the Laboratory of the Department of Chemical Samarinda State Polytechnic. In the Laboratory of Chemical Engineering, Samarinda State Polytechnic, namely the Laboratory of Operations and the Waste & Utilities Laboratory, sedimentation equipment is needed to support the smooth running of the practicum and it is arguably still limited. Therefore, through this Basic Research (Research and Development Prototype Scheme), researchers try to make sedimentation devices in batch modification which is expected to help the implementation of experiments or experiments for the smooth sedimentation practicum in Samarinda State Polytechnic, especially in the Chemical Engineering Department and specifically in the Laboratory of Operations Unit and Waste & Utilities Laboratory.

2. Research Methods

The method used in this research is analysis and analysis. To support the implementation of this research, research subjects are needed. The subjects in this study are as follows: 1). Media: As the media in this study are solids in the form of precipitation materials and liquid in the form of water. 2). Sample: What is taken as a sample is a mixture of media that is water and depositional material (solids) where the depositional material has previously been sieved and cleaned of impurities. 3). Variables: The variables of this study are deposition time, deposition height, suspension concentration and concentration of flocculant addition

2.1. Tools and materials

The tools used in a batch sedimentation tool include: Caliper, analytical balance, Stop watch, beaker, mortar & pestle, screen (screening), spatula, plastic spoon and bucket While the material used: PDAM water, Flocculant Material: Alum (Aluminum Sulfate (Al2 (SO3) 4)) , Precipitating material: Lime (CaCO3), Wheat Flour, Soil (Dry), and Activated Carbon

2.2. Work procedures

Making batch sedimentation equipment is carried out at Samarinda State Polytechnic campus in the Laboratory of Operations Unit of the Chemical Engineering Department. As for the following Work Procedures as follows:

1. Trial sedimentation without flocculants
   a. Sift the settling material to be used and weigh the settling material as much as 50 grams, 100 grams, 150 grams, 200 grams, and 250 grams, then transfer it to the settling tube.
   b. Add water until it reaches a height of 70 cm
   c. Shake the mixture by turning the 900 tube 10 times then let it sit and remove the tube cover
   d. Observe and record the height of the suspension every 5 minutes until it is constant.
2. Trial Sedimentation experiments with the addition of flocculants
   a. Sift the settling material to be used to obtain a more uniform particle diameter of the settling material
   b. Weigh the depositor material as much as 50 grams, 100 grams, 150 grams, 200 grams, and 250 grams then transfer it to the settling tube.
   c. Add water to reach a height of 80 cm then cover the tube with each lid.
   d. Weigh the flocculant each 1 gram, then transfer it to the settling tube which contains a mixture of water-settling material and the tube must be closed.
   e. Stir the mixture by flipping the tube so that the concentration of suspense is evenly distributed.
   f. Then set aside the cylinder and remove the cylinder lid and run the stopwatch.
   g. Observe and record the height of the suspension every 5 minutes until it is constant.
   h. Repeat the above steps by adding 2 grams of flocculants, 3 grams, 4 grams and 5 grams

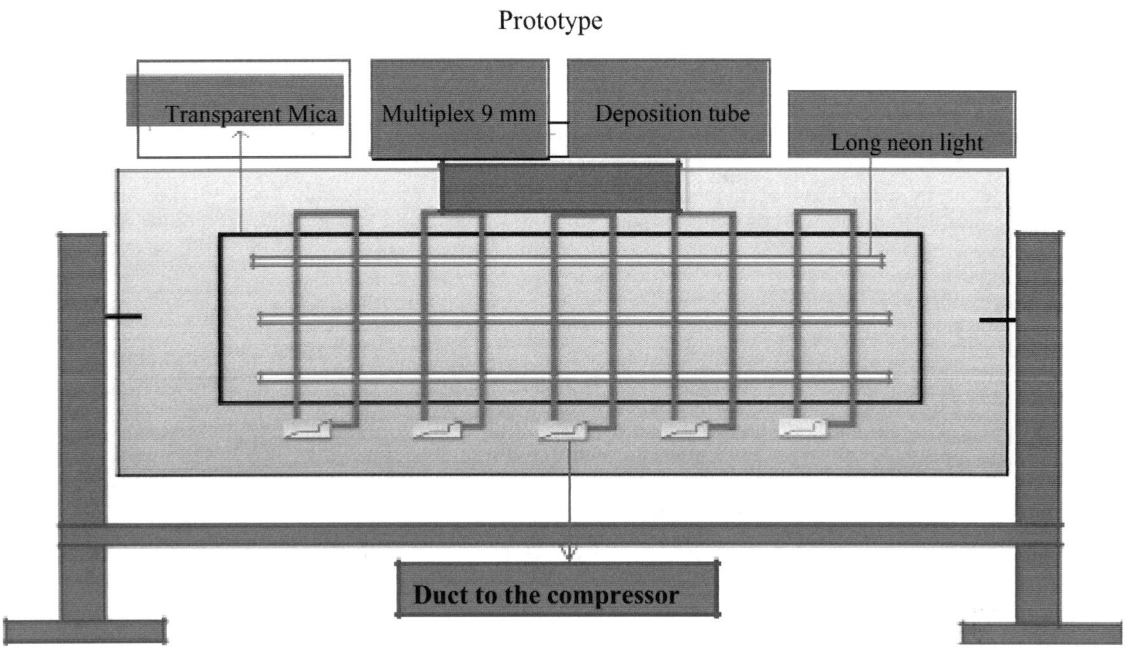

Figure 1. Prototype tool

3. Results and discussion

3.1. $CaCO_3$ Residual materials

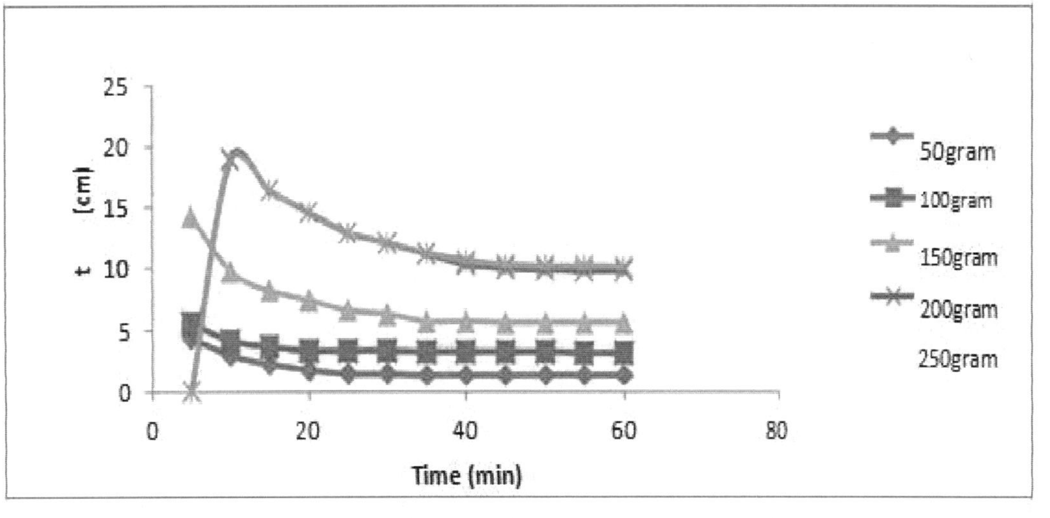

Figure 2. Without Addition of Coagulants $CaCO_3$

In determining the deposition speed of CaCO3 without the addition of coagulant can be determined by the graph of the relationship between the height of the compressible zone Vs of time. In the graph it can be seen that the more mass of solids added to the tube, the longer the time needed for the floces formed to settle.

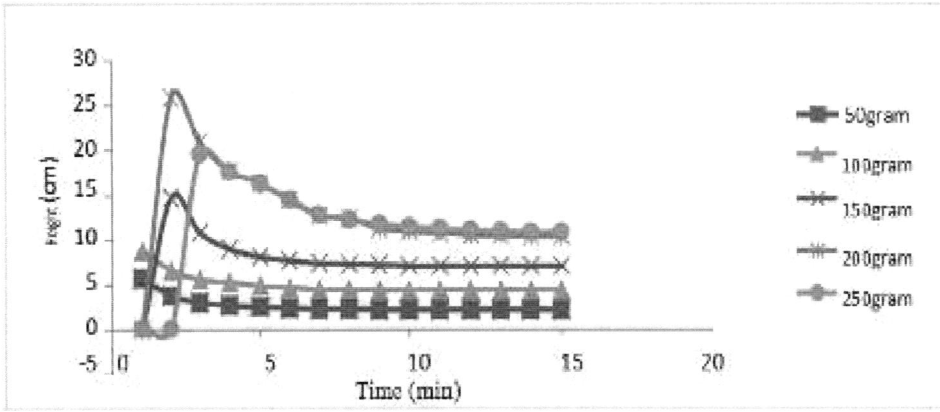

Figure 3. With Addition of Coagulants $CaCO_3$

Furthermore, the deposition of CaCO3 with alum was added as a coagulant. The added coagulant mass is 1 gram into each tube. The function is to add alum as a coagulant so that clumping occurs between alum and floc in the water so that the flocks become larger and relatively heavier, causing suspended particles to settle faster with the addition of alum.

3.2. Tapioca Flour Residual Materials

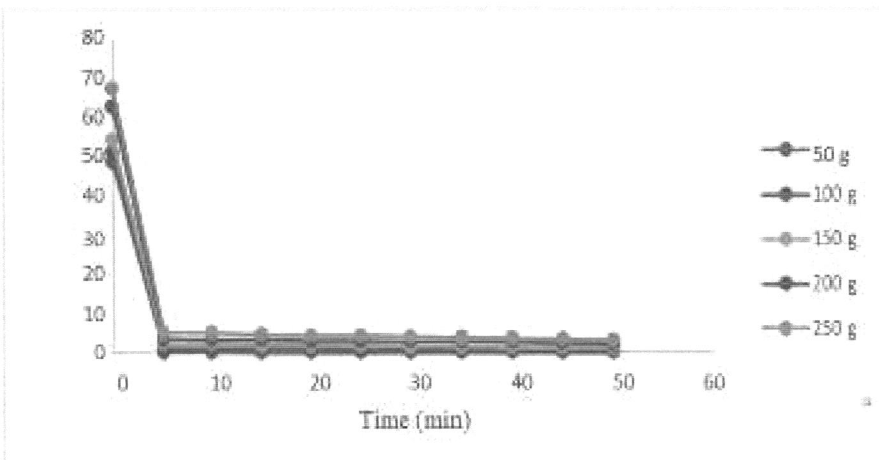

Figure 4 .Time vs height of Tapioca Flour without flocculant

In determining the deposition speed of tapioca flour without the addition of coagulant can be determined by the graph of the relationship between the height of the compressible zone Vs of time. In the graph it can be seen that the more mass of solids added to the tube, the longer the time needed for the floces formed to settle.

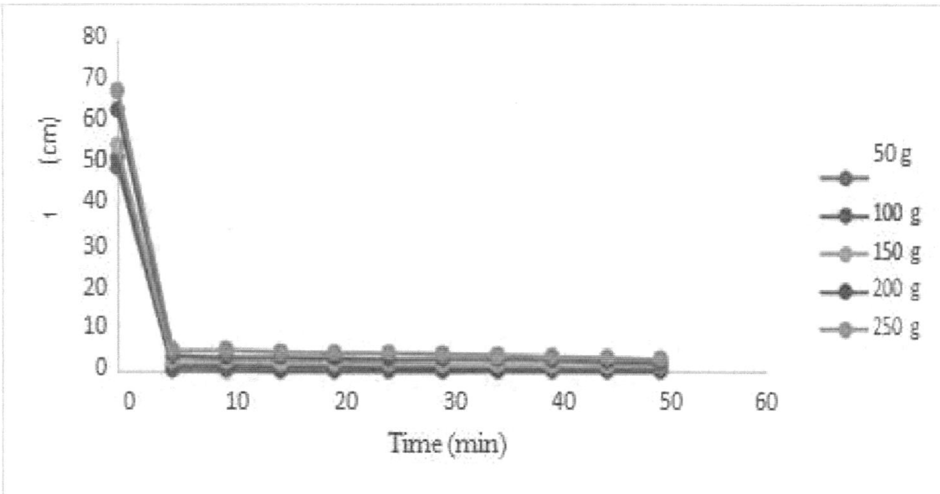

Figure 5 . Time vs height of Tapioca Flour with flocculant

Furthermore, the deposition of tapioca flour with alum was added as a coagulant. The added coagulant mass is 1 gram into each tube. The function is to add alum as a coagulant so that clumping occurs between alum and floc in the water so that the flocks become larger and relatively heavier, causing suspended particles to settle faster with the addition of alum.

3.3. Dry Soil Residual Materials

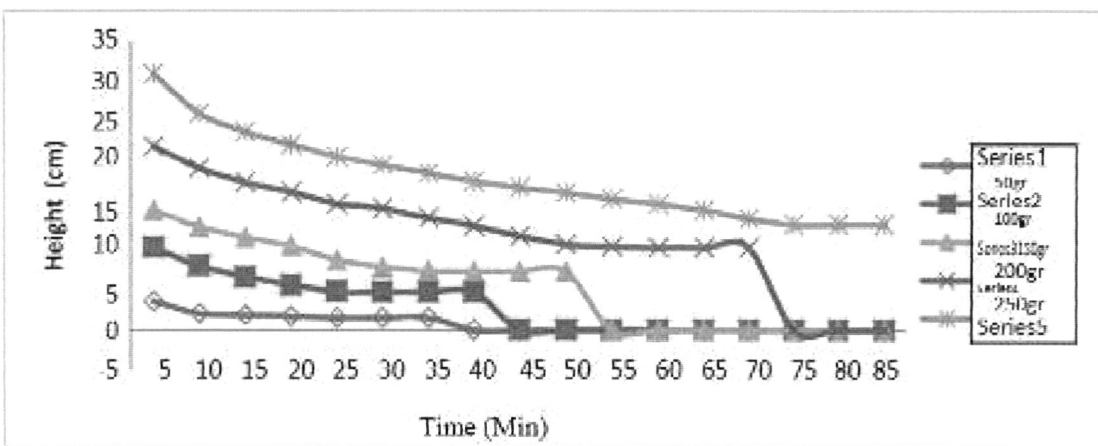

Figure 6. Time vs height of Dry Soil without flocculant

In determining the deposition speed of dry soil without the addition of coagulant can be determined by the graph of the relationship between the height of the compressible zone Vs of time. In the graph it can be seen that the more mass of solids added to the tube, the longer the time needed for the floces formed to settle.

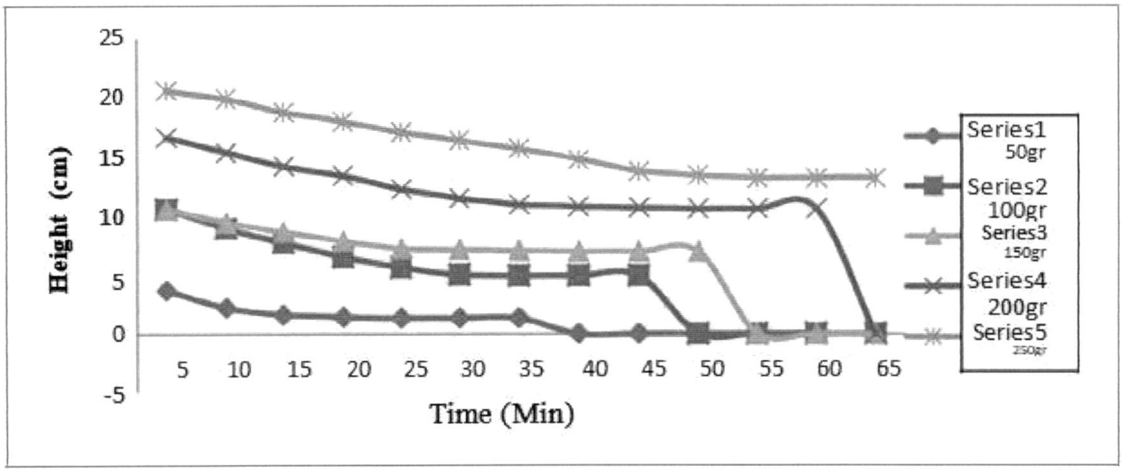

Figure 7. Time vs height of Dry Soil with flocculant

Furthermore, the deposition of dry soil with alum was added as a coagulant. The added coagulant mass is 1 gram into each tube. The function is to add alum as a coagulant so that clumping occurs between alum and floc in the water so that the flocks become larger and relatively heavier, causing suspended particles to settle faster with the addition of alum.

3.4. Activated Carbon Residual Materials

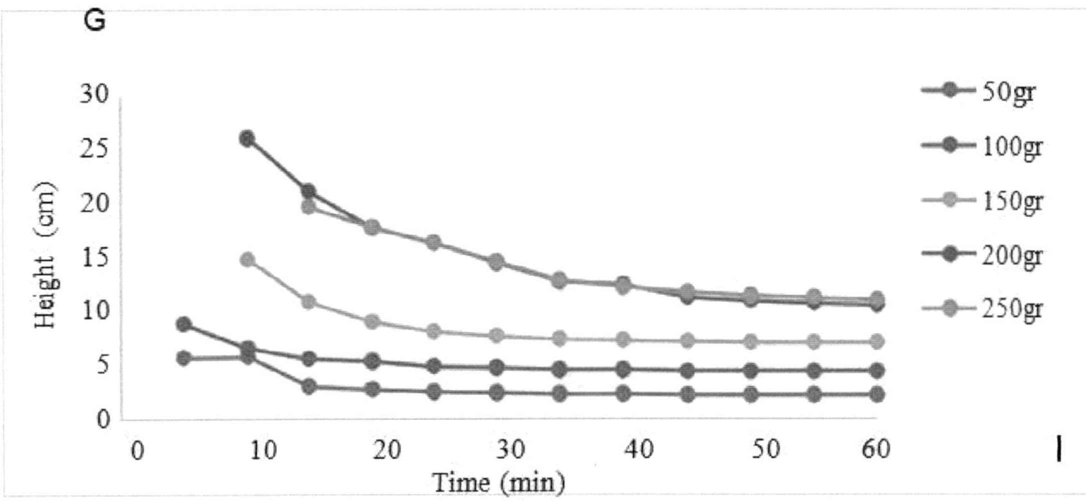

Figure 8. Time vs height of Dry Soil with flocculant

Furthermore, the deposition of dry soil with alum was added as a coagulant. The added coagulant mass is 1 gram into each tube. The function is to add alum as a coagulant so that clumping occurs between alum and floc in the water so that the flocks become larger and relatively heavier, causing suspended particles to settle faster with the addition of alum.

3.5. Activated Carbon Residual Materials

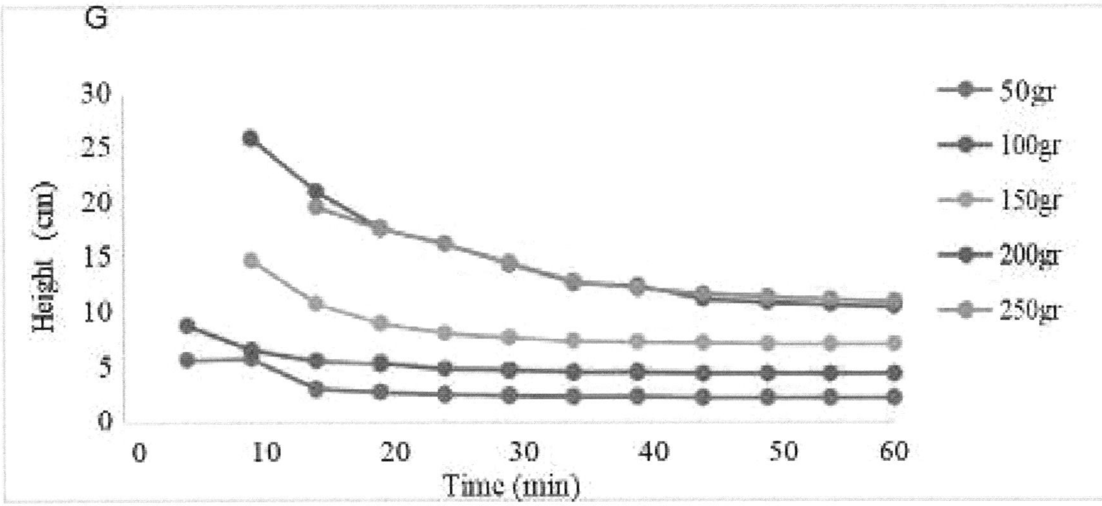

Figure 9. Time vs height of Activated Carbon (CA) without flocculant

In determining the deposition speed of activated carbon without the addition of coagulant can be determined by the graph of the relationship between the height of the compressible zone Vs of time. In the graph it can be seen that the more mass of solids added to the tube, the longer the time needed for the floces formed to settle

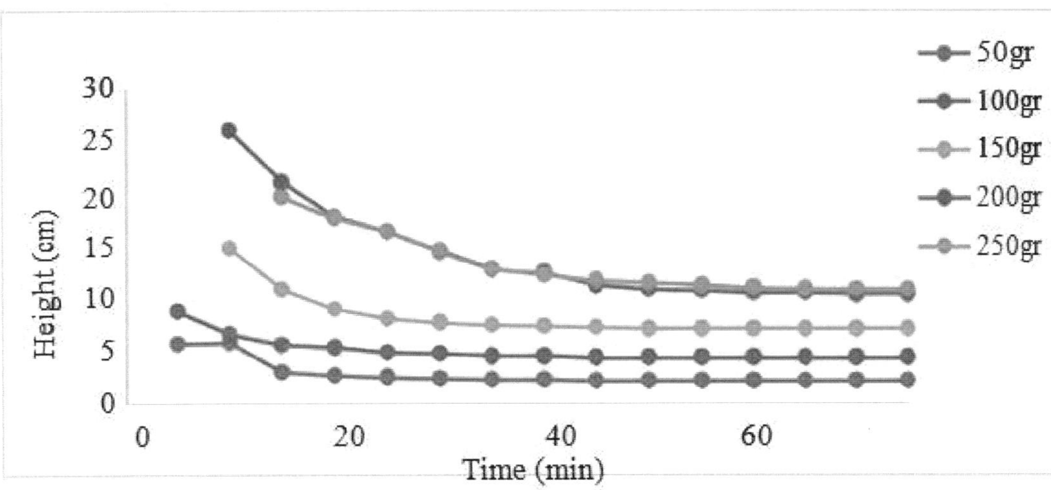

Figure 10. Time vs height of Activated Carbon (CA) with flocculant

Furthermore, the deposition of activated carbon with alum was added as a coagulant. The added coagulant mass is 1 gram into each tube. The function is to add alum as a coagulant so that clumping occurs between alum and floc in the water so that the flocks become larger and relatively heavier, causing suspended particles to settle faster with the addition of alum.

4. Conclusion

From the experiments that have been carried out it can be concluded that The more alum used, the more time is needed to precipitate the formed flocks. The greater the concentration of solids, the smaller the settling speed. The greater the concentration of solids, the greater the sediment obtained

Acknowledgement. The researcher would like to thank the Research and Community Service Center of the Samarinda State Polytechnic Community (P3M POLNES) for providing the opportunity to research, and obtain research funding with a scheme: research and development prototypes and all those who have supported for the smooth running of this research.

References

[1] Goel A, Rana G and Rastogi R 2010 Knowledge management as a process to develop sustainable competitive advantage *South Asian J. Manag.* **17** 104

Comparison of *freeze drying* and f*oam mat drying* effects on characteristics of inulin from gembili *(dioscorea esculenta)*

N Indah, Zainal, D Ganesa

Departmen of Agricultural Technology, Faculty of agriculture, Hasanuddin University, Indonesia.

Email: zainal.itp@gmail.com

Abstract. Gembili (Dioscorea esculenta) is of Dioscorea spp. which contains high inulin. Several factors can affect the physico-chemical characteristics of inulin, one of which is drying method. The purpose of this study was to compare the freeze drying and foam mat drying methods in producing of inulin from gembili (Dioscorea esculenta) based on physicochemical characteristics. Based on the analysis of inulin content of day the best gembili obtained, the samples dried using foam mat drying contained about 9,38% inulin while those dried using freeze contained about 8,66%inulin. The colors of dried samples were relatinglight as indicated by high L values 95.9% for freeze drying and 91,3% for foam mat drying.

1. Introduction

Inulin is a polymer of fructose units with glucose terminal groups. The fructose units in inulin are linked by β- (2-1)-glycosidic bonds. Nearly every inulin linear chain fructose has a GFn structure (G = glucose unit, F = fructose unit and n = number of fructose units that bind to each other). The nature of inulin as soluble dietary fiber is very beneficial for digestion and health of the body [1]. Inulin can dissolve in water but cannot be digested by enzymes in the digestive system of mammals. In the large intestine, inulin is fermented by bacteria found in the large intestine, so that it has a positive effect on health. Some types of Bifidobacteria can use inulin as an energy source because it produces an extracellular inulinase enzyme that can hydrolyze β- (2- 1) -D-fructose bonds to fructose [2]. Inulin is very widely used in the food industry, both in Europe, the USA, Canada and Indonesia as a component (ingredient) of various types of food products.

Inulin is used in various food products. For example, instant milk for children and adults or the elderly. In children's instant milk, inulin provides the benefit of increasing the body's absorption of calcium which supports the growth of bones and teeth. In adult and elderly instant milk, the ability of inulin increases calcium absorption which will prevent osteoporosis or bone loss. Some industries also include it in ice cream and yogurt products [3].

Freeze dryer is a dryer that uses a freezing method in which it dries materials by removing water and solvents by sublimation. The advantages of freeze drying in maintaining the quality of the results of drying, especially for products that are sensitive to heat, among others, can maintain product stability (avoid changes in aroma, color, and other organoleptic elements), can maintain the stability of the material structure (shrinkage and shape changes after drying is very small) and the results of drying in the form of physiological, organoleptic and physical form are almost the same as before drying) [4–6].

Foam mat drying is a way of drying liquid in the form of a previously made foam or foam first by adding a foam agent by stirring or shaking, then dried. The material which is dried by foam mat drying method has a characteristic namely crumb structure, easy to absorb water and easily dissolved in water

The addition of foaming agents and fillers to the drying of inulin with the foam mat drying method can affect the properties of inulin. therefore it is necessary to test the physicochemical characteristics of the tuberous gembili obtained by the method of freeze drying and foam mat drying so that it can reduce the import of inulin by utilizing functional food and increasing its economic value [7].

2. Material and Methods

2.1*Preparation of samples*

The Gembili tubers are washed and cut into small pieces, the sample is then blended with the addition of hot water temperature of 80-90 °C, the filtrate is then discharged using a water temperature of 90 °C for 1 hour, afterwards it is centrifuged at a speed of 5000 rpm for 15 minutes until a white precipitate is obtained. The white precipitate obtained was then divided into two parts 1 part dried using an oven with a foam mat drying method of temperature 60 °C for 7 hours and the second part was dried using a freeze drying temperature of -52 °C for 36 hours, after drying inulin then milled and sieved.

2.2 *Analysis inulin of gembili*

The concentration / purity of inulin was measured using HPLC with Aminex Colom HPX-87C (250mm x 4mm), water index refraction detector model 410 and LCHE Waters pump model M-45. Water is used as a car phase at a speed of 0.3 ml / min, injection volume of 20 µl. The column temperature is set at 80 ° C and the detector is 40 ° C.

2.3 Analysis of lightness

Color is measured by a digital color meter (T-135) for the measurement of the color values of L, a and b, and the color quantification of the indicator is done using a digital camera (Sony PowerShot S110) to take indicator pictures.

3. Results and discussion

Inulin extraction conducted in this study is inulin which is sourced from gembili tubers. Selected tuber tubers are tubers whose age is more than 8 months. 200 grams of tubers produce 24 gr (b / k) for drying foam mat and produce 26 gr for freeze drying. This is caused by the addition of a foam mat agent which affects the water content and inulin texture. The results of the study can be seen in Figure 1 for inulin content of gembili and Figure 2 for lightness of gambili.

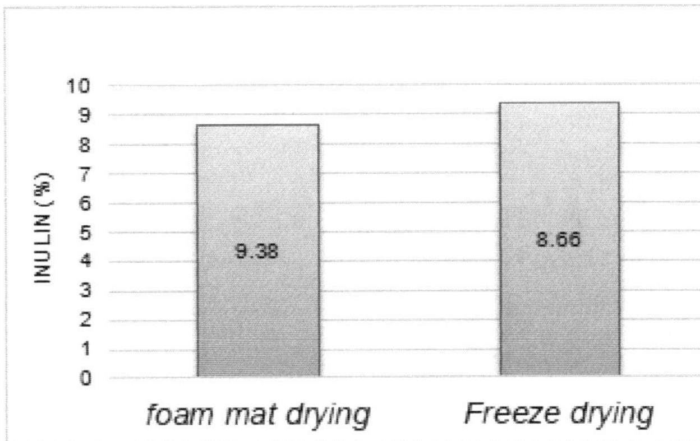

Figure 1. Inulin of Gembili *foam mat drying* dan *freze drying*

Based on Figure 1 above, it can be seen that the level of inulin foam mat drying is greater than freeze drying, which is 9.38% and 8.66%, this is thought to occur due to the glucose reaction contained in inulin with air. states that one of the local raw materials containing inulin is the tuber Dioscorea spp. which has inulin levels vary between 2.88% -14.77%, and the highest is Dioscorea esculenta (gembili) of 14.77% [8]. Inulin research from plant tubers has been conducted since 2006, including; The potential of inulin as a functional food component of dahlia tubers has been investigated . Isolation and purification of inulin from dahlia flower bulbs by Suleman et al in 2008 obtained inulin levels of 7.50% using ethanol and diethyl ether solvents [9]. received the highest yield of inulin in tubers of uwi from gembili with a BEP value of 24.17% with a deposition temperature of -30 °C. Inulin extraction from dahlia plant tubers was studied getting a yield of 6.87% for peeled tubers and 4.9% for tubers that were not peeled using 90% ethanol solvent [10]. High and low of inulin obtained largely depends on the inulin extraction process, extraction using ethanol produces higher levels of inulin than water solvents [11].

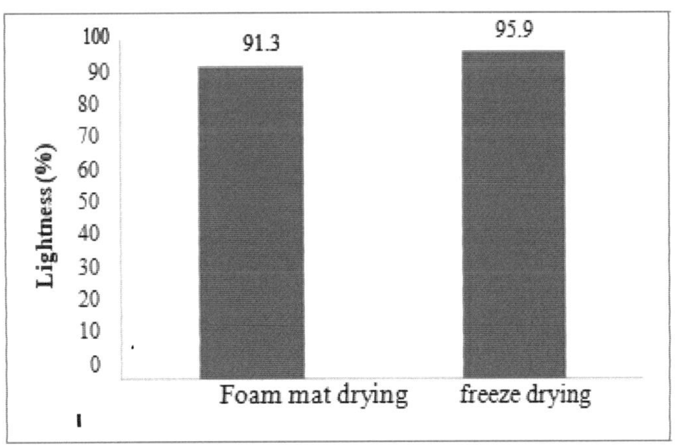

Figure 2. Lightess of inulin gembili foam mat drying and frezze drying

The lightness of gembili is caused by white starch, non-crystalline powder which is insoluble in cold water, and has no sweet taste. The color of dry food products is strongly influenced by the drying temperature. From the results of the study on dry asparagus, obtained materials for dry samples, the higher the temperature or the smaller the temperature below zero, the greater the level of lightness (L *), for freeze drying leading to green color, while the others are yellow color. Better color in the freeze drying method at a minimum temperature - 10 °C.

Appearance of a product will affect consumer interest in the product. Where if food can not be accepted aesthetically (appearance), food will not get the opportunity to play a role in meeting the nutritional needs of a person . The color of the product is not liked or considered to deviate from the color that should be, then consumers are usually no longer interested in giving a good assessment of other quality attributes. Appearance on each treatment given produces different values. The temperature and duration of drying affect the appearance of gembili tuber powder produced, the higher and longer the drying temperature can change the appearance of flour to brownish. However, the low temperature and duration of drying can produce a white powder appearance [12].

4. Conclusion

Gembili tubers with foam mat drying method results in higher levels of inulin than drying using the freeze drying method as well as with lightness foam mat drying producing lightness colors.

References

[1] Sardesai V 2011 *Introduction to clinical nutrition* (CRC Press)
[2] Roberfroid M 2007 Prebiotics: the concept revisited *J. Nutr.* **137** 830S-837S
[3] Kumalaningsih S and Beni Y M W 2005 Membuat Makanan Siap Saji *Trubus Agrisarana. Surabaya*
[4] Toneli J T C L, Mürr F E X, Martinelli P, Dal Fabbro I M and Park K J 2007 Optimization of a physical concentration process for inulin *J. Food Eng.* **80** 832–8
[5] Widjaja B, Setyawan D and Moechtar J 2013 Development of piroxicam orally disintegrating tablets by freeze drying method *Int. J. Pharm. Pharm. Sci.* **5** 795–8
[6] Indriani Y and Hikmawati D 2019 Fabrication of Chitosan-Chondroitin Sulfate/Hydroxyapatite Composite Scaffold by Freeze Drying Method. *J. Int. Dent. Med. Res.* **12**
[7] Toneli J, Park K J, Ramalho J R P, Murr F E X and Fabbro I M D 2008 Rheological characterization of chicory root (Cichorium intybus L.) inulin solution *Brazilian J. Chem. Eng.* **25** 461–71
[8] Winarti S, Harmayani E and Nurismanto R 2011 Karakteristik dan profil inulin beberapa jenis uwi (Dioscorea spp.) *Agritech* **31**
[9] Widowai Y, Mulyani K and S. H 1998 *Potential of Cassava Plants as Carbohydrate* (UNSOED)
[10] Istianah N 2010 *Variety on production of Umbi Uwi (Dioscorea spp.).* (Universitas Pembangunan Nasional)
[11] Ronkart S N, Paquot M, Fougnies C, Deroanne C and Blecker C S 2009 Effect of water uptake on amorphous inulin properties *Food Hydrocoll.* **23** 922–7
[12] Tranggono S, Haryadi A, Murdiati S, Sudarmadji S, Rahayu K, Naruki S and Astuti M 1989 Bahan Tambahan Pangan (Food Additives) *Pus. antar Univ. Pangan dan Gizi. Univ. Gadjah Mada. Yogyakarta*

IOP Conf. Series: Materials Science and Engineering **885** (2020) 012047 doi:10.1088/1757-899X/885/1/012047

Effect of knowledge management on performance: A case study at PT. Eastern Pearl Flour Mills Makassar

A Rapi[1,2], A V Baharuddin[1] and M P Shabrina[1]
[1] Department of Agro Industrial Engineering, Polytechnic of ATI Makassar, Makassar, 90211, Indonesia
[2] Department of Industrial Engineering, University of Hasanuddin, Makassar, 91711, Indonesia

Email : amrin_rapi@yahoo.com, andi

Abstract. This study aims to determine the effect of Knowledge Management on the performance of maintenance department employees at PT. Makassar Eastern Pearl Flour Mills and to find out which Knowledge Management indicators are most influential on employee performance. This study uses the statistical application IBM SPSS 21. The population in this study is employees at PT. Eastern Pearl Flour Mills Makassar, amounting to 78 people. The collection of data is done by using a questionnaire. 78 questionnaire were handed out with 65 questionnaire are valid. The data were processed using multiple linear regression analysis technique. The results of the study indicate that there is the influence of the variable Knowledge Management which consists of Personal Knowledge (X1), Job Procedure (X2), and Technology (X3) against the performance of the employee (Y) with the value of the coefficient of determination of 0.286. Analysis of the data do indicate that the variable Personal Knowledge of the results of the t test for 0.3562 is the most giving effect compared with other variables.

1. Introduction

Knowledge as capital has the effect of which is very large in the progress of an organization. In a neighborhood that is very quickly changing it then knowledge usually will undergo obsolescence by because it needs to be kept constantly updated through a process of learning.

Knowledge which is owned by an organization is an asset that is very valuable [1] and is an asset that is not naked eye or intangible assets [2], knowledge is also a resource internal company the most valuable, unique, difficult to replace, and difficult to imitate [3].

Science and technology as a factor of power of competitiveness that is very important to be able to survive and compete so that Knowledge Management is required so that employees, groups , or organizations are constantly developing its capacity to achieve result is optimal, which is commonly called the performance [4]

According to Khammarnia et al. [5], knowledge management is an important component in organizational success, including innovation, strategic, economics, and management. Knowledge management emerges as an important aspect of achieving excellent organization and has been approved and used as an effective organizational management method in a variety of companies.

The maintenance department at PT. Eastern Pearl Flour Mills recognizes the importance of a consistent and ongoing learning process. Then from the department of maintenance do the management of knowledge as a system of management that acts as a means of learning and doing activities transfer or sharing of knowledge one of them to follow training or training in accordance with the field and perform activities of brainstorming. PT. Eastern Pearl Flour Mills particular department maintenance of implementing knowledge management in the form of a briefing every morning and do brainstorming every three months.

The maintenance department of PT. Eastern Pearl Flour Mills applies the knowledge management method as a tool for managing employee performance in the company's management process and empowering employees to get superior results from the maintenance department's employees to improve company performance, so that it is expected that each employee can continue to explore his knowledge and not just depend on or fixate on the existing system so that each maintenance department employee has a role in improving the quality of the company.

The study is to investigate the construct of Knowledge Management, which consists of three variables, namely personal knowledge, job procedures, and technology by using a method of multiple linear regression analysis. This is a method that is used to determine how large the influence of the management of knowledge on the performance of employees of the department of maintenance, so it can be known variables that most influence on the performance of employee of maintenance department.

2. Methods

This research is using primary data obtained by asking the respondent a set of questions in the form of a questionnaire. The respondents of this study were all employees at the Maintenance Department of PT. Eastern Pearl Flour Mills. The employees of the maintenance department that are this research respondents are oddly all male. From the 78 respondents, 45% respondents has education level as high as high school, 31% hold diploma degree or college graduates, 23% respondents hold bachelor degree, and 2% of the respondents hold master degree. Of the 78 questionnaires were distributed as much as 65 questionnaires collected back and can be processed much further. The data analysis technique used is multiple linear regression analysis. The conceptual framework of this study can be seen in Figure 1.

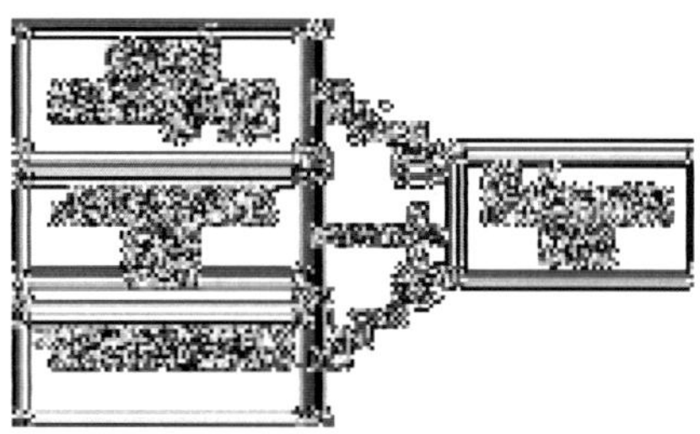

Figure 1. Research conceptual framework

The study is concluded the effect of variable independent of the variable dependent by comparing the value of the significance of models in statistics. Constructs Knowledge Management was measured by using three indicators, namely Personal Knowledge (X1), Job Procedure (X2), and Technology (X3). The third indicator is used to measure the effect of Knowledge management of the variable performance Employees (Y). Indicator variables Performance Employees in research is adopted from research Bernardin and Russel [6] which consists on Quality, Quantity, Timeliness, Need for supervision, and interpersonal impact.

3. Results

By using the multiple linear regression analysis technique, obtained the results of the processing of data such as that shown in table 1.

3.1. Multiple Linear Regression Analysis

Table 1. Coefficient of multiple linear regression test

Model		Unstandardized Coefficients		Standardized Coefficients		
		B	Std. Error	Beta	t	Sig.
1	(Constant)	12.286	4.434		2.771	.007

personal knowledge	.426	.120	.669	3.562	.001
job procedure	-.004	.090	-.007	-.041	.976
technology	-.067	.080	-.139	-.838	.405

Table 1 is the result of data processing that shows the coefficients of the multiple linear regression analysis. Table 2 shows the R values of the multiple linear regression analysis. In table 1 can also be known the value of t-calculate.

Table 2. R table of multiple linear regression test

Model	R	R Square	Adjusted R Square	Std. Error of the Estimate
1	.565[a]	.320	.286	301.848

Based on the results of the data that then can be compiled equation multiple linear regression as follows: M

$$Y = 12.286 + 0.426_{x1} - 0.004_{x2} - 0.067_{x3}$$

The regression equation above implies that:

a. The constant value is 12.286. This constant value indicates that if there are no personal knowledge variables (X1), work procedures (X2) and technology (X3), so employee performance is valued as 12.286.
b. It is proven that the variable of personal knowledge (X1) is 0.426 which is mean that if there is a one-unit increase, the employee's performance (Y) will also increase by 0.426.
c. The job procedure variable (X2) is -0.004 which is mean that if there is a one-unit increase, it will also lose the performance of employees (Y) of 0.004.
d. The technology variable (X3) value is -0.067 which is mean that if there is a one-unit increase, it will also lose employee performance (Y) of 0.067.
e. R value of 0.549 shows that relationship between the performance of employees with the variables of personal knowledge, work procedures and technology are sufficient.
f. Adjusted R square value of 0.286 which also shows the relationship between employee performance with personal knowledge variables, work procedures, and technology affect positively or sufficiently on employee performance. The variables of personal knowledge, work procedures and technology on the performance of maintenance employees amounted to 28.6%, and the remaining 71.4% was kept by other variables outside this study.

3.2. Partial t-test

Partial t-test was conducted to gain the understanding of the influence of knowledge management to performance. The results of the t-test are:

a. t calculate value for variable X1 is 3.562, meanwhile t table value is 1.999 which is imply that X1 influence variable Y significantly.
b. t calculate value for variable X2 is 0.127, meanwhile t table value is 1.999 which is imply that X2 does not has significant influence on variable Y.
c. t calculate value for variable X3 is 1.158, meanwhile t table value is 1.999 which is imply that X3 does not influence variable Y significantly.

3.3. F-test

F test is done to understand the influence of variables X1, X2, and X3 simultaneously influencing variables Y. The results of the F test can be seen in Table 3.

Table 3. F value

Model		Sum of Squares	df	Mean Square	F	Sig.
1	Regression	261.230	3	87.977	9.557	.000[b]
	Residual	555.785	61	9.111		

Total	817.015	64

a. Dependent Variable: performance
b. Predictors: (Constant), technology, job procedure, personal knowledge

4. Discussion

Based on the table coefficient test of multiple linear regression shows that personal knowledge has an effect that is significant to the performance of the employee. In practice, at PT. Eastern Pearl Flour Mills Makassar employees will attend training activities (training) in accordance with their fields so that each employee's knowledge can be increased to support his work . Employees also usually perform discussion or sharing of knowledge among co- working , so that employees of the department of maintenance can work much better and new knowledge.

Based on the results of the conducted interview, maintenance department employees have more than one year's work experience. With experience that is owned , employees agree that the work that they do will be easy and able to finish the job more quickly than on the other . With so variable personal knowledge or knowledge of personal one only experience will greatly impact on performance.

As shown in the table coefficient test of multiple linear regression analysis indicates that the job procedure does not have the influence that is significant to the performance of the employee. In this study SOP (standard operating procedure) is similar to job procedure. SOP applied by PT. Eastern Pearl Flour Mills Makassar particular department maintenance implement the SOP with quite well. However, according to some employees in companies that do not agree that the SOP in the department have meet the standard because sometimes still often happens implementation of activities that leave behind in the SOP as operators still driving a forklift in which the operator has SIO (operator permit letter) who has passed the limit driving license, this is not in accordance with the SOP.

The third variable used in this research to measure knowledge management is Technology. Based on the table coefficient test of multiple linear regression shows that the technology does not have the influence that is significant to the performance of the employee. In practice, some maintenance department employees do not really use technology in the form of the internet because employees only focus on improving technology in the form of machines.

5. Conclusion

Based on the results of the test regression showed that knowledge management impact positively on the performance of employees of the department of maintenance by 28.6% with the variable of personal knowledge as a variable having influence largest compared to job procedure and technology and the rest of 71.4% is influenced by variables other outside this research. Other similar research as this research are encouraged to including another variables that has been proved that could affect employee performance and not include in this research model such as organization culture to explain more of the influence of knowledge management to employee performance in companies.

References

[1] Delgado-Verde M, Martín-de Castro G and Navas-López J E 2011 Organizational knowledge assets and innovation capability *J. Intellect. Cap.*
[2] Goel A, Rana G and Rastogi R 2010 Knowledge management as a process to develop sustainable competitive advantage *South Asian J. Manag.* **17** 104
[3] Bonner D 2000 Knowledge: from theory to practice to golden opportunity *Am. Soc. Train. Dev.* **5** 12–3
[4] Purnama M and Budihardjo A 2012 Peran budaya pembelajaran dan knowledge management terhadap kinerja perusahaan: studi kasus PT XYZ *Int. Res. J. Bus. Stud.* **1**
[5] Khammarnia M, Shahsavani F, Shahrakipour M and Barfar E 2015 Relationship Between Knowledge Management and Quality of Working Life in Nursing Staff of Zahedan Teaching Hospitals, 2014 *Heal. Scope* **4**
[6] Bernardin R 2011 Human Resource Management. New Jersey: International Editions

IOP Conf. Series: Materials Science and Engineering **885** (2020) 012048 doi:10.1088/1757-899X/885/1/012048

Preliminary Analysis of Electricity Energy Efficiency in SMEs

Y S Akil[1], H Miyauchi[2], S Mangngenre[3], and K Amar[3]

Department of Electrical Engineering, Universitas Hasanuddin, Gowa, Indonesia
Department of Computer Science and Electrical Engineering, Kumamoto University, Kumamoto, Japan
Department of Industrial Engineering, Universitas Hasanuddin, Gowa, Indonesia

E-mail: yusakil@unhas.ac.id

Abstract. In general, industrial sector consumes high energy. Therefore, efforts to manage the energy usage in this sector are useful as it can reduce significantly total energy consumption mainly for electricity. This research aims to observe consumers' perceptions and success factors to improve efficiency of electricity energy usage for industry particularly for manufacturing small-medium enterprises (SMEs) in Makassar as a representative big city in Indonesia. A questionnaire for survey is designed to get data from 127 respondents and analysed using statistical approach. Observed variables related to energy efficiency in SMEs include human resources, production technology, rainfall, and tax. Results shown the developed questionnaire is reliable (Cronbach's alpha value exceeds 0.6). Perception of consumers about energy saving is quite good. From aspect of demographic characteristic, there is a difference in users' perceptions on energy saving. Other results show the significant relationship between human resources, production technology, and rainfall condition to the energy efficiency level at SMEs. Knowledge or findings of this study provide valuable information in relation to strategy to improve efficiency of electricity usage in industries.

1. Introduction

Electricity load in many countries increase over the time. To assure the balancing between electricity production and load growth, efforts such as building renewable based power plants at certain times and energy saving are needed. The balancing is an important aspect to keep stability electricity service to consumers in power systems.

As in other places, industrial sector in Indonesia consumes high energy. Based on this, energy saving actions or energy efficiency improvement (EEI) in this sector such as for SMEs is meaningful as a strategy to reduce energy consumption mainly electricity. However, its implementation is not an easy task as several factors can influence the success of energy saving actions or EEI in one place. One of barriers for some consumers to practice energy saving optimally is the lack of information [1]. Therefore, determining the best way to promote energy saving practice for enhancing consumers' knowledge and observing main factors affecting EEI are the challenge for researchers. Consumers that have enough information or have positive perceptions are normally easier to practice energy saving.

They will utilize energy efficient equipments and use electricity energy wisely in their daily activities as a habit.

There are a number of references which focus on the energy efficiency analysis for different sectors. Barris et al. [2] analyses end user to increase electricity energy efficiency for households in Latvia using logistic regression. The authors perform a survey, and goal-framing theory is adopted to measure the deployment of electricity efficiency concept in society. Rahman et al [3] analyses energy saving potential for households particularly in suburban area in Malaysia. Awareness of consumers and type of electrical appliances are some indicators used in the study to understand factors in reducing energy consumption. Soepardi et al. [4] investigates Indonesian steel industry concerning the barriers to increase energy efficiency. Policy, financial-economic factor, technology and three other variables are examined in relation to EEI. Fleiter et al. [5] analyses factors influencing energy efficiency for SMEs in German. Data from energy audit and statistical approach are used in their empirical analysis. Ackah et al. [6] studies energy efficiency practice of the rural SMEs in Ghana. Product generational dematerialization approach is adopted to investigate electricity and fuel energy consumption. Another reference, by Apeaning et al. [7] for Ghana context observed the barriers to increase energy efficiency especially for large industries. Exploratory and qualitative research are applied to get specific knowledge in enhancing efficiency of energy usage. Recently, study in [8] performs a literatures review on energy policies to increase energy efficiency in SMEs, and in Yonemoto et al. [9], the authors develop physical simulator to assess energy efficiency related to productivity for manufacturing industry.

This study aims to measure perceptions of consumers and to observe factors affecting efficiency of electricity energy usage in industry particularly for manufacturing SMEs in Makassar. For this purpose, survey questionnaire is designed, and regression model is proposed as a tool analysis. In general, increasing efficiency is not only a key strategy to maintain electricity availability for consumers, but also to reduce greenhouse gas emissions in short to medium term or to increase carbon neutral energy system [5][8]. SMEs business avoids lack of energy supply as it may lead to lost profits and customers [10]. However, more advocating of electricity saving is needed as energy efficiency level is not too clear in SMEs, budget for efficiency improvement program is quite low, and often lack of resources which can be an obstacle in adopting innovation [11][12].

To the authors' knowledge, study on energy efficiency for Indonesian SMEs is very limited. This is in line with literature review on energy efficiency reported in [8] which shown no study for Indonesian context. Majority of the studies are from European countries. Therefore, this present work is expected can fill the gap. The structure of paper consists of five parts. After introduction, this paper continues with typical Indonesian SMEs, followed by research methodology and then results. The last part is conclusion including future research.

2. Indonesian SMEs Consumers

Typical industrial consumers in Indonesia are classified into micro, small, medium and large scale which can be grouped into micro-small-medium industry (MiSMEs) and large industry. According to BPS [13], employee number for micro scale is 1 to 4, small scale is 5 to 19, medium scale is 20 to 99, meanwhile for large industry is 100 and above. Among the groups, majority of the consumers is coming from MiSMEs group. Figs. 1 and 2 show Indonesian MiSMEs and their number of employees for year 2012 and year 2014 [14]. As seen from Fig. 1, micro industry is highly dominant compared to other classifications (SMEs). Particularly for SMEs, the number of small and medium sized industries in year 2012 are 602,195 and 44,280, respectively and in year 2014 increased becomes 654,222 and 52,106, respectively. Similar tendency to the number of employees, which is from 3.9 million in year 2012 and becomes 5.5 million in year 2014 for small industry. Meanwhile for medium scale, it is increased from 2.8 million in year 2012 and becomes 3.9 million employees in year 2014. The same situation in South Sulawesi province which Makassar city is located, the largest consumers group for industrial sector is also coming from MiSMEs.

As Indonesian industries and employee absorption are very high, moreover the number of industries in the future may increase higher which proportional to energy consumption, therefore, it is important to explore industrial characteristic to improve efficiency of energy usage mainly for manufacturing SMEs. This is in line with the concern of Indonesian government to increase SMEs performance.

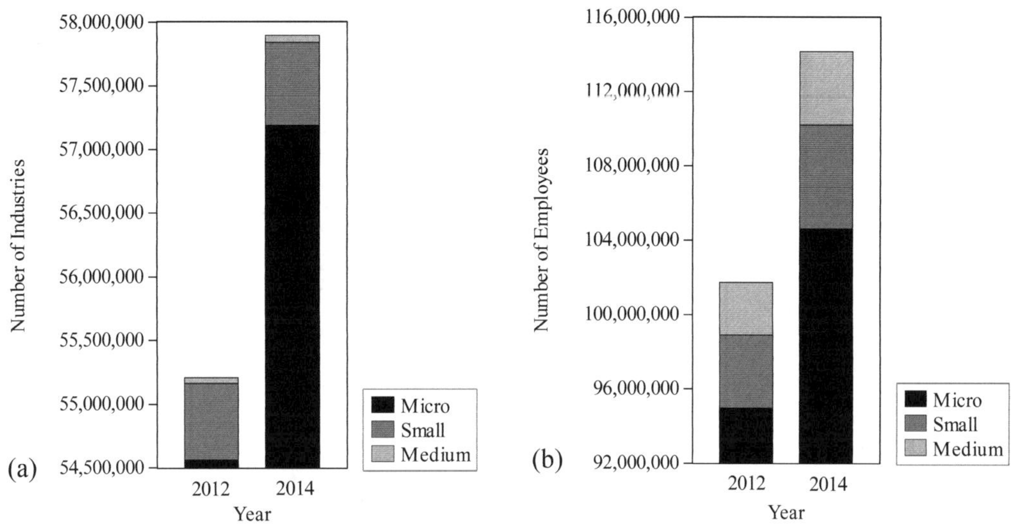

Figure 1 (a)(b) Indonesian MiSMEs and number of their employees for year 2012 and 2014.

3. Research Methodology

The target respondent selected in this study is SMEs particularly manufacturing industry in Makassar area in South Sulawesi province. Required data are collected using a questionnaire survey. The questionnaire is developed based on the related prior studies [15], and it is organized into three main parts, A to C. In Part A is questions about general information of respondents, e.g. age, education level, position in SMEs, and category of industry. Part B is users' perceptions about electricity energy saving, the statements include e.g. "Utilization of low watt electricity equipments in SMEs supports energy saving program in industry", and "Energy saving behaviour can support the sustainability of electricity energy in the future". Meanwhile in Part C is influencing factors related to implementation of energy efficiency. The statements include e.g. "Employees have good enough education", "Used machine is new, it can minimize obstacles in the production process", "Rain with high intensity often occurs in Makassar". Items on the questionnaire is evaluated by using 1 'strongly disagree' to 7 'strongly agree' rating scale (7-point Likert scale).

Purposive sampling technique is used during face-to-face survey that conducted in August to October 2019. The respondents are business owner, manager, or employee who knows energy usage in their organization including electricity equipments and production process. To get more information from collected data after reliability test, perception of consumer is measured using mean score analysis, and regression approach is applied to identify success factors for implementation of energy efficiency in manufacturing SMEs. Here, the examined factors include human resources, production technology, rainfall, and tax.

4. Results

4.1. Reliability test and validity

For reliability test, Cronbach's alpha value is applied to evaluate items in the questionnaire. Forty respondents are used as a pilot survey. From testing, obtained Cronbach's alpha values are above 0.6 for all dimensions (perception, human resources, production technology, rainfall factor, tax, and efficiency energy). This means items in the designed questionnaire have internal consistency or reliable and it can be used for further analysis (main survey). For validity, this study adopts the expert validity approach.

4.2. Demographic of respondents

Figure 2 shows descriptive statistic for all participants (127 data sample) in this study. The data sample is around 24% of the total number of manufacturing SMEs in the observed area for year 2017. The descriptive data include gender, educational level, age, position of respondents, category of industry, and installed power capacity (IPC). From the figure, 66.14% of the respondents is male and dominantly is business owner (62.2%). Majority of the participants has small industry (92.12%), only 7.87% of them has medium industry. Meanwhile for IPC, 69.3% of respondents have IPC 2,200 – 6,600 VA in their SMEs.

4.3. Perceptions of consumers

Analysis for all participants in this study shown perception of SMEs' consumers concerning energy saving is generally good (5.68 out of 7 Likert scale). Normally, people with good perceptions or have enough knowledge are easier to practice energy saving. They tend to use electricity energy wisely in their daily activities either at home or in workplace.

Further analysis shows a significance difference in the users' perceptions on energy saving from perspective of demographic characteristics, namely based on the category installed power capacity (IPC). The SMEs respondents for IPC 2,200 – 6,600 VA show more positive perception (Mean = 5.76, SD = 0.91) than respondents who have IPC below 2,200 VA (Mean = 5.61, SD = 1.02) and IPC above 6,600 VA in their SMEs (Mean = 4.75, SD = 1.88).

4.4. Success factors of energy efficiency improvement

Regression analysis is performed to examine the relationship between observed variables (human resources, production technology, rainfall, and tax) and implementation of energy efficiency in SMEs as seen in Table 3. From results, variables in the regression model can explain around 42% of the energy efficiency level which well validated. Three of variables in the model namely human resources, production technology, and rainfall have significance at 10% associated with energy efficiency with expected signs. The significance of the variables is indicated by obtained p-value for each variable below 0.1. Among the factors, production technology gives the highest effect associated to EEI in SMEs as indicated by its coefficient value (0.4975) is higher than others.

The result implied that human resources, production technology, and rainfall are some factors to enhance energy efficiency. The better of human resources and production technology, the higher energy efficiency in SMEs. Based on this, efforts include providing information and giving training to increase knowledge of employees including management level are important to implement energy efficiency concept. For production technology, some aspects such as renew production equipments at particular time and utilize efficient-based support devices are useful to achieve optimal efficiency in energy usage. This is in line with previous studies such as in [15]. For rainfall variable, under severe condition such as high intensity rainfall during peak period of rainy season can make consumption higher in SMEs due to additional lighting and the longer production process at certain industries. This can stimulate consumers to use electricity more effective to minimize additional energy consumption.

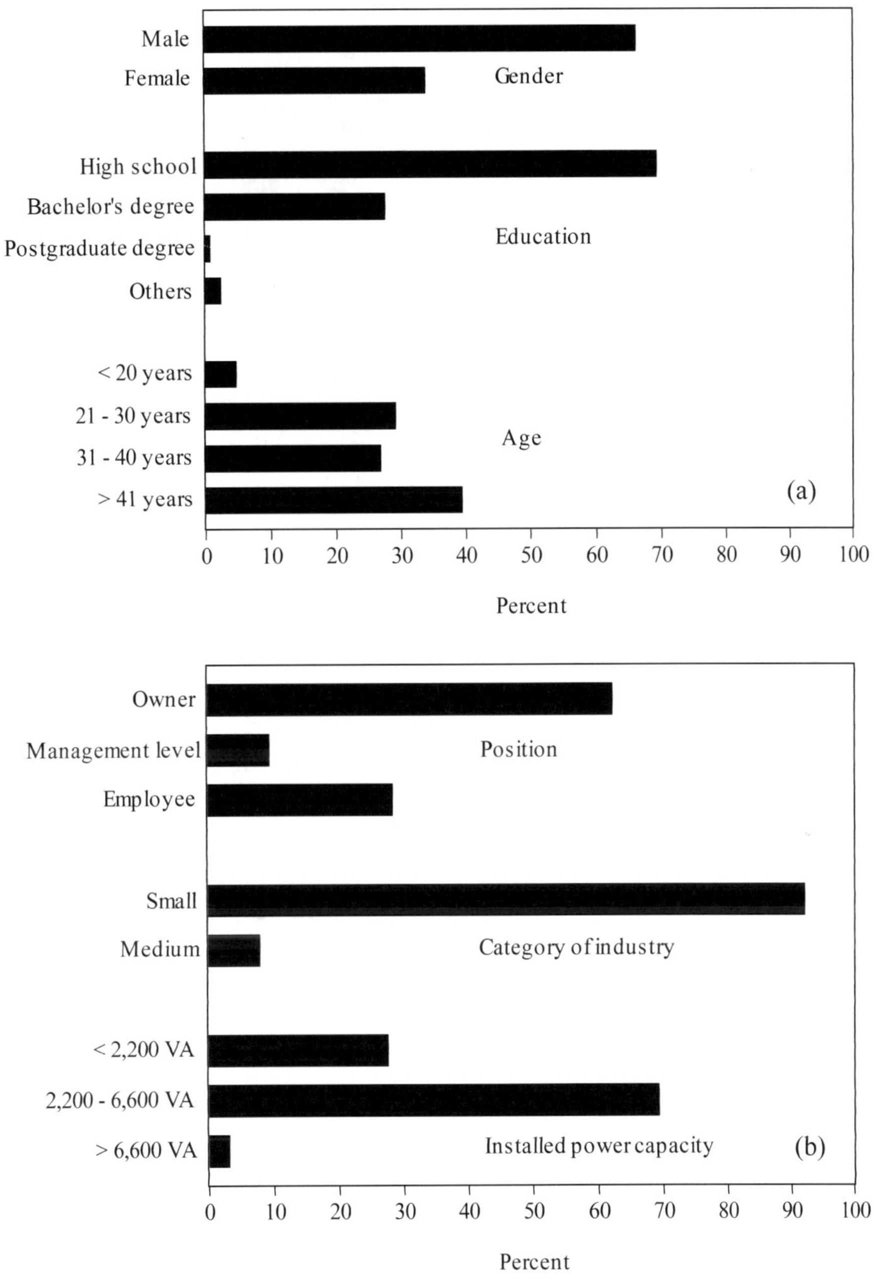

Figure 2. (a)(b) Demographic of respondents

Table 1. Regression result

Variables	Energy Efficiency		
	Coef.	p-value	t-statistic
α_0	1.3473	0.0093	2.6453
Human resources	0.1480	0.0762	1.7886
Prod. technology	0.4975	0.0000	5.6050
Rainfall	0.0957	0.0733	1.8070
Tax	0.0344	0.5693*	0.5705
AR(1)	0.2516	0.0077	2.7124
R^2		0.4392	
$R^2_{adj.}$		0.4158	
Prob. (F-stat.)		0.0000	
Durbin-Watson		2.0464	

*Non significance variable

As implications of the study, policy maker should determine more suitable strategies in promoting EEI to industrial consumers. They can be designed based on the perception level such as by giving priority according to power capacity level in SMEs as IPC affects perceptions of consumers. For example, IPC above 6,600 VA or below 2,200 VA show lower perceptions on energy saving than consumers with IPC 2,200 – 6,600 VA. Similar perspective for designing strategies based on finding for success factors. For example, according to human resources which may refers to educational background of consumers as an indicator. Respondents' background is dominantly junior / senior high school (around 69%). Based on this, education in the form of training or others should more concern to the SMEs consumers.

Enhancing energy efficiency in industry is a pivotal aspect to increase SMEs performance, and it is more important in global competition era. Concerning this, industry business in all scales should implement more energy efficiency concept for effectiveness and competitiveness.

5. Conclusions

This study observed consumers' perceptions and success factors for electricity energy efficiency in industry particularly for manufacturing small-medium enterprises (SMEs) in Makassar. A questionnaire for survey is designed to get data from consumers and analysed using statistical approach. From analysis, it is concluded that the developed questionnaire is reliable. Perception of consumers on energy saving is generally good. Based on the demographic characteristic, there is a difference in users' perceptions on energy saving. Other results show strong relationship between human resources, production technology, and rainfall condition to the energy efficiency level at SMEs. It is expected information from this study can contribute in relation to strategy to improve efficiency of electricity usage in industries. Future research will observe other variables to get more detail information regarding success factors of energy efficiency for industry in Indonesia.

Acknowledgments

The authors would like to thank the Indonesian government for support given in this research (under program SAME 2019). Special thanks go to all who participated during data collection.

References

[1] Sweeney J C, Kresling J, Webb D, Soutar G N and Mazzarol T 2013 Energy saving behaviours: Development of a practice-based model *Energy Policy* **61** 371–81

[2] Bariss U, Dandens A, Timma L, Blumberga A and Blumberga D 2015 How to assess involvement of electricity end user in energy efficiency improvement-Analysis of survey results *Energy Procedia* **72** 270–7

[3] Rahman K A, Leman A M, Mubin M F, Yusof M Z M, Hariri A and Salleh M N M 2017 Energy consumption analysis based on energy efficiency approach: a case of suburban area *MATEC Web of Conferences* vol 87 (EDP Sciences) p 2003

[4] Soepardi A, Pratikto P, Santoso P B, Tama I P and Thollander P 2018 Linking of barriers to energy efficiency improvement in Indonesia's steel industry *Energies* **11** 234

[5] Fleiter T, Schleich J and Ravivanpong P 2012 Adoption of energy-efficiency measures in SMEs—An empirical analysis based on energy audit data from Germany *Energy Policy* **51** 863–75

[6] Ackah I and Kizys R 2018 Analysis of energy efficiency practices of SMEs in rural Ghana: an application of product generational dematerialization method *Energy Effic.* **11** 1359–74

[7] Apeaning R W and Thollander P 2013 Barriers to and driving forces for industrial energy efficiency improvements in African industries–a case study of Ghana's largest industrial area *J. Clean. Prod.* **53** 204–13

[8] Johansson I, Mardan N, Cornelis E, Kimura O and Thollander P 2019 Designing Policies and Programmes for Improved Energy Efficiency in Industrial SMEs *Energies* **12** 1338

[9] Yonemoto R and Suwa H 2019 Evaluation of Energy Efficiency and Productivity in Scheduling by Using Physical Simulator *Trans. Inst. Syst. Control Inf. Eng.* **32** 185–91

[10] Shahbaz M 2015 Measuring economic cost of electricity shortage: current challenges and future prospects in Pakistan

[11] Khalifa A, Caporin M and Di Fonzo T 2019 Scenario-based forecast for the electricity demand in Qatar and the role of energy efficiency improvements *Energy Policy* **127** 155–64

[12] Amar K and Davis D 2015 Ready for Six Sigma?-a survey of Indonesian SMEs *Int. J. Product. Qual. Manag.* **16** 281–97

[13] BPS Statistics Indonesia

[14] Anon Kementerian KUKM

[15] Akil Y S, Mangngenre S, Mawar S and Amar K 2018 Preliminary Study of Perception and Consumer Behaviour Towards Energy Saving for Household Appliances: A Case of Makassar *Journal of Physics: Conference Series* vol 979 (IOP Publishing) p 12091

Biohydrogen production from palm oil mill effluent by dark fermentation using *clostridium butyricum*

M Pasaribu[1], F S P Purba[2], E Y R Sitompul[2] and E A Nainggolan[2]

[1]Department Mineral Chemical Engineering, Politeknik ATI Makassar, 90211 South Sulawesi, Indonesia
[2]Department Bioprocess Engineering, Institut Teknologi Del, 22381 North Sumatera, Indonesia

E-mail: monitapasaribu@gmail.com

Abstract. The process of producing oil from fresh fruit bunches produces palm oil effluent (POME). Good processing of POME is needed to prevent emissions that can damage the environment. POME can be used as a substrate to produce biohydrogen gas. This study aims to treat palm oil liquid waste-producing biohydrogen by dark fermentation using Clostridium butyricum. The initial stage is the addition of NaOH to POME which functions to degrade lignin and stop methanogenic bacteria. Then the hydrolysis method was carried out during dark fermentation using Clostridium butyricum with varying concentrations of NaOH (0.3; 0.5; 1% w/v POME) and hydraulic retention time. The results showed the highest biohydrogen production was obtained at a concentration of NaOH 0.3% w/v with a hydraulic retention time of 32 hours with an amount of 2824.2 mg H_2/L POME.

1. Introduction

Palm oil is one of the important industrial plants. The largest amount of vegetable oil is produced from crude palm oil (CPO). CPO is an industrial raw material that is widely used for various industrial sectors, including the food industry, cosmetics industry, pharmaceutical industry, steel industry and others. The increase in the palm oil industry in addition to having an impact on meeting the needs of the community for palm oil consumption, also impacts the environment. From an environmental perspective, the palm oil processing industry produces waste in the form of solid waste and liquid waste. One type of waste produced by the CPO industry is palm oil mill effluent (POME). Processing every 1 ton of oil palm produces more than 50% of POME [1]. The large amount of POME produced is an environmental problem for the CPO industry [2]. This is due to the high organic content of POME which can reduce dissolved oxygen levels which are known from the COD parameter values of 50 gr / L and BOD 25 gr / L [3,4] One type of processing carried out is the conversion of POME into a more valuable product using the help of microorganisms. One product that can be produced from POME processing is biohydrogen.

Biohydrogen is a renewable energy that is environmentally friendly. Therefore biohydrogen has the potential as alternative energy that can be used to overcome the depletion of fossil oil reserves and reduce the impact of emissions produced on the environment [5]. The great demand for hydrogen in the future prompted a number of researchers to optimize biohydrogen production. Parameters that affect biohydrogen production include residence time and substrate concentration. This research aims to produce biohydrogen from POME processing by dark fermentation using Clostridium butyricum. The variables used in this study are variations in residence time and variations in NaOH concentration in the pretreatment of POME substrates.

2. Experimental Methods

2.1. Pretreatment
POME storage was carried out in a 4^0C freezer. POME was centrifuged for 7 minutes with 21,000 rpm before adding NaOH. Solids obtained were collected and taken as much as 40 gr to the flask. There are 3 flask with NaOH concentrations variations 0.3% (w/v), 0.5% (w/v) and 1% (w/v) which was added as much as 400 ml. The result of mixing was heated at a temperature of 30^0C for 4 hours. The POME solid was then filtered and washed with distilled water to remove the base. POME was centrifuged again for 7 minutes with a speed of 21,000 rpm to remove the water content.

2.2. Bacteria cultivation
Cultivation of Clostridium butyricum was carried out using RCM as much as 10 mL in a 50 mL ampoule bottle for one week. Then the results of cultivation were transferred in 40 mL RCM media to obtain new 50 mL inoculums

2.3. Dark fermentation
1 L POME of the enzymatic hydrolysis was put into the 2 L flask. Then nitrogen gas has flowed for 5 minutes in the flask. Furthermore, bacterial culture in RCM media as much as 100 mL was inserted into the flask. Fermentation was carried out for 4 days by taking and analyzing samples and products in 32 hours, 64 hours and 96 hours.

2.4. Analysis of biohydrogen products in gas chromatography
Biohydrogen products were analyzed after fermentation had completed in 32 hours, 64 hours and 96 hours sample variations in each NaOH pretreatment concentration substrate 0.3: 0.5 and 1% w/v POME. Biohydrogen products were analyzed using gas chromatography

2.5. Glucose analysis in the fermentation process at HPLC
HPLC was used to analyze glucose content in palm oil liquid waste substrate. Theoretically, the glucose content in POME hydrolysis is enzymatically around ± 1-5%. The concentration of glucose standard solution used is 1%, 2%, 5% and 8% with a ratio of acetonitrile: water (80 : 20% v/v). The samples which to be analyzed are filtered to remove noise particles.

3. Results and Discussion
The sugar content obtained from the hydrolysis of POME is used by Clostridium butiricum as an energy source to increase the rate of cell growth. Initial glucose concentration has an important role in the yield and rate of hydrogen production during fermentation. The relatively low initial glucose concentration causes the fermentation rate to be also low [6].

(Figure 1) shows that the highest concentration of hydrogen gas occurs at a retention time of 32 hours, while at 64 and 96 hours, the concentration of hydrogen decreases. This is inversely proportional to the graph of glucose concentration. Longer retention time causes greater hydrolyzed glucose content.

(Figure 2) with 0.5% NaOH and (figure 3) 1% NaOH concentration show that at a retention time 32 and 64 hours increase glucose concentration result but decreases at retention time 96 hours and 0.5% NaOH concentration. Biohydrogen production has an increased concentration at a retention time of 32 hours to 64 hours, but the decrease that occurs is much lower than in the condition of 0.5% NaOH concentration

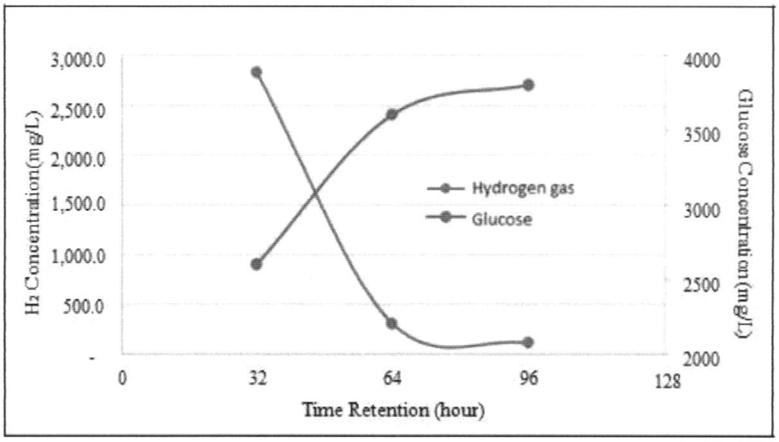

Figure 1. Effect of glucose concentration to hydrogen gas production in the addition of 0.3% NaOH

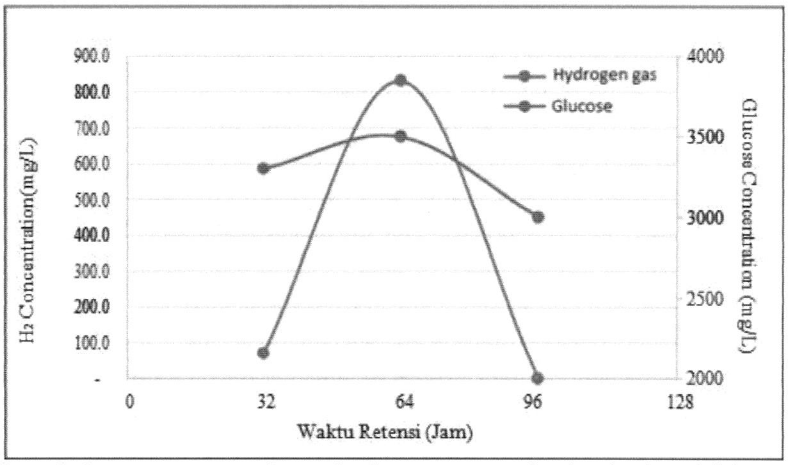

Figure 2. Effect of glucose concentration to hydrogen gas production in the addition of 0.5% NaOH

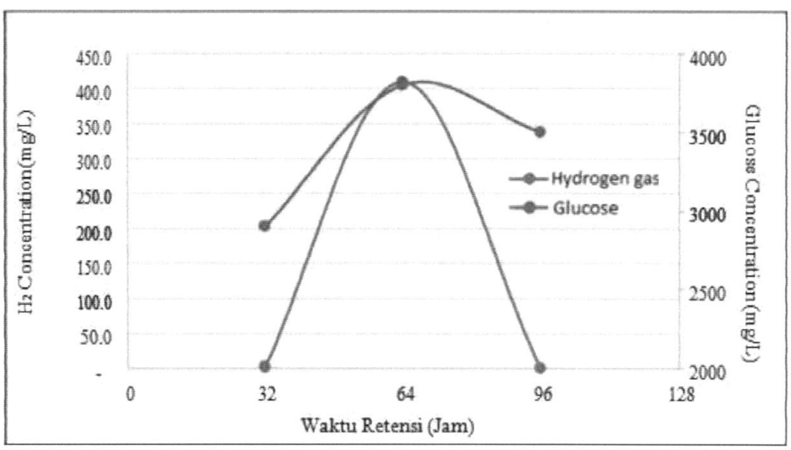

Figure 3. Effect of glucose concentration to hydrogen gas production in the addition of 1% NaOH

Glucose as an inhibitor occurs when glucose inhibits the action of the hydrogenase enzyme in which the active center of this enzyme in the form of ionized ion groups cannot maintain the conformation of the active site to bind the substrate and catalyze the reaction [6]. This results in a decrease in the acquisition of biohydrogen. Decreased glucose concentration is influenced by the initial pH value. This happens because of an increase in the concentration of VFA (Volatile Fatty Acid) which causes a decrease in pH thereby inhibiting hydrogen production [7].

Decreasing glucose substrate concentration causes increasing partial pressure in the fermentation process. This will divert acidogenesis in the process to solventogenesis, thereby inhibiting hydrogen production. The process of hydrogen production by fermentation bacteria also accompanies the formation of volatile fatty acids (propionic acid) and alcohol as metabolic byproducts. This product formation reduces equivalents like NADH. This reduced NADH equivalent supports the metabolism of glucose into ethanol and the process of propionogenesis. As a result, the production of ethanol and propionic acid will cause a decrease in the acquisition of hydrogen products so that the ratio of $NADH/NAD^+$ in cells must be maintained.

Several research reports show that biohydrogen production which initially obtains high production yields, will tend to decrease to low levels further processing due to the formation of propionic acid and the emergence of partial hydrogen pressure in the reactor [8]. If the concentration of propionic acid is more, it will reduce the production of biohydrogen. Conversely, if the ratio of acetic acid and butyric acid formation increases, the concentration of biohydrogen produced increases as well. Failure to maintain the level of carbon source (glucose) in the form of a liquid phase in the reactor also optimally affects the growth of microorganisms, the level of substrate utilization, enzyme activity and the overall results of the process itself. It is concluded that the hydraulic retention time affects the amount of glucose hydrolysis produced.

4. Conclusion

The conclusions obtained from this research are :

1. Biohydrogen can be produced from palm oil liquid waste using the bacterium Clostridium butyricum
2. The highest level of biohydrogen production is the concentration of NaOH 0.3%, which is 2824.2 mg H_2 / L POME.
3. The best fermentation retention time in producing the highest levels of biohydrogen is at 32 hours at a 0.3% NaOH concentration.

References

[1] Mohammadi P, Ibrahim S and Annuar M S M 2012 Effects of biomass, COD and bicarbonate concentrations on fermentative hydrogen production from POME by granulated sludge in a batch culture *Int. J. Hydrogen Energy* **37** 17801–8

[2] Amira R D, Roshanida A R, Rosli M I, Zahrah M F S F, Anuar J M and Adha C M N 2011 Bioconversion of empty fruit bunches (EFB) and palm oil mill effluent (POME) into compost using Trichoderma virens *African J. Biotechnol.* **10** 18775–80

[3] Tangahu B V, Ningsih D A, Kurniawan S B and Imron M F 2019 Study of BOD and COD removal in batik wastewater using scirpus grossus and Iris pseudacorus with intermittent exposure system *J. Ecol. Eng.* **20**

[4] Chaisri R, Boonsawang P, Prasertsan P and Chaiprapat S 2007 Effect of organic loading rate on methane and volatile fatty acids productions from anaerobic treatment of palm oil mill effluent in UASB and UFAF reactors *Warasan Songkhla Nakharin (Sakha Witthayasat lae Technol.*

[5] Sompong O, Prasertsan P, Intrasungkha N, Dhamwichukorn S and Birkeland N-K 2007 Improvement of biohydrogen production and treatment efficiency on palm oil mill effluent with nutrient supplementation at thermophilic condition using an anaerobic sequencing batch reactor *Enzyme Microb. Technol.* **41** 583–90

[6] Fabiano B and Perego P 2002 Thermodynamic study and optimization of hydrogen production by Enterobacter aerogenes *Int. J. Hydrogen Energy* **27** 149–56

[7] Khanal S K, Chen W-H, Li L and Sung S 2004 Biological hydrogen production: effects of pH

and intermediate products *Int. J. Hydrogen Energy* **29** 1123–31
[8] Wang J and Wan W 2008 The effect of substrate concentration on biohydrogen production by using kinetic models *Sci. China Ser. B Chem.* **51** 1110–7

Analysis of FMEA and FTA methods in identifying the causes of decreased quality of refined sugar products (case study: PT. Makassar Tene)

N Khaerani B, Arminas, and R Hendrawan

Agro-Industrial Engineering Study Program Polytechnic of ATI Makassar

E-mail: rhany_busri@yahoo.com

Abstract. PT. Makassar Tene is one of the manufacturing industries that produce refined sugar from raw sugar as basic materials. Most of the refined sugar production process has been carried out automatically, but it still requires regular supervision and monitoring from the operator to ensure the smoothness and accuracy of the sugar processing steps in reducing the number of defect products significantly. Failure Mode and Effect Analysis (FMEA) is a method to identify product failures based on potential causes. FMEA is used to determine the value of Risk Priority Number (RPN), then further identified as the main focus in determining the root of the potential cause using the Fault Tree Analysis (FTA). The application of FTA can provide suggestions for improvements and controls in order to help companies reduce the risk of production failure. Based on the results of the research using FMEA method, it can be seen that the process stage with the greatest risk of failure lies in the stages of sugar evaporation/ crystallization and the packaging stage in the packaging bin. The next step is to identify the type of problem at each stage of the process selected using Pareto diagram so that the results show that the evaporation/crystallization stage CV deviation often happened which can produce sugar dust, while at the packaging stage the problem that often occurs is metal detection, torn sack, and sweeping sugar. Then proceed with the use of the FTA method to identify the underlying causes of each failure that arises at the chosen process stage, and the results show that most of the failures that arise are caused by operator negligence, the operation process is still done manually, and the design of operating equipment is less optimal.

1. Introduction

The current rapid development of industry happened. The rapid development of the industry followed by competition increasement among industrial companies to attract the attention of customers, both in product pricing approach as well as by improving the quality/grade products [1,2]

Product pricing are not be the only factor that can influence customer interest in the products offered. Few customers are even willing to pay for the expensive price and become loyal to certain products simply because the quality of the products are better than the others [3]. Unfortunately, not all industry give an urge to this factor. They are stick to the principle of "price is everything" to attract buyers. Therefore, not a few of the industries that went bankrupt because of lags to the new industries that offer competitive quality products at affordable prices.

Based on data from the Ministry of Industry estimates that the national requirements for refined sugar reached up to 3.6 million tons for industrial purposes, which represents an increment of 6%

compared with the level of sugar consumption in 2017 reached 3.4 million tons. Projected growth is largely influenced by the growth of the food and beverage industry as much as 7% to 8% [1].

PT. Makassar Tene as the only refined sugar company industry in Eastern Indonesia. In carrying out the production activities of PT. Makassar Tene proved that it still has defects on the quality of produced sugar products such as the creation of sugar dust, metal detection, and unstablility of quality on the crystallization process. This phenomenon brings disadvantageous for both consumers and for the company. Besides, as the competition increase, sugar industries that offering sugar products with higher quality could be a threat if this condition continued over time and did not receive serious attention.

Based on this problem, consideration of methods selected in order to improve the quality/grade of refined sugar products with the adoption of FMEA (Failure Mode and Effect Analysis) and the method of FTA (Failure Tree Analysis). FMEA is a method that can be used to evaluate the probability of failures or accidents within the certain system [4]. While FTA is an analytical method that can analyze system failure, look for aspects of the system involved in the main failure, and find the cause of product defects in the production process [5]. The successful application of both the FTA and FMEA techniques has increased their adoption in several reliability and safety engineering analyses studies [6]. Therefore, this paper aim to identify the cause of decreasing refined sugar quality at. Makassar Tene assisted with analysis tools such as Pareto diagram.

2. Research methods

2.1. Identification of the type of failure that occurred

Researchers conducted a study on the possibility of failures in each stage of the production process with the aim of improving the quality of refined sugar products and minimized failures in the process of PT. Makassar Tene using Failure Mode and Effect Analyze (FMEA).

2.2. Identify the cause of the failure

At this stage, the causes of the failure will be identified and the percentage of failures projected using Pareto diagram.

2.3. Determine Rate of Severity, occurrence and Detection

Severity is the first step to analyze risks, which calculates how much impact or intensity of incident affect the final outcome of the process. Those impacts on the rating scale ranging from 1 to 10, where 1 is the most solve-able impact while 10 is the worst impacts and the determination of the rating. If the rating already specified on the severity factor, the next step is to determine the rating of the occurrence factor. Occurrence is a possibility of caused failure during the production of the product. After the value of occurrence obtained, the next step is to determine the value of the detection factor. Detection value estimates how well the control parameter can detect the cause of the failure when occurred but before the consumer knows it. Value detection judged on a scale of 1 to 10, where 1 means that the control are specifically made to detect occurring failures and the number of 10 means a special control for not detecting the problem of the failure (or no control at all).

2.4. Calculation and Sortment of RPN (Risk Priority Number) Value

Having obtained the rating value of severity, occurence and detection, the next step is to multiply the value of the rating is to get the value of RPN then continued by the score sorted based on the highest to lowest RPN value. Once RPN value sorted, it determines the percentage of the cumulative value of the RPN and presented in the form of the Pareto diagram. The cumulative percentage of the value formed in the scale of priorities based on data clustering 80-20%.

2.5. Analysis of Defect Cause with FTA Method

The next step after the obtained priority grouping 80-20% on the Pareto diagram are analyze the cause of the defect using FTA method. The use of FTA method serves to identify the root of the occurring defect.

2.6. Proposed Improvements and Control

Proposed improvements and control in prevention phase to the product performed as a corrective measure that is able to do in the next production process.

2.7. Conclusions and suggestions

From the results of all calculations and analysis using FMEA and FTA method can be attached to conclusions and suggestions that are beneficial to the company.

3. Collection and processing of data

3.1. Identification of the failure type that occurred

Based on the process flow of sugar processing, ranging from raw sugar to the final product obtained some kind of failure with the percentage of reject for most of the sugar product caused by metal detection at the level of 8.78%, followed by rejection due to sacks torn of 6.81%, and reject caused by sugar sweep of 3.15%.

Table 1. Percentage of evaporation with a value of CV does not meet standards

No	Sugar Grade	Overall Number Evaporation Vacuum Pan	Number Evaporation Rated by CV (Not meet Standards)	Incompatibility Percentage CV Value (%)
1	Smooth	766	155	20.23
2	Rough		26	3.39
		Total	181	23.63

In (table 1) above shows that the amount of sugar fondant and percentage of evaporation that the CV value is not in accordance with established standards. The standard of CV values assigned to each process in a vacuum pan is no more than 32 (<32), when the value of CV over 32 consequently has an impact on the creation of sugar dust (dust sugar).

In smooth sugar grade there are 155 dishes that do not meet the standard of CV value with a percentage of the overall number of dishes is 20.23% while in rough sugar grade there are 26 dishes that do not meet the standard of CV value with a percentage of the overall number of the cook was 3.39%, a total of 23.63% of cook not meet the standards of CV value.

3.2. Identify the cause of the failure

Results of identification as a result of the failure can be seen in (table 2).

Table 2. Processing the results identifying the causes of failure of Refined Sugar

Process Stages	Requirement	Potential Failure Mode	Potential Effect (s) of Failure
Raw Sugar	Transportation and equitable distribution of raw sugar for tunnel	Raw sugar pile up at a certain point	Raw sugar in uneven distribution conveyor
Affinity	The temperature of the hot water and steam that is appropriate	Improper hot water the temperature of 70 ° C	Raw sugar and melting process is not running optimally decolorizes
Carbonation	The reaction was perfect on carbonator	Further reaction occurs at carbonator	The process of carbonation the reaction takes longer than it

IOP Conf. Series: Materials Science and Engineering **885** (2020) 012050 doi:10.1088/1757-899X/885/1/012050

			should
Curing	Liquar clear sugar	Liquar still there are deposits	Liquar can not be crystallized
IER Operation	Impairment ICUMSA sugar	ICUMSA value is still above the standard	ICUMSA value is still relatively high and the crystallization process in the vacuum pan is not running normally
Evaporation / Crystallization	Sugar crystal sugar according to the standard specifications set by the quality control department	Sugar crystals do not meet the standards	Crystal formation conglomerate and the standard CV discrepancy resulting sugar dust
Screening	Filtering sugar MA size according to specifications consumer demand	Filtering sugar MA size does not match the specifications of consumer demand	Crystal size mixed
Bin Packing	Sugar packed neatly and ready to be stored in warehouses of materials so	Sacks of sugar were torn and did not pass metal detection time has been packed	Sugar can not be saved and need to undergo a process of reprocessing
Packed Sugar	Sugar stored neatly on each Lot of storage per specifications sugar	Sugar pile collapse	Sugar can not be saved and need to undergo a process of reprocessing

3.3. Calculation and Sortment of RPN (Risk Priority Number) Value

Here are the results of the calculation of the RPN value (Severity x Occurrence x Detection), as shown in (table 3).

Table 3. Calculation result RPN value (top 2)

Process Stages	Requirement	Potential Failure Mode	Potential Effect (s) of Failure	S	O	D	RPN
Evaporation / Crystallization	Sugar crystal sugar according to the standard specifications set by the quality control department	Sugar crystals do not meet the standards	Crystal formation conglomerate and the standard CV discrepancy resulting sugar dust	7	7	5	245
Bin Packing	Sugar packed neatly and ready to be stored in warehouses of materials so	Sacks of sugar were torn and did not pass metal detection time has been packed	Sugar can not be saved and need to undergo a process of reprocessing	7	8	2	112

Based on the calculation of the RPN value at every stage of the sugar processing, it can be arranged into the shape of Pareto diagram as follows:

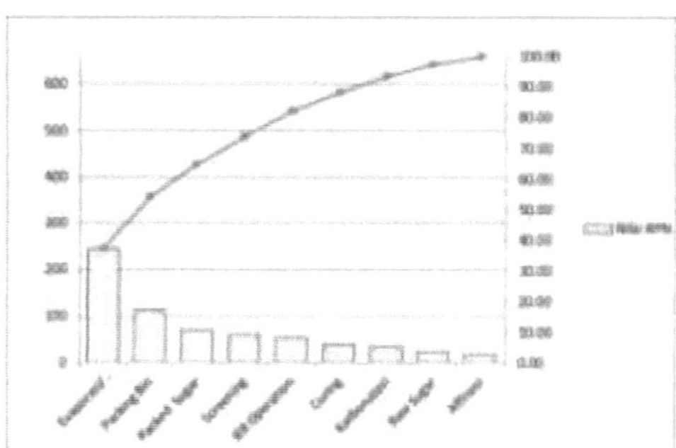

Figure 1. Pareto diagram of earned RPN value

IOP Conf. Series: Materials Science and Engineering **885** (2020) 012050 doi:10.1088/1757-899X/885/1/012050

3.4. Analysis of defect cause with FTA method

The election of the two stages of the above process is then translated into the potential risks that occur in each stage of the process using FTA method, the following stages of the implementation of the FTA:

3.4.1. The evaporation/crystallization process. In the cooking process of the evaporation stage using a vacuum pan, it was found that the sources of the occurring problem often arises as follows:

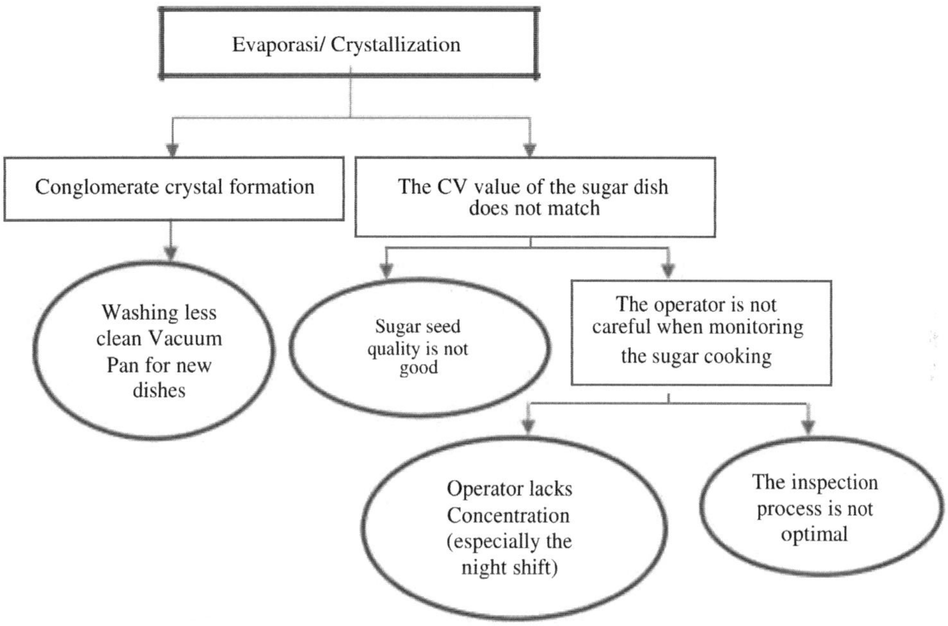

Figure 2. FTA process of evaporation/crystallization phase

3.4.2. Packaging Stages on Packaging Bin. At the stage of the packaging process, sugar-packed in polypropylene sacks by the operator found that the sources cause of the problem that often arises is as follows:

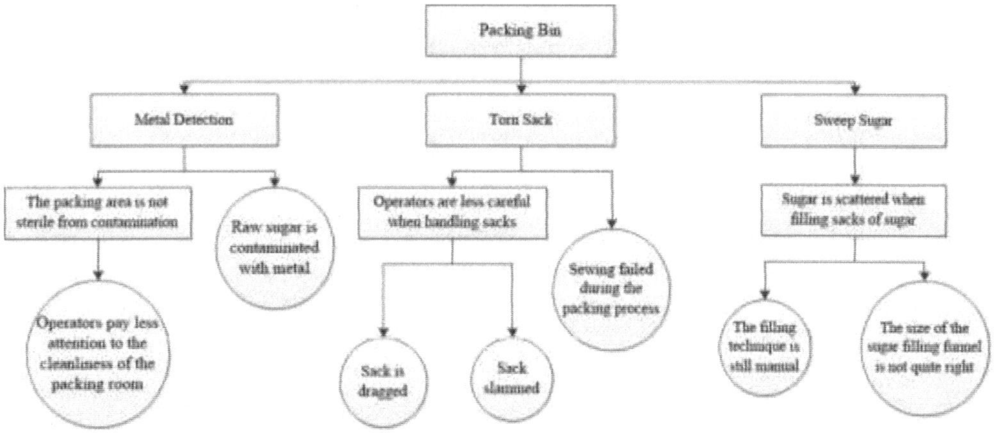

Figure 3. FTA of Packaging Stage on Packaging Bin

4. Discussion result

4.1. FMEA (Failure Mode and Effect Analysis)
Based on calculations using FMEA method by looking at the value of the RPN can be seen that the stages of the process to the value of the RPN largest to smallest in a row is a stage evaporation/crystallization with RPN 245 points, the stage of packing bin with RPN 112 points, the stage of packed sugar with RPN 70 points, RPN screening stage with 60 points, IER stage operation with RPN 54 points, carbonation stage with RPN 36 points, phase RPN raw sugar with 24 points, and phase RPN affinity with 18 points. So was taken two stages with the highest RPN value that stage evaporation/crystallization and bin packing stage.

4.2. Pareto diagram
Based on the stage of the process that has been selected using FMEA method before, when described again at each stage has a series of problems that often occur. At this stage Pareto diagram is used as a tool to determine the priority of the type of damage from the highest to the lowest, the following description:

4.2.1. Phase Evaporation / Crystallization. The stage evaporation/crystallization kind of damage is divided into 2 types of crystal formation conglomerate and CV values mismatch, but the case is very rare crystal conglomerate that deviation CV values on the type of refined sugar with 155 deviations (85.63% of total damage) into focus problem.

4.2.2. Phase Packaging Bin. At the packing stage bin, Pareto diagram shows that there are three types of problems that often occur that metal detection of 2934 units (46.85% of total damage), torn sacks of 2278 units (36.36% of total damage), and sugar sweep of 1051 units (16.79% of total damage).

4.3. FTA (Fault Tree Analysis).
After the percentage and type of damage have been known through the application of the Pareto diagram, then the implementation of the FTA method is intended to detect the root cause of the problem, the following description:

4.3.1. Stage evaporation / crystallization. As for the discrepancy evaporation stage CV value in cooking, sugar is generally caused by sugar beet quality is not good or those less scrupulous operators while monitoring the cooking process sugar. Poor seed quality sugar tends to rarely happen since PT Makassar Tene stop producing sugar beet seeds manually and import sugar from Germany while the lack of rigor operator is usually caused by the lack of concentration in particular operator on duty on the night shift, as well as the process of assessing the dishes still manually by checking the tube.

4.3.2. Pha*se Packaging Bin.* At this stage of packaging bin, metal breakage detection usually occurs due to metal contamination in early stages, namely raw sugar, but is most often caused by not sterile packaging space of contamination due to less operator attention to the cleanliness of room packaging. Karung tear is usually caused by two things: operators who are less careful when handling sacks of sugar such as slamming or dragged repeated or failed sewing during sewing sacks entering the stage in an automatic sewing machine using a belt conveyor, while sugar strokes caused by sugar scattered during the process of filling bags of sugar due to the filling technique of sugar are still manually by the operator so charging intermittently between one sack to sack more and funnel filling sugar does not quite fit in terms of design that allows the sugar scattered upon sugar sack filling process takes place.

5. Conclusion

Based on the discussion of the results obtained from PT. Makassar Tene, as well as a discussion of the causes of the decline in the quality of the analysis of sugars by using FMEA and FTA, the authors conclude from the results of the discussion that:

1. In the FMEA method, the two-stage process has a high risk of failure is the stage evaporation/crystallization with the acquisition value of the RPN 245 points than with the packaging stage in the packaging bin with the acquisition value of the RPN 112 points.
2. In the implementation of the FTA method, the basic causes (basic causes) of the onset of problems/failures are:
   a. Stages evaporation / crystallization
      1) The checking process and sugar cook are not optimal (not fully automatic).
      2) Lack of rigor for the operator during the sugar cooking process, including cooking temperature, cooking time, as well as monitoring through a control panel.
   b. Phase Packaging Bin
      1) Lack of implementation of SOP procedure
      2) Lack of implementation of hygiene
      3) Handling sacks are not careful (slammed and dragged)
      4) Failure in sack sewing.
      5) Processes and techniques of refined sugar filling still done manually by the operator.
      6) Lack of strict material quality control to prevent contamination of raw sugar.

References

[1] Ansori munib 2017 Proyeksi Kebutuhan Gula Industri 3,6 Juta Ton di 2018
[2] Hadia N H, Zakib I and Fatihinc M K 2020 The Preference and Prospects of Sugar Needs in Micro, Small and Medium Enterprise Industries of Food and Beverage in Surabaya City *Int. J. Innov. Creat. Chang.* **10**
[3] Fanani Z, Hartono B and Nugroho B A 2015 The Influence Of The Marketing Mix (Product, Price, Promotion, Place, Process, Entrepreneurs And Physical Evidence) To Customer Satisfaction And Loyalty In Buying Shredded Beef In Palu City, Indonesia *Int. J. Econ. Res.* **12**
[4] Aulady M F N, Nuciferani F and Wicaksono S 2018 Application of Failure Mode Effects Analysis (FMEA) Method and Fault Tree Analysis (FTA) Towards Health and Occupational Safety on Jetty Project, Gresik, Indonesia *JACEE (Journal Adv. Civ. Environ. Eng.* **1** 100–8
[5] Hidayat A A and Kholil M 2018 The Implementation of FTA (Fault Tree Analysis) and FMEA (Failure Mode And Effect Analysis) Methods to Improve the Quality of Jumbo Roll Products *IOP Conference Series: Materials Science and Engineering* vol 453 (IOP Publishing) p 12019
[6] Shafiee M, Enjema E and Kolios A 2019 An integrated FTA-FMEA model for risk analysis of engineering systems: a case study of subsea blowout preventers *Appl. Sci.* **9** 1192

Application of quality function deployment for the development of sulam tumpar motif design

Dwi Cahyadi, Etwin Fibrianie Soeprapto, Andi Farid Hidayanto and Ditha Nizaora

Design Departement, Politeknik Negeri Samarinda

Email : dwicahyadi@polnes.ac.id

Abstract Regional crafts from East Kalimantan Province is a special embroidery, that's call sulam tumpar. Sulam tumpar is a decoration made on cloth or other materials with sewing needles and thread. Craft originating from the Dayak tribe in the East Kalimantan Province has a variety of patterns with bright colors. The problem in this research is how to develop a new motif of sulam tumpar by not leaving the local wisdom characteristic in its design in order to raise the name of this East Kalimantan craft in national and international. The purpose of developing this sulam tumpar motif is to design and develop a new motif that can be applied to all handicraft products, including fashion products for men and women. The results of this research activity are contributing new ideas in the development of the sulam tumpar motif in accordance with customer desires using the Quality Function Deployment (QFD) method, which is a blend of modern and traditional concepts followed by unisex motif designs and universal designs.

1. Introduction

Regional crafts from East Kalimantan Province, one of which is a special embroidery, that's call sulam tumpar. Sulam tumpar is a decoration made on cloth or other materials with sewing needles and thread fig 1. Craft originating from the Dayak tribe in the East Kalimantan Province has a variety of patterns with bright colors. Not only in the form of cloth, but sulam tumpar is also widely applied to various products such as bags, clothes, to wall hangings. Generally, sulam tumpar is used for handicraft and clothing products, which tend to be used by women. This is inseparable from the motif and colors that tend to be feminine.

Figure 1. Sulam tumpar.

The current development demands the development of fashion not only for women but also for men. Fashion trends that bring local wisdom can be combined with the development of fashion for all genders.

Sulam tumpar motif that tends to be feminine provides limits for men to wear this embroidery. The embroidered motif that is applied in most fashion products is for women, such as women's clothing, bags, necklaces, and others.

The problem in this research is how to develop a new motif of sulam tumpar by not leaving the local wisdom characteristic in its design in order to raise the name of this East Kalimantan craft in national and international. The purpose of developing this sulam tumpar motif is to design and develop a new motif that can be applied to all handicraft products, including fashion products for men and women.

2. Literature Review

2.1. Embroidery As Craft and Fashion Products

Culture is a lifestyle that develops continuously and is passed down hereditary [1]. Sulam tumpar is typical embroidery handicrafts from the Dayak culture in East Kalimantan Province. In its development, embroidery is widely used as a craft and fashion product. In developing of sulam tumpar, there is a need for market identification through input from customers' needs [2].

Each region has a unique cultural and ethnic characteristic, as well as the kerawang embroidery is one of Gorontalo Province's mainstay products. Naturalist and geometric motifs from embroidery cannot stand alone, so the distillation motif becomes a decorative motif in order to add to the beauty of embroidery. The development of other embroidery motif is done by making new designs that are more creative and innovative as their characteristics without ignoring the techniques applied previously [3] [4]. The embroidery can also be visualized for interior products [5]. The development of handicraft products is not only focused on one fashion but can be expressed as an identity of the city where the craft originated [6].

2.2. Quality Function Deployment (QFD)

To develop the motif of the embroidery, it needs input from the consumer or that calls the voice of costumers. One method that departs from the needs of consumers is Quality Function Deployment, commonly abbreviated as QFD. It controls the quality of a product based on the wants and needs of costumers. This method has four stages, where each stage produces one matrix. The four stages are product planning, design planning, process planning, and production planning. The identification of customer needs is very useful in the product development process to get closer to the target [7, 8]. It is really directed to the client, not the product, where the efforts of all the departments involved allow a match between client requirements and the product offered [9].

QFD as a method for quality design with consumer expectations, then translates it into the target design and critical quality points, so that it can be used in the development products [10]. QFD, as a method for quality design with customer expectations, then translates it into the target design and critical quality points so that it can be used in the development of the motif. QFD itself not only fulfills as many customer expectations as possible but also tries to exceed customer expectations [11].

Identification of customer needs is very useful in the product development process to get closer to the target of how exactly the attributes or products desired by consumers [12]. It is a quality tool that helps translate Voice of the Customer (VoC) into new products that truly satisfy their needs [13].

It's important to understand and meet customer needs to increase customer satisfaction. House of Quality (HOQ) is used as an interface tool with QFD. HoQ includes six main sections are 1. Customer's requirements, 2. Quality elements, 3. Relationship matrix, 4. Correlation matrix, 5.Competition Analysis, and 6. Priority [14].

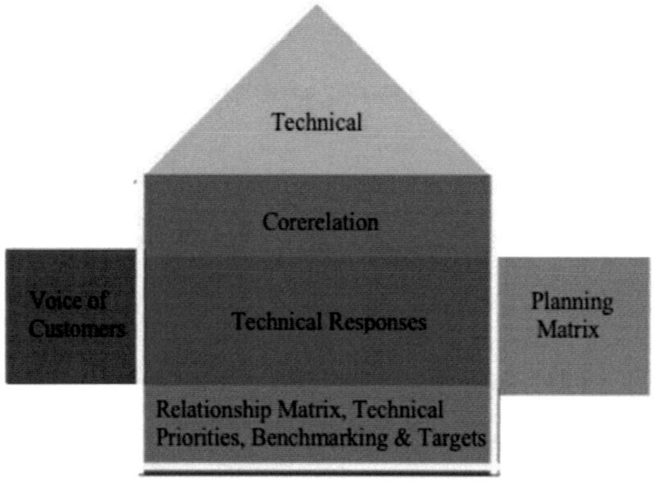

Figure 2. Matrix House of Quality (HoQ).

QFD, with the main use in product development, is expected to assist in determining the direction of the development of the design of the motif in accordance with consumer needs through questionnaire results. It is also useful in developing products with the customer's sound application for fashion products [15, 16]. In fact, it is expected to be able to improve quality management and is designing new products to be made to meet customer requirements [17, 18].

3. Method

The method used in developing the Sulam Tumpar motif is the application of the use of Quality Function Deployment (QFD), which begins with the identification of customer needs through questionnaires. Questionnaires were given to 100 respondents who came from the community, academics, artisan, and the government, and it will be used as a basis for the use of QFD as a basis for determining the final priority of what attributes should be developed from this new Sulam Tumpar motif.

4. Results and Discussion

In developing the motif of the sulam tumpar, a survey was conducted on 100 respondents consisting of the community, government, artisan, and academics in the East Kalimantan Province. The results of the survey will get input from consumers or often referred to as the voice of the customer (VoC). Some stages that will be carried out in the QFD method are:

4.1. Determine Voice of Customer (VoC)

At this stage, information gathering related to the needs and desires of costumers will be carried out for the development of the design of the sulam tumpar motif. Information gathering can be done through a survey through a questionnaire, interviews, and even direct observation of costumers. The data that will be obtained from costumers are the attributes of the sulam tumpar products needed by costumers.

4.2. Making a House of Quality (HoQ)

House of Quality (HoQ) is a form of table or picture of quality houses that will be compiled to meet the needs of the desired attributes of costumers and the technical needs of companies or designers and their work teams. The steps in preparing this quality house are:

4.2.1. Matrix attributes customer needs. From the results of a survey conducted in the process of developing the motif of a sulam tumpar obtained the following data:

- The motif is not crowded

- The motif characterizes the east Kalimantan
- Variation in the size of the motif
- Form of renewable / latest motif
- Color is inconspicuous
- The combination of colors
- Motives give the impression of masculine
- The motif is used for fashion/craft products

Attributes of the desired needs of customers, along with the level of importance, can be seen in table 1.

Table 1. Attributes of customer importance.

Attributes	Customer Importance
The motif is not crowded	3.66
The motif characterizes the East. Kalimantan	4.09
Variation in the size of the motif	3.88
Form of renewable / latest motif	4.05
Color is inconspicuous	4.06
The combination of colors	3.70
Motives give the impression of masculine	3.29
The motif is used for fashion / craft products	3.29

4.2.2. Technical requirements. It contains the identification of product characteristics that can be measured to meet the attributes of customer needs regarding the development of the sulam tumpar motif. The technical requirements are:

- Geometrically shaped design
- Design Motif Shaped organic
- Unisex Motif Design
- Proportion of shapes
- A combination of modern and traditional concepts
- Universal design
- color composition

4.2.3. Relationship Matrix. The matrix that illustrates in table 2 and table 3 are the relationship between technical requirements and customer needs of the sulam tumpar motif.

IOP Conf. Series: Materials Science and Engineering **885** (2020) 012051 doi:10.1088/1757-899X/885/1/012051

Table 2. Relationship matrix.

	Technical Requirement						
Costumers Need	Geometrically shaped design	Design Motif Shaped organic	Unisex Motif Design	Proportion of shapes	A combination of modern and traditional concepts	Universal design	color composition
The motif is not crowded.							
The motif characterizes the East Kalimantan							
Variation in the size of the motif							
Form of renewable / latest motif							
Color is inconspicuous							
The combination of colors							
Motives give the impression of masculine							
The motif is used for fashion/craft products							
The motif is not crowded							
The motif characterizes the East Kalimantan							

Relationship Symbol Information:
= 9
= 3
= 1
(empty) = 0

IOP Conf. Series: Materials Science and Engineering **885** (2020) 012051 doi:10.1088/1757-899X/885/1/012051

Table 3. Calculation of relationship matrix.

Attribute	Target Value	Existing Value	Customer Importance	Row Weight	% Weight
The motif is not crowded	5.00	4.00	3.66	6.56	13.15
The motif characterizes the east Kalimantan	5.00	5.00	4.09	9.17	18.38
Variation in the size of the motif	4.00	4.00	3.88	4.12	8.27
Form of renewable / latest motif	4.00	4.00	4.05	5.93	11.88
Color is inconspicuous	4.00	4.00	4.06	3.94	7.90
The combination of colors	5.00	4.00	3.70	6.49	13.00
Motives give the impression of masculine	5.00	3.00	3.29	6.84	13.71
The motif is used for fashion/craft products	5.00	3.00	3.29	6.84	13.71

Figure 3. House of Quality (HoQ).

4.3. House of Quality (HoQ)

In fig 3 is an HoQ that is made based on the customer's needs and technical requirements from the HoQ can be seen in the relationship between the matrix and the results in the form of technical priority, benchmarking, and targets of what will be developed from the motif.

Based on the House of Quality (HoQ), the ranking of the development of the sulam tumpar motif focuses primarily on the blend of modern and traditional concepts followed by unisex motif designs and universal designs. From this, the development of the design sulam tumpar motif will be oriented to the three main things.

5. Conclusion

The results of this research are contributing new ideas in the development of the sulam tumpar motif in accordance with customer desires using the Quality Function Deployment (QFD) method, which is a blend of modern and traditional concepts followed by unisex motif designs and universal designs.

6. Acknowledgment

The authors wish to thank the Ministry of Technology Research and Higher Education Indonesia, which provided financial assistance in this research and also the National Craft Council of East Kalimantan Province (Dekranasda Kaltim), which has made it to find supporting data.

References

[1] Cahyadi D, Fibrianie E and Shyafary D 2018 *Ilmu Pengetahuan Populer Desain Produk* (Jakarta: PNJ Press)

[2] Cahyadi D, Fibrianie E and Hidayanto A F 2019 Development Of Embroidery Motif Designs Using The Principle Theory Of Gestalt *Int. J. Sci. Technol. Res.* **8**

[3] Hariana and Trifandi L 2012 Analisa Proses Produksi Sulaman Kerawang Khas Gorontalo *Prosiding Seminar Nasional "Peningkatan Kompetensi Guru dalam Menghadapi UKG* (Gorontalo: Universitas Negeri Gorontalo) pp 80-86

[4] Ranelis and Kendall M 2018 Pengembangan Desain Dan Motif Produk Sulam Koto Gadang, Sumatera Barat *Idealogy* **3** 23-46

[5] Nuri M R and Panggabean R 2014 Penerapan Citra Wayang Dengan Teknik Tusuk Jelujur Untuk Produk Interior *Jurnal Tingkat Sarjana Senirupa dan Desain* **2**

[6] Akbar A A 2018 Analisis Deskriptif Pesan Motif Belang Hatta Pada Sarung Tenun Samarinda *J. Ilmu Komun. Univ. Mulawarman* **6** 173-187

[7] Canan S, Fatma K and Aslı A 2015 Internal Benchmarking Methodology and Its Implementation on Apparel Retil Industry *TEKSTİL ve KONFEKSİYON* **4**

[8] Ginting R, Batubara T Y and Widodo 2017 Desain Ulang Produk Tempat Tissue Multifungsi Dengan Menggunakan Metode Quality Function Deployment *J. Sist. Tek. Ind.* **19**

[9] André S D, Silvia J C M and Pereira J A 2017 On the Use of the Quality Function Deployment Matrix for Flexible and Quantitative Prioritization *J. Adv. Manag. Sci.* **5**

[10] Susantoa B P and Wurjaningrumb F 2019 Service Blueprint and Quality Function Deployment in Designing Service Quality Improvement in Hospital *Int. J. Innov. Creat. Chang.* **9**

[11] Yustian O R 2015 Analisis Pengembangan Produk Berbasis Quality Function Deployment (QFD) (Studi Kasus Pada Produk Susu PT MSA) *J. Ekon. dan Bisni* **18**

[12] Prabowo R and Zoelangga M I 2019 Pengembangan Produk Power Charger Portable dengan Menggunakan Metode Quality Function Deployment (QFD) *J. Rekayasa Sist. Ind.* **8**

[13] Eshan S and Jaiswal A 2012 Case Study on Quality Function Deployment (QFD) *IOSR J. Mech. Civ. Eng. (IOSR-JMCE)* **3** 27-35

[14] Kuo H M and Chen C W 2011 Application of Quality Function Deployment To Improve The Quality of Internet Shoping Website *Interface Des. Int. J. Innov. Comput. Inf. Control* **7**

[15] Kabukcu E 2016 A Research on QFD-"House of Brand" in Fashion Industry *Chinese Business Review* **15** 207-213
[16] Mafazah N and Sujoko H 2014 Penerapan Metode Quality Function Deployment (QFD) Untuk Pengembangan Desain Motif Batik Khas Kalimantan Timur *J@ti Undip J. Tek. Ind.* **9**
[17] Kunecka D 2016 Quality Function Deployment as Method Used in The Development of Quality od Education for The Nursing Practice *Gen. Prof. Educ.* **1** 42-45
[18] Premkumar D and Balamurugan M 2014 Implementation of Quality Function Deployment in Pump Industry *Int. J. Innov. Res. Sci. Eng. Technol.* **3**

Implementation of The Rought Cut Capacity Planning (RCCP) Methods for The Planning of Bottled Drinking Water Production Capacity: A study case

A Rapi, I S Muthalib, S Asmal

Department of Industrial Engineering, Universitas Hasanuddin, Makassar, Indonesia

Email: irwan@tiunhas.net

Abstract. Syifa Ltd is a manufacturing company that produces bottled drinking water with several types of products produced from their production floor. The main objective of this research was to predict the total quantity production of the 220 ml bottled mineral water to meet the consumer demand as scheduled time. In forecasting the demand for production of the 220 ml glass bottle packaging, the methods used were the Weighted Moving Average (WMA) method and the Single Exponential Smoothing (SES) method by comparing the size of forecasting errors using Mean Absolute Deviation (MAD), Mean Square Error (MSE), and Mean Absolute Percentage Error (MAPE). To determine the supply of glass bottles used the EOQ (Economy Order Quantity) method to determine the purchase of the most economical glass bottles. The Rought Cut Capacity Planning (RCCP) analysis result shows that there was insufficient capacity/lack of capacity in the filling and packing work center. Alternative capacity management used to optimize production utility to 95% and make capacity adjustments (re-adjustment). So that alternative capacity management was able to meet the capacity shortages that occur

1. Introduction

The development of the Industrial Revolution that continues and seems to be endless, requires every company to continue to improve and requires companies to respond carefully to the changes that occur [1]. The main problem faced by companies today is how the company be able to create the best strategy so that the company can survive and continue to grow and minimize the risk of this uncertainty, and can be fulfilled the best forecasting method to determine how much demand for an item in the future [2]. Knowledge of future needs will greatly support the planning process on the production floor.

In production planning, as in Gyulai et al. [3], the forecast results in the form of production plans for future needs sometimes not in line with the available capacity on the production floor. The demand for fluctuating products results in the available capacity sometimes being able to produce more than the results of forecasting needs and or even lacking for each period. Therefore capacity planning needs to be done at first, so that the demand was in line with an available capacity [4].

Syifa Ltd. is a company that has several types of products produced at its production floor, including the 220 ml glass packaging, the 600 ml bottle packaging, and 19-liter gallon packaging. However, in general, the production process takes place intuitively, where the company cannot anticipate the occurrence of low or high needs of existing demand. The research objectives expected from this study was to forecast the future demand and determining the best outcome forecast as a production plan, as well as validating the production plan using.

2. Research method

The data were observed and processed according to Gyulai et al. [3] and [5] sequence:

1. Determination of the data using demand and actual data. As mention by Najy [6], the time series analysis is very appropriate to be used to predict demand whose demand patterns in the past are quite consistent over a long period [7][3].
2. Quantitative forecasting methods can be defined in a number of ways [8]:
   a. Single Moving Average (SMA)

$$F_t = \frac{A_{t-1} + A_{t-2} + \cdots + A_{t-n}}{n}$$

   b. Weighted Moving Weight (WMW)

$$F_t = \frac{(w1A_{t-1}) + (w2A_{t-2}) + \cdots + (wnA_{t-n})}{w1 + w2 + \cdots + wn}$$

   c. Single Exponential Smoothing (SES)

F_t = F_ (t-1) + α (A_ (t-1) -F_ (t-1))

3. Measuring Forecasting Accuracy. In determining forecasting accuracy, there are some commonly used calculations:
   a. Mean Absolute Deviation (MAD), The MAD value can be calculated using the following formula.

$$MAD = \frac{\sum|A_t - F_t|}{n}$$

   b. Mean Square Error (MSE), is another method for evaluating forecasting methods. This method produces moderate errors which are probably better for small mistakes, but sometimes make a big difference.

$$MSE = \frac{\sum(A_t - F_t)^2}{n}$$

   c. Mean Absolute Percentage Error (MAPE), indicates how much error in predicting compared to the real value.

$$MAPE = \frac{\sum(\left(\frac{|A_t - F_t|}{A_t}\right)}{n} x100$$

4. Determining the efficiency and utility factors on the production floor, and determine the standard time used in the production process
5. Validateingforecast results with Rough Cut Capacity Planning (RCCP) calculations
6. Determining the Master Production Schedule (MPS).

3. Result

The selection of forecasting methods is based on the data patterns that are formed in scatter diagrams as presented in Figure 1. The data patterns formed are random data patterns or random data patterns, this happens because of irregular fluctuating requests. Historical conditions of actual demand that exist allow the most appropriate forecasting method is the method of forecasting the WMA and the SES method.

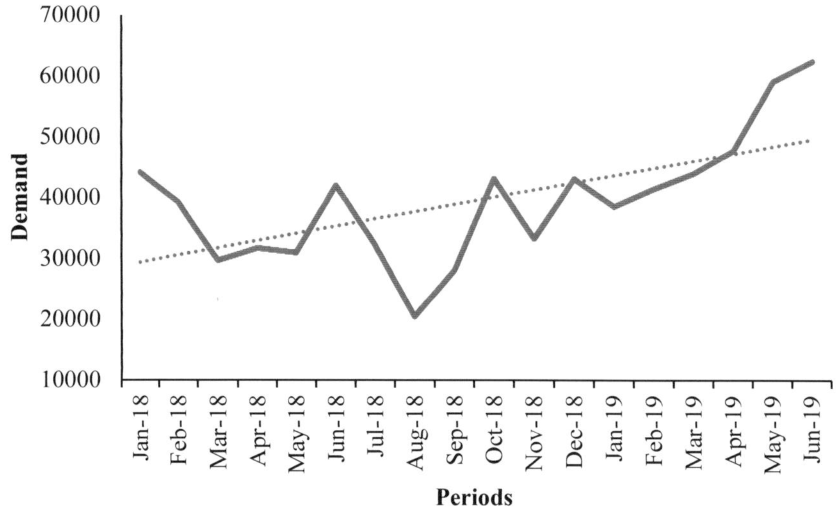

Figure 1. Scatter diagram of demand historical data

The next step is to perform forecasting needs using forecasting methods by referring to the data patterns that are formed. Forecasting using the weighted moving average method is carried out using weighting value. The choice of weights value is intuitive because there is no formula that sets them. Calculation of weighted moving average forecasting is done with an average of three and five months. Forecasting using the weighted moving average method is done using a smoothing constant (α), where the value of α is a value with a range of $0<\alpha<1$. The determination of constants is intuitive because there is no formula that sets them. Single exponential smoothing forecasting calculations are done using three constants: 0.3; 0.5; and 0.9. The comparison of MAD, MSE, and MAPE for each forecasting method is presented in Table 1.

Table 1. Comparison of MAD, MSE and MAPE

METHOD	MAD	MSE	MAPE (%)
WMA 3	5.393.92	57,560,046.59	0.759
WMA 5	5,405.87	60,327,815.32	0.761
SES 0,3	7,478.69	80,738,201.74	1.052
SES 0,5	6,467.63	66,301,854.65	0.910
SES 0,9	6,521.47	62,213,669.42	0.918

The efficiency factor is a condition of how far the work station can use the available capacity efficiently. At the filling work center, the actual average production output was 37.5 cartons per hour, and production output 72 per hour. The utility factor is a measure of the ability of a work station to utilize available capacity effectively. That is the percentage ratio of actual processing time and effective processing time. At filling the work center, the average available work hours are 15 hours of work per day, and only an average of 10 hours per day is used as the actual work hours. The utility and efficiency factors for each work center are presented in Table 2.

Table 2. Utility and efficiency factor

Work Centre	Efficiency (%)	Utility (%)
Filling	52.1	66.7
Packing	72.5	66.7
Display	75.0	66.7
Average	66.5	66.7

Based on the data uniformity test and the calculation of standard time that has been done, it can be seen the standard time needed at each work center to produce one carton of the 225 ml is as shown in Table 3. The Rough Cut Capacity Planning (RCCP) is shown in Table 4.

Table 3. Processing time of each work center

No	Work Centre	Processing Time	
		Second	Hour
1	Filling	50	0.0139
2	Packing	89	0.0247
3	Display	21	0.0058
	Total	160	0.0444

Table 4. Rough Cut Capacity Planning

No	Period (2019-2020)	Forecasting	Rough Cut Capacity Planning
1	October	26,606	31,988
2	November	34,072	40,059
3	December	35,163	36,854
4	January	40,107	40,059
5	February	38,814	40,059
6	March	40,881	41,661
7	April	42,107	41,661
8	May	45,305	41,661
9	June	52,609	41,661
	Total	355,664	355,664

4. Conclusion
The production plan is validated using the RCCP approach by means of CPOF method and the BOLA method. The validation carried out to determine whether the capacity available is able to meet the required capacity. Capacity management by optimizing production utilities was able to meet capacity shortages in the filling work center periods on the period of October 2019, November 2019, December 2019, February 2020, and March 2020 as well as meeting all capacity shortages in overall work center packing. Meanwhile, the remaining capacity deficiencies can be fulfilled with the re-adjustment approach MPS made from the production plan RP which is the result of the forecast. Furthermore, it was validated with the RCCP approach to finding out capacity shortages. Capacity shortages were met by optimizing production utilities and making adjustments (re-adjustments) so that a master production schedule is feasible to run.

References
[1] Tran L 2019 *Supply chain losses in the brewing industry* (JAMK University of Applied Science)
[2] Coetzer M 2016 *Development of planning rules and an optimized Master Production Schedule for new water and ice tea bottling plant* (University of Pretoria)
[3] Gyulai D, Pfeiffer A, and Monostori L 2017 Robust production planning and control for multi-stage systems with flexible final assembly lines *Int. J. Prod. Res.* **55** 3657–73
[4] Matswaya A, Sunarko B, Widuri R and Indriati S 2019 Analisis Perencanaan Kapasitas Produksi dengan Metode Rought Cut Capacity Planning (RCCP) pada Pembuatan Produk Kasur Busa (Studi pada PT Buana Spring Foam di Purwokerto) *Performance* **26** 128–42
[5] Joshi A D 2017 *Optimal end-of-life decision-making strategies for products with design alternatives* (Northeastern University Boston)
[6] Najy R J 2014 Rough Cut Capacity Planning (RCCP)-(Case Study). *Adv. Theor. Appl. Mech* **7** 53–66
[7] Kempf K G, Keskinocak P and Uzsoy R 2011 *Planning Production and Inventories in the Extended Enterprise* (New York, NY: Springer)
[8] Heizer J, Render B and Munson C 2016 *Principles of operations management: sustainability and supply chain management* (Pearson Higher Ed)

IOP Conf. Series: Materials Science and Engineering **885** (2020) 012053 doi:10.1088/1757-899X/885/1/012053

Model development of risk assessment of risk-taking relationship with integrated approach

A B R Indah, S Mangngenre[1] , R D Mudiastuti, N I Syamsul and Nilda

Department of Industrial Engineering, Universitas Hasanuddin, Makassar, Indonesia

Mail: a.besseriyani@gmail.com

Abstract. Each organization or company must always be dealing with risk. The decision-makers strive to be able to avoid the impact of such risks. Those generated risks may affect the achievement of the objectives, therefore risk management is needed. Risk management is a solution to avoid or to minimize the occurrence of the risk. Risk management consists of several steps, one of them is risk assessment. An appropriate risk assessment would support the achievement of the objectives. In this study, a risk assessment model is developed with *PLTSa Kota Surabaya* as a (research) object, which was selected because there were still constraints that affect the energy produced and waste volume used as fuel in this *PLTSa*. This study aims to develop risk assessment models integrated with the Delphi method, Fuzzy Cognitive Maps, and Fuzzy Synthetic Evaluation, then implementing these models in real cases and proposing risk mitigation steps by considering the relationship of the risks. The Delphi method was used to identify risks, FCM was used to identify the relationship among risks and the relationship between risks with the objectives of PLTSa, and FSE was used to assess the level of risks to find risks with the critical category. Delphi method identified 15 risks as a result and reached the consensus based on statistical analysis, FCM generated positive value of the relationship among all of the identified risks and negative value of the relationship between identified risks with the objectives of PLTSa. Meanwhile, FSE method obtained six types of critical risk, which were the machine brake all of a sudden (RC = 3,082), the blower did not work (RC = 3,041), congestion in fuel flow (RC = 2,954), clogged filter (RC = 2.915), lack of machine maintenance knowledge by the operator (RC = 2,872), and the energy produced had not maximum. (RC = 2.828). Those six types of risk used a scenario to run the FCM simulation and meet the dynamic behavior of the risk relationship value. The mitigation proposals are to perform maintenance activities programmatically, to improve the management system related to human resources, and to do the fuel examination according to the standard.

1. Introduction

In carrying out activities both in business and not business oriented will always face problems or obstacles. The problem is unpredictable, such problems could happen beyond human control. If the problem continues to occur then there will be something unexpected commonly referred to risk. Risk is an uncertainty that is likely to occur in a case causing harm. In risk management, several steps must be passed. One of them is risk assessment. Several previous studies have developed models of risk assessment on several objects. Andric and Lu [1] conducted a risk assessment on the bridge construction process with the Fuzzy Analytic Hierarchy Process (FAHP). Cao and Song [2] developed a risk assessment approach to rough group and Analytic Network Process (ANP) to conduct a quantitative risk assessment. Cheng and Lu [3] developed a risk assessment on construction projects jacking pipes with fuzzy inference method and failure mode effect analysis (FMEA). Taylan et al. [4] conducted a study with the selection of construction projects and conducted risk assessments using fuzzy AHP and Fuzzy TOPSIS. Wu et al. [5] made a risk assessment model on the offshore pipeline project by integrating methods Interpretive Structural Modeling (ISM) and Bayesian Network.

This research will conduct development models of risk assessment which take into account the relationship between risk and risk relations with the objectives to be achieved to make risk mitigation measures. The development of valuation models is done by integrating the Delphi method, Fuzzy Cognitive Maps, and Fuzzy Evaluation Synthetic for. Delphi was used for risk identification, FCM was used to identify the relationship of risk, and FSE was used to conduct a risk assessment to find critical. Critical risk obtained will be used as the basis of the scenario FCM to run simulations of FCM to see the dynamic behavior of the risk relationship.

2. Risk assessment

Risk assessment consists of several main stages such as risk identification, risk analysis, and risk evaluation.

2.1 Risk identification

This phase will conduct a risk identification process, which first identified is the potential of risk. Identification of potential risks carried out using the Delphi method. Delphi method is a method of decision-making process involving multiple experts [6]. The aim is to avoid domination by another expert opinion and minimize bias. The stages of the Delphi method are explained in Figure1.

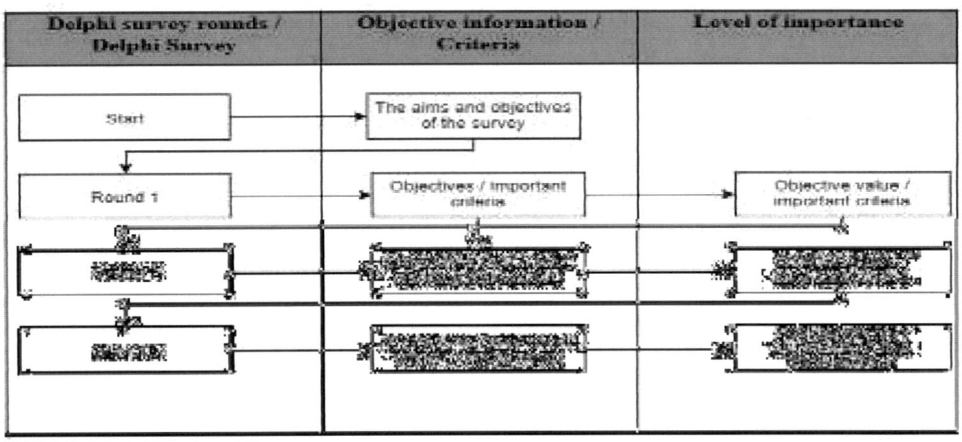

Figure 1. Stages of the Delphi Method [7]

Having obtained any potential risks that may occur by using the Delphi method, the next step is obtaining the relationship between the risk potentials. To find the relationship between the potential risks we use Fuzzy Cognitive Maps. Fuzzy cognitive maps are soft computing techniques and knowledges-based method that involves human reasoning and human decision-making in its application. The purpose of this phase is to determine whether a potential risk can be occurred due to other potential risks. In this method, extracting information will be made to the expert by interviews directly related to the relationship between the potential risks categorized by risk variables. Participants will fill out the questionnaire by assigning the value of the relationship between the potential risks.

a. If participants give value > 0, then there is a positive relationship between the potential risks (proportional).
b. If participants give value = 0, then there is no relationship between the potential risks.
c. If participants give value < 0, then there is a negative relationship between the potential risks (opposite).

2.2. Risk analysis

After gaining the relationship between the risks, the next step is conducting a risk analysis process. Fuzzy Synthetic Evaluation (FSE) is a branch of the fuzzy logic developed by Zadeh [8] to represent and manipulate the vagueness of the term [9]. The risk analysis performed using Fuzzy Synthetic Evaluation, in analyzes performed consists of several stages.

2.3 Risk Evaluation

Based on the results of risk assessment, therefore, we can make proposed mitigation measures that should be conducted immediately. By looking at the critical risk category, we can determine the priority of risk mitigation. Also, we understand the terms of the relationship between risks, the relationship may be based on the results of FCM method that has been used to describe the risks that influence the critical risks by simulating FCM conducted by using a critical risk scenario.

3. Case Study

The research was conducted on a small scale PLTSa objects located in the Bratang landfill and Wonorejo landfill of Surabaya. PLTSa Surabaya was an object selected in this study because city officials until now made a target to build PLTSa in a small scale in 35 polling stations located in Surabaya, therefore the government of Surabaya can make the results of this study as a reference to face the potential risks to the development of next PLTSa. Researchers will look at the enforceability of the development of the proposed risk assessment model on that object. The data collection was conducted by using interviews and questionnaires addressed directly to the experts/stakeholders. Data collected consists of primary data and secondary data. Primary data was the data taken directly from the results of surveys, interviews, and questionnaires to the expert. The number of experts who were taken in this study is four

3.1 Research design method

This research conducted an integrated approach (Figure. 3) by creating a model of risk assessment by integrating several methods such as the Delphi method, Fuzzy Cognitive Maps, and Fuzzy Synthetic Evaluation.

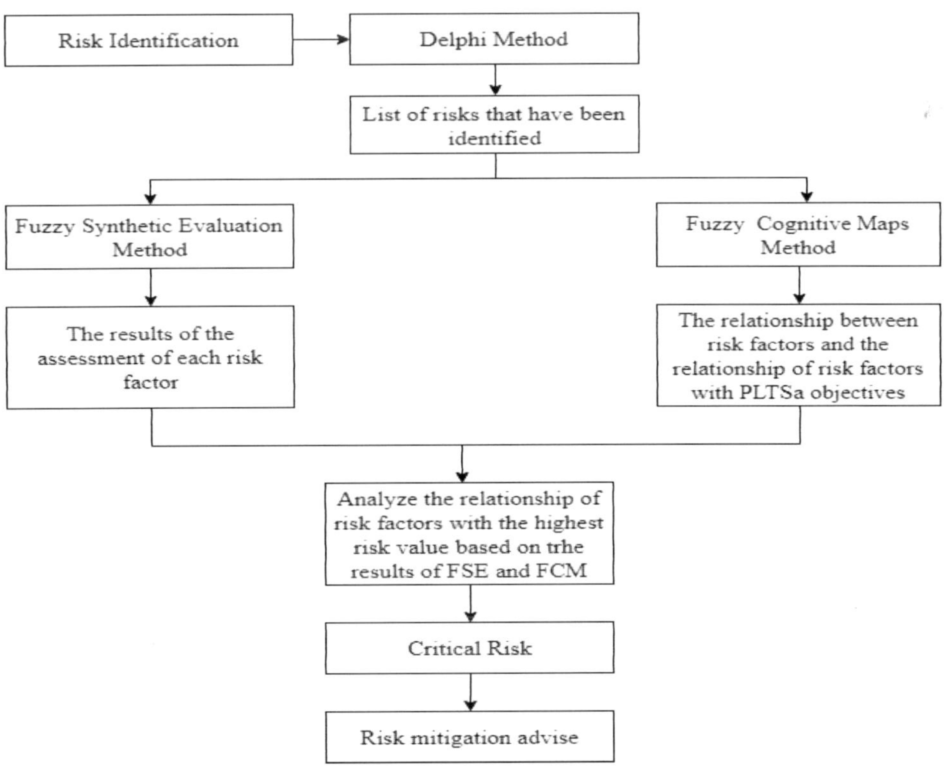

Figure. 2 Integrated Approach Model

3.2 Data Analysis

3.2.1 Delphi method

In this research, the Delphi method was conducted by distributing Delphi questionnaires with three turns. From the results of collection from each respondent obtained it summarized possible risk factors that occur are as many as 26 risk factors. From those 26 risk factors can be classified of financial risk, operational risk, and maintenance.

Delphi's second round conducted an assessment of risk factors that have been identified in the Delphi rounds I. Assessment is based on a Likert scale to express statements of agreeing or disagree related to 26 risk factors that have been summarized in previous rounds of Delphi. The results of questionnaires showed that 11 risk factors that otherwise have an average value of 3.5 to 3.5 over indicate that respondents agree with these risk factors. Besides, there are 15 risks with the total value averaging below 3.5 indicating respondents undecided even disagree with these risk factors. Because there are still some potential risks to the value of 3 is doubtful therefore re-division of the questionnaire was conducted. It is called the Delphi questionnaire round III. However, in the second round of the Delphi questionnaire, there are additional risk factors, namely the explosion of the reactor. In Delphi III because the respondents remain consistent with questions of risk factors which have been summarized by the first round Delphi questionnaire, the respondents reported that they have reached an understanding uniformity through statements risk factor that has been written in the questionnaire. So the modified Delphi II questionnaire produce a questionnaire Delphi III only form of additional risk factors that originally consisted of 26 risk factors assessed by each respondent to 27 risk factors assessed at Delphi questionnaire round III. From the results obtained by the Delphi method, 15 risks were obtained to reach a consensus until the third round (Table 1)

Table 1. Risks were obtained to reach a consensus until the third round

No.	Code	Risk	No.	Code	Risk
1	(R1)	Levels of methane gas produced are still inconsistent	9	(R9)	The blower is suddenly not working
2	(R2)	There are no tar waste processing methods	10	(R10)	No determination of process schedule operation of PLTSa
3	(R3)	Delay operation process of PLTSa	11	(R11)	Filter clogged
4	(R4)	The energy produced is not maximized	12	(R12)	The workforce who lack discipline
5	(R5)	The number of engine units is still lack	13	(R13)	The lack of skill of the operator to operate PLTSa
6	(R6)	The machine was broken suddenly	14	(R14)	Knowledge of the operator to maintain the machine is still lacking
7	(R7)	PLTSa process in the rainy season is not maximum	15	(R15).	There was an explosion in the reactor
8	(R8)	There was congestion fuel flow			

3.2.2 Method of fuzzy cognitive maps

In this study, besides, to identify risk factors that occurred or likely to occur also conducted the process of identifying the relationship of risk to see the correlation relationship between one risk to other risks. It also identified the relationship of risk to the objectives of PLTSa. The objectives to be achieved on a pilot project PLTSa in Surabaya consists of two main objectives, namely:

1. Anticipate an accumulation of trash by reducing the volume of waste through the operation of PLTSa (P1)
2. Produce renewable energy by using waste in Surabaya (P2)

As for the data processing with the FCM is:

1. Establish experts/stakeholders who have understood about PLTSa to determine the relationship of risk.

2. FCM questionnaire (Appendix D) for each expert to give values to the relationships between these risk factors.
3. Each expert will give values to the relationship between risk and risk relations with the objectives to be achieved by giving more values to the value range -1 to 1. Each expert determines the amount of value between these ranges generously. If there is no relationship then the assigned value of 0.
4. Each value of the relationship obtained from each expert will form an adjacency matrix. Valuing is conducted by each expert forming matrix adjacency respectively. In this study, because it uses four experts consequently there is four adjacency matrix. Below is an example adjacency matrix:

$$A = \begin{bmatrix} 0 & 0 & 0 & w_{14} \\ 0 & 0 & 0 & w_{24} \\ w_{31} & w_{32} & 0 & 0 \\ 0 & 0 & 0 & 0 \end{bmatrix}$$

Adjacency matrix obtained from four experts then processed into adjacency augmented matrix by using equation (1):

$$W_{ij}^{Aug} = \frac{\sum_{i=1}^{m} W_{ij}^{k}}{m} \quad (1)$$

The pattern of risk relationships later is depicted in the form of a cognitive map based on the augmented adjacency matrix. Values obtained from the expert relationship can be seen in its dynamic behavior. If you want to see the dynamic behavior of the risk relationship therefore we should make a scenario to simulate FCM based on the equation (2) computing scenarios, namely:

$$C_j^{t+1} = f(C_j^t + \sum_{\substack{i=1 \\ i \neq j}}^{n} C_i^t \; x \; W_{ij}) \quad (2)$$

f is the activation function, C_j^{t+1} is the value impact (effect) node *j* at time *t + 1*. C_j^t is the value impact (effect) at time *t*. C_i^t is a cause value (causal) node *i* at time *t*. And wij is to show the intensity of the relationship between causal effect node *i* and *j*. The tangent function consists of two: the hyperbolic tangent function and sigmoid. The formula for the hyperbolic tangent function according to Lopez and Salmeron (2012), equation (3):

$$f(x) = \tanh(\lambda x)$$

Hyperbolic tangent is used on the value of connectedness with the value range [-1,1]. At this stage only limited identify relationships based on expert assessment.

3.2.3 method of fuzzy synthetic evaluation

The risk assessment is carried out by using FSE. After identifying the relationship of risks using the FSE method derived from the opinion of the expert, then any identified risks are analyzed quantitatively using FSE. The steps of data processing, risk assessment with the FSE method are:

1. Appoint experts to conduct a risk assessment.
2. The list of risks assessed was risks identified using the Delphi method which consists of 15 types of risk factors.
3. Furthermore, questionnaires are distributed to each expert. Parameter of risk assessment consists of two types of consequences (impact) and likelihood (likelihood of risk occurrence). Each of these parameters is assessed on a Likert scale with values 1-5.
4. Calculating process of FSE method uses the following equation:

IOP Conf. Series: Materials Science and Engineering **885** (2020) 012053 doi:10.1088/1757-899X/885/1/012053

The matrix equation for consequences (impact) type of risk R6 (broken machine suddenly) is (0.00,0.00,0.00,0.00,1.00), which answers with a value of 1 = 0%, value of 2 = 0%, value of 3 = 0%, value of 4 = 0%, and the value of 5 = 100%.

$$\left(R_i^C\right)_{1x5} = \left(r_{i1}^C, r_{i2}^C, r_{i3}^C, r_{i4}^C, r_{i5}^C\right) \quad (3)$$

$$\left(R_i^C\right)_{1x5} = (0.00, 0.00, 0.00, 0.00, 1.00)$$

$$C_i = \sum_{j=1}^{5}(s_j\, x\, r_{ij}^C) \quad (4)$$

$$C_i = (1 \times 0.00) + (2 \times 0.00) + (3 \times 0.00) + (4 \times 0.00) + (5 \times 1)$$

$$C_i = 5$$

The matrix equation for Likelihood (Probability) type of risk R6 (broken machine suddenly) is as follows: (0.00,0.00,0.00,0.50,0.50). Here is the calculation:

$$\left(R_i^L\right)_{1x5} = \left(r_{i1}^L, \quad r_{i2}^L, r_{i3}^L, r_{i4}^L, r_{i5}^L\right)$$

$$\left(R_i^L\right)_{1x5} = (0.00, 0.00, 0.00, 0.50, 0.50)$$

$$L_i = \sum_{j=1}^{5}(s_j\, x\, r_{ij}^L)$$

$$L_i = (1 \times 0.00) + (2 \times 0.00) + (3 \times 0.00) + (4 \times 0.50) + (5 \times 0.50)$$

$$L_i = 4.50$$

Sj is the value given to the risk factors, namely *Sj* = 1,2,3,4,5.

$$RC_i = \sqrt{L_i\, x\, C_i} \quad (5)$$

$$RC_i = \sqrt{5 + 4.5}$$

$$RC_i = 3.082$$

The results of the calculations by the FSE method can be seen in Table 1. Based on the calculations conducted there are six types of risk included in the category of critical value with the value of RC is between 2.8 to 3.1:

a. Engine broken abruptly with the value of RC is 3.082
b. The blower is not worked suddenly with the value of RC is 3.041
c. Fuel flow congestion occurred with the value of RC is 2.958
d. The filter clogged with the value of RC is 2.915
e. Knowledge of machine Maintenance operator is still lacking with the value of RC is 2.872
f. The energy generated is not maximized with the value of RC is 2.828

3.2.4 Results of fcm simulation by critical risk

This section will conduct an FCM analysis of simulation results based on the scenario built that is critical risks that consist of six types of risk.

Simulation 1

In the first simulation, we have to understand what if scenario 1 run so if the risk factors occurred are broken engine suddenly (R6) with the value 1 on FCM simulation means active. The results obtained are affecting some of the risk factors and objectives of PLTSa. As for the risk factors associated with the R6 is:

1. R1 (levels of methane produced is still inconsistent) = 0.9434 (very powerful positive correlated)
2. R3 (delay on operation process of PLTSa) = 0.9464 (very powerful positive correlated)
3. R4 (energy produced is not maximized) = 0.9165 (very powerful positive correlated)
4. R7 (PLTSa process is not maximum in rainy season) = 0.8812 (very powerful positive correlated)
5. R8 (fuel flow congestion occurs) = 0.8946 (very powerful positive correlated)
6. R9 (blower suddenly not working) = 0.9295 (very powerful positive correlated)
7. R11 (filter clogged) = 0.9295 (very powerful positive correlated)
8. R12 (undiscipline labors) = 0.5647 (very powerful positive correlated)
9. R15 (explosion in reactor) = 0.8946 (very powerful positive correlated).

If R6 occurs it is likely that the risk of R1, R3, R4, R7, R8, R9, R11, and R15 will occur because they have a very powerful positive correlation. Meanwhile, as for relations with the aim of PLTSa R6 is negatively correlated where P1 (reduce the volume of waste) with relationship value = -0.9612 (very powerful negative correlation), and P2 (produce renewable energy) = -0.9612 (very powerful negative correlation), which means if R6 occurs then the purpose of P1 and P2 are not achieved in full potential.

Simulation 2

The second simulation conducted is running the second scenario that is R9 (blower suddenly not worked) if it happens then how is their relationship to other risk factors and objectives of PLTSa. The results of the scenario conducted are if R9 is enabled by providing value 1 on FCM simulation therefore risk factors affected are:

1. R1 (levels of methane produced is still inconsistent) = 0.9211 (very powerful positive correlated)
2. R3 (delay in operation process of PLTSa) = 0.9464 (very powerful positive correlated)
3. R4 (energy produced is not maximized) = 0.9295 (very powerful positive correlated)
4. R6 (faulty engine suddenly) = 0.9115 (very powerful positive correlated)
5. R7 (PLTSa process is not maximum in rainy season) = 0.7078 (powerful enough positive correlated)
6. R8 (fuel flow congestion occurs) = 0.8812 (very powerful positive correlated)
7. R11 (filter clogged) = 0.7861 (powerful enough positive correlated)
8. R15 (an explosion in the reactor) = 0.7516 (powerful enough positive correlated).

The relationship between risk factor R9 with the aim of PLTSa is negative with a value of P1 = (– 0.9369), P2 = (– 0.9403). If R9 occurs, therefore, the aims of P1 and P2 are not achieved in full potential. Meanwhile with the positive relationship occurred in R9 with other risks, if R9 happens then likely R1, R3, R4, R6, R7, R8, R11, and R15 also have a positive relationship.

Simulation 3

In the third conducted what if scenario 3 is executed if risk factors occurred is R8 (fuel flow congestion) with the value 1 on FCM simulation means active. The risk factors influenced by results of FCM simulation is:

1. R1 (levels of methane produced is still inconsistent) = 0.9369 (very powerful positive correlated)
2. R3 (delay in operation process of PLTSa) = 0.9434 (very powerful positive correlated)
3. R4 (energy produced is not maximized) = 0.9211 (very powerful positive correlated)
4. R6 (faulty engine suddenly) = 0.9165 (very powerful positive correlated)
5. R7 (PLTSa process is not maximum in rainy season) = 0.7861 (powerful enough positive correlated)
6. R9 (blower suddenly not working) = 0.9211 (very powerful positive correlated)
7. R11 (Filter clogged) = 0.9006 (very powerful positive correlated)
8. R15 (explosion in reactor) = 0.9006 (very powerful positive correlated).

The relationship between risk factors R8 with aim of PLTSa is negative with a value of P1 = (– 0.9295), P2 = (–0.9295). If R8 happens then likely R1, R3, R4, R6, R7, R9, R11, and R15 happen because they have a positive relationship. Meanwhile, if R8 occurred, therefore, PLTSa goals can not be achieved in the maximum because they have a negative relationship between the objective and the R8.

Simulation 4

Simulation 4 is run by third scenario if the risk factors occurred are R11 (filter clogged) to give a value of 1 on FCM simulation which means active, therefore risk factors influenced by the results of the FCM simulation are:

1. R1 (levels of methane produced is still inconsistent) = 0.9434 (very powerful positive correlated)
2. R3 (delay in operation process of PLTSa) = 0.9403 (very powerful positive correlated)
3. R4 (energy produced is not maximized) = 0.9165 (very powerful positive correlated)
4. R6 (faulty engine suddenly) = 0.9115 (very powerful positive correlated)
5. R7 (PLTSa process is not maximum in rainy season) = 0.6810 (powerful positive enough correlated)
6. R8 (fuel flow congestion) = 0.8812 (very powerful positive correlated)
7. R9 (blower suddenly not working) = 0.9165 (very powerful positive correlated)
8. R13 (lack of skill of the operator in operating PLTSa) = 0.6496 (powerful positive enough correlated)
9. R14 (lack of knowledge of operator to perform machine maintenance) = 0.6910 (powerful positive enough correlated)
10. R15 (explosion in reactor) = 0.6496 (powerful positive enough correlated).

The relationship between risk factors R11 with the aim of PLTSa is negative with a value of P1 = (–0.9568), P2 = (–0.9612). Values of relationships indicates that if R11 occurred then the risk of potentially happen is R1, R3, R4, R6, R7, R8, R9, R13, R14, and R15 because they have positive relationships, the value of the relationship between risk assessed from expert assessment directly, changing the value of the relationship after simulation of dynamic behavior based on scenarios built. While the value of the relationship between R11 with the aim of PLTSa is negative, which means that if R11 occurred it will hamper the achievement of predetermined PLTSa.

Simulation 5

To run a simulation 5 we use scenario 5 to determine the value of the relationship of risk and PLTSa purposes. In simulation 5 we run scenario 5 that is R14 (knowledge of operator to maintain machine is still lacking) if it occurred then how they relate to other risk factors and objectives of PLTSa. The results of the scenario run are if R14 activated with value 1 on the FCM simulation of risk factors affected are:

1. R1 (levels of methane produced is still inconsistent) = 0.5647 (powerful positive enough correlated)
2. R3 (delay in operation process of PLTSa) = 0.7861 (powerful positive enough correlated)
3. R4 (energy produced is not maximized) = 0.9006 (very powerful positive correlated)
4. R6 (faulty engine suddenly) = 0.9369 (very powerful positive correlated)
5. R7 (PLTSa process is not maximum in rainy season) = 0.5647 (powerful positive enough correlated)
6. R8 (fuel flow congestion) = 0.9211 (very powerful positive correlated)
7. R9 (blower suddenly not working) = 0.9434 (very powerful positive correlated)
8. R11 (filter clogged) = 0.9434 (very powerful positive correlated)
9. R15 (explosion in reactor) = 0.9434 (very powerful positive correlated)

The relationship between risk factors R11 with the aim of PLTSa is negative with a value of P1 = (–0.9493), P2 = (–0.9493), which means that if R11 occurred then PLTSa goal can not be realized in maximum. Meanwhile, the relationship between R11 to other risks associated with the positive cause most likely risk factors R1, R3, R4, R6, R7, R8, R9, R11, and R15 will occur.

Simulation 6

In simulation 6 conducted is if scenario 6 run that is if the risk factors occurred is R4 (energy produced is not maximum) with the value 1 on FCM simulation means active. The risk factors influenced by the results of the FCM simulation are:

1. R1 (levels of methane produced is still inconsistent) = 0.9454 (very powerful positive correlated)
2. R7 (PLTSa process is not maximum in rainy season) = 0.5647 (powerful positive enough correlated).

Based on the value of relationship risk factor R4 there are only two risk factors that have a positive correlation, which means that if R4 occurred then likely R1 and R7 are affected. The relationship between risk factor R8 with the aim of PLTSa is negative with a value of P1 = -0.9464, P2 = -0.9519, a negative value indicates that if risk factors R4 occurred, therefore, we will make PLTSa goal not achieved in maximum.

3.3. Proposed Risk Mitigation

Perform maintenance activities in the program and schedule on the machine/equipment of PLTSa

Programmed maintenance activities can be conducted based on the systematic scheme of maintenance activities proposed by Muhtad (2009), as well as on Figure 4.

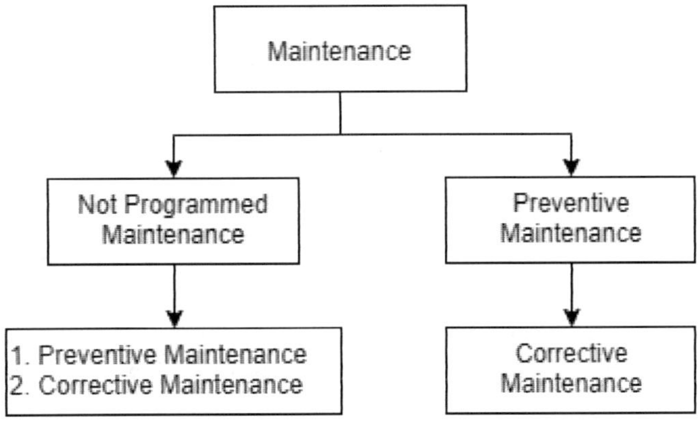

Figure 4. Systematic scheme of maintenance activities

Improving management system relating to human resources.
Some efforts should be made to overcome the lack of qualified human resources according to in Widiasih [10] are:

a. Training (education and training)
b. Job design

Verify the fuel adjusted with fuel quality standards for the gasification process.
Some standards that need to be concerned related to the energy content of the fuel, fuel moisture, shape and size of the fuel, the uniformity of fuel, heavy fuel per cubic meter, volatile substance (easily evaporated) on fuel and ash substance. Those characteristics must be considered to generate maximum energy.

4. Conclusion

In this research, we conducted the development models of risk assessment by considering relationship aspects between risks and risk relations with the objectives to be achieved to make risk mitigation measures, with a case study on power plant waste of Surabaya. The methods integrated are the Delphi method, Fuzzy Cognitive Maps, and Fuzzy Synthetic Evaluation. Delphi method is used as a method to identify risks in small-scale PLTSa in Surabaya conducted by distributing questionnaires three times round. The results of the questionnaire in reaching consensus obtained fifteen risks managed to reach a consensus. Fuzzy Cognitive Maps (FCM) is the method used to identify the relationship between risk and risk the relationship with the project objectives. The results of value identification of risk relationship showed that the relationship between the risks of all is positive (one-way), and risk relationship with PLTSa goal of all is negative (opposite-way). From the results of FSE methods successfully obtained six critical risks that is R6 (faulty engine suddenly), R9 (blower suddenly not working), R8 (fuel flow congestion), R11 (filter clogged), R14 (the lack of operator's knowledge to maintain machine), and R4 (the energy produced is not maximized). The proposed mitigation measures that need to be addressed are as follows:

a. Conduct maintenance activities programmed and scheduled on machinery/equipment of PLTSa.
b. Improve the management system related to human resources.
c. Examine the fuel adjusted fuel quality standards for the gasification process.

References

[1] Andrić J M and Lu D-G 2016 Risk assessment of bridges under multiple hazards in operation period *Saf. Sci.* **83** 80–92
[2] Cao J and Song W 2016 Risk assessment of co-creating value with customers: A rough group analytic network process approach *Expert Syst. Appl.* **55** 145–56
[3] Cheng M and Lu Y 2015 Developing a risk assessment method for complex pipe jacking construction projects *Autom. Constr.* **58** 48–59
[4] Taylan O, Bafail A O, Abdulaal R M S and Kabli M R 2014 Construction projects selection and risk assessment by fuzzy AHP and fuzzy TOPSIS methodologies *Appl. Soft Comput.* **17** 105–

16
[5] Wu W-S, Yang C-F, Chang J-C, Château P-A and Chang Y-C 2015 Risk assessment by integrating interpretive structural modeling and Bayesian network, case of offshore pipeline project *Reliab. Eng. Syst. Saf.* **142** 515–24
[6] Tambun B S, Jananuraga I G, Mahatmavidya P A and Soesanto R P 2014 Perancangan Sistem Informasi Berbasis Geografis untuk Mendukung Keputusan Pemilihan Tempat Makan di Kawasan Pendidikan Telkom University dengan Menggunakan Metode Factor Rating dan Metode Delphi *Seminar Nasional IENACO 2014* (Universitas Muhammadiyah Surakarta) pp 474–83
[7] Udisubakti C 2001 Integrasi Metode Delphi dan Prosedur Analisis Hierarkhis (AHP) untuk Identifikasi dan Penetapan Prioritas Objektif Kriteria Keputusan *J. Iptek* **12** 37–42
[8] Zadeh L A 1965 Fuzzy Sets *Inf. Control* **8** 338–53
[9] Ameyaw E E and Chan A P C 2015 Evaluation and ranking of risk factors in public–private partnership water supply projects in developing countries using fuzzy synthetic evaluation approach *Expert Syst. Appl.* **42** 5102–16
[10] Widiasih W 2015 *Pengembangan Model Risiko pada Implementasi Lean Manufacturing Di PT. Dirgantara Indonesia (Indonesian Aerospace) Dengan Pendekatan Terintegrasi* (Institut Teknologi Sepuluh November)

IOP Conf. Series: Materials Science and Engineering **885** (2020) 012054 doi:10.1088/1757-899X/885/1/012054

A Hybrid Firefly and Local Search Algorithms to Solve Orienteering Problems

C Sitompul, R Pramudi and F Setiawan

Department of Industrial Engineering, Parahyangan Catholic University, Bandung, Indonesia

E-mail: carles@unpar.ac.id

Abstract. A satisfied customer is a goal for every tourism agency who put efforts designing a tour route for its customers. The resulting problem often called an orienteering problem attempts to design a route that maximizes customer's value for all destinations. We propose a firefly algorithm combined with a hybrid of local search algorithms to solve the problem. Our finding however suggests that firefly algorithms cannot outperform Ant Colony (AC) and Genetic Algorithm (GA).

1. Background

For Indonesia, tourism has been one of the biggest contributors to its gross domestic products. In 2018, Indonesia attracted more than 15.8 million foreign tourists, reflecting a 12% growth from the previous year [1]. However, Indonesia still runs behind Thailand, Singapore and Malaysia which forces the Indonesian Government to ramp up its tourism sector. Tour and travel agencies have the potential to support the industry through designing routes that maximize customers' satisfaction. This routing problem is often called as an orienteering problem [2]. For example, a travel agency provides a tour package in Bali for a day tour visiting places of interest such as Pura Tanah Lot, Pura Uluwatu, Ubud Monkey Forest, Jati Luwih Rice Fields, Bedugul Lake, and Kuta Beach. A customer's satisfaction can be described using rating systems for those places, e.g. given in *TripAdvisor*.

The orienteering problem has been discussed for decades starting with Tsiligirides first attempt to solve the problem using a heuristic method [3]. Since its introduction, the problem has evolved to many variations for example Time Dependent Orienteering Problem (TDOP), Orienteering Problem with Time Windows (OPTW), and Team Orienteering Problem (TOP). The orienteering problem can be seen as a combination of two classical combinatorial problems, i.e. Knapsack Problem and Travelling Salesmen Problem [2]. Furthermore, the orienteering problem is categorized as an NP-hard problem [4] where solutions can only be found in a non-polynomial time or solutions cannot be found. This orienteering problem can be formulated as in Vansteenwegen [2] as follows. Suppose N is the number of nodes or destinations, where p_i denotes profit of visiting i, and c_{ij} as costs of transportations from i to j, $i, j\{1..N\}$. Variable x_{ij} equals to 1 if a route exists from i to j and variable $\mu_i \geq 0$ denotes node i in the route assuming that the route starts at node *1* and ends in node N.
Maximize

$$\sum_{i=1}^{N-1}\sum_{j=2}^{N} p_i x_{ij} \tag{1}$$

Subject to:

$$\sum_{j=2}^{N} x_{ij} = \sum_{i=1}^{N-1} x_{iN} = 1 \tag{2}$$

$$\sum_{i=1}^{N-1} x_{ir} = \sum_{j=2}^{N} x_{rj}\,, \forall r = 2, \cdots, N - 1 \tag{3}$$

$$\sum_{i=1}^{N-1}\sum_{j=2}^{N} p_i\, x_{ij} \le B, \tag{4}$$

$$2 \le u_i \le N, \forall i, j = 1, 2, N \tag{5}$$

$$u_i - u_j + 1 \le (n - 1)\left(n - x_{ij}\right), \forall r = 2, \cdots, N \tag{6}$$

$$x_{ij} \le \{0,1\}, \forall i, j = 2, \cdots, N \tag{7}$$

2. A Firefly Algorithm

Firefly algorithms are introduced by Xin-She Yang in 2010 [5] inspired by fireflies and their conspicuous use of bioluminescence during twilight to attract mates or prey. Some assumptions used in the firefly algorithms are:

1. Fireflies are unisex
2. Attractiveness is proportional to their brightness.
3. The brightness of a firefly is determined by the objective function.

Based on these three rules, the basic steps can be summarized as pseudo code as in Figure 1.

```
Firefly Algorithm
Objective function f(x), x = (x_1, ..., x_d)^T
Generate initial population of fireflies x_i (i = 1,2, ..., n)
Light intensity I_i at x_i is determined by f(x_i)
Define light absorption coefficient γ
while (t< MaxGeneration)
for i = 1:n all n fireflies
    for j = 1:n all n fireflies (inner loop)
    if (I_i < I_j), Move firefly i towards j; end if
        Vary attractiveness with distance r via exp[-γr]
        Evaluate new solutions and update light intensity
    end for j
end for i
Rank the fireflies and find the current global best g *
end while
Post-process results and visualization
```

Figure 2. Pseudo code of the firefly algorithm (Yang, 2010)

The firefly algorithm can be implemented to solve the orienteering problem as follows. A firefly's location is analogous to a solution in orienteering problems, hence finding a better solution can be achieved through a movement of their locations. An exploitation is considered when fireflies are attracted to other fireflies, otherwise a random movement is considered as an exploration. Encoding is a process of translating a solution into a firefly's location. A vector of firefly's location has $(N-2)$ dimensions where entries are generated using a random number $U(0,1)$. A decoding is used to translate a vector into solutions, i.e. (1) determining the sequence without constraints, (2) inserting each destination in the sequence into the solution taking into account the constraints, (3) calculating the objective function which is seen as an intensity level. The light intensity $I(r)$ varies according to the inverse square law and the absorption law following Gaussian form:

$$I(r) = I_0 e^{-\gamma r^2}$$

The attractiveness β can be defined by:

$$\beta = \beta_0 e^{-\gamma r^2}$$

where β_0 is the attractiveness at $r = 0$. The attractiveness can also be approximated as:

$$\beta = \frac{\beta_0}{1 + \gamma r^2}$$

The distance between two fireflies x_i and x_j can be defined as a Cartesian d–dimension:

$$r_{ij} = \sqrt{\sum_{k=1}^{d} \left(x_{i,k} - x_{j,k}\right)^2},$$

where x_i and x_j denotes vectors of two fireflies.

Fitness value of each firefly is calculated and then compared one another. A firefly i is attracted to firefly j if firefly j has a higher fitness function. If there exist no other fireflies with higher fitness function, then firefly i will move randomly. As input parameters, we use the following: randomization parameter (α), attractiveness (β), light absorption coefficient (γ), number of population and number of iterations.

The movement of a firefly is determined by:

$$x_1 = x_1 + \beta_0 e^{-\gamma r_{ij}^2 (x_i - x_j)} + \alpha \epsilon_i$$

where ϵ_i is random error generated from a uniform distribution.

3. Local Search Algorithms

We also propose some local search algorithms to improve the quality of the fitness function provided by firefly algorithm. The fitness function is improved using inner swap algorithm, outer swap algorithm, insertion algorithm, and 2-opt algorithm. Inner swap algorithm is used to swap two nodes within the route, while an outer swap algorithm exchanges a node in the route from a node outside the route. An insertion algorithm is also used to add a node into existing route expecting that it will improve the solution. Lastly, a 2-opt local search is used to reduce the number of crossovers among routes.

4. Computational Results

We implement the proposed algorithm using JAVA Net Beans IDE 8.2 and use it to solve three benchmark problems as suggested in Tsiligirides [3]. The following parameters are used, i.e. number of iterations 10.000 to avoid failures finding the solution. The number of fireflies is experimented using 3 levels: 15, 25 and 40. The randomization parameter (α) and attractiveness (β) both has three levels: 0.1, 0.5 and 1. We use 0,10 and 50 as light absorption coefficient (γ). Using one factor at a time (OFAT) experimental design, we have the following parameter combination as given in Table 1.

Table 1. Parameter combination

Combination	n	α	β	γ
A	10	0.1	0.1	0
B	25	0.1	0.1	0
C	40	0.1	0.1	0
D	10	0.5	0.1	0
E	10	1.0	0.1	0
F	10	0.1	0.5	0
G	10	0.1	1.0	0
H	10	0.1	0.1	10
I	10	0.1	0.1	50

From all three datasets, the firefly algorithm performs worse than any best-known solution averaging, i.e. 5%, 8% and 10% worse in the first, second and third datasets respectively. Even when compared with Ant Colony Algorithm [6] and Genetic Algorithm [7], the firefly algorithm also performs worse.

However, when combined with local search, the algorithm performs better achieving almost 60% of best-known solutions for all datasets. When compared with Ant Colony (AC) and Genetic Algorithm (GA), the proposed algorithm differs less than 2% worse in dataset 1 and dataset 3. In dataset 2, the algorithm can reach the same level with both AC and GA algorithms.

5. Conclusion

In general, firefly algorithm in itself cannot outperform Ant Colony (AC) and Genetic Algorithm (GA). However, when combined with local search algorithms the algorithm may reach the same level as those algorithms. We propose that further research should be directed to a more effective way to implement the algorithm within the JAVA programming environment.

ACKNOWLEDGMENTS

We would like to thank Directorate General of Learning and Student Affair, Ministry of Research Technology and Higher Education for giving us the financial support in this publication as implementation of grant program in the postgraduate improvement quality by the contract number (No.036/B4/PPK-SPK/VII/2019).

References

[1] Post J 2019 Indonesia welcomed 15.8 million foreign tourists last year: BPS
[2] Vansteenwegen P, Souffriau W and Van Oudheusden D 2011 The orienteering problem: A survey *Eur. J. Oper. Res.* **209** 1–10
[3] Tsiligirides T 1984 Heuristic methods applied to orienteering *J. Oper. Res. Soc.* **35** 797–809
[4] Golden B L, Levy L and Vohra R 1987 The orienteering problem *Nav. Res. Logist.* **34** 307–18
[5] Yang X-S 2010 *Nature-inspired metaheuristic algorithms* (Luniver press)
[6] Liang Y-C and Smith A E 2006 An ant colony approach to the orienteering problem *J. Chinese Inst. Ind. Eng.* **23** 403–14
[7] Tasgetiren M F and Smith A E 2000 A genetic algorithm for the orienteering problem *Proceedings of the 2000 Congress on Evolutionary Computation. CEC00 (Cat. No. 00TH8512)* vol 2 (IEEE) pp 910–5

IOP Conf. Series: Materials Science and Engineering **885** (2020) 012055 doi:10.1088/1757-899X/885/1/012055

Reducing variability in process and the risk of product defects: A case study in SME's

A Darmawan, R Hanafi and A Asqarye

Industrial Engineering Department, Faculty of Engineering, Universitas Hasanuddin, Makassar

E-mail: darmawanarmin@gmail.com

Abstract. The study aims to determine the quality control level of mineral water packaging. Six Sigma is recognized as an essential tool used in this study in determining the quality level by measuring variability in the process, reducing product defects (packaging), and developing the improvement comprehensively. The implementation of Six Sigma has been implemented in many advanced companies. The implementation was conducted in one small-medium enterprise (SME) that produced the mineral water. However, the implementation of Six Sigma in the small-medium level of industries has relatively new challenges. The result showed that identified four defects: bottle cover defect, bottle defect, label defect, label, and process defect. The process variability was highly fluctuated and identified about 28 data out from control lines. Based on the capability process was measured and gained sigma level in 3.76. Some recommendations were set to reduce the variability of the process.

1. Introduction

The advancement of technology and business competitiveness pushed many companies to compete to produce a high-quality product. In a competitive environment, the fundamental goals of the company are to survive in the market for the long term. The companies survived in this competitiveness era, only those that can satisfy the customer by improving the quality product, enhancing cost efficiency, and increasing productivity and delivery.

Quality had become a critical factor in determining the quality of products and affected the success of the industry in the national market and international market as well. Quality control effectiveness depends based on raw material, resources, transformation process, methodology, and supported equipment/machine integrated into a system. Maintaining product quality means ensuring product met the specifications, fitted requirements, and customer satisfaction. It broadly influences the effectiveness of the process, output productivity, and efficiency in resources used. Some methods

Six Sigma is defined as one tool in integrating a systematic way in improving the quality of the production process and satisfy given needs. Customer satisfaction has a positive relationship significantly to financial benefit, company growth, equality, and productivity [1]. Furthermore, it is essential to the industry to consider the implementation of six Sigma in supporting sustainability company growth. However, the application of quality control by using scientific approach (Six Sigma) is not yet implemented comprehensively and consistently in developing countries particularly in the context of Small Medium Enterprises (SME), so that there were huge losses due to high defects in production process [2][3][4]. Instead of the latest economic development, SME contributed to the resilience and economic growth of the country and even the world. Besides that, SME was expected to enhance the quality of production to meet demand and satisfy customers [3]. For that, the development of quality studies in the SME sector is an important thing to investigate implement.

Effective quality control will produce products that can meet quality standards that are based on individual specifications and set by the company. Besides, the factors that cause product failure will be minimized, and consumer confidence and satisfaction can be maintained and enhanced. To support effective quality control, various quality methods have been developed both partially and thoroughly. Various types of methods are developed by companies to produce better quality products. Six Sigma is a term that was first developed by the Motorola Company in the mid-1980s, which emphasizes process improvement to reduce variability and make general improvements. The Six Sigma, quality improvement process includes the Define, Measure, Analyze, Improve, Control (DMAIC) process. Then the control is carried out by analyzing the causes of disability and seeking continuous improvement with the Kaizen approach.

MKM company is a small to mid-size enterprise (SME) engaged in the production of bottled drinking water. The primary material used by the company is water from wells in the hills. Several industries use water as the primary raw material, such as pharmaceuticals and food and beverage products [5][6]. The material is then processed to meet the requirements of the Indonesian National Standard (SNI) using specific tools so that the water is suitable for drinking, where the water is clear, colorless, odorless, not turbid and there are no other foreign objects such as stones or other foreign objects. In addition to water as the primary raw material, packaging plays an essential role in a product. The packaging functions as a container to hold and place the primary material so that it can be distributed. Besides, packaging must be able to build uniqueness and attractive to consumers. A good packaging must be able to meet several factors, namely: 1) must be able to protect the product from potential that can damage the product such as the environment and others, 2) economical design, 3) distribution, packaging is easy to distribute, 4) communication, packaging must be able to communicate product, 5) ergonomic, easy to carry, hold, open, and easily take the product, 6) esthetic, the packaging must look attractive, 7) identity, show the uniqueness of different product identities. Product packaging design gives important meaning to brand values and influences consumer purchasing decisions. [7]. Besides aesthetics and identity, it can be concluded simply from several other factors related to the durability of the packaging during the production process until the product is consumed. Furthermore, there is a substantive relationship related to perception values (brand values/awareness, brand association, and brand loyalty) and purchasing decisions. The higher the perceived value, the higher the potential for consumers to buy the product [8].

Bottled drinking water production activities carried out by MKM has been established since 2008. There are four types of packaging sizes for drinking water products manufactured and marketed by the company, namely 220 ml glass, 600 ml bottles, 1.5-liter large bottles, and 19-liter gallon packaging. There are four production machines: 2 machines for the production of 220 ml glass packaging sizes, one machine for 660 ml and 1.5-liter packaging bottle sizes, and one machine for 19-liter gallon sizes. The production capacity of machines for 600 ml bottle packaging is capable of producing approximately 10,000 pcs in one working day. However, the production results often have several products that experience packaging defects in the production process. By implementing Six Sigma in this company, it is expected that it can improve and enhance product quality and can reduce inefficient costs due to defective products.

Related to the above background, research and discussion to improve the quality of processes need to be carried out by applying an integrated and effective quality control method. For this reason, It is aimed that the company can identify the type of defect and the causes of defects in the product during the production process. Also, it is intended to identify and reduce the variability in the packaging process to minimize the risk of product defects. Furthermore, to improve the quality process.

2. Methodology

Data was collected by getting information about the company's historical data, observations, and interviews. Based on the information and data collected, the product that needs to be examined is the production of 600 ml of bottled mineral water. Then, data collected was processing by a simple statistical method to identify, map variability, and analyze the quality control level of the production process

Quality control is a sub-system of quality management (Quality Management/QM). Quality control activities include sampling, testing, monitoring, and inspection, as well as reporting the results of activities. The results of testing, monitoring, and inspection are managed as a basis for quality assurance (QA). They are used as a basis for product release to the market or quality improvement (QI). Quality control in the bottled water industry starts from the water source, each stage of the process, to the release of the product to the market. The Bottle Drinking Water (BDW) processing scheme, in principle, must meet and comply with the national regulation [9].

Six Sigma is an essential tool for production management to maintain, improve, maintain product quality, and especially to achieve quality improvement towards zero defects. In this study, the application of quality control used is the Six Sigma method that should go through five stages. But in this study, only four stages of analysis are used, namely Define, Measure, Analyze, Improve at SME. The study was carried out at the Production Department, where the production process of 600 ml bottled mineral water is very concerned about the quality of its production before it is packaged and distributed. As an object of research in the process of improving quality, at this stage, data collected from 600 ml bottled mineral water produced for 45 days.

The Head of Production is the person who is fully responsible during the production process. There are several stages during the production process, starting from the water quality checking stage to the packaging process stage. The first stage of the production process is the QC section checks the pH of water in a raw water tank by taking samples and testing them in the laboratory. After that, the production department checks and settings all production machines to meet the standards. After everything is up to standard, raw water is channeled to the sand filter tank. At this stage, raw water undergoes three processes of filtering foreign matter, carbon and cartridge. After passing through the screening process, raw water flows for the ozonization process. Furthermore, raw water enters the mixing process and finally into the finished water tank.

The next stage is the production process. The water is flowed into the machine through a UV lamp along with other materials. Then the filling process will be carried out to the packaging bottle by the filling machine, and then several samples will be taken to be tested in the laboratory. If the laboratory results meet the standard, then the next process of bottled water enters the lid (labeling) process, then goes into quality testing. The final stage is the finishing process. Products that passed quality testing will be directly packed into a box, then arranged in a pallet, then stored in a warehouse and ready to distribute.

3. Result

3.1. Define

Define is the stage of defining quality problems in the production of 600 ml of bottled mineral water. At this stage, the defective product is defined as the cause. At the stage of identifying the problem, there will be explained several causes of defects, which are potential causes in producing 600 ml bottled mineral water products. The types of defects are identified through interviews with the head of the production, employees, and direct observation in the production area. The types of defects can be seen in Figure 1 and describe as follows:

Defect bottle (A)

In this type of defect, the body of the bottle is dented and damaged on the head of the bottle. This is caused by the wrong machine flanking the head of the bottle so that the head of the bottle is damaged, and the poor quality of the bottles imported by the supplier so that during the production process takes place, the bottle becomes dented.

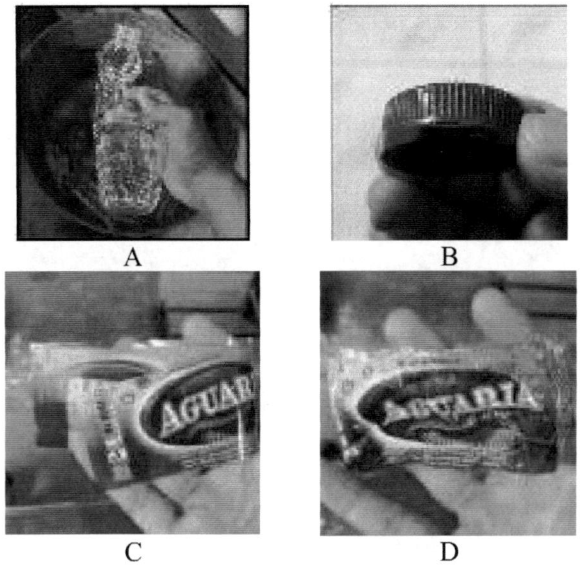

Figure 1. Defects of packages

Defect bottle cup (B)
Bottle caps are not appropriately attached to the bottle body during the sealing process. This is usually caused by the machine or the position of the bottle cap, which is not quite right during the process of sealing the bottle.
Defect label supply (C)
In this type of defect, glue from the label comes off. This is caused by the quality of the labels that are not good. Other causes are also common because workers are not careful enough to put labels.
Defect label process (D)
In this type of defect, the label shrinks and almost melts. This is because the heat in the engine is too high and also due to negligent workers inputting the label in the right position.

3.2 Measure

At the measurement stage, it is useful to validate problems, measure or analyze problems from existing data. The measure is an action that aims to measure the dimensions of product performance, processes, and other activities. In this stage, Critical to Quality (CTQ) priorities and process stability and process capability will be determined and sorted.

The output of this stage was resulted in measuring the quality level by involving DPMO (Defect Per Million Opportunities) as a base in identifying industry performance on the quality side. In this stage, data were collected from 45 subgroup samples in 45 days. Data then plotted to the Pareto diagram to determine the critical to the quality of defect as in Figure 2. From this Pareto diagram, it shows that the highest defect is a bottle cap defect of 17,596 pcs with a percentage of 47.9%. Then it is followed by bottle defects of 17,504 pcs with a percentage of 47.7%. Next, the label defects were supplied with 814 pcs, and the percentage was 2.2%, and the last was the process label defects with 798 pcs and the percentage of 2.2%. Therefore, repairs that must be done first are defective bottle caps. This is because the defect is the most significant defect, among other types of defects.

3.3 Capability Process

From Figure 3 above, it appears that 28 data are outside the control limit. Therefore, it must be eliminated so that the process is stable so that new revision 1 data is obtained. To get new revision 1 data, first, the process stability is recalculated until the process data is within the control limits.

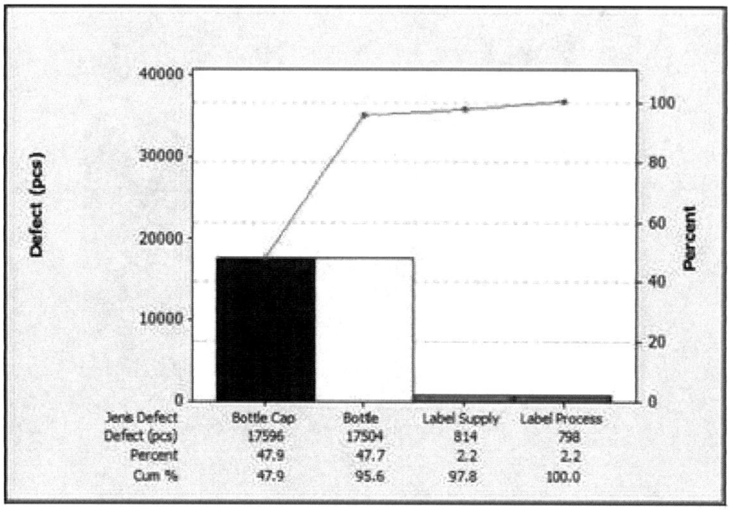

Figure 2. Critical to the quality

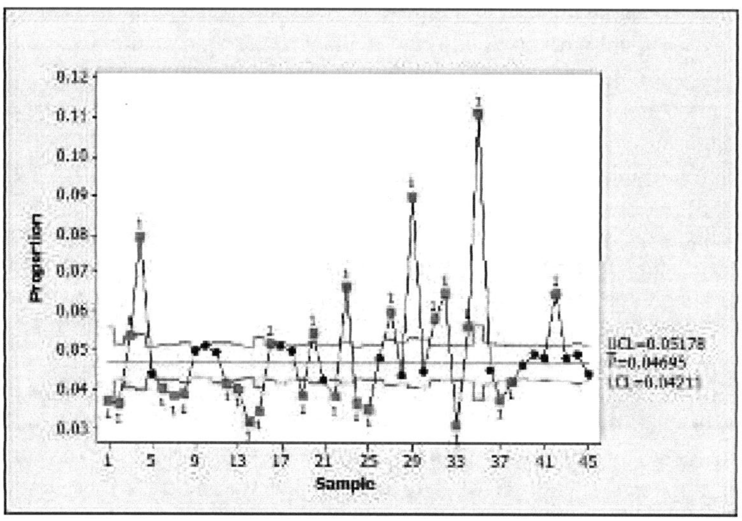

Figure 3. P chart of the defect

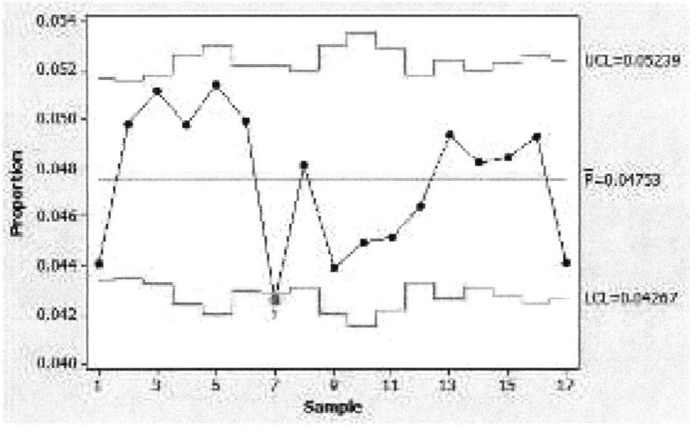

Figure 4. P chart of the defect (II)

From Figure 4, it can be seen that there is still data that is outside the control limits, which is on the 21st day (February). Therefore, it must be eliminated so that the process is stable so that new data can be revised II. To get revised II new data, the process stability is recalculated first to get the process data that is within the control limits.

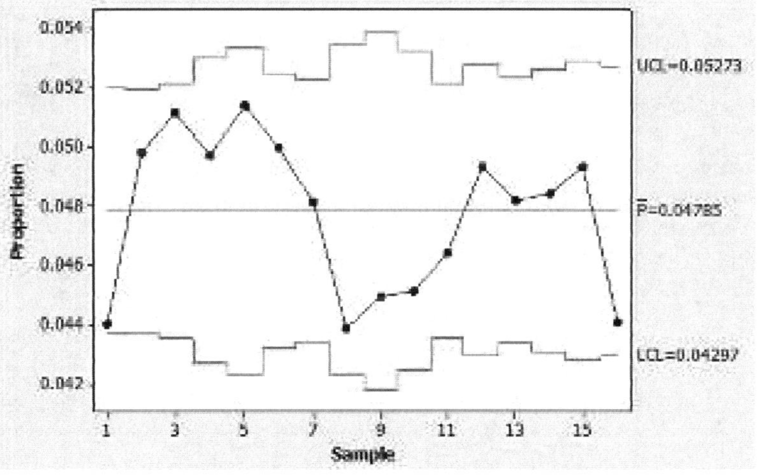

Figure 5. P chart of the defect (III)

From the results of calculations and Figure 5, it can be seen that the production of 600 ml bottled mineral water has a sigma level of 3,756 or 3.76, with possible damage of 11,937 pcs for a million materials. The calculation shows the DPMO and Sigma levels are quite good. With good results in the industry average in Indonesia, it still needs to be controlled continuously so that it will show the DPMO pattern of the failure of the production of 600 ml bottled mineral water that continues to fall and the capability of Sigma that continues to increase.

3.4 Analyse
Causes and effects were analyzed using the fishbone diagram in particularly for data that existed jumped out from LCL and UCL. The identification process in clarifying causes and effects was conducted by observation and interviewing the personnel and key people involved in the process directly. Five factors caused the defect of raw material. Some of the causes of defects in the packaging process in terms of several factors, namely:

1. Human: Lack of knowledge of workers related to the packaging and production process in detail, especially related to production machines. Besides that, the fatigue factor on workers has an impact on decreased concentration in the process of installing labels and packaging. This also becomes a dominant factor in the packaging process.
2. Machine: Engine adjustment is unstable. This is usually because the bottle gripper rinser makes the transfer of the bottle not right, so the bottle head is damaged. The engine pump is not stable. This makes the volume of water when charging is not up to standard.
3. Method: Bottle caps are put into the machine without being sorted, and the packaging process was still done manually by the operator.
4. Material: The quality of the material does not meet qualifications such as defective bottle caps from the supplier, so it makes a dent bottle during the production process.

3.5. Improve and Control
The variability that occurs during the production and packaging processes is generally caused by product defects during the process. Furthermore, reducing variability can be carried out by reducing the number of product defects that occur during the production and packaging processes. After identifying the causes and impacts that occur, several proposals are developed to emphasize variability or increase the improvement of the production process. Some of the recommendations are as follows:

1. Giving special assistance for workers during the production and packaging process and provide intensive training to improve their knowledge and skills related to the production and packaging process.
2. Conducting a minimum material checking sampling test by random sampling of the total material ordered periodically.
3. Replacing bottle rinser gripper, which is not feasible for use or damaged so that the measurement of water content can be accurate.
4. Conducting preventive maintenance to machines regularly and replace parts that do not meet the standard.
5. Increasing attention to the sorting process of packaging, labels, and bottle caps during the receiving process so that it can detect early the process failure in the production and packaging process.

4. Discussion
The empirical results show the picture of what happened in SMEs related to huge losses along the process caused by uncontrolled variability with four major defects. Production of 600 ml bottled mineral water has a sigma level of 3,756 or 3.76, with a possible defect of 11,937 pcs for a million productions. The measurement shows the DPMO and Sigma levels are relatively good. Besides, it still needs to be controlled continuously to ensure continuous improvement.

Weaknesses in quality control generally caused by a lack of knowledge related to effective quality control methods. Therefore, tools and six sigma methods as an effective quality methodology in reducing variability and six Sigma is still needed to popularize in SMEs [10]. This is in line with the empirical finding in this study that actually in SMEs, human error still dominated in high variability along the process [11]. Besides that, mechanical maintenance had become the other concern to improve in reducing the variability. Uncontrolled variability, generally caused by specific cause (assignable cause) such as failure modes in the process, human error, and lack of knowledge.

Philosophically, Six Sigma is an approach to study, analyze, and reduce process variations. The concept of reducing variation and using statistical control tools can be applied in many areas where real understanding is needed by making direct observations on the actual process [12] [13].

The company is expected to implement or consider the Six Sigma method to reduce the variability due to defective products in the next production process. The use of the Six Sigma method needs to be carried out for the quality improvement process because the company can find out the value of DPMO and its Six Sigma level as reference material to determine the feasibility of the ongoing production process.

Aside from the strengths of six Sigma, there is certainly no perfect methodology. Likewise, this methodology has limitations where this method is so dependent on central evaluation and critical testing of the entire capability process, such as requires firm statistical experimentation [14]. In addition, it requires total employee engagement, and of course, the most important is related to costs.

5. Conclusion

A possible defect rate of 11,937 products per million productions indicates that it needs ongoing and continuous control. The main thing that requires attention in reducing the number of defects is to eliminate the potential for human error in the production and packaging process. Because human error is the leading cause in causing uncontrolled variability. By eliminating human error, it is expected to reduce uncontrolled variability in the production and packaging processes. Also, on the other hand, it can maintain the level of production quality and reduce the cost of product defects.

This study is still limited to specific case studies in SMEs and without considering work environment factors as developed previous work. For this reason, environmental considerations and expansion of application are opportunities for future research development.

References

[1] N. Venkatesh and C. Sumangala, "Success of manufacturing industries – Role of Six Sigma," vol. 05002, pp. 1–15, 2018.

[2] T. K. Chakrabortty, Ripon Kumar; Biswas and I. Ahmed, "Reducing Process Variability by Using DMAIC Model: A Case Study," vol. 7, no. 1, pp. 127–140, 2013.

[3] T. Y. M. Zagloel, R. Ardi, and W. Poncotoyo, "Six sigma implementation model based on critical success factors (CSFs) for Indonesian small and medium industries," vol. 04017, 2018.

[4] D. A. Purnama, R. C. Shinta, and V. N. Helia, "Quality improvements on creative industry by using Six Sigma : a study case," vol. 01088, 2018.

[5] D. Rimantho and D. M. Mariani, "Penerapan Metode Six Sigma Pada Pengendalian Kualitas Air Baku Pada Produksi Makanan," 2013.

[6] W. Soedarmadji, S. Surachman, and E. Siswanto, "Penerapan Konsep Green Manufacturing Pada Botol Minuman Kemasan Plastik," *J. Eng. Manag. Industial Syst.*, vol. 3, no. 2, pp. 76–81, 2015.

[7] S. A. Immawati, "The Product Packaging Design And Advertisement Attraction On Brand Awareness And Its Impact On Repurchasing Teh Botol Sosro Among Young Children At Tangerang City," *Int. J. Entrep. Bus. Dev.*, vol. 2, no. 1, pp. 145–159, 2018.

[8] A. O. Oke, P. Kamolshotiros, O. Y. Popoola, M. A. Ajagbe, and O. J. Olujobi, "Consumer behavior towards decision making and loyalty to particular brands," *Int. Rev. Manag. Mark.*, vol. 6, no. 4, pp. 43–52, 2016.

[9] M. Perindustrian, D. A. N. Perdagangan, and R. Indonesia, "Persyaratan Teknik Industri Air Minum Dalam Kemasan dan Perdagangannya." 2003.

[10] P. Grudowski and P. Waszczur, "Characterization of Small and Medium Enterprises (SMES) of Pomeranian Region in Six Sigma Methodlogy Application," vol. 7, 2011.

[11] D. P. Sulistyowati and S. Anggarini, "Pengendalian Kualitas Pengalengan Jamur dengan Metode Six Sigma di PT Y , Pasuruan , Jawa Timur Quality Control of Mushrooms Canning using Six Sigma Method at Company Y , Pasuruan , East Java," vol. 6, no. 1, pp. 1–7, 2017.

[12] V. Narula and S. Grover, "Application of six sigma DMAIC methodology to reduce service resolution time in a service organization," vol. 1, pp. 43–50, 2015.

[13] S. Habib, A. Rahmati, A. Kazemi, and M. Saidi-mehrabad, "A New Fuzzy Method for

Assessing Six Sigma Measures," vol. 13, pp. 39–47, 2013.
[14] K. Jenab, C. Wu, and S. Moslehpour, "Design for Six Sigma: A review," vol. 8, pp. 1–18, 2018.

IOP Conf. Series: Materials Science and Engineering **885** (2020) 012056 doi:10.1088/1757-899X/885/1/012056

Logistic regression model for user preference of online parking: A study case of Sidoarjo

E Widodo[1], R N Harnaningrum[1], Suparno[1] and A Santoso[2]

[1] Department of Industrial Engineering Institut Teknologi Sepuluh Nopember (ITS), Surabaya, Indonesia
[2] Department of Industrial Engineering University of Surabaya, Surabaya, Indonesia

Email: erwin@ie.its.ac.id,

Abstract. This paper investigated the factors affecting user preference of online parking in Sidoarjo Regency of Indonesia, using a binary logistic regression model approach. This model could predict user preference between online and offline parking system. Through a nonprobability convenience sampling technique, structured questionnaire was used to collect primary data from 384 respondents in three different parking location of Sidoarjo. The result of the binary logistic regression model revealed that gender, age, parking frequency, and parking duration are the four main factors that are significantly affect user parking decision. This will be definitely helpful to the government of transportation department as parking policy maker to evolve strategies in encouraging people to use online parking.

1. Introduction

The number of motor vehicles in Indonesia is increasing significantly from year to year and caused many problems such as traffic jams and parking problems. In a review of 16 studies conducted in 11 cities in the United States, an average 30% of the causes of traffic congestion occurred were parking spaces [1]. This happens because of the difficulty of finding parking spaces when the parking area is crowded, driver must search for available parking lot when they want to park their vehicles, this causes frequent queues of vehicles. The good governance of parking system is needed to deal with these problems.

Parking policies are one of the most effective ways that urban planners and government as policy makers to manage travel demand and traffic in the city center. In many countries, governments use parking policies as a way to reduce urban road traffic where an effective parking system is one that can reduce congestion. Local governments in Indonesia have begun to think about how to overcome parking problems in their areas by making parking policies that can provide optimal regional income and can also solve parking problems in Indonesia. Parking policies that has been developing in the industrial revolution 4.0 is an online vehicle parking reservation system. This system can provide smart parking management solutions by providing information on the availability of parking facilities and parking reservation systems [2]. The description of an online parking system where this system runs on a mobile platform and provides parking location information and available parking slots to users, making it easier for parking users to find an empty spot to park a vehicle. User can make a reservation of parking slot through the application and choose any payment method including cashless. There are certain things that must be research by parking policy makers when implementing an online parking system in Indonesia. Analysis of user preference should be performed to determine the factors that influence the user in choosing online parking. So that parking policy makers can consider the strategies that must be taken to encourage people to switch from offline parking systems to online parking systems.

The case study in this research was conducted in Sidoarjo regency as in 2019 the Sidoarjo government had conducted a feasibility study on the Online Parking System as a parking system solution based on mobile apps. In 2018, parking income contributed Sidoarjo revenue only IDR 28,176,793,500 of the expected revenue IDR 102,146,595,652 [3]. This indicate that there is a leakage of parking revenue in Sidoarjo. To overcome this problem the government of Sidoarjo proposed an online parking system which is trusted to be able to facilitate the public in accessing local government parking facilities and could increase regional revenue since the parking payments go directly to the local treasury.

A certain previous research has explored the factors that influence parking user decisions. Most studies have shown that user parking decision is influenced by demographic variables. For example, Shiftan and Burd-Eden [4] found that younger individuals were more likely to change travel behavior in response to parking restrictions such as increased parking costs or reduced parking availability. Research conducted by Golias et al. [5] show that parking costs have the most important impact on alternative choices of on-street or off-street parking, besides all the other variables that have a significant impact on time-related parking choices. Zong et al. [6] shows that an increase in parking fees will shorten the duration of parking; tourists for work purposes will have a longer parking duration; and the greater the number of passengers, the higher the probabilities of on-street parking. But there is no research that has explored the factors that influence users in choosing online parking, where this research will analyze that problem. The results of previous research are used as a reference in determining factors that are expected influencing user's decision in choosing a parking system.

Logistic regression is one model for estimating the relationship between categorical response variables with one or more continuous or categorical predictor variables. In cases where the response variable is a qualitative variable which is binary then binary logistic regression is used. The binary response variable consists of two categories namely "yes (success)" and "no (failure)" and denoted 1 = "success" and 0 = "fail [7]. Therefore, binary logistic regression is used to analyze the problem in this research, which is to investigate the factors affecting user preference of parking system where the response variable is binary namely "online parking" and "offline parking". This will be definitely helpful to the government of transportation department as parking policy maker to evolve strategies in encouraging people to use online parking.

2. Literature Review

2.1 Online Vehicle Parking Reservation System

The online vehicle parking reservation system is an online-based application for parking lot reservations for vehicles and parking data processing. This system can provide smart parking management solutions by providing information on the availability of parking facilities and parking reservation systems [2]. Vehicle drivers who really need a parking lot in a short time can access this system without any other vehicles blocking. This is quite the opposite of conventional parking systems which take quite a long time. In addition, the existence of online and cashless payment patterns makes it easier for drivers to process parking briefly.

2.2 Factors Influencing Parking Decisions

A review of the various factors that influence the parking decision that has been identified in the literature will be carried out in this section. Demographic variables such as age, gender, parking frequent, parking duration, and level of education is demographic variables that expected affect the parking decisions. Shiftan and Burd-Eden [4] found that younger individuals were more likely to change travel behavior in response to parking restrictions such as increased parking costs or reduced parking availability. Mo et al. [8] found that men were more likely to park for short periods (<1 hour) than women, even though both sexes parked most often between 1 and 3 hours. Thompson et al. [9] found that tourists with high parking frequencies to an area tend not to need information about waiting times and parking location. Golias et al. [5] who examined the parking duration, parking costs, search time and egress time, and found an increase in preferences for off-street car parking. The influence of an individual's level of education on parking choice was investigated by Anastasiadou et al. [10] who found university graduates were willing to pay, on average, higher parking charges than non-university educated individuals.

2.3 Logistic Regression Approach

Logistic regression is used if the response variable is categorical (nominal or ordinal) with continuous and categorical predictor variables [11]. Logistic regression is a regression method that describes the relationship between a response variable with one or more predictor variables, where the response variable is dichotomous (nominal or ordinal scale with two categories) or polychotomous (having a nominal or ordinal scale with more than two categories). In cases where the response variable is a qualitative variable which is binary or dichotomous, then binary logistic regression is used. Dichotomous variables are variables that have only two possible values, for example success and failure. Binary logistic regression is a logistic regression analysis between predictor variables and response variables consisting of two categories (dichotomous). The response variables consisted of two categories namely "yes (success)" and "no (failed)" and denoted 1 = "success" and 0 = "failed". The response variable (y) follows the Bernoulli distribution with the probability function as follows [7].

$$f(y_i) = \pi(x_i)^{y_i}\left(1 - \pi(x_i)\right)^{1-y_i} \text{ with } y_i = 0,1$$

$$If\ y_i = 0, then\ f(0) = \pi(x_i)^{0}\left(1 - \pi(x_i)\right)^{1-0} = 1 - \pi(x_i)$$

$$If\ y_i = 1, then\ f(1) = \pi(x_i)^{1}\left(1 - \pi(x_i)\right)^{1-1} = \pi(x_i)$$

The general form of the Logistic Regression model with p independent variables is as follows:

$$\pi(x) = \frac{exp\left(\beta_0 + \beta_1 x_1 + \beta_2 x_2 + \cdots + \beta_p x_p\right)}{1 + exp\left(\beta_0 + \beta_1 x_1 + \beta_2 x_2 + \cdots + \beta_p x_p\right)}$$

By using logit transformation, the binary logistic model can be written as follows:

$$g(x) = \beta_0 + \beta_1 x_1 + \beta_2 x_2 + \cdots + \beta_p x_p$$

3. Research Methodology

The purpose of this study is to predict the user preference of an online parking system when his/her exact demographic profile is known. At this stage the data was collected using primary data by conducting a survey directly to 384 respondents in three different on-street parking location of Sidoarjo which is managed directly by the city government. Respondents in this study were user who had parked their motorcycle on-street in Sidoarjo Regency. A nonprobability convenience sampling technique has been adopted for identifying samples from the population. A specific questionnaire was developed for this study and the same was used to collect data from the respondents.

The researcher has used the binary logistic regression as a part of model building where the dependent variable is a dichotomous variable like preference of parking system (online or offline parking system). The logistics regression analysis could identify the relation between user preference of parking system and the factors that influence user preferences to choose online or offline parking system. The dependent variable is analyzed with respect to the 5 independent variables presented in Table 1. The response variable is user decision to choose parking system which assigns the value "0" to the "offline parking" answer and the value "1" to the "online parking" answers. Hence, the results are interpreted with respect to these values in the following section. The predictor variables used in this study can be seen in Table 1.

Table 1. Predictor variables

Predictor Variable (X)	References	Categories
Age(X_1)	Shiftan and Burd-Eden [4]; Golias et al. [5]; Van der Waerden et al. [9]; Anastasiadou et al. [10]; Tsamboulas [12]; Teknomo and Hokao [13]	1. 20-29 years 2. 30-39 years 3. 33-42 years 4. 43-52 years 5. > 53 years
Gender(X_2)	Golias et al. [5]; Mo et al. [8]; Tsamboulas [12]	1. Male 2. Female
Parking Frequent(X_3)	Golias et al. [5]; Tsamboulas [12]; Van der Waerden et al. [13]	1. < 30 times/month 2. 30-45 times/month 3. 46-60 times/month 4. >60 times/month
Parking Duration(X_4)	Shoup [1] ; Golias et al. [5]; Tsamboulas [12]; Van Ommeren et al. [14]	1. < 1 hours 2. 1-2hours 3. 2-4hours 4. 4 hours
Education Level(X_5)	Anastasiadou et al. [10]; Salomon [15]	High School Diploma bachelor's degree

4. Result and Discussion

4.1. Demographic Characteristics of Respondents

The demographic characteristics of respondents in this study were classified by age, gender, parking frequency, parking duration and education level. This demographic is cross tabulated with user preference on parking system. The results of this cross-tabulation are summarized in Table 2.

Table 2. Crosstab preferences with participant demographics

			Parking System Preference		Total
			Offline	Online	
Age	<19 years	Count	13	30	43
		%	30.2%	69.8%	11.2%
	20 - 29 years	Count	34	80	114
		%	29.8%	70.2%	29.7%
	30 - 39 years	Count	31	39	70
		%	44.3%	55.7%	18.2%
	40 - 49 years	Count	36	27	63
		%	57.1%	42.9%	16.4%
	>50 years	Count	49	45	94
		%	52.1%	47.9%	24.5%

Table 2. Crosstab preferences with participant demographics (*continued*)

			Parking System Preference		Total
			Offline	Online	
Gender	Male	Count	101	115	216
		%	46.8%	53.2%	56.3%
	Female	Count	62	106	168
		%	36.9%	63.1%	43.7%
Parking Frequency (/month)	<30	Count	56	69	125
		%	44.8%	55.2%	32.6%
	30-45	Count	81	130	211
		%	38.4%	61.6%	54.9%
	46-60	Count	14	13	27
		%	51.9%	48.1%	7.0%
	>60	Count	12	9	21
		%	57.1%	42.9%	5.5%
Parking Duration	< 1 hours	Count	41	66	107
		%	38.3%	61.7%	27.9%
	1 - 2 hours	Count	32	51	83
		%	38.6%	61.4%	21.6%
	2 - 4 hours	Count	28	54	82
		%	34.1%	65.9%	21.4%
	> 4 hours	Count	62	50	112
		%	55.4%	44.6%	29.2%
Education Level	High School	Count	134	175	309
		%	43.4%	56.6%	80.5%
	Diploma	Count	5	7	12
		%	41.7%	58.3%	3.1%
	Bachelor	Count	24	39	63
		%	38.1%	61.9%	16.4%
	Total	Count	163	221	384
		%	42.4%	57.6%	100.0%

Table 2 provides information that out of the 384 respondents in this study, based on classification of age groups '20-29 years' (29.7%), followed by age '>50 years' (24.5%), 30-39 years (18.2%), 40-49 years (16.4%) and '<19 years' (11.2%). 216 respondents (56.3%) were male and the remaining 168 respondents (43.7%) were female. Based on the classification of parking frequency in a month, 54.9% of respondents generally parking 30-45 times and only 5.5% respondents parking more than 60. The findings for parking duration indicated that the highest number of responses occurred for the longest duration '>4 hours' with 82 respondents (29.2%) selecting this option. This was followed by the shortest duration of '< 1 hours' with 107 respondents (27.9%). Classification by education level indicates that the majority of participants are at education level of high school (80.5%), followed by bachelor (16.4%), and diploma (3.1%).

Table 2 generally indicates that the majority of participants in this study had a tendency to choose to online parking system (61% of participants). Moreover, it also indicates that younger user, on contrary with older user in this study prefer to choose online parking.

4.2. Logistic Regression Results

The researcher has developed a logistic regression model for predicting user preference based on demographic factors that influence user to choose online or offline parking system. For selecting the significant independent variables, the researcher has used backward likelihood ratio method of logistic regression analysis. Backward likelihood ratio has been selected that include all candidate variables at initial step and ended with the most significant variables. This method is removal testing which based on the probability of the likelihood-ratio statistic based on the maximum partial likelihood estimates. The final iteration results of the logistic regression model are given in Table 3.

Table 3. The results of binary logistic regression

Parameter	B	S.E.	Wald	df	Sig.	Exp(B)
Age_X1			17.183	4	0.002	
Age_X1(1)	0.905	0.411	4.849	1	0.028	2.472
Age_X1(2)	0.843	0.303	7.739	1	0.005	2.323
Age_X1(3)	0.174	0.332	0.274	1	0.601	1.19
Age_X1(4)	-0.365	0.343	1.135	1	0.287	0.694
Gender_X2(1)	-0.468	0.226	4.285	1	0.038	0.626
Parking_Frequency_X3			9.067	3	0.028	
Parking_Frequency_X3(1)	0.393	0.502	0.614	1	0.433	1.482
Parking_Frequency_X3(2)	1.003	0.496	4.091	1	0.043	2.726
Parking_Frequency_X3(3)	0.292	0.619	0.223	1	0.637	1.339
Parking_Duration_X4			12.814	3	0.005	
Parking_Duration_X4(1)	0.968	0.307	9.936	1	0.002	2.632
Parking_Duration_X4(2)	0.972	0.324	8.989	1	0.003	2.644
Parking_Duration_X4(3)	0.638	0.323	3.903	1	0.048	1.892
Constant	-1.036	0.548	3.569	1	0.059	0.355

Table 3 shows the logistic regression results, Wald's test and the odd-ratio ratio of each predictor variable in predicting user preferences of parking system. The significance level used in this study is 5%. At this level, age, gender, parking frequency and parking duration have a significant effect in predicting user preferences of parking system since the p-value (sig) is less than 0.05 while the variables of education level have no significant effect in the proposed model.

Table 3 also displays odd ratio value which is an important information to explain the influence of the independent variables on the increasing or decreasing of probability to occur the event measure by the dependent variable. For example, based on odds ratio of variable X2(1) (that is predictor variable, category 1: gender is male) is 0.626. These results indicate that under conditions where other variables in this study are the same, male user who prefer to choose online parking system application is 0.626 times more than female user. Table 3 shows the odd-ratio value of age decrease by the increasing age. It can be concluded that the increasing of user age causes a shrinkage of user preference to choose online parking system.

5. Conclusion

This study results demonstrate that user preferences of a parking system both online and offline are influenced by the demographic variables of a parking user. The logistic regression was created to predict user preference of online parking system. Through this model, the factors affecting the user preference of parking system were evaluated. Factors influenced by user age, gender, parking frequency and parking duration, were found to be significant. The study concluded that female more prefer to park online than male. Moreover, the study reveals that the younger user is, the more the chance they will choose to park online. Furthermore, no significant difference in education level group is revealed. As implications, this study identifies user parking system preferences and their tendency toward online parking. Government of transportation department as parking policy maker can be better prepared to evolve strategies to encouraging people using online parking based on which factors have a significant influence on the preferences of users choosing online parking.

Acknowledgement
The authors would like to thank Directorate General of Learning and Student Affair, Ministry of Research Technology and Higher Education for giving us the financial support in this publication as implementation of grant program in the postgraduate improvement quality by the contract number (No.036/B4/PPK-SPK/VII/2019).

References

[1] Shoup D C 2006 Cruising for parking *Transp. Policy* **13** 479–86
[2] Sadhukhan P 2017 An IoT-based E-parking system for smart cities *2017 International Conference on Advances in Computing, Communications and Informatics (ICACCI)* (IEEE) pp 1062–6
[3] Azta F C 2019 *Strategy Analysis of Parking Management By Using Game Theory Approach (Case Study: Kabupaten Sidoarjo)* (Institut Teknologi Sepuluh Nopember)
[4] Shiftan Y and Burd-Eden R 2001 Modeling response to parking policy *Transp. Res. Rec.* **1765** 27–34
[5] Golias J, Yannis G and Harvatis M 2002 Off-street parking choice sensitivity *Transp. Plan. Technol.* **25** 333–48
[6] Zong F, Yu P, Tang J and Sun X 2019 Understanding parking decisions with structural equation modeling *Phys. A Stat. Mech. its Appl.* **523** 408–17
[7] Menard S 2002 *Applied logistic regression analysis* vol 106 (Sage)
[8] Mo Y, Zhang B and Yan K 2008 A study of parking behavior and parking information requirements in Shanghai CBD *Plan, Build, and Manage Transportation Infrastructure in China* pp 673–82
[9] Thompson R G and Bonsall P 1997 Drivers' response to parking guidance and information systems *Transp. Rev.* **17** 89–104
[10] Anastasiadou M, Dimitriou D J, Fredianakis A, Lagoudakis E, Traxanatzi G and Tsagarakis K P 2009 Determining the parking fee using the contingent valuation methodology *J. Urban Plan. Dev.* **135** 116–24
[11] Agresti A 2018 *An introduction to categorical data analysis* (John Wiley & Sons)
[12] Tsamboulas D A 2001 Parking fare thresholds: a policy tool *Transp. Policy* **8** 115–24
[13] Van Der Waerden P, Borgers A and Timmermans H 2006 Attitudes and behavioral responses to parking measures *Eur. J. Transp. Infrastruct. Res.* **6**
[14] Teknomo K and Hokao K 1997 Parking behavior in central business district a study case of Surabaya, Indonesia *Easts J.* **2** 551–70
[15] Van Ommeren J N, Wentink D and Rietveld P 2012 Empirical evidence on cruising for parking *Transp. Res. part A policy Pract.* **46** 123–30

IOP Conf. Series: Materials Science and Engineering **885** (2020) 012057 doi:10.1088/1757-899X/885/1/012057

Recommendations of ergonomic checkpoints and total ergonomics intervention in the pempek kemplang palembang industry

H Setiawan[1] and M Rinamurti[2]

[1)]Industrial Engineering Department, Faculty of Science and Technology Universitas Katolik Musi Charitas, Palembang, Indonesia
[2)]Management Department, Faculty of Business and Accounting Universitas Katolik Musi Charitas, Palembang, Indonesia

E-mail: heri_setiawan@ukmc.ac.id

Abstract. This study evaluates the potential industrial working environment of the Pempek & Kemplang Palembang. Observation of the work environment found a lot of un-ergonomics impact that has not optimal productivity and exceeds the threshold value of aspects of human health. In order to optimize productivity, the production line is required to work quickly in an ECSHE (Effective, Comfortable, Safety, Health, Efficient) ergonomics for workers. Evaluate the work environment by implementing Ergonomics Checkpoints to see whether the working environment has been running well or not. Ergonomics Checkpoints used in this study include material handling aspects, safety, workstation design, lighting, development of human resources and sources of danger contribute to the value of most incompatibility which becomes the top priority in determining suggestions for improvement. While the Total Ergonomics intervention study includes; work posture, un-ergonomics design tools, un-physiological work, classification of subjectivity level of musculoskeletal system is 55-76 (high risk), classification of subjectivity level of fatigue is 66-79 (high category), physical workload at workload level is 125-150 bpm (weight category), and the %CVL category is 78-85% (rather heavy). To be more holistic the 5S method and participatory training in human resources are also used as a method of evaluating the work environment in this industry.

1. Background

Pempek & Kemplang as a typical Palembang food has become one of the 17 mainstay non-oil and gas export commodities in South Sumatera Province. Palembang is famous as a city of pempek & kemplang potentially developing a traditional food industry. It has been able to penetrate the markets of ASEAN countries such as Malaysia, Singapore and Thailand. They are very valuable because from the field data of production and sales centers in Palembang every day as many as 14 tons a day they are sent to domestic and foreign markets [1]. On the other hand, increasing industrial competitiveness is an important issue that needs attention from various parties. Competitiveness about standardization of quality and work systems in an ergonomic production process. Increasing civilization through civilizing the work system that is ECSHE (Effective, Comfort, Safety, Healthy, Efficient) and improving the welfare of the community through participatory empowerment of all workers involved in the industry. Work systems that are not ergonomically designed that can impact ECSHE, enable workplace accidents which actually cause industrial and physical losses.

Figure 1. shows how the working environment conditions. Seen in Figure 1, the working environment conditions that do not support the creation of an ergonomic work environment. The observation of the work environment that has not been ergonomic and ECSHE is therefore an evaluation that is useful to assess the extent and completeness of the work environment. The results of Ergonomic Checkpoints (EC) are synergized with the recommendations from the results of measurement of Total Ergonomics (TE) Intervention. To ensure the research can run well, it is necessary to conduct a literature study to study and answer theoretical questions around research problems where in this study will discuss EC, TE Intervention, and 5S methods.

(a) (b)

Figure 1. (a). The Process of Making Pempek; (b). The Kemplang Drying Process Figure 1. The Proses of Making Pempek and Kemplang

2. Literature Review

2.1. Ergonomic Checkpoints

EC are methods used to assess the current work environment and apply a work environment that is in accordance with working conditions [2],[3],[4]. Easy implementation solutions in practice for improving security, health and working conditions. The stages of how to use EC, are; a) knowing the workplace, b) defining work area to be checked, c) walk-through initials, d) writing your check results, e) selecting priorities, and f) group discussion about the check results. The reviewed checkpoints incorporated in these checklists have common structures as indicated in Table 1. [5], [2], [3], [4].

Table 1. Common Structure of the Checkpoints Incorporated in The Checklists of the Typical Participatory Programs [4]

WISE checklist (44 items)	**WIND checklist (42 items)**	**Construction site checklist (45 items)**	**ISO/TS 20646 checklist (25 items)**
For small enterprise	**For farmers**	**For small sites**	**For muscular load**
Materials handling (8)	Materials handling (8)	Materials handling (7)	Handling work (6)
Machine safety (6)	Workstations (7)	Work at height (7)	Workstations (9)
Workstations (8)	Machine safety (5)	Work posture (5)	Work organization (2)
Lighting (5)	Work environment and control of Hazards (9)	Machine safety (5)	Heat and lighting (2)
Control of hazard sources (6)	Welfare facilities (8)	Physical environment (7)	Design process (1)
Premises (4)	Work organization (5)	Welfare facilities (5)	Resting facilities (1)
Welfare facilities (4)		Emergency preparedness (3)	Work time (2)
Work organization (3)		Work organization (3)	
		Safety organization (3)	

2.2. Total Ergonomics Intervention

TE Intervention is a methodology of intervention studies based on TE, evaluating and analyzing areas of the work environment through stages of 8 ergonomic problems, SHIP approach (systemic, holistic, interdisciplinary, participatory) and appropriate technology (AT). The application empowers all workers in SMEs Pempek and Kemplang Palembang by conducting workshops to identify the conditions and the old work environment, determine priority scale, design tools, environment and work organization, conduct evaluations and improvements/ finishing.

2.3. The Concept of 5S or 5R

Often we find irregularities in the environment in which we are today. When looking at these irregularities, what we feel is discomfort. When the disorder turns into a habit, it affects our work culture going forward. The solution to this irregularity problem is to apply the 5S or 5R concept. The concept of 5R [5-8] stands for RINGKAS, RAPI, RESIK, RAWAT, AND RAJIN is a way to manage the area of the work environment to be a better place. By applying the 5R concept, a work culture will be formed that can improve our efficiency and productivity in work. *a) RINGKAS (SEIRI)* is to distinguish between items that are needed and those that are not needed and then get rid of items that are not needed. *b) RAPI (SEITON)* is arranging items needed in a place that makes it easy to immediately take the item. *c) RESIK (SEISO)* is keeping the work environment and all items in good condition and clean. *d) RAWAT (SEIKETSU)* is to make a standard / condition that supports maintaining the 5R components, which are Ringkas, Rapi, and Resik. And *e) RAJIN (SHITSUKE)* is making the previous 4R a habit for sustainable development.

3. Materials and Methods

This study uses the synergy of EC, TE Intervention and the 5R concept [5], [2], [3], [4], [9] to evaluate and provide recommendations on whether the working environment in the SMEs Pempek and Kemplang Palembang industries has meet the rules of ergonomics that are good or not based on ECSHE related to the conditions of the work environment in the production process area.

Evaluation of current working environment conditions uses a synergy method between EC with stages [3]; a) knowing the workplace, b) defining work area to be checked, c) walk-through initials, d) writing your check results, e) selecting priorities, and f) group discussion about the check results. Assessment aspects include; storage and handling materials, hand tools, machinery design, lighting, premises, hazardous substances and agents, welfare facilities, and work organizations. and TE Intervention with the identification of 8 ergonomic problems (nutrition & calories, posture, muscle power utilization, environmental conditions, time conditions, social conditions, information conditions, and human and machine interaction), simultaneously applying the SHIP approach and applying AT (technical, economical, ergonomic/ healthy, socio-cultural, energy efficient, and not damaging to the environment) [10], [11], [12]. Based on the results of EC, TE Intervention then identification of problems, priority scale of problem solving, implementation of 5R, design of tools, improvement of the environment and work organization, which are then evaluated accordingly or not. The Synergy EC and TE Intervention models applied in this study are presented in Figure 2.

4. Results and Discussion

Data collection on the condition and the old work environment were examined using EC as one of the tools to correct the standards that have been implemented in SMEs Pempek and Kemplang. EC detect at least 11 items, can be seen in Table 2.

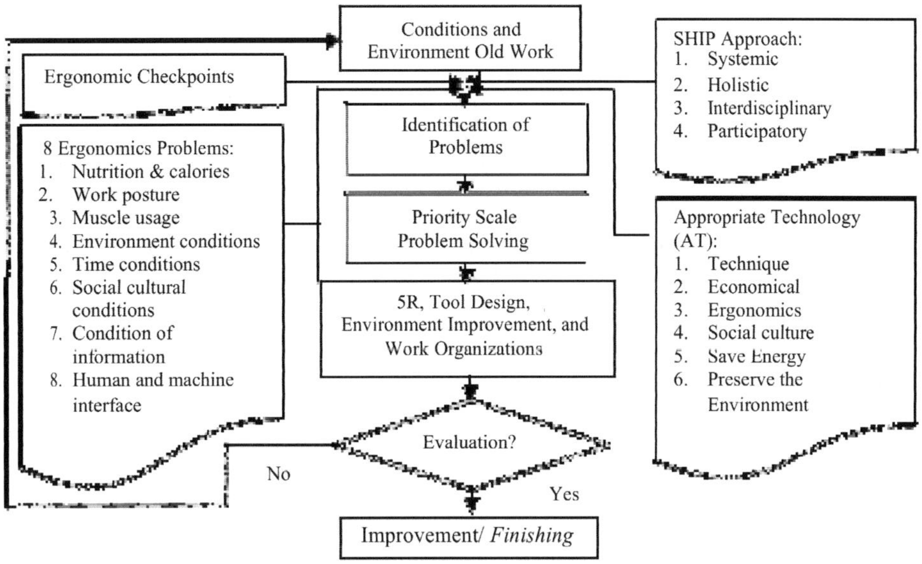

Figure 2. Synergy Model of Ergonomic Checkpoints and Total Ergonomics Intervention

Table 2. Table of Ergonomic Checkpoints aspects

Correction Aspects	Items
Storage & materials handling	8 item
Machine safety	6 item
Workstations	9 item
Heat & lighting	2 item
Work posture	5 item
Physical environment	7 item
Resting facilities	1 item
Work time	2 item
Premises	4 item
Welfare facilities	4 item
Work organization	5 item

Table 3. Ergonomic Checkpoint Observation Results

Aspect	Sub Aspect	Assignment	
		Good	No
Storage & materials handling	8	3	5
Workstations	9	2	7
Heat & lighting	2	1	1
Work posture	5	1	4
Physical environment	7	2	5
Work time	2	1	1
Premises	4	3	1
Welfare facilities	4	2	2
Work organization	5	1	4
Total	**46**	**16**	**30**

From 11 aspects of correction in EC in (Table 2), there were 9 aspects used in this study. These aspects are; storage & materials handling, workstations, heat & lighting, work posture, physical environment, work time, premises, welfare facilities, work organization. The results of filling in EC show in Table 3. The reason for choosing these 9 aspects is because this study only examined the problem of physical conditions and work environment in SMEs Pempek and Kemplang Palembang. While the 2 aspects that were not included in this study, such as the machine safety aspect, it is caused that in this work area workers only use simple machines without having to pay attention to machine safety. And the resting facilities aspect is also not included in the research because this aspect is not very urgent needed in SME, since resting facilities simply exist.

From the 46 items, 30 of them need repairs. While of the 30 items, there are priority scale problems, namely; storage and material handling (urgent), work posture (urgent), physical environment (essential), and work organization (important). Synergy of EC with TE Intervention harmonizes ILO 2010 rules with interventions to improve working conditions and environments based on TE. The application of the TE Intervention method, begins with observation and evaluation of old working conditions and environments with an approach to 8 ergonomics problems and AT to identify problems, determine priority scale of problem solving, then conduct design studies of tools, environmental improvement studies and work organization studies. One improvement in the un-ergonomic problem that is urgent related to storage & material handling and work posture is to design a saving rack trolley tool ergonomic that can minimize workload due to incorrect posture work.

4.*1. Results of Application of the 5R Concept*

While improving the conditions of the work environment by applying 5R, are as follows:
a) *RINGKAS* is the activity of sorting out items that are needed and not needed (to be disposed of) based on Figures 3 & 4, piles of wood and materials that cannot be used anymore are immediately cleaned because they will become a nest of disease and damage the work environment. The recommendation is to place the pile of wood and material into a more representative container or warehouse. b) *RAPIH* is how workers arrange a good layout so that the work environment is ECSHE. Warehouse area, production process, scrap, and room arrangement are still low, which disrupts transportation routes. The work environment area integrated with the owner's residence, is very narrow and is not available between human lines and workstations. The recommendation for this problem is to make road lines and road dividers so that the items are placed more in accordance with their purpose and look neat. c) *RESIK* is an activity to maintain the cleanliness of the work environment by removing garbage and foreign goods so that the workstation becomes cleaner. Scraping the kneading process and crumbs of the frying process will interfere with health. Not yet available representative bins. Recommendations for cleaning problems by cleaning regularly before and after the production process. d) *RAWAT* is an action to care for items in the work environment. Seiketsu will be effectively implemented if the 3 main aspects of 5R (Ringkas, Rapi, Resik) have been implemented properly. In SMEs Pempek and Kemplang do not have SOP's to maintain simple machines used, safety procedures, etc. e) *RAJIN* is the advanced stage of care. Recommendations about PPE that are good, simple and relevant to the task, such as hair covers, masks, etc. are highly recommended.

4.*2 Results of Application of EC*

The results of the EC (Table 2 and Table 3) show that the problems of storage and materials handling, workstations, and physical environments have high complaints so that the physical work environment also impacts workstations and storage and material handling which are not optimal in the production process. The recommended work environment with the application of the 5R concept can be seen in section 3.1.

4.*3 Results of the Implementation of Total Ergonomics Intervention*

Ergonomics intervention studies at the SMEs Pempek and Kemplang Palembang, in terms of the aspects of Task, Work Organization, and Environment. This review of 3 aspects is synchronized with ECSHE.

Figure 3 shows the kneading process & pempek shape, pempek submarine egg shape process, pempek adaan shape process, pempek frying process, and pempek cutting process. ***Task:*** no physiological work attitude (sitting with a chair that is not anthropometric, bending, standing). Muscle energy use is large because it forms a manual pempek by bending, the NBM results show musculoskeletal pain/ complaints 78-85% of the wrist, arm, shoulder, and back with a high risk 55-76, the classification of fatigue level is high 66- 79, the category of physical workload weighs 125-150 bpm, and the % CVL category is 78-85% (rather heavy). ***Work Organization:*** work without regulating work breaks, without providing additional nutrition. ISHOMA Time (Rest, Prayer, Lunch), 12.00 - 13.00 a.m. ***Environment:*** hot workspace 38-42°C, dark, moist, lacking ventilation so that air circulation is less optimal. ***Recommendation:*** it is necessary to apply 5R, redesign of the ergonomic work system (cadaver, printer of pempek, boilers, drums, etc.) and increase productivity. Necessary increase in productivity in order to fulfill the demand for pempek which has not been optimal with the application of TE interventions. Identification of 8 ergonomic problems, to implement SHIP Approach, AT and 5R through workshops on conditions and working environment of SMEs so far, to determine the priority scale of problems and improve tool design, work organization and involving participatory of all SMEs workers and owners.

Figure 3. Work System and Work Environment in SMEs Pempek Palembang: Shaping

Figure 4 shows the kneading of material process, boiling, drying, roasting, frying, kemplang crackers. ***Task:*** the work attitude is not physiological (bending and pressing repeatedly). Muscle energy use is large because the cadre process is still manual, the NBM results show musculoskeletal pain / complaints 50-62% of the wrist, arm, back and shoulder with high risk 52-74, classification of fatigue level in the high category 65-72, the physical workload category weighs 125-150 bpm, and the %CVL category is 70-88% (rather heavy). ***Work Organization:*** work without regulating work breaks, without providing additional nutrition. ISHOMA Time (Rest, Prayer, Eat), 12.00 - 13.00 a.m. ***Environment:*** hot workspace 38-40°C, moist, lack of ventilation so that air circulation is not optimal. ***Recommendation:*** the application of 5R is required, working system redesign (cadaver, boilers, ergonomic trolley, drying, frying) and AT. The results of various types of kemplang and crackers, there are still many who have not met food standards. The environment and work system that have not been ergonomic have an impact on the rapid level of fatigue experienced by workers, heavy physical burden and finally the level of productivity is not optimal. It is necessary to increase productivity in order to fulfill the demand have not been optimal with the application of TE interventions. Figure 5, here is one of the drying tool designs and the storage of a rack saving trolley which is useful for drying and storing kemplang before the cooking process or frying. The work system is more ergonomic than before it was designed. Proven total ergonomic interventions are indications of increasing the quality of life of workers and increasing productivity. Quality of life, aspects of workload, musculoskeletal complaints rate, fatigue rate, work stress rate, boredom numbers have decreased indications, while numbers of comfort and satisfaction indicate an increase. Ultimately it has an impact on increasing productivity [11], [12], [13].

Figure 4. Work System and Work Environment at SMEs Kemplang Palembang: Processing

Figure 5 and 6, here is one of the drying tool designs and the storage of a rack saving trolley which is useful for drying and storing Kemplang before the cooking process or frying. The work system is more ergonomic than before it was designed. Proven total ergonomic interventions are indications of increasing the quality of life of workers and increasing productivity. Quality of life, aspects of workload, musculoskeletal complaints rate, fatigue rate, work stress rate, boredom numbers have decreased indications, while numbers of comfort and satisfaction indicate an increase. Ultimately it has an impact on increasing productivity [11], [12], [13].

Figure 5. The Work System Un-Ergonomics

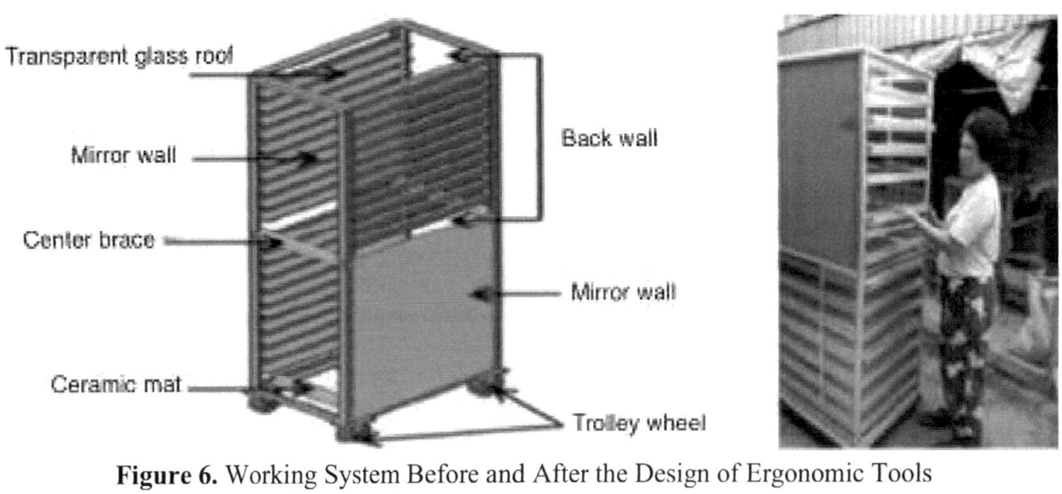

Figure 6. Working System Before and After the Design of Ergonomic Tools

5. Conclusion

Recommendations of EC and TE Interventions in The Pempek & Kemplang Palembang industry, can be summarized as follows:

1. Application of EC tools (9 aspects), the results show that 16 items are appropriate and 30 items are not in accordance with ILO 2010 standards. The aspects with the most incompatibility are less ergonomic or ECSHE, namely; storage & materials handling, workstations, and physical environment. Warehouses that are not available, routes that are not safe and there are no lines markings, poor workstations because there are still many stations that are integrated with the owner's residence and many posture work is not ergonomics at work. In the production process at SMEs Pempek has not been ergonomics in the process of kneading & pempek shape, egg submarine shape, pempek adaan shape, pempek frying pan, and pempek slicing. And in the production process in the SMEs Kemplang it is un-ergonomics in the kneading of material process, boiling, drying, roasting, frying, there are still many un-physiological working posture (sitting with chairs that are not anthropometric, bending, standing). Muscle energy use is large because it forms a manual pempek by bending, the NBM results show musculoskeletal pain on the wrist, arm, shoulder and back with high risk, high fatigue classification, heavy physical workload category, and category % CVL is rather heavy.
2. Implementation of TE Intervention and 5R concept in SMEs Pempek and Kemplang, has not been fully implemented because of constrained infrastructure investment costs. Where the selection of goods (RINGKAS) is still not well implemented by accumulating unused items and materials. Arrangement (RAPI) in the workstation area of the production process is also not good, because there are still many packages of scrap that accumulate in any place, and cleanliness (RESIK) work area that is not good, especially in the parts of a simple tool/ machine. Synergy model of TE intervention analysis and EC results of a nine aspects assessed; storage & material handling, workstation, heat & lighting, work posture, physical environment, work time, premises, welfare facilities, work organization contribute to the value of most incompatibility which becomes the top priority in determining suggestions for improvement. While the TE intervention study includes; work posture, un-ergonomics design tools, un-physiological work, classification of subjectivity level of musculoskeletal system is 55-76 (high risk), classification of subjective fatigue level is 66-79 (high category), physical workload at workload level is 125-150 bpm (weight category), and the % CVL category is 78-85% (rather heavy). To be more holistic the 5S method and participatory training in human resources are also used as a method of evaluating the work environment in this industry.
3. Recommendations on the conditions and working environment of SMEs Pempek and Kemplang Palembang are still found in work system inefficiencies, which have an impact on productivity not optimal and exceed the threshold value of human health aspects; limitations, ability and capability. The role of ergonomics interventions is still very much needed in designing work systems (task, work organization and environment), given the interrelationships between ideas, people, and the tools practiced in SMEs. The study of ergonomic interventions in the context of a suitable idea is to conduct ergonomics analysis (work posture, muscle usage, nutrition, conditions of information, time conditions, environment conditions, socio cultural conditions and man-machine interface) and AT. The adjustment of suitable work equipment is a tool that must be user friendly and sustainable. It is necessary to apply 5R, redesign the work system (cadaver, pempek printer, frying pan, boilers, drying trolley, waster, etc.) which is ergonomics and appropriate technology.

References

[1] Palembang K B 2018 5 Alasan Pempek Layak Mendunia

[2] Helali F 2009 Using ergonomics checkpoints to support a participatory ergonomics intervention in an industrially developing country (IDC)-a case study *Int. J. Occup. Saf. Ergon.* **15** 325–37

[3] Association I E 1996 *Ergonomic checkpoints: practical and easy-to-implement solutions for improving safety, health and working conditions* (International Labour Organization)

[4] KOGI K 2007 Action-oriented use of ergonomic checkpoints for healthy work design in different settings *J. Hum. Ergol. (Tokyo).* **36** 37–43

[5] Gaspersz V and Fontana A 2011 Integrated Management Problem Solving Panduan bagi Praktisi Bisnis dan Industri *Vinchristo Publ.*
[6] Santoso G 2004 *Ergonomi manusia, peralatan dan lingkungan* (Jakarta)
[7] Suma'mur 2009 *Higiene Perusahaan dan Keselamatan Kerja* (Jakarta: CV Sagung Seto)
[8] Setiawan H 2019 Internalization of the champion core values in work system design and ergonomics learning *IOP Conference Series: Materials Science and Engineering* vol 508 (IOP Publishing) p 12088
[9] Wigjosoebroto S 2003 *Ergonomi Studi Gerak dan Waktu - Teknik Analisis untuk Peningkatan Produktivitas Kerja* (Guna Widya)
[10] Setiawan H 2018 Redesigning the work system of rubber industries based on total ergonomics and ergo-micmac integration *Journal of Physics: Conference Series* vol 953 (IOP Publishing) p 12025
[11] Setiawan H 2017 Edukasi Pendekatan Ergonomi Total Dalam Praktik Kearifan Lokal Keilmuan Teknik Industri *J. Saintek* **1** 1–8
[12] Setiawan H 2017 Rekomendasi Intervensi Ergonomi Pada UKM Unggulan Provinsi Sumsel *Log. J. Ranc. Bangun dan Teknol.* **17** 86–91
[13] Setiawan H 2012 Technology Innovation Roadmap To Industrial Development of Rubber-Raw Material In South Sumatera (Ergonomics SHIP Approach & Appropriate Technology Point of View) *Procedia Econ. Financ.* **4** 255–63

IOP Conf. Series: Materials Science and Engineering **885** (2020) 012058 doi:10.1088/1757-899X/885/1/012058

Drying of extracted betacyanin from red dragon fruit (*Hylocereus Costaricensis*) as natural colorant by using spray dryer in various operating condition

I Susanti, F Hasanah , H Wijaya, S Heryani and E S Hartanto

Researcher at Center for Agro-Based Industry, Ministry of Industry, Republic of Indonesia

E-mail: fitrihasanah@yahoo.com

Abstract. One of the natural dyes that can be used in food is betacyanin from red dragon fruit. The betacyanin extracted from the red dragon fruit potentially be used for industrial ingredients such as a natural dye that is appropriate and can be applied further for food colouring. The drying process is one of the important stages in producing betacyanin extract from dragon fruit. The objective of this research was to study drying process of extract betacyanin from red dragon fruit using spray dryer in various operation condition. There materials from red dragon fruit: whole fruit, fruit flesh and fruit skin were used for extraction process with water as a solvent. Maltodextrin (30 %) was added as filler to the extracts. The inlet temperature that used in spray dryer condition were: 140 °C, 160 °C, 180 °C. The results showed the highest betacyanin content was found in whole dragon fruit extracts used spray dryer with Inlet temperature 140°C (411.80 mg / kg).

1. Introduction

Fruit is a rich plant in natural colourants resources. More than thousands of plants could be possible used as natural colourant on food. The plant has been used for thousands of years as a source of ingredients to provide attractive colours and appearance to food and drinks. However, in order to introduce in the worldwide market, more advance research should be conducted to study the active compounds that produce natural colour with the extraction techniques on those plants.

One of the potential plants as a source of natural food colourant is Red dragon fruit. Dragon fruit is the fruit of several cactus species that have been classified as white (*Hylocereus undatus*), red (*H. polyrhizus*), and yellow (*Selenicereus megalanthus*) [1][2]. Several studies have mentioned that dragon fruit contains high moisture and rich in fiber, organic acids, protein, phosphorus, vitamin C and other minerals like potassium, magnesium, calcium [3][4]. Furthermore, research investigated dragon fruit identified at least seven betacyanins, namely: betanin, isobetanin, lhyllocactin, isophyllocactin, betanidin, isobetanidin and bougainvillein-R-I where all have identical absorption spectra that contribute to the deep purple coloured pulp [5] [6][7].

The dragon fruit has been used as a food, natural dye, craft, tribal ceremony and medicine [8][9][10]. Dragon fruit is considered as a heavenly fruit on the earth with high nutritive and medicinal values. Various studies on the potential of red dragon fruit to treat some degenerative diseases such as cervical cancer, breast cancer, colon cancer, antidiabetic, anti-inflammatory has been done in recent years [11][12]. Swarup et al. [13] investigated the antidiabetic of dragon fruit. Results had shown that dragon fruit extract revealed oxidative damage inhibition activities and decreased the aortic stiffness by pulse wave velocity in streptozotocin-induced diabetes in rats. Widyaningsih et al. [14] investigated that effect of red dragon fruit juice on pregnant women has been able to increase haemoglobin and erythrocyte levels.

Commercialized natural dyes come in two forms, namely concentrate and powder. The concentrate can be obtained by evaporation under vacuum until a 40-60% solid is reached, while the powder dosage form is obtained by drying the extract using a spray dryer. In general, powder forms are preferred over concentrates because they are easier to handle, stable, and have a longer shelf life. The use of dragon fruit as a food colouring simply can be done by blending with the addition of water, and directly used or filtered first then applied to food. However, the colouring produced from this simple method has a short shelf life so that it is only suitable to be applied on a home industry scale where this product is directly used as a mixture of dough. One method that can be used to obtain colouring powder extract from dragon fruit is drying using spray dryer. Drying with a spray dryer will produce a product with a low water content that has a long shelf life and is easily applied on an industrial scale. Therefore, it is necessary to study the effect of drying using a spray dryer on the content of betacyanin in natural dye extracts from dragon fruit. The results of the literature study show that the purplish red colour in plants can also be obtained from the active ingredient of betacyanin substances [15][16][17][18].

2. Materials and methods

2.1 Materials and instruments

Materials used in this research are red dragon fruit (*Hylocereus costaricensis*) that obtained from farmers in Subang West Java, aquades as solvent, maltodextrin as filler. Red dragon fruit will be collected into 3, namely whole dragon fruit (meat and fruit are used), fruit flesh, and fruit skin. The chemical materials for analysis are betacyanin standard, n-butanol, acetic acid was purchased from Sigma Aldrich. The equipment used in this study were spray dryer, UV-Vis Evolution 201 spectrophotometer, blender, filter press.

2.2 Methods

Betacyanin pigments of red dragon fruits (whole dragon fruit, fruit flesh, and fruit skin) were extracted by blending using water as a solvent in a ratio of 1 : 3. Furthermore, the characterization of betacyanin extracts of red dragon fruit included water content analysis with oven method and total Betacyanin analysis (spectrophotometric method). The result of extraction was dried using spray dryer with variations of inlet temperature of 140 °C, 160 °C and 180 °C.

In order to extract the betacyanin pigmen of red dragon fruit, the fruits were washed and disinfected, and prickles were removed manually. Red dragon fruit was collected into 3 parts those are whole dragon fruit (meat and fruit), fruit flesh, and fruit skin. The fruit for the extraction was crushed using knife to reduce the size. Extraction was carried out by boiling at a temperature of 98°C for 30 minutes using water as solvent with ratio of red dragon fruit and water was 1:3. The filtrate extracted was filtered using a filter cloth to obtain a liquid extract. Then concentration and concentrated extract were obtained whole dragon fruit (meat and fruit are used), fruit flesh, and fruit skin. The concentrated extract is dried with a spray dryer which aims to produce a product with a low water content that has a long shelf life, maintain the quality and stability of the extract. The drying of betacyanin extract using spray dryer was the drying of extracts stage from whole dragon fruit, fruit flesh, fruit skin using a spray dryer. The extract was added with maltodextrin (30%) and then dried with a spray dryer. Spray Dryer conditions was set as follows:

- Carrier agent /filler : maltodextrin 30%
- Spray nozzle size: 0.5 mm
- Compressed air pressure: 3 bar
- Feed temperature: 30oC
- Feed flow rate: 22.9 ml / min
- Inlet air temperature: 140 °C, 160 °C, 180 °C.

Analysis of total Betacyanin was carried out using Standard of Betacyanin to create a standard curve. Calibration curves was created as follows, a solution stock was prepared by transferring 0.2g of the

Betacyanin standard to a 10 mL volumetric flask and making up to volume with acidified water (1% acetic acid/water v/v). The concentration of betacyanin was calculated by the extinction coefficient for Betacyanin (ε=60,000). The absorption maxima at 538, the equation used was [mg/l]= [(A*DF*MW*1000)/(ε*L)], where A is the absorption, DF is the dilution factor, MW and ε are the molecular weight and extinction coefficient and L the pathlength of the 1-cm cuvette [18] [3].

The concentration of betacyanin pigment in extracts from whole fruit, flesh fruit and skin fruit was determinated using a calibration curve. The absorbance of the samples was measured at wavelength of 540 nm using UV-Vis Thermo Scientific Evolution 201 spectrophotometer. The linear equation derived from the calibration curve was used to determine the concentration of the pigment extraction from whole dragon fruit, fruit flesh, and fruit skin of red dragon fruit. The absorbance obtained by spectrophotometric measurements was included in the linear equation of the standard curve so that the concentration of the sample solution in mg / 10 ml is obtained.

3. Result and Discussion

3.1. Pigment powder extract appearance of red dragon fruit

The natural dye extraction process has been carried out with water as solvents (aquadest). The treatment consisted whole dragon fruit, fruit flesh, and fruit skin. Each of them was cut into small pieces, then crushed using blender, and added with water (ratio of 1: 3) respectively. Each of treatment were filtered with filter clothes then added with maltodextrin (30%) and dried with spray dryer. Process condition of spray dryer was carried out at varying of inlet temperature: 140°C, 160°C and 180°C. This operating condition was setup to know the effect of the inlet temperature toward the betacyanin content which gave purplish red pigment in red dragon fruit. Pigment powder extract from red dragon fruit dried using spray dyer can be seen in Figure 1.

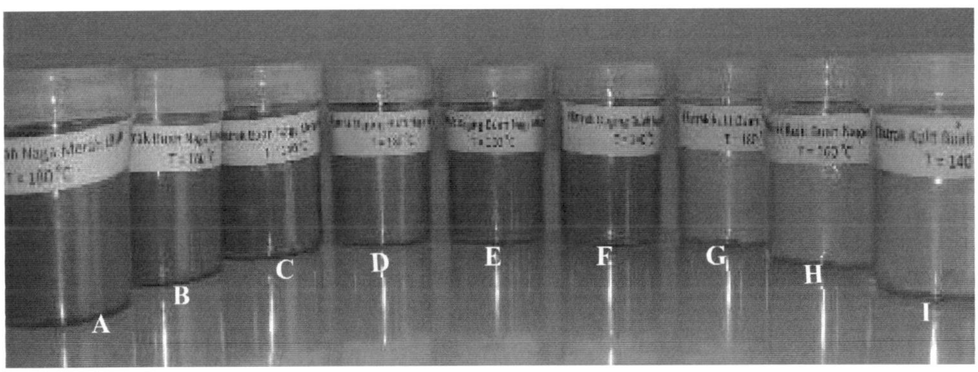

A : Whole fruit, inlet temperature of 180 °C
B : Whole fruit, inlet temperature of 160 °C
C : Whole fruit, inlet temperature of 140 °C
D : Fruit flesh inlet temperature of 180 °C
E : Fruit flesh inlet temperature of 160 °C
F : Fruit flesh inlet temperature of 140 °C
G : Skin fruit inlet temperature of 180 °C
H : Skin fruit inlet temperature of 160 °C
I : Skin fruit inlet temperature of 140 °C

Figure 1. Pigment powder extract from red dragon using spray dyer

From Figure 1, it can be seen that the results of the extraction of whole fruit and flesh of the dragon fruit were more concentrated (more purple visually) compared to the fruit skin (pink). It indicated that the betacyanin contained in whole fruit and fruit flesh are more than fruit skin.

3.2. Total betacyanin in pigment extract from red dragon fruit using spray dryer

Freshly extracted and concentrated of pigmen from red dragon fruit was dried using spray dryer. Due to the nature of food which was sticky and formless, drying techniques to increase the active compound of pigment producing plant become a challenge for researcher [19]. Besides the inlet temperature, the others important parameter of spray dryer was spray angle, outlet temperature, air humidity, residence time, etc [20]. Because the inlet temperature gives significant impact to atomized feed droplets to their wet-bulb temperature [21], this has a direct correlation to the wet-bulb temperature of the surrounding hot air which cause different concentration of pigment recovered.

In this research, there is a variation of inlet temperature treatment in 3 temperature conditions, namely 140, 160 and 180 so that it can be seen the effect of inlet temperature on the content of betacyanin in natural dye extracts from dragon fruit. Total betacyanin in Powder pigment extract from red dragon fruit were calculated and presented in in Figure 2.

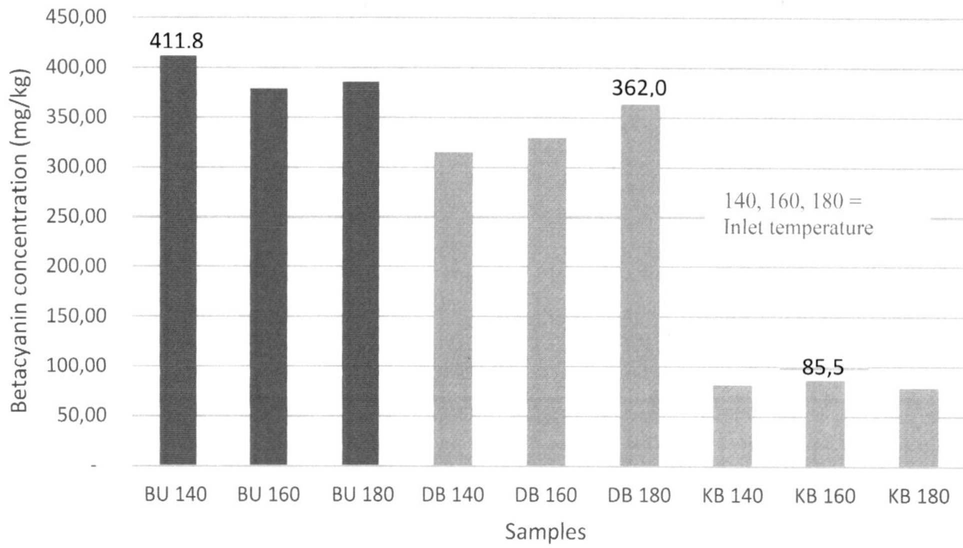

Figure 2. Total betacyanin in powder pigment extract from whole fruit (BU), fruit flesh (DB), skin fruit (KB) of red dragon fruit

The highest betacyanin concentration is in whole fruit with inlet temperature of 140 °C amounting to 411, 80 mg/kg. Betacyanin is water-soluble and therefore suitable when extracted with water. Some studies investigated that the increasing inlet-air temperature more than 200°C caused less pigment recovery [22]. The other studies investigated that the addition of maltodextrin can maintain the anthocyanin and betacyanin [23][24]. The degradation of active compounds during spray drying become main concern in production of natural pigment therefore the inlet temperature must be set at the lower temperature if the wet-bulb temperature surrounding hot air to be reduced [25][26][27]. The trade-off between the aforementioned circumstances is important in deciding an optimal inlet temperature for the spray drying process.

3.3. pH Value at pigment extract from red dragon fruit using spray dryer
The pH value of the red dragon fruit colour pigment powder can also be used as a distinguishing parameter between anthocyanin and betacyanin. According to Harborne [26], Betacyanin also functions as a natural colouring agent in plants, and has properties not susceptible to changes in pH. The results of the pH analysis of dragon fruit natural colouring powder can be seen in Figure 3.

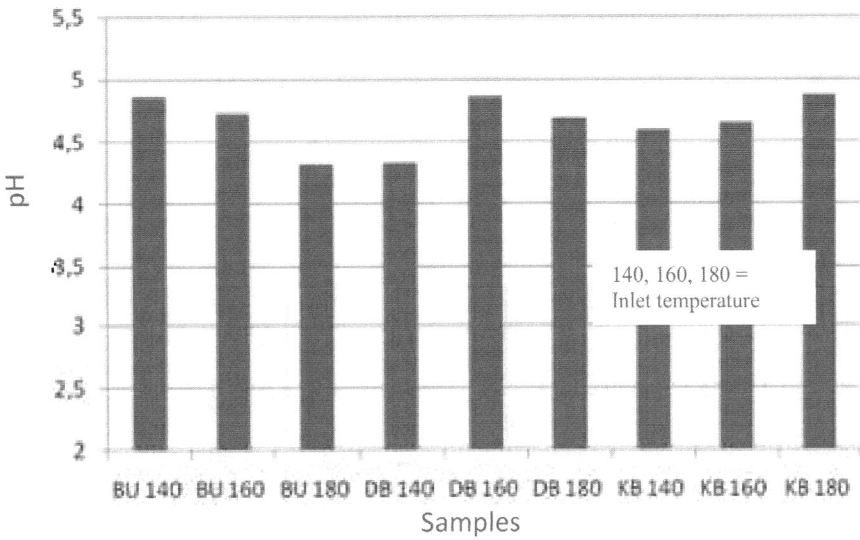

Figure 3. pH Value of whole fruit (BU), fruit flesh (DB), skin fruit (KB) natural colouring powder

From Figure 3 shows that the range of the pH value of the red dragon fruit dye powder is 4.3 to 4.7. This is in line with the optimum pH of betacyanin, which is around 4.6 to 4.7 [27][28][29].

3.4 Effect of pH changes on colour stability
The stability against pH is one of differentiation indicators between betacyanin and anthocyanin. Betacyanin is more stable to pH (the colour does not differ much in a certain pH range) and the red purple will turn yellow when the pH gets higher (base). The results of stability testing for changes in pH can be seen in Figure 4.

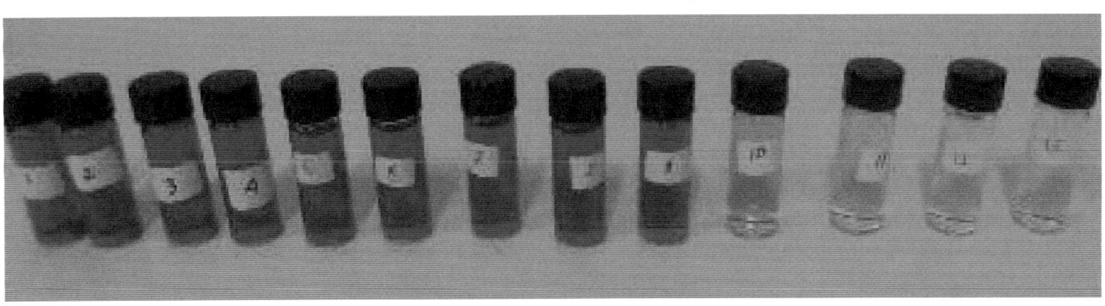

Figure 4. The stability of red dragon fruit colouring extract to changes in pH (pH 1 - 13)

From Figure 4 it can be seen that at pH 1-9, the colour of the solution is relatively the same, namely purplish red, but when the pH reaches 10 the colour of the solution turns yellow. This is in accordance with one of the distinguishing characteristics of anthocyanins and betacyanin is the higher the pH (base), the red colour will turn yellow [26][27][28]. Betacyanin pigments produce pink / pink to red in the pH range of 4-8. These results provide important information for the application of dyes from red dragon fruit in food products and others in the future.

3.5 Water content of dragon fruit natural colouring powder

Water content is a parameter that shows the drying conditions of the colouring powder. The lower the water content, the longer the shelf life will be. The range of water content of the colouring powder of red dragon fruit is 4.3% to 5.1% (Figure 5). The water content is low, so it is expected that with good storage conditions, the shelf life of the colouring powder will be long.

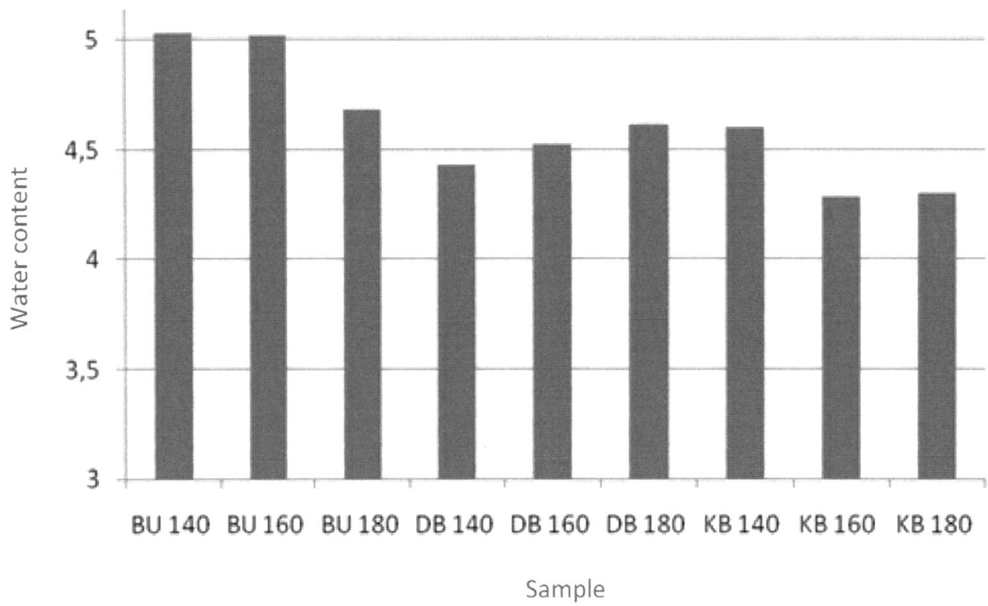

Figure 5. Water content of dragon fruit natural colouring powder

The results showed that the higher the inlet temperature used when drying, the lower the water content obtained both whole fruit, fruit flesh and skin fruit of red dragon fruit pigment extract (figure 5). The others studies observe that maltodextrin in high quantity can result better stability of powder [30][31][32]. Reducing reactant mobility and moisture content might be the reason of these. Other research show that higher inlet temperature results low water content and hygroscopicity, and high solubility and stable colour [33].

4. Conclusion

According to this research, inlet temperature of the spray dryer has an effect on the content of betacyanin, pH and water content of the natural dyes of the red dragon fruit. The highest concentration of betacyanin in whole fruit with an inlet temperature amount of 140° C.

Acknowledgment
The authors would like to thank Center for Agro Based Industry, Ministry of Industry of Republic Indonesia for funding this research activity. Thank you also to Mrs. Fince and Mr. Indera Wirawan for their technical assistance during the study

Bibliography

[1] Nerd A, Sitrit Y, Kaushik R A and Mizrahi Y 2002 High summer temperatures inhibit flowering in vine pitaya crops (Hylocereus spp.) *Sci. Hortic. (Amsterdam).* **96** 343–50
[2] Hoa T T, Clark C J, Waddell B C and Woolf A B 2006 Postharvest quality of Dragon fruit (Hylocereus undatus) following disinfesting hot air treatments *Postharvest Biol. Technol.* **41** 62–9
[3] Stintzing F C, Schieber A and Carle R 2003 Evaluation of colour properties and chemical quality parameters of cactus juices *Eur. Food Res. Technol.* **216** 303–11
[4] Delgado-Vargas F, Jiménez A R and Paredes-López O 2000 Natural pigments: carotenoids, anthocyanins, and betalains—characteristics, biosynthesis, processing, and stability *Crit. Rev. Food Sci. Nutr.* **40** 173–289
[5] Stintzing F C, Schieber A and Carle R 2002 Betacyanins in fruits from red-purple pitaya, Hylocereus polyrhizus (Weber) Britton & Rose *Food Chem.* **77** 101–6
[6] Wybraniec S, Platzner I, Geresh S, Gottlieb H E, Haimberg M, Mogilnitzki M and Mizrahi Y 2001 Betacyanins from vine cactus Hylocereus polyrhizus *Phytochemistry* **58** 1209–12
[7] Rebecca O P S, Boyce A N and Chandran S 2010 Pigment identification and antioxidant properties of red dragon fruit (Hylocereus polyrhizus) *African J. Biotechnol.* **9** 1450–4
[8] Raveh E, Weiss J, Nerd A and Mizrahi Y 1993 Pitayas (Genus Hylocereus): a new fruit crop for the Negev Desert of Israel *New Crop.* 491–5
[9] Tel-Zur N, Abbo S, Bar-Zvi D and Mizrahi Y 2001 Hybridization between species and genera of fruit-crop vine cacti of the genera Hylocereus and Selenicereus *HortScience* **36** 441
[10] Nerd A, Gutman F and Mizrahi Y 1999 Ripening and postharvest behaviour of fruits of two Hylocereus species (Cactaceae) *Postharvest Biol. Technol.* **17** 39–45
[11] Abd Hadi N, Mohamad M, Rohin M A K and Yusof R M 2016 Effects of Red pitaya fruit (Hylocereus polyrhizus) consumption on blood glucose level and lipid profile in type 2 diabetic subjects *Borneo Sci.* **31**
[12] Nurliyana R d, Syed Zahir I, Mustapha Suleiman K, Aisyah M R and Kamarul Rahim K 2010 Antioxidant study of pulps and peels of dragon fruits: a comparative study *Int. Food Res. J.* **17**
[13] Swarup K R L A, Sattar M A, Abdullah N A, Abdulla M H, Salman I M, Rathore H A and Johns E J 2010 Effect of dragon fruit extract on oxidative stress and aortic stiffness in streptozotocin-induced diabetes in rats *Pharmacognosy Res.* **2** 31
[14] Wahdaningsih S, Wahyuono S, Riyanto S and Murwanti R 2018 Antioxidant Activity of Red Dragon Fruit Peel (Hylocereus Polyrhizus (F.A.C. Weber) Britton And Rose) Isolates Using 2,2-Diphenyl-1-Picrylhydrazyl Method *Asian J Pharm Clin Res* **11** 124–8
[15] Grotewold E 2006 The genetics and biochemistry of floral pigments *Annu. Rev. Plant Biol.* **57** 761–80
[16] Vaillant F, Le Bellec F and Imbert E P 2006 A new fruit crop, a market with a future *Fruits* **61** 37–250
[17] Biswas M, Dey S and Sen R 2013 Betalains from Amaranthus tricolor L. *J. Pharmacogn. Phytochem.* **1**
[18] Wong Y-M and Siow L-F 2015 Effects of heat, pH, antioxidant, agitation and light on betacyanin stability using red-fleshed dragon fruit (Hylocereus polyrhizus) juice and concentrate as models *J. Food Sci. Technol.* **52** 3086–92
[19] Woo M W, Daud W R W, Tasirin S M and Talib M Z M 2009 Controlling food powder deposition in spray dryers: wall surface energy manipulation as an alternative *J. Food Eng.* **94** 192–8
[20] Anandharamakrishnan C 2015 *Spray drying techniques for food ingredient encapsulation* (John

Wiley & Sons)

[21] Oldfield D J, Taylor M W and Singh H 2005 Effect of preheating and other process parameters on whey protein reactions during skim milk powder manufacture *Int. Dairy J.* **15** 501–11

[22] Goula A M and Adamopoulos K G 2005 Stability of lycopene during spray drying of tomato pulp *LWT-Food Sci. Technol.* **38** 479–87

[23] Ruri A S, Norziah M H, Iswandi, Uthumporn and Najla L 2017 Spray Drying of Enzymatic Clarification Red Purple Pitaya (Hylocereus Polyrhizus) Pigment Extract *IJSR* **6** 386–91

[24] Goula A M, Adamopoulos K G and Kazakis N A 2004 Influence of spray drying conditions on tomato powder properties *Dry. Technol.* **22** 1129–51

[25] Chong P H, Yusof Y A, Aziz M G, Nazli N M, Chin N L and Muhammad S S 2014 Effects of spray drying conditions of microencapsulation of Amaranthus gangeticus extract on drying behaviour *Agric. Agric. Sci. Procedia* **2** 33–42

[26] Harborne J B 2013 *The flavonoids: advances in research since 1980* (Springer)

[27] Prudencio I D, Prudêncio E S, Gris E F, Tomazi T and Bordignon-Luiz M T 2008 Petit suisse manufactured with cheese whey retentate and application of betalains and anthocyanins *LWT-Food Sci. Technol.* **41** 905–10

[28] Stintzing F C and Carle R 2007 Betalains–emerging prospects for food scientists *Trends Food Sci. Technol.* **18** 514–25

[29] Cai Y and Corke H 1999 Amaranthus betacyanin pigments applied in model food systems *J. Food Sci.* **64** 869–73

[30] Loksuwan J 2007 Characteristics of microencapsulated β-carotene formed by spray drying with modified tapioca starch, native tapioca starch and maltodextrin *Food Hydrocoll.* **21** 928–35

[31] Koul V K, Jain M P, Koul S, Sharma V K, Tikoo C L and Jain S M 2002 Spray drying of beet root juice using different carriers

[32] Grabowski J A, Truong V and Daubert C R 2006 Spray-drying of amylase hydrolyzed sweetpotato puree and physicochemical properties of powder *J. Food Sci.* **71** E209–17

[33] de Souza V B, Thomazini M, de Carvalho Balieiro J C and Fávaro-Trindade C S 2015 Effect of spray drying on the physicochemical properties and color stability of the powdered pigment obtained from vinification byproducts of the Bordo grape (Vitis labrusca) *Food Bioprod. Process.* **93** 39–50

IOP Conf. Series: Materials Science and Engineering **885** (2020) 012059 doi:10.1088/1757-899X/885/1/012059

Analysis of time acceleration using Critical Path Method (CPM) to increase motorcycle maintenance in authorized service station

I Y Marit[1], E Nursanti[2] and P Vitasari[2]

[1]Master Student, Postgraduate Program of Industrial Engineering, National Institute of Technology, Malang, Indonesia
[2]Industrial Engineering, Postgraduate Program, National Institute of Technology, Malang, Indonesia

E-mail: iinmarit06@gmail.com

Abstract. Numerous studies have shown that maintenance requires proper scheduling by using preventive and corrective maintenance scheduling to maintain the state of the machine. The fact is the number of Honda motorcycle products has increased. This has resulted in a long motorcycle queue. Motorcycle queues cannot be prevented only by scheduling preventive and corrective maintenance, so scheduling of maintenance activities is necessity to speed up motorcycle maintenance time. This study uses the Critical Path Method (CPM) activity scheduling method with the help of the POM-QM program to increase the acceleration of maintenance time. Scheduling is made by calculating the logic of dependency and time of each maintenance activity. Activity scheduling is carried out on 7 types of maintenance, namely carburetor light service, carburetor and CVT light service, injection light service, injection and CVT light service, oil change, spare part replacement, oil and spare part replacement and final inspection of each type of maintenance. After that, the analysis of the scheduling results is done by comparing the results before and after the proposed improvement. Based on the results of research conducted it can be seen that the maintenance time at the authorized service station can be accelerated by using the CPM scheduling method. The efficiency of acceleration of maintenance time is 16.666% to 57.889%. The preparation of maintenance activities and processing of heavy service data is required if there is heavy service maintenance in the authorized service station.

1.Introduction

Maintenance activities are undertaken to maintain the condition of the equipment and its components to be ready for operation [1]. Millions of motorbike populations need routine maintenance to function properly, this results in queues at workshops, eventually requiring maintenance scheduling to speed up maintenance time. Some research on scheduling planning using several methods, namely bar chart method, network method (CPM, PERT, PDM), LOB and Time Chainage Diagram. Where Bar Charts are easy to make and understand, but cannot recognize the critical path, determine the interrelationships between these activities and are not effective on large projects [2]. The network method such CPM, PDM, and PERT, it has the reliability to show specifically the logic of dependency relationships between activities and determine the critical trajectory of project activities so that the priority activities in the event of delays can be identified, but cannot directly detect activities that experience disturbances in scheduling projects [3][4]. LOB is simple and easy to understand because as a line that shows the productivity of a job. However, LOB cannot show specifically the relationship between the logic of dependency and activities. LOB has the advantage of being able to directly detect activities that experience disturbances in project scheduling by looking at the presence of intersecting bar charts [5]. Another name for Time Chainage Diagram is Space-Time Diagram. Time Chainage Diagrams are another variation of the LOB. This method is also known as Time Distance Chart which is a simple extension of the Bar Chart method which is widely known by users of the planning system [6]. The scheduling needs to determine the relationship of each activity, identify relationships that should take precedence over other activities, help the use of labor and other resources in an optimal way on a project. In the CPM diagram, it can be showing specifically for dependency logic relationship used on all work items is Finish to Start (FS). Likewise, the maintenance completion time can be estimate by mathematics calculation. Also, the CPM method can also be seen a critical trajectory in a project schedule so that if there is a delay in project work, the priority of the work to be evaluated becomes easier to do. Work items traversed by the critical track monitored, the other hand overall maintenance delays cause of experience delays as well [7].

Long queues at official service stations can be described as a good level of customer loyalty [8]. However, this can also be a serious problem because they can lose up to 10 customers a week. Therefore, the Critical Path Method (CPM) is a suitable method to solve critical trajectory of maintenance activities in the authorized service station. This study uses CPM method to propose priority work on maintenance activities and increasing time acceleration for motorcycle maintenance.

2. Method

The scheduling method used in this study is CPM using a POM computer program. Sample taken is the type of motorcycle maintenance such as carburetor-type light service, carburetor and CVT light service, injection-light service, injection-light service and CVT, oil change, replacement of spare parts, oil and spare parts replacement and inspection of each type of maintenance. Sample total is taken based on the type of maintenance served during the duration of data collection. The research design stage showed in Figure 1

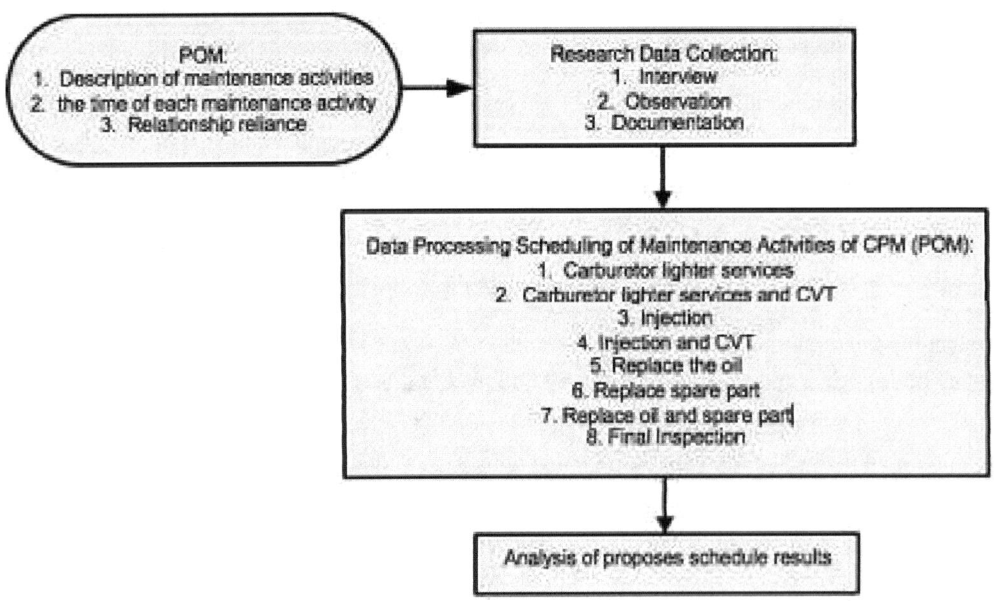

Figure 1. Research design stage

IOP Conf. Series: Materials Science and Engineering **885** (2020) 012059 doi:10.1088/1757-899X/885/1/012059

CPM has six basic steps [9], namely:

1. Identify the project and prepare a fractional work structure.
2. Building relationships between activities, deciding which activities should go first and which ones follow.
3. Describe the network that connects the whole activity.
4. Establish an estimated time for each activity
5. Calculate the longest time path through a network called the critical path.
6. Using networks to help with project planning, scheduling, and control.

Data processing using the POM program can be formulated into 4 steps, namely [10]:

1. In the Project management module select a single time estimate with network immediate predecessor.
2. Enter the number of numbers based on the number of work activities.
3. Input by existing data, ranging from tasks (activities) to activities that precede (Predecessor).
4. Then solve.

3. The results

Reliable logic of each activity item that is protected for scheduling maintenance activities, one of the results of scheduling using the CPM method can be seen in Table 1. Early Start is the earliest start time of an activity while Early Finish is the earliest time to finish the activity. Late Start shows the slowest start time of the activity can be started without delaying maintenance. time. Late Finish shows the end time of the activity can be ended at the latest without an extended period of maintenance. The Slack is the difference in time, if slack is zero then the activity is called critical activity.

Table 1. The logic of carburetor type motorcycle light service maintenance dependency

Activity Code	Activity	Activity Time (Second)	Pre-deccessor	Early Start	Early Finish	Late Start	Late Finish	Slack
1	Light Check	30		0	30	1650	1680	1650
2	Horn Check	30		0	30	1650	1680	1650
3	Cover Body Removal	420		0	420	0	420	0
4	Engine Oil Inspection	60		0	60	1620	1680	1620
5	Oil Filter Check	60	3	420	480	960	1020	540
6	Inspection and Adjustment	60	3	420	480	960	1020	540
7	Carburetor Cleaning and Adjustment	600	3	420	1020	420	1020	0
8	Cleaning and Replacement of Fuel Filters	60	3	420	480	960	1020	540
9	Cleaning and Replacement of Air Filters	60	3	420	480	960	1020	540
10	Valve Check and Adjustment	300	3	420	720	720	1020	300
11	Clutch Check and Adjustment	120	3	420	540	900	1020	480
12	Chain Adjustment	120			120	1560	1680	1560
13	Brake Check and Adjustment	60			60	1620	1680	1620
14	Brake Light Switch Adjustment	60			60	1620	1680	1620
15	Battery Checkup Examination	120	3	420	540	1560	1680	1140
16	Addition of Tire Wind Pressure	120			120	1560	1680	1560
17	Shock Beaker examination	60			60	1620	1680	1620
18	Steering Handlebar Checks and Adjustments	60	3	420	480	1620	1680	1200
19	Wheel Bearing Inspection	60			60	1620	1680	1620
20	Mounting Cover	600	4,5,6,7,8, 9,10,11	1020	1620	1020	1620	0
21	Nuts and Bolts Inspection and Tightening	60	20	1620	1680	1620	1680	0
	Total Before Proposed Improvement = 3120			Total After Proposed Improvements (CPM) = 1680				

Based on the results of scheduling the overall type of maintenance using the CPM method, it is known that the speed of time before and after the proposed improvement can be seen in Table 2.

Table 2. Comparison results of scheduling activities before and after proposed improvements

No	Type of Maintenance	Before Proposed Improvement	After Proposed CPM Repair	Time Saving	Equivalent Efficiency
1	Carburetor	3120 second (52 minute)	1680 second (28 minute)	1440 second (24 minute)	46,153 %
2	Carburetor and CVT	3900 second (65 minute)	1680 second (28 minute)	2220 second (37 minute)	43,076 %
3	Injection	3120 second (52 minute)	1680 second (28 minute)	1440 second (24 minute)	46,153 %
4	Injection and CVT	3900 second (65 minute)	1680 second (28 minute)	2220 second (37 minute)	57,894 %
5	Oil	720 second (12 minute)	600 second (28 minute)	120 second (2 minute)	16,666 %
6	Spare part	2760 second (46 minute)	2700 second (45 minute)	60 second (1 minute)	2,173 %
7	Oil and Spare part	660 second (11 minute)	510 second (8,5 minute)	150 second (2,5 minute)	22,727 %
8	Final Inspection	270 second (4,5 minute)	120 second (2 minute)	150 second (2,5 minute)	55,555 %

The graph of the comparison of the scheduling of activities before and after the proposed improvement can be seen in Figure 2. The graph of the results of the comparison of the scheduling of the activities before and after the proposed improvement shows that there is an acceleration of maintenance time after the proposed improvement of the activity schedule using the critical path method.

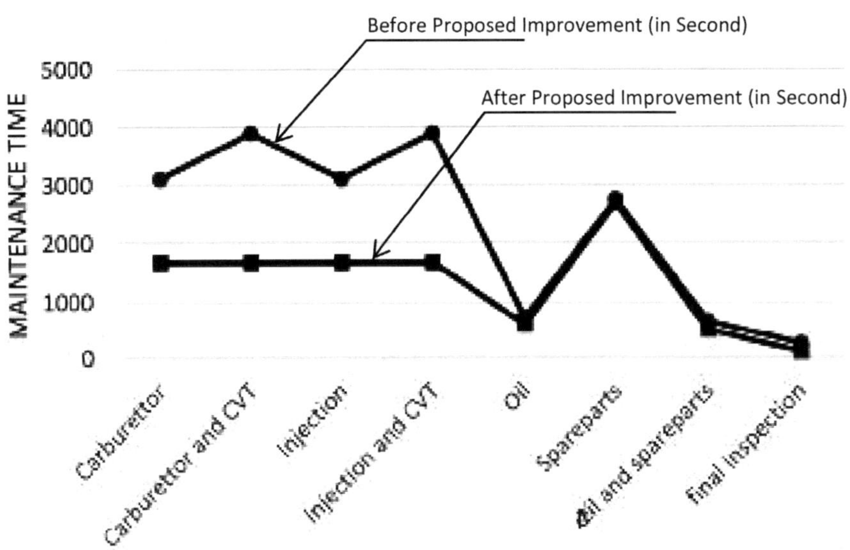

Figure 2. Comparison results of scheduling activities before and after proposed improvements

4. Conclusion

Maintenance of a light carburetor motorcycle service accelerates for 1440 seconds, with an efficiency of 46.153%. Maintenance of the carburetor and CVT lightweight service motorcycle accelerates for 2220 seconds, with an efficiency of 43.076%. Maintenance of light injection service motorbikes accelerates for 1440 seconds, with an efficiency of 46.153%. Maintenance of light injection and CVT service motorbikes accelerates for 2220 seconds, with an efficiency of 57.894%. Motorcycle oil changes obtain acceleration for 120 seconds with an efficiency of 16.666%. Motorcycle spare parts replacement gained acceleration for 60 seconds with an efficiency of 2.173%. Oil and motorcycle spare parts changes have accelerated for 150 seconds, with an efficiency of 16.666%.

Acknowledgement

We would like to thank Directorate General of Learning and Student Affair, Ministry of Research Technology and Higher Education for giving us the financial support in this publication as implementation of grant program in the postgraduate improvement quality by the contract number (No.036/B4/PPK-SPK/VII/2019).

References

[1] Kandemir C, Celik M, Akyuz E and Aydin O 2019 Application of human reliability analysis to repair & maintenance operations on-board ships: the case of HFO purifier overhauling *Appl. Ocean Res.* **88** 317–25

[2] Rolfsen C N and Merschbrock C 2016 Acceptance of construction scheduling visualizations: bar-charts, flowline-charts, or perhaps BIM? *Procedia Eng.* **164** 558–66

[3] Hajdu M and Isaac S 2016 Sixty years of project planning: history and future *Organ. Technol. Manag. Constr. an Int. J.* **8** 1499–510

[4] Nursanti E, Avief R M, Sibut S and Kertaningtyas M 2018 Peningkatan Efisiensi Waktu dan Biaya Pemeliharaan Overhaul Pesawat Tempur *J. Teknol. dan Manaj. Ind.* **4** 7–15

[5] Badukale P A and Sabihuddin S 2014 Line of balance *Int J Mod Eng Res* **4** 45–7

[6] Olivieri H, Seppänen O and Denis Granja A 2018 Improving workflow and resource usage in construction schedules through location-based management system (LBMS) *Constr. Manag. Econ.* **36** 109–24

[7] Hegazy T and Menesi W 2010 Critical path segments scheduling technique *J. Constr. Eng. Manag.* **136** 1078–85

[8] Khadka, Kabu and Maharjan S 2017 *Customer satisfaction and customer loyalty* (Centria University of Applied Science)

[9] Barry R, Hezser J and Chu M 2014 *Operations Management* (Pearson)

[10] Weiss H J 2004 *POM/QM for Windows* (Prentice Hall)

IOP Conf. Series: Materials Science and Engineering **885** (2020) 012060 doi:10.1088/1757-899X/885/1/012060

Hygiene Factors of Outsourcing Workers: A Literature Review

I Ramdhani and I Bakri

Department of Industrial Engineering, Universitas Hasanuddin, Makassar, Indonesia

E-mail: irin.ramdhani@gmail.com

Abstract. In a large company, the human resources consist of permanent employees and temporary employees. Both in industry, service, banking, etc., workers outsourcings also one of the supporting factors in the company's business. Therefore, we need a study to identify reviews of their satisfaction and dissatisfaction at work. This study aims to find out the working conditions of outsourced workers in various fields of work and relate them to the two-factor theory of Herzberg. This research was carried out based on a literature review from various previous studies on outsourcing workers, work motivation and also a factor of two theories by Herzberg. The results of this study show that outsourcing workers both in Indonesia and abroad have been protected by law from the government. So that some companies have implemented outsourcing labour regulations following applicable regulations. The results of several previous studies show that the two-factor theory by Herzberg influences the job satisfaction of outsourcing workers. Both that is influenced by motivational factors and hygiene factors.

1. Introduction

Employees are an important resource for the company because it has a role as a driver of the entire activity of the company [1], however, the stiff competition has led many organizations to rethink the use of its human resources. The problem that arises is the fact that not all of the human resource competencies expected. In fact, not infrequently employee-owned makes the cost becomes higher. This makes the company should improve efficiency in various ways according to the condition of each company. One strategy that began to be applied by the company to create efficiency is the use of contract Labour or known as outsourcing [2], Outsourcing or outsourcing is the transfer of responsibilities and work - the work of the business process which was performed in the company to a third party as service providers where the agency service providers do the administration and management based on the definitions and criteria agreed to by both parties (users and providers services) [3], As a result of the increased presence of the worker or workers are not permanent or contract status, then comes a few complaints relating to the welfare of workers or labour [2], So that the company can continue to grow, one of the major problems is to motivate employees to continue to work well [4], Job satisfaction is the positive attitude shown by someone happy and proud of the work [5], Satisfaction and dissatisfaction of employees in the work described in the theory of two factors, namely the hygiene factors and motivators. Something which causes dissatisfaction of employees at work is a hygiene factor, but when filled with not enough to make employees feel satisfied. While motivator is something that leads to the nature of the work, giving satisfaction and higher motivation can be achieved [6],

These hygiene factors theory comes from the evaluation of the work sequence, which proved to have a real effect on work attitudes. This is done by asking respondents to think of a time and to describe the situation when they feel very good or very bad about their work, whether their current job or other jobs they have. Following these questions, there are many special investigations designed to clarify the nature of the event and their personal reactions to the event. Typically, each respondent was asked to provide one or more sequences of events for the period of the satisfied and dissatisfied [7],

In theory, hygiene factors and motivation showed that various factors influence the satisfaction and dissatisfaction in work. Something that makes satisfied in the work leading to the employee's job description, especially the achievements, awards, interesting work, career advancement, growth, and progress. Meanwhile, something that makes the employee is not satisfied with the work is not about what they do but how they are treated. The main factors associated with employee dissatisfaction is the company's policy and administration, supervision, relationships with superiors or colleagues, working conditions, income, status and security [8].

a. Interpersonal relationships.
   This relationship is limited to personal and working relationships between co-workers, superiors or subordinates. These include work-related interaction and social discussion in the workplace and during leisure time in office [9].
b. Salary.
   This includes all forms of compensation in a person's workplace, such as increases in wages or salary, or the expectation of an increase or decrease of wages or salaries are not met [9],
c. Policy and administration of the company.
   In this case, describe the policies and rules of the organization and management of the company sufficient or insufficient. This factor involves organizational policy is good or bad that affects employees. For example, the lack of delegation of authority, communication or lack of procedure [9].
d. Supervision.
   The oversight here pertains to competence in terms of the ability or inability and the justice or injustice of supervisors. These include the willingness of employers to delegate their responsibilities or to teach, justice and knowledge of the job. Supervisory good, or access to supervision, it is important to improve the level of employee satisfaction [9].
e. Working conditions.
   These factors are related to the physical environment of the work and its facilities and infrastructure work. The work environment can create feelings of employees satisfied and proud [9],

2. Methods

Search a literature review conducted by the search term hygiene factors, outsourcing, motivation, theory of two factors, and other literature relating to hygiene factors in outsourcing is Labour. Searches were conducted using google scholar search engine, ScienceDirect, and others. All kinds of research included regardless of the types of companies as research objects. The purpose of this paper is to identify the hygiene factors in outsourcing workers.

3. Result

3.1. Research on Hygiene Factors

Several theories explain the motivation, one of which is by Herzberg's Two Factor Theory, namely Motivation and Hygiene Factors [4], Table 1 shows some of the research conducted related to the Hygiene Factors.

IOP Conf. Series: Materials Science and Engineering **885** (2020) 012060 doi:10.1088/1757-899X/885/1/012060

Table 1. Research of hygiene factors

Author (Year)	Research Title	Results and Conclusion
Tesavrita C and Suryadi D, 2012 [4]	Identification of Herzberg's Motivator-Hygiene Factors for SME's Workers: Case Study of SME in Bandung, Indonesia	This study showed that motivation factors are grouped into six categories based on model canoe. From this model obtained any factors that motivate workers, keeping the satisfaction of the workers and the things that make them survive in his job
Juariyah L and Rizky Saktian I, 2018 [5]	Does Motivators Determine Employees' Job Satisfaction? Testing Herzberg Theory of Motivation in Indonesian Café and Restaurant Context	The results of this paper are research and Hygiene Factors motivating factors to have a significant positive effect on job satisfaction. However, Hygiene Factor has a greater influence than the motivation
Feizal R, 2015 [10]	Hygiene Influence Factors and Motivating Factors on Job Satisfaction and Its Impact on Organizational Citizenship Behaviour (OCB) in Indonesia	From the results of hypothesis testing can be concluded that the Hygiene Factors direct effect on job satisfaction and job satisfaction directly influence the OCB
Nirmalasari Y F, 2018 [11]	Factors Influence Analysis Hygiene and Motivator Factors on Employee Satisfaction (*in Bahasa*)	The results of this study are Hygiene Factors and motivational factors influencing job satisfaction. However, the motivation to work has a greater influence on job satisfaction than hygiene factors
Amoako G K and Baah K D, 2011 [6]	Application of Frederick Herzberg's Two-Factor Theory in Assessing and Understanding Employee Motivation at Work: A Ghanaian Perspective	This study describes the factors that motivate workers of Ghana and evaluate the workers thought that the working arrangement motivating factor is higher than the hygiene factor
SH Shaikh et al., 2019 [12]	The Impact of Job Satisfaction and Job Dissatisfaction on Herzberg Theory: A Case Study of Meezan Bank Limited and National Bank Limited	This research was conducted at Meezan Bank and the National Bank, where the result is obtained that job satisfaction has a positive influence on employee performance and job dissatisfaction negatively affect the performance of employees
Hilmi A et al., 2016 [13]	Herzberg's Motivation- Hygiene Theory Applied to High School Teachers in Turkey	This research was conducted by distributing questionnaires to a high school teacher in turkey, where the result is obtained that the Hygiene Factors and the motivational factors affect job satisfaction Hygiene Factors teachers but have a greater influence

IOP Conf. Series: Materials Science and Engineering **885** (2020) 012060 doi:10.1088/1757-899X/885/1/012060

Table 1. Research of hygiene factors (*count.*)

Author (Year)	Research Title	Results and Conclusion
Bryant O A, 2018 [14]	Employee Turnover in the Long-Term Care Industry using Herzberg's Motivation-Hygiene Theory	This study was analysed using multiple regression tests, where the result is obtained that no statistically significant relationship between salaries, employee involvement, job satisfaction, work environment, and employee turnover intention. Whereas no significant effect between employee motivation and employee turnover intention in the industry in the long term
Anah C I, 2015 [15]	Re-Evaluating the Relevance of Herzberg's Motivation Hygiene Theory to The Nigerian Worker: A Political Economy Perspective	This research was conducted at the faculty and staff of the Federal University of Technology Owerri, the results of the questionnaire showed that in addition to his salary as Hygiene Factors, but also improve the recognition and dignity of workers
de Waal A and Jansen P, 2013 [16]	The Bonus as a Hygiene Factor: The Role of Reward Systems in The High-Performance Organization	Results from this study are that the bonus system or certain benefits do not have a positive or negative influence on the performance of the company

IOP Conf. Series: Materials Science and Engineering **885** (2020) 012060 doi:10.1088/1757-899X/885/1/012060

3.2. Outsourced Workers

Outsourcing labour regulations can only cover workers in the supporting processes (non-core business units) or in practice all the working parts can be transferred as an outsourcing unit. The number, area, and shape in the outsourced work is growing very fast, not only the work of factories but also more advanced work. Outsourcing is a way to get the skilled labour as well as reducing the burden and costs of companies in improving corporate performance to continue to compete in the face of global economic and technological development to give up some activities on third-party companies [3],

The practice of outsourcing in Indonesia provided for in Article 64 of Law No. 13 of 2003 about Employment which states that: "The Company may subcontract part of the work to another company through an agreement contract of work or the provision of worker/Labourer made in writing". Furthermore, in 2011 the Constitutional Court issued Decision No 27 / PPUIX / 2011 which granted the judicial review of Law No. 13 of 2003 on employment submitted by the Chairman of the Alliance Officer Electricity Meter Readers. Of the decision explained that the position of outsourcing in the Act - Act No. 13 of 2003 on employment is still valid and remain in force unless the provisions contained in Article 65 paragraph (7) and Article 66 paragraph (2) related to employment agreement specified the time. Then in 2012, the Minister of Manpower issued Regulation of the Minister of Manpower and the Transmigration Republic of Indonesia Number 19 the Year 2012 About Terms - Terms of Delivery Part of Work Implementation for Service-Oriented Others mention that the work is biased handed over to the enterprise IT services provider of workers/Labourers shall constitute service activities -supporting services. The activities supporting services in question is a cleaning service, catering, security, mining, and petroleum chartering and transport [17]. This theme was chosen in consideration of the growing practice of outsourcing in Indonesia that are considered to support the efficiency and effectiveness of a company. But, on the other hand, this practice is also considered to cause major problems in the field of employment [17], Table 2 lists some studies about outsourcing workers who have done well in the field of the metal industry, banking, telecommunications, electricity, etc.

3.3. Research on Work Motivation for Outsourcing

Recruitment outsourcing workers in this era is one way to win the competition and support the efficiency of the company. However, some things must be considered in imposing the system of outsourcing in companies [2], Here are some of the research that has been done concerning outsourced workers in terms of working motivate.

IOP Conf. Series: Materials Science and Engineering **885** (2020) 012060 doi:10.1088/1757-899X/885/1/012060

Table 2. Research of outsourced workers

Author (Year)	Research Title	Results and Conclusion
Tjandraningsih I et al., 2010 [18]	Job Training Outsourcing Contracts and Metal Workers in the Industrial Sector in Indonesia	1. Employment practices and outsourcing give effect fragmented, degrading, discriminatory and exploitative of workers. 2. The practice occurs because of differences of thought and various violations of the Act and regulations implementing the employment relationship and outsourcing contracts amid the weakening of competition, the role and function of oversight by Manpower in the framework of regional autonomy. 3. Employment practices and outsourcing show the violation of core Labour standards in the ILO Convention No. 87, 98, 100, 102 and 111
Indrayani and Sukmawati, 2018 [19]	Overview The use of Personal Protective Equipment Power Distribution Outsourcing in PT PLN (Persero) Rayon Wonomulyo Polewali Mandar, Indonesia (*in Bahasa*)	1. There are still outsourced workers who have low knowledge about the use of PPE 2. Distribution of outsourced workers there who feel less comfortable with the PPE to be used 3. There are still some outsourced workers who do not use PPE properly
Ishak A H S et al., 2018 [20]	Employee Performance Outsourcing do PT Telkom Regional Division VII of Makassar, Indonesia (*in Bahasa*)	From the research results that outsourced workers have good performance following procedures established by the company
Safira and Hamidah, 2018 [3]	Employee Performance Difference Analysis Outsourcing and Permanent Employees (Studies in Bank BRI Branch Office Employees Kawi Malang) In Indonesia (*in Bahasa*)	Based on the analysis of the comparative approach showed that no significant difference to the performance of the permanent employees with outsourced employees

IOP Conf. Series: Materials Science and Engineering **885** (2020) 012060 doi:10.1088/1757-899X/885/1/012060

Table 2. Research of outsourced workers (*cont.'*)

Author (Year)	Research Title	Results and Conclusion
Belton B et al., 2018 [21]	Mechanization Outsourcing Services in Myanmar's Dry Zone	The business outsourcing company has a considerable influence in the growth of the agricultural economy in rural Myanmar in the last five years
Dean E et al,. 2010 [22]	Consequences of Outsourcing Strategies on Employee Quality of Work Life, Attitudes, and Performance	These results indicate that the system is outsourced workers have a negative effect on the quality of the working environment. Although outsourcing has certain advantages for companies
Waluyo S, 2019 [23]	Employment Rights and Industrial Relations Systems Implementation: The Case of the Outsourcing Policy in Banking Industry	Outsourced workers at several banks in the object of this study have adopted regulations based on outsourcing Labour legislation and government regulation
Siew-Chen S and Seow-Voon Y, 2016 [24]	Exploring Human Resource Outsourcing Trends in Malaysia	This research was conducted by distributing questionnaires to 145 HR managers from several companies. The result show that manufacturing and service companies employ about 40% of outsourced workers. Two things that encourage outsourcing is employing workers from factors of resource requirements and cost factors
Kuruvilla S and Ranganathan A, 2010 [25]	Globalization and outsourcing: human resource confronting new challenges in India's Business Process Outsourcing Industry	This study shows that outsourced workers can reduce poverty in India, but the lack of training and knowledge of the workers outsourcing became an obstacle in the growth of the company
Mutua A N and Gichure J M, 2017 [26]	Assessment of Factors That Are Strategic to Guarantee Quality Outsourced in Aviation Ground Services: A Case of Strami Limited	This study shows that knowledge work, motivation, remuneration, and the competence of outsourced employees have a significant impact on the quality of service to customers on the flight sector
Elmuti D, 2010 [22]	Consequences of Outsourcing Strategies on Employee Quality of Work Life, Attitudes, and Performance	These results indicate that the system is outsourced workers have a negative effect on the quality of the working environment. Although outsourcing has certain advantages for companies

IOP Conf. Series: Materials Science and Engineering **885** (2020) 012060 doi:10.1088/1757-899X/885/1/012060

Table 3. Research on work motivation for outsourcing

Author (Year)	Research Title	Results and Conclusion
Robinson J R, 2016 [27]	Role of Outsourcing in Stress and Job Satisfaction of Information Technology Professionals	The result of the 20 workers in the enterprise IT outsourcing is that the outsourced workers have special attention to job stress and job dissatisfaction. However, the results of research in the IT companies showed that most of the workers were satisfied with their jobs as outsourcing
Lee G R et al., 2019 [28]	Outsourcing and Organizational Performance: The Employee Perspective	This study found that there is a negative correlation between outsourcing with companies where increased outsourcing decrease performance and job satisfaction. Besides the cost incurred also even greater because the cost of the outsourcing company agent is also greater than the benefits provided by outsourced workers
Muzanenhamo G N et al., 2016 [29]	The Relationship Among Change Implementation, Job Satisfaction and Organizational Citizenship Behaviour in the Business Process Outsourcing Industry in South Africa	There is a positive relationship between: 1. Implementation of changes to the Citizen Organizational Behaviour (OCB) 2. Implementation of changes to job satisfaction 3. OCB to job satisfaction 4. Job satisfaction over the implementation of changes and OCB
Sinuhaji E, 2014 [30]	Influence of Personality, Ability to Work and Work Motivation Performance Outsourcing of Human Resources at PT. Catur Karya Sentosa Medan (*in Bahasa*)	These results indicate that the variable personality, ability, and motivation have a significant effect on employee performance outsourcing
Tusholihah M et al., 2019 [31]	Effect of Motivation and Work Discipline Employee Performance Against Outsourcing	This research was conducted at PT Citus Branch in West Java where the results of this study showed that the variables of motivation and discipline of employees affect the performance of employees
Horwitz J, 2019 [32]	Exploring the Ethics of Outsourced in Post Apartheid South Africa	Findings from this study that outsourced workers at one of the existing University of South Africa's existence is under-recognized so that there are rules that do not fit. Therefore, the government is expected to be more concerned with outsourced workers

4. Conclusions and Suggestions

In a large company, Human Resources (HR) is one of the important resources in the sustainability of a company. In its activities, the company - large companies split the employees are permanent employees and outsourcing. Employee outsourcing in this study discusses the prevailing regulations on Labour outsourcing and the things that influence performance and employee motivation outsourcing.

From the results of several studies indicate that some companies have implemented a system of outsourcing the appropriate laws that apply. The company is also treated with either outsourcing Labour so that workers have the motivation and good performance of the company.

So that the company can create a good working environment and comfortable, good for permanent employees and for employees of outsourcing requires the application of two-factor theory by Herzberg were able to explain about the things that affect employee satisfaction and dissatisfaction in work.

In this research, there is still a shortfall in which the authors only describe some previous research on work motivation on employee outsourcing but do not focus on factors that influence job satisfaction and dissatisfaction with outsourced employees. So, expect further research to undertake specific research to identify the factors of satisfaction and dissatisfaction with outsourced workers in a company.

References

[1] Andriani M and Widiawati K 2017 Penerapan Motivasi Karyawan Menurut Teori Dua Faktor Frederick Herzberg Pada PT Aristika Kreasi Mandiri *J. Adm. Kant.* **5** 83–98
[2] Bakhri S, Buchori and Ardi R P 2018 Praktik Pengelolaan SDM Melalui Outsourcing: Konsep dan Kenyataan *J. Visi Manaj.* **3** 203–11
[3] Safira W and Hamidah N U 2018 Analisis Perbedaan Kinerja Karyawan Outsourcing dan Karyawan Tetap (Studi Pada Karyawan Bank BRI Kantor Cabang Malang Kawi) *J. Adm. Bisnis* **60** 96–103
[4] Tesavrita C and Suryadi D 2012 Identification of Herzberg's Motivator-Hygiene Factors for SME's Workers: Case Study of SME in Bandung, Indonesia *Procedia Econ. Financ.* **4** 299–303
[5] Juariyah L and Saktian I R 2018 Does Motivators Determine Employees' Job Satisfaction? Testing Herzberg Theory of Motivation in Indonesian Café and Restaurant Context *KnE Soc. Sci.* **3** 482–90
[6] Dartey-Baah K and George K A 2011 Application of Frederick Herzberg ' s Two -Factor theory in assessing and understanding employee motivation at work : a Ghanaian Perspective *Eur. J. Bus. Manag.* **3** 1–8
[7] Herzberg F 1964 The motivation to work among Finnish supervisors *Pers. Psychol.* **18** 393–402
[8] Herzberg F 1974 *Motivation-hygiene profiles: Pinpointing what ails the organization*
[9] Alshmemri M, Shahwan-Akl L and Maude P 2017 Herzberg's Two-Factor Theory *Life Sci. J.* **14** 12–6
[10] Feizal R 2015 Pengaruh Hygiene Factors dan Motivating Factors Terhadap Kepuasan Kerja dan Dampaknya Terhadap Organizational Citizenship Behavior (OCB) *Mix J. Ilm. Manaj.* **5** 481–99
[11] Nirmalasari Y F 2018 Analisis Pengaruh Faktor Hygiene dan Faktor Motivator Terhadap Kepuasan Kerja Karyawan *Parsimonia* **5** 199–210
[12] Shaikh S H, Shaikh H and Shaikh S 2019 The Impact of Job Satisfaction and Job Dissatisfaction on Herzberg Theory: A Case Study of Meezan Bank Limited and National Bank Limited *Int. J. Bus. Soc. Sci.* **10** 143–7
[13] Hilmi A, Ali C and Nihal C 2016 Herzberg's Motivation- Hygiene Theory Applied to High School Teachers in Turkey *Eur. J. Multidiscip. Stud.* **1** 90–7
[14] Bryant O A 2018 Employee Turnover in the Long-Term Care Industry using Herzberg's Motivation-Hygiene Theory *Int. J. Acad. Res. Bus. Soc. Sci.* **8** 57–76
[15] Anah C I 2015 Re-Evaluating The Relevance of Herzberg ' S Motivation-Hygiene *Int. J. Educ.*

Res. **3** 731–40
[16] de Waal A and Jansen P 2013 The bonus as hygiene factor: the role of reward systems in the high performance organization *Evidence-based HRM* **1** 41–59
[17] Wulansari C D 2019 Outsourcing: Dilema antara Strategi Peningkatan Profitabilitas dan Tanggung Jawab Sosial
[18] Indrasari T, Rina H, Sriwulan F F and Suhadmadi 2010 *Praktek Kerja Kontrak dan Outsourcing Buruh di Sektor Industri Metal di Indonesia* (Bandung)
[19] Indrayani I and Sukmawati S 2018 Gambaran Penggunaan Alat Pelindung Diri Tenaga Outsourcing Distribusi Di PT. PLN (Persero) Rayon Wonomulyo Kabupaten Polewali Mandar *J-KESMAS* **4** 59–71
[20] Akhmad H S I, Adys A K and Haerana 2018 Kinerja Pegawai Outsourcing di PT . Telkom Divre VII Kota Makassar *KOLABORASI J. Adm. Publik* **4** 268–81
[21] Belton B, Fang P and Reardon T 2018 *Mechanization Outsourcing Services in Myanmar's Dry Zone*
[22] Elmuti D, Grunewald J and Abebe D 2010 Consequences of Outsourcing Strategies on Employee Quality of Work Life, Attitudes, and Performance *J. Bus. Strateg.* **27** 177
[23] S Waluyo Y 2019 *Employment Rights Implementation and Industrial Relations Systems : The Case of the Outsourcing Policy in Banking Industry*
[24] Siew-Chen S and Seow-Voon Y 2016 Exploring Human Resource Outsourcing Trends in Malaysia *Procedia - Soc. Behav. Sci.* **224** 491–8
[25] Kuruvilla S and Ranganathan A 2010 Globalisation and outsourcing : confronting new human resource challenges in India ' s business process outsourcing industry *Ind. Relat. J.* **41** 136–53
[26] Mutua A N and Joseph Mwangi G 2017 Assessment Of Factors That Are Strategic To Guarantee Quality Outsourced Ground Services In Aviation : A Case Of Strami Limited *IJIRAS* **4** 155–60
[27] Robinson J R 2016 *Role of Outsourcing in Stress and Job Satisfaction of Information Technology Professionals* (Walden University)
[28] Lee G R, Lee S, Malatesta D and Fernandez S 2019 Outsourcing and Organizational Performance: The Employee Perspective *Am. Rev. Public Adm.* **49** 973–86
[29] Muzanenhamo G N, Allen-lle C, Adams A and Iwu C G 2016 The relationship among change implementation, job satisfaction and organizational citizenship behavior in the Business Process Outsourcing industry in South Africa *Bus. Perspect.* **14** 473–82
[30] Sinuhaji E 2014 Pengaruh Kepribadian, Kemampuan Kerja dan Motivasi Kerja terhadap Kinerja SDM Outsourcing pada PT. Catur Karya Sentosa Medan *J. Ilman* **1** 11–22
[31] Tusholihah M, Hasyim A N, Novitasari A, Oktavia P P, Lestari F I, Fadli M and Sobari A M 2019 Pengaruh Motivasi dan Disiplin Kerja Terhadap Kinerja Karyawan Outsourcing *UNEJ e-Proceeding* vol 5
[32] Horwitz J 2019 *Exploring the Ethics of Outsourced Labour in Post- Apartheid South Africa by* (Stellenbosch University)

Overall Equipment Effectiveness (OEE) analysis and Failure Mode and Effect Analysis (FMEA) on Packer Machines for minimizing the *Six Big Losses* - A cement industry case

I S. Muthalib, M Rusman and G L Griseldis

Industrial Engineering Department, Universitas Hasanuddin, Makassar - Indonesia

E-mail: irwan@tiunhas.net

Abstract. Bosowa Maros Cement Company (BMC) is a company engaged in the production of cement. There are seven packer machines for the packaging process to fulfill the demand in BMC. The most problems are many breakdowns that occur on packer machines, and the amount of production, which not maximum, causes a lack of Overall Equipment Effectiveness (OEE) value. The research aimed to identify the performance of packers based on production data using OEE and Failure Mode Effect Analysis values. The results of the six big losses and the Pareto diagram show that the most affect losses in causing the low achievement of OEE values on both machines are reduced speed loss, idling and minor stoppages, equipment failure (breakdown loss) for Packer 3 is 94.27% and Packer 7 is 90.41%. The average OEE value of the Packer 3 machine is 53%, and the Packer 7 machine is 36%. OEE values on the two machines have not reached the OEE ideal standard by the Japan Institute of Plant Maintenance (JIPM), which is 85%.

1. Background

The strict competition in the era of globalization requires industries to be more productive with the best quality for each product produced. Bosowa Maros Cement Company (BMC) is an industry located in Maros Regency, South Sulawesi, Indonesia. BMC itself has 7 Packer machines that are operated for the cement packaging process, but not all packer machines are able to perform optimally.

Based on company data for the period January 2018 to July 2019 Packer 1 operated for 442 days, Packer 2 operated for 483 days, Packer 3 operated for 521 days, Packer 4 operated for 516 days, Packer 7 operated for 374 days, Packer 8 operated for 385 days day, and Packer 9 operated for 401 days. There was no Packer that could operate 100% during that period due to some failures or damages on the equipment, or the packer was only idle [1]. Based on the results of interviews in the Production Planning and Inventory Control (PPIC) and Production Engineering (PE) sections, the packer has experienced to some of the same types of failures, including the spout not functioning or being clogged by cement solids, the machine dies due to various equipment malfunctions, bag filling which exceeds the maximum time from the target, and so on that those will result in unplanned downtime on the machine.

In the world of engine maintenance, there are six type of losses that must be avoided by every company that can reduce the level of effectiveness of a machine, known as the Six Big Losses [2]. As mention in Almeanazel [3], the six big losses are usually categorized into 3 main categories based on aspects of loss: downtime, speed losses and defects. A comprehensive calculation method to identify the level of productivity and machine performance is called Overall Equipment Effectiveness (OEE). The purpose of this research were to find out how much the achievement of the packer's OEE values includes availability rate, performance efficiency, and quality rate at BMC, identify the losses that most affect the Packer engine's evectivity, and provide suggestions for improving engine performance from the measured OEE values, which will be proposed as consideration for improvements to the Packer engine.

2. Literature Review

2.1. Overall Equipment Effectiveness (OEE)

Overall Equipment Effectiveness (OEE) is a calculation carried out to determine the extent of the effectiveness of an existing machine or equipment [3]. OEE is one of the methods available in Total Productive Maintenance (TPM) according to [4]. Generally, OEE is used as an indicator of the performance of a machine or equipment. The purpose of OEE is as a measure of the performance of a maintenance system, by using this method it can be seen the availability of machines or equipment (availability), production efficiency (performance), and the quality of the output of the machine/ equipment (quality) [5].

2.2. Six Big Losses

There are six equipment losses that cause low performance of machines and equipment. the six losses are known as six big losses which are divided into 3 categories of losses: downtime losses (breakdown, set up and adjustment) speed losses (idling and minor stoppages losses, reduce speed), and quality losses (rework and quality defects, start up reject) [6].

2.3. Failure Mode and Effect Analysis (FMEA)

Failure Mode and Effect Analysis (FMEA) is a structured procedure to identify and prevent possible failure modes [7] and [4]. The factors are defined are: Severity, Occurance, and Detection. The Risk Priority Number (RPN) is get by multiplying Severity, Occurance, and Detection.

3. Research Methodology

Data is taken from PPIC department of BMC. By used the OEE method as written in [2], there were three stages of the process, including the calculation of availability ratio (down time losses), performance ratio (speed losses), and quality ratio (quality losses).

Also mentioned in [8] and [9], after the effectiveness of the engine was known, the calculation of the magnitude of the ratio of the six components of the six big losses was equipment failure, set up and adjustments, idling and minor stoppages, reduced speed, scrap and rework and startup losses to find out what categories of losses are causing low OEE value of the machine.

Then the research continued by filling in the FMEA questionnaire as a method of identifying problems. FMEA questionnaire data was obtained from one respondent namely Production Engineering employees who were responsible for managing the activities of the Packer machine. How to calculate FMEA questionnaire data by using RPN, i.e. multiplying severity, occurance and detection [4]. The results to be obtained were expected to be a reference for decision making to achieve a good maintenance management system. The data processed was the downtime of packer 3 and packer 7 machines from January 2018 to June 2019 where later the processing results are compared with international OEE standards.

4. Results

4.1. Calculation of OEE.

The ABC inventory management system is a technique for managing inventory control by dividing inventory systems that need to be controlled more. [10][11] mention that the ABC category is based on the price of the type of inventory and the amount of company demand. From the results of data processing obtained spare parts grouping as in Table 1.

Table 1. Calculation of OEE results of *Packer* 3 dan *Packer* 7

Period	Avaliability Rate (%)		Performance Rate (%)		Quality Rate (%)		OEE (%)	
	Packer 3	Packer 7	Packer 3	Packer 7	Packer 3	Packer 7	Packer 3	Packer 7
1	76	78	84	51	99	99	64	40
2	82	80	75	51	99	99	61	40
3	67	48	72	66	99	99	47	31
4	55	64	94	95	98	99	51	60
5	84	84	72	75	99	99	60	63
6	84	84	55	55	99	100	46	45
7	88	88	68	74	99	99	59	65
8	100	100	64	37	99	99	63	37
9	65	57	70	63	98	99	44	35
10	70	28	80	77	99	99	56	21
11	75	65	72	43	99	99	53	27
12	74	83	76	37	99	99	56	30
13	85	22	62	78	98	99	51	17
14	85	79	64	17	98	99	54	13
15	77	69	76	44	98	99	58	30
16	78	77	60	42	99	99	47	33
17	68	58	69	46	99	99	46	27
18	84	72	53	50	99	99	44	36
Median	78	69	70	56	99	99	53	36

4.2. Calculation of Packer Machine Losses

This calculation is to identify the six components of six big losses, which component has the highest ratio or influence on the low OEE value that has been obtained.

Table 2. Cumulative recapitulation of Six Big Losses Packer 3 and Packer 7

Group Losses	Six Big Losses	Time Losses (hour)		Time Losses (%)		Cumulative	
		Packer 3	Packer 7	Packer 3	Packer 7	Packer 3	Packer 7
Downtime Losses	Equipment Failure	2,112.58	1,620.76	17.71	22.55	18	23
	Set Up & Adjusment	560.71	660.05	4.76	8.84	22	31
Speed Losses	Idle & Minor Stoppages	2,673.29	2,280.81	22.47	31.39	45	63
	Reduce Speed	6,491.00	3,049.00	54.09	36.47	99	99
Quality Losses	Scrap & Rework	59.28	25.16	0.00	0.00	100	100

IOP Conf. Series: Materials Science and Engineering **885** (2020) 012061 doi:10.1088/1757-899X/885/1/012061

Startup Losses	0,00	0.00	0.00	0.00	100	100

4.3. Failure Mode and Effect Analysis (FMEA)

From the results of the total time loss on the Packer 3 engine and the Packer 7 engine, there was three losses factors, namely Reduced Speed Loss, Idling and Minor Stoppages, and Equipment Failure. These three losses factors analyzed using Failure Mode and Effect Analysis. The FMEA analysis was made based on the results of interviews and discussions with the head of production and Production Engineering of the Packer machinery section at BMC.

Table 3. Questionaire filling data

Factor Losses	Failure Mode	Failure Cause	Failure Effect	Initial Conditions			
				S	O	D	RPN
Equipment Failure (breakdown)	Breaker motor trip	Dust conceded contactor	The spout does not work during the cement filling process	7	5	4	140
	Trouble instrumen	Slip ring encountered a problem (carbon brush was used up / worn out)	The engine is off for a moment, and repair is needed	8	4	5	160
Idling and Minor Stoppages	Flow attempt failed	The spout is blocked by solid cementa	The spout lights up but does not release material (cement)	6	5	6	180
	Bag can't released	Rubber Buffer aus	Couldn't remove bag	4	5	4	80
	Lost work time	Power Blackout	Working time is reduced, production targets may not achieve than targeted	6	4	4	96
		Workers (laborers / operators) are not available	Work is pending to wait	6	3	3	54
Reduced Speed Loss	Beyond maximum filling time	Material discharge is not optimal	Bag filling exceeds the maximum amount of time targeted	7	5	6	210
		Aerasi Cylinder Bag Holder leaking out	Cannot remove the bag according to the filling limit and release towards the irregular conveyor belt	3	5	4	60
	Actual and Ideal production time doesn't match	Workship takes too long, workers tend to chat while working	Work completion time sometimes does not match the target	4	3	3	36

From three losses factors, the reduced speed loss factor, which is the failure mode beyond the maximum filling time, is the largest RPN, which means that these losses must get more attention to be overcome.

5. Discussion

In Packer 3 it appears that in the 9th and 18th periods the OEE value of the machine was very low by 44%. In packer 7 it appears that in the 13th period the OEE value of the machine was very low by 17%. This is due to the low level of engine performance ratio in that period. Overall, the OEE value on Packer 3 machines was 56% and in Packer 7 it was 36% below the Japan Institute of Plant Maintenance (JIPM) standard of 85%. Of the three components, the value that affects OEE is the value of a very low performance ratio. Those were stated in [12] and [13]. So, it can be stated that the overall performance or productivity of the two Packer machines is not maximal, still far below ideal conditions.

From the OEE result, it can be seen that in Packer 3, the value of the biggest loss ratio is caused by reducing speed losses by 54.09%, idle & minor stoppages losses by 22.47%, and Equipment Failure Losses by 17.71% which is 94, 27% so it becomes a major factor in the low level of effectiveness of the Packer 3 engine. Based on the cumulative recapitulation of six big losses packer 7, it can be seen that the value of the largest loss ratio is caused by reducing speed losses by 36.47%, idle & minor stoppages losses by 31.39%, and Equipment Failure Losses by 22.55% which amounts to 90, 41% so it becomes a major factor in the low level of effectiveness of the Packer 7 engine. The three other factors: set up & adjusment losses, scarp & rework and start up losses as a whole only affect 4.76% for Packer 3 and 8.84% for Packer 7 so that they do not significantly influence the low OEE value on the machine, then the three factors are not a priority for improvement.

Based on the RPN from the Worksheet FMEA of the Packer 3 and Packer 7 machines before, the biggest cause of failure was known to affect the low OEE value in the reduced speed loss factor, the same emntion in [6], which is beyond the maximum filling time with an RPN of 210 due to the material discharge from the Packer machine which is not optimal which results in the bag filling exceeding the limit maximum time than targeted. The next biggest cause of failure is the idling and minor stoppages factor, namely the flow attempt failed with an RPN of 180 caused by a blockage in the spout so that the bag is not filled with material. In the equipment failure factor (breakdown loss) the biggest cause of failure was trouble instrument with an RPN of 160 because the engine was running and then turns itself off. The three losses factors based for the reduced speed loss factor, which is the failure mode beyond the maximum filling time, was the largest RPN, which means that these losses must get more attention to be overcome.

6. Conclusion

The OEE value on the Packer 3 machine were 53%, and Packer 7, which was 36%, and still far from the world standard OEE standard set by the Japan Institute of Plant Maintenance (JIPM) by 85%. So, it can be stated that the effectiveness of the two machines, namely Packer 3 and Packer 7 during low production. This was due to the large amount of loading time that absorbed due to Reduce Speed, Idle and Minor Stoppages, and Equipment Failure (Breakdown Loss) that occured on the engine.

It was known that the biggest factors that cause the low achievement of OEE values on Packer 3 and Packer 7 machines were Reduced Speed Loss, Idling and Minor Stoppages, and Equipment Failure (Breakdown Loss). On the Packer 3 engine, the Reduce Speed Loss factor results in an inefficient time of 54.09% or 6491 hours, Idling and Minor Stoppages results in an inefficient time of 22.47% or 2,673.29 hours, and Equipment Failure (Breakdown Loss) resulting in inefficient time of 17.71% or 2,112.58 hours during January 2018 - June 2019.

On the Packer 7 engine the Reduce Speed Loss factor results in an inefficient time of 36.47% or 3049 hours, Idling and Minor Stoppages results in an inefficient time of 31.39% or 2280.81 hours, and Equipment Failure (Breakdown Loss) results inefficient time of 22.55% or 1620.76 hours during January 2018 - June 2019. Therefore, the higher the total time loss, the less the effectiveness of the Packer engine in producing products.

Proposed design improvements proposed by researchers to the company in minimizing the Six Big Losses that occur were to make changes in preventive parts and determine the life of the parts life, perform routine scheduled cleaning every month so that no more blockages occur in the spout during the filling process, and preventive maintenance measures taken before the machine operates and can prevent damage to the machine while operating for the sake of smooth production.

References

[1] Corder A and Hadi K 1988 *Teknik Manajemen Pemeliharaan* (Erlangga)
[2] Anggraini M, Marcelly Widya W and Kujol Edy F B 2017 Implementasi Metode Overall Equipment Effectiveness Dalam Menentukan Produktivitas Mesin Rotary Car Dumper *Seminar dan Konferensi Nasional IDEC 2017* pp 78–87
[3] Almeanazel O T R 2010 Total productive maintenance review and overall equipment effectiveness measurement *Jordan J. Mech. Ind. Eng.* **4** 517–22
[4] O'Brie M 2015 *TPM and OEE* (Limerick, Ireland: About LBSPartners)
[5] Gaspersz V 2002 *Total Quality Management* (Jakarta: PT Gramedia Pustaka Utama)
[6] Mayangsari D F, Adianto H and Yuniati Y 2015 Usulan Pengendalian Kualitas Produk Isolator dengan Metode Failure Mode And Effect Analysis (FMEA) dan Fault Tree Analysis (FTA) *Reka Integr.* **3** 81–91
[7] Press D 2003 *Guidelines for failure mode and effects analysis (FMEA), for automotive, aerospace, and general manufacturing industries* (Florida: CRC Press)
[8] Muhtadi M Z Z 2009 Manajemen Pemeliharaan Untuk Optimalisasi Laba Perusahaan *J. Pendidik. Akunt. Indones.* **VIII** 35–43
[9] Sayuti M, Juliananda, Syarifuddin and Fatimah 2019 Analysis of the Overall Equipment Effectiveness (OEE) to Minimize Six Big Losses of Pulp Machine: A Case Study in Pulp and Paper Industries *IOP Conf. Ser. Mater. Sci. Eng.* **536** 1–7
[10] Venkatesh J 2007 An Introduction to Total Productive Maintenance (TPM) *Plant Maint. Resour. Cent.*
[11] Nakajima S 1988 *Introduction to TPM: Total Productive Maintenance (Translation)* (Productivity Press, Inc.,)
[12] Ahuja I P S and Khamba J S 2008 Total productive maintenance: literature review and directions *Int. J. Qual. Reliab. Manag.* **25** 709–56
[13] Hansen R C 2001 *Overall equipment effectiveness: a powerful production/maintenance tool for increased profits* (New York: Industrial Press Inc.)

IOP Conf. Series: Materials Science and Engineering **885** (2020) 012062 doi:10.1088/1757-899X/885/1/012062

International Joint Venture (IJV) in emerging economies: Case Study of Automotive Industry

K Nadiyah[1], V Vojisavljevic[1], K Amar[2]

[1] School of Engineering, RMIT University, Melbourne, Australia
[2] Department of Industrial Engineering, Faculty of Engineering, Universitas Hasanuddin, Indonesia

E-mail: nadiyahkhairun@gmail.com

Abstract. International Joint Venture (IJV), a popular form of Foreign Direct Investment (FDI), is a favourite mode for automobiles multinational companies (MNCs) to expand into new markets, especially emerging economies. However, many things are considered by MNCs in selecting investment destinations, including industry growth and the quality of destination countries. Therefore, emerging economies need to show their attractiveness to attract MNCs to invest there. This research has analysed the growth of the automotive industry in two emerging economies, India and Indonesia, and its economic impacts, and provides information on the contribution of the IJV to the host country and explains the attractiveness of the two countries, particularly in the automotive field. The findings on the research are useful in giving consideration to MNCs to invest in both countries, especially in the automotive field. Furthermore, it also provides information to emerging economies about the contribution of the IJV company to the host country.

1. Introduction

International Joint Venture (IJV), a popular form of Foreign Direct Investment, has become a business growth strategy favoured by MNCs to enter emerging economies since the 1980s [1]. The increase has been sharp in the last few decades [2] and has succeeded in becoming an expansion mode favoured by MNCs to enter the global market [3].

Experts agree that the IJV has a strong appeal for MNCs because it opens wide opportunities to global markets [4][5][6]. The existence of local partners provides great benefits for the IJV, including accelerating the expansion process, transferring technology, resources and skills, minimizing risk and reducing costs [7][8] hence it opens up greater opportunities for success if done in a way correct and appropriate.

The surprising fact is that the IJV has a high failure rate of 70% [9] and the basic trigger is the wrong partner selection [10], causing a lot of conflict and ultimately leading to IJV instability. Therefore, companies must reckon the process of selecting partners if they desire to increase the chances of the IJV success. In addition, companies also need to take into account internal factors (technological capability, cultural compatibility) and external (business environment and industry growth) factors of the potential partners [7].

The high failure rate of IJV means that IJV has a high level of complexity. This complexity will be even greater when IJV operates in the automotive industry, a large industry that involves many business components, from upstream to downstream, bringing them together from different cultural backgrounds and understandings. Emerging economies are the most profitable market for automotive industry [11]. As a result, IJV research in the automotive sector is crucial to be carried out in order to encourage the success of IJV in the automotive sector, especially in emerging economies.

Indonesia and India, two of emerging economies, are well aware of the large contribution of foreign investment in developing the country hence both countries are currently trying to attract foreign investors to invest, especially in several priority sectors, one of which is the automotive sector. To response that condition, this paper is focused on analysing India and Indonesia's automotive industry.

2. Methodology

This research has used quantitative and qualitative methods. Quantitative method was used to analyse the growth of the automotive industry in India and Indonesia while qualitative method was used to analyse the attractiveness and contribution of IJV companies to both countries. The data used were secondary obtained from several published corporate, government and private institutions data namely SIAM (Society of Indian Automobile Manufactures), DIPP (Department of Industrial Policy and Promotion), IBEF (India Brand Equity Foundation), BKPM (Indonesia Investment Coordinating Board Official), GAIKINDO (The Association of Indonesia Automotive Industries), AISI (Indonesian Motorcycles Industry Association), Maruti Suzuki Reports, Toyota Motor Manufacturing Indonesia Reports etc. The numerical data that have been used were automotive production, FDI, Export volume, GDP and domestic sales of the two countries. Furthermore, numerical data have been analysed with some statistical tools such as AAGR (Average Annual Growth Rate), CAGR (Compound Annual Growth Rate), and correlation. Case study design was used in this study to narrow discussions about IJV in both countries. Maruti Suzuki and Toyota Astra Motor are two venture companies which became the object of IJV case studies in this research.

3. Results and Discussions

3.1. India and Indonesia automobile production's growth

The automotive industry has an important role in driving the country's economic growth. This industry supports many related industries, both upstream and downstream, such as steel, aluminium, rubber, battery, software and so on.

Table 1. Automotive Production

Country (Period)	Growth	Passenger Vehicles	Commercial Vehicle	Three Wheelers	Two Wheelers	Grand Total
India	AAGR	3,9%	5,7%	6,9%	8,0%	7,2%
(2010-19)	CAGR	3,8%	4,9%	5,9%	7,9%	7,1%
Indonesia	AAGR					
(2010-18)	CAGR	9,9%	4,3%			

Table 1 shows the growth trend of automotive production in India and Indonesia. Overall, India and Indonesia were experiencing the positive growth with AAGR and CAGR value above 3,8% in all of vehicle's types over the given period.

3.2. Foreign direct investment growth in India and Indonesia automotive industry

The automotive sector is ranked in the top 10 largest attractors of FDI in India [12] and Indonesia [13]. The increase in FDI gives a positive effect on the growth of related industries.

Table 2. FDI Inflows in Automotive Industry

	India	Indonesia
Year	2010-19	2010-18
AAGR	23%	30%
CAGR	15%	11,90%

The result of data processing in Table 2 shows that both countries were experiencing a positive growth of FDI Inflows. India shows AAGR of 23% and CAGR of 15% in FY 2010-19 while Indonesia registered an AAGR of 30% and CAGR of 11,9% during 2010-18.

Table 3. Correlation analyses between FDI inflows in automobile industry and total production

Country (Period)	r
India (2010-19)	0.764
Indonesia (2010-18)	0.512

Table 3 presents that the value of r in India is +0,764 which means that there is a sufficient correlation between FDI Inflows in Automotive industry and Total Production while the correlation both variables in Indonesia is rather weak with r value of +0.512.

3.3. Export by India and Indonesia automotive sector

The noble export performance will add to Indonesia and India tax revenue. Below are the export analyses in India and Indonesia automotive industry (Table 4)

Table 4. Export trend

	Growth	Passenger Vehicles	Commercial Vehicle	Three Wheelers	Two Wheelers
India (2011-19)	AAGR	4.5%	1.7%	10.1%	8.1%
	CAGR	4.2%	1.1%	6.6%	7.7%
	Growth	Complete Built Up (CBU)	Complete Knock Down (CKD)	Component Part	Two Wheelers
Indonesia (2011-18)	AAGR	15.3%	7.5%	160.9%	101%
	CAGR	13.7%	-0.3%	8.7%	54%

The data above shows that there has been an increase in export volume in every vehicle's category in India and Indonesia. If we look at both sides (AAGR and CAGR), overall types have a substantial positive value.

3.4. Domestic sales pattern

Table 5 is the Domestic Sales Pattern of India and Indonesia Automotive industry

Table 5. Domestic sales

Country	Growth	Passenger Vehicles	Commercial Vehicle	Three Wheelers	Two Wheelers	Grand Total
India	AAGR	6,20%	10.10%	8.30%	7.50%	8.03
(2013-19)	CAGR	6.20%	9.70%	7.90%	7.40%	7.80
Indonesia	AAGR	11.40%			4.00%	
(2007-18)	CAGR	9.30%			3.00%	

The data above shows the domestic sales of four kind of vehicles in India and two kind of data (four wheelers and two wheelers) in Indonesia. It can be seen clearly from AAGR and CAGR values, domestic sales in both countries were experiencing a great growth in all of vehicle types over the given period.

3.5. Employment in Automotive Industry

Along with renewed government policies that are friendly to foreign investors, rising FDI in the automotive sector, expanding domestic and export markets, and increasing automotive production, have opened up great opportunities for increased absorption of the automotive sector workforce in India and Indonesia. Based on data from the Indonesian Ministry of Industry [14], the automotive industry was able to absorb more than 1 million workers. On the other side, in India, the automotive industry has employed more than 30 million people according to data from SIAM [15].

3.6. GDP per capita and automobile domestic sales

Table 6 is the correlation analyses between GDP Per Capita and Automobile Domestic Sales.

Table 6. Correlation analyses between GDP Per capita and automobile domestic sales

Country	r
India	0.995
Indonesia (4-wheeler)	0.923
Indonesia (2-wheeler)	0.557

Table 6 presents that the value of r in India is +0,995 which means that there is very strong correlation between GDP Per Capita and Automobile Domestic Sales. Similarly, the result shows that the correlation between both variables in Indonesia (4-wheeler) is very strong (r = +0.932) while the correlation in 2-wheelers is rather weak with r value of +0,570.

3.7. GDP contribution by automotive sector in India and Indonesia

Automobile Industry in India had been contributing more than 7% of total GDP recently [15] while Indonesia is slightly larger at an average of 9.4% in the last seven years [16].

3.8. IJV contribution to the host country

Table 7 is the contribution of Maruti Suzuki and TMMIN to the host country.

Table 7. Maruti Suzuki and Toyota Motor Manufacturing contribution

	MARUTI SUZUKI (India)	TMMIN (Indonesia)
ENVIRONMENT [17][18]	Adopt and promote the ISO14001 standard	Adopt and promote ISO 14001 standard Toyota Eco Youth and Toyota Forest
ECONOMY	• 33.180 Employee (until 30 March 2019) [17] • Export (AAGR of 10% and a CAGR of 6% over 2004-19 period) and Sales (CAGR and AAGR of 10% over 2004-19 period)	• Income Generating Society [18] • More than 8.800 employee (until January 2019) [18] • Biggest contributor in the export volume • Biggest domestic market share
SKILL DEVELOPMENT	Training program; Employee Union; Schools and Learning Centre [17]	Training program; Manufacturers Club; Learning Centre [18]
COMMUNITY DEVELOPMENT [17-18]	Has some Schools • WATER ATM :21 in 20 villages • Cleanliness: Waste treatment program; household toilets (4.345 in 26 villages); collecting and disposing household waste. • Paved Road Renovations • Road Safety Campaign	Has an Education Foundation • Scholarship: > 99.000 recipients since 1974 • Donation: Rp.4.306.000.000 in 2017 Health Sector: fogging and *posyandu* (Integrated Healthcare Centre) Infrastructure development Social Program Technology Transfer and Management [19]
ADDITIONAL	• Road Safety Technology [17] • Technology Transfer and Management [20]	

3.9. India and Indonesia Attractiveness

Table 8 shows several attractiveness of India and Indonesia as destination for investing, particularly in automotive sector:

Table 8. India and Indonesia attractiveness

	INDIA	INDONESIA
ECONOMY	• Projected to become the fourth largest economy in the world by 2020 [21] • By 2030, about 600 million people will live in cities [22]	• The largest economy in Southeast Asia [23] • GDP stability [16] • More people will live in urban area [16]
LABOUR	• Large working-age population, projected will reach 869 million by 2020, constituting 28% of the global workforce [22]	• Large working-age population (67.6%)[24] • Monthly wage average: 2,83 million Rupiah [24] • Literacy rate: 95,7% in 2018 [24]
RESOURCES	• Literacy rate: 74% [22] • Second largest steel producer and combined aluminum production stood at 1,12 million tones in 2019 [25]	• World largest producer of nickel in 2018 and currently aggressively focused on battery production development for Electrical Vehicle [26]
POLICY RENEWAL MARKET	• 100% FDI with a simple entry procedure [27] • Total population: 1,37 billion people [22] • Ranks fourth as the country with the most vehicle sales at present [27] • In 2018, out of 1.000 people, only 27 had cars in India [28]	• 100% foreign ownership; tax holidays or tax incentives; OSS System [14] • Total population: above 265 million people [22] • The most vehicle sales in ASEAN [29] • In 2016, out of 1.000 people, only 59 people owned cars, and 67% of car owners consider that their car is a symbol of their success [16]
PROGRAM ADDITIONAL	• Make in India [30] • EODB Ranking: 77 [31] • English is the second language [32]	• Making Indonesia 4.0 Program [33] • EODB Ranking: 73 [31]

4. Conclusion

India and Indonesia are two emerging economies with noble performance in the automotive industry. It can be seen from the results of the analysis that both countries' automotive industry experienced positive growth hence it had a positive impact on export volumes, GDP, FDI, and employers in both countries. As an FDI earner, export producer, and GDP contributor, the automotive industry is vital in determining both countries' economy. IJV is a popular mode used by MNCs to expand into emerging economies. Two examples of successful IJV companies, Maruti Suzuki in India and TMMIN in Indonesia, have had many positive impacts on host countries, in the areas of environment, economy, skill development and community development. India and Indonesia have a strong appeal as a destination for investment. In addition, the understanding of the two countries on the positive impact of foreign investment, one of which is in the form of the IJV, makes the two countries more eager to increase their attractiveness to become the chosen destination for investors to invest, especially in the automotive sector.

Acknowledgment
The author would like to express a genuine gratitude to LPDP (Indonesia Endowment Fund for Education) for its generous financial support in finishing this research.

References

[1] Ali M 1995 *Performance of international joint ventures (IJVs) in developing countries: a study of IJVs in Bangladesh* (University of Wollongong)
[2] Kabiraj T and Sengupta S 2018 A theory of joint venture instability under inter-partner learning *Res. Int. Bus. Financ.* **46** 363–72
[3] Makino S, Chan C M, Isobe T and Beamish P W 2007 Intended and unintended termination of international joint ventures *Strateg. Manag. J.* **28** 1113–32
[4] Kogut B 1988 Joint ventures: Theoretical and empirical perspectives *Strateg. Manag. J.* **9** 319–32
[5] Yan A and Zeng M 1999 International joint venture instability: A critique of previous research, a reconceptualization, and directions for future research *J. Int. Bus. Stud.* **30** 397–414
[6] Shenkar O 1990 International joint ventures' problems in China: Risks and remedies *Long Range Plann.* **23** 82–90
[7] Schillaci C E 1987 Designing successful joint ventures *J. Bus. Strategy* **8** 59
[8] Gaur A S, Ma X and Ding Z 2018 Home country supportiveness/unfavorableness and outward foreign direct investment from China *J. Int. Bus. Stud.* **49** 324–45
[9] Lowen A and Pope J 2008 Survival analysis of international joint venture relationships *J. Bus. Econ. Stud.* **14** 62
[10] Fatehi K and Choi J 2019 International communication and negotiation *International Business Management* (Springer) pp 109–43
[11] Mohr D, Müller N, Krieg A, Gao P, Kaas H-W, Krieger A and Hensley R 2013 The road to 2020 and beyond: What's driving the global automotive industry *McKinsey Co Automot Assem. Latest Think* **28** 2014
[12] DIPP 2019 FDI Statistics
[13] BKPM 2019 FDI Statistics
[14] RI M of I 2018 *Kebijakan Sektor Industri Otomotif Dalam Rangka Implementasi Roadmap Industri 4.0*
[15] Muzzafar K, Singh R, Bhat S A, Haseeb M and Ghani A 2017 Early result of pediatric shaft femur fractures treated by hip spica cast in children upto 10 years of age *Int. J. Res. Med. Sci.* **5** 2430
[16] Indonesia M A S 2017 *Automotive: Too big to fail* (Jakarta)
[17] Limited M S I 2019 *Integrated Report 2018-19* (New Delhi)

[18] Indonesia P T M M 2018 *Sustainability Report 2018* (Jakarta)
[19] Irawati D and Charles D 2010 The involvement of Japanese MNEs in the Indonesian automotive cluster *Int. J. Automot. Technol. Manag.* **10** 180–96
[20] Motohashi K 2015 Suzuki Motor's Expansion in India *Global Business Strategy* (Springer) pp 223–41
[21] Budhwar P S, Varma A and Kumar R 2019 *Indian Business: Understanding a rapidly emerging economy* (Routledge)
[22] Bliss S 2019 India: Population and urbanisation *Geogr. Bull.* **51** 78–98
[23] The World Bank The World Bank in Indonesia *2019*
[24] BPS-Statistics Indonesia 2019 *Statistical Yearbook of Indonesia 2019* (Jakarta)
[25] IBEF 2019 *Metals and Mining*
[26] Young L E 2019 *Nickel and the Battery Revolution. A New Dawn for Nickel in Batteries* (Singapore)
[27] IBEF 2019 Automobile Industry in India
[28] McGee P and Mundy S India displaces Germany to become fourth-largest auto market Financial Times *2019*
[29] AAF 2019 Statistics
[30] Rajesh K 2019 Make in India: Chalengges and Opportunities *IJSSER* **4** 2075–95
[31] Bank W 2019 *DOING BUSINESS 2019: Training for Reform* (New York)
[32] Graddol D 2010 *English Next India: The future of English in India*
[33] Satya V E 2018 Strategi Indonesia Menghadapi Industri 4.0 *Info Singk.* **10** 19–24

IOP Conf. Series: Materials Science and Engineering **885** (2020) 012063 doi:10.1088/1757-899X/885/1/012063

Analysis of Total Productive Maintenance (TPM) and Failure Mode And Effect Analysis (FMEA) to improve machine effectiveness: A study on Indonesia's sugar mills

Lukmandono, R Prabowo, E Sulistyowati

Adhi Tama Institute of Technology, Surabaya, Indonesia

Email: lukmandono@itats.ac.id

Abstract. Many companies experienced a halt of operation because their production machines are damaged. As a result, customer demand is not fulfilled; such productivity decreases. This study conducted at PT. PG Candi Baru Sidoarjo.They often experience a suffered engine damage on their rolling and grinding station. The purposes of this study were to improve their machine performance at the grinding station by using the Total Productive Maintenance (TPM) and Failure Mode and Effect Analysis (FMEA), to measure the value of Total Productive Maintenance (TPM), and to determine the type of damage that often occurs using the Age Replacement method in order to improve engine maintenance schedule. As the results, the Overall Equipment Effectiveness (OEE) at the grinding station decreased from 2015 to 2016 in the amount of 74.6% to 69.7%. The most frequent types of damage were the IMC chain abrasion, IMC chain breaking, dropping/off track and slip mill. For the type of squashed IMC damage, treatment was carried out on the 12th day at the cost of IDR 3,806,425.92. The dropped/exited IMC from the operation line was repaired on the 13th day at the expense of IDR 5,614,466.88For the type of rolling slip damage, treatedment done on the 10th day at the cost of IDR 34,272,482.34.

1. Introduction

The machine is one of the crucial things in the manufacturing world because in production activities, it will directly require a device, whether it is used as a tool or as the main production machine. According to Wakjira et al. [1], the engine is an input in the production process that requires energy to carry out the activities of the production process, the energy in question is energy in the form of fuel, lubricating oil, electric power, water for factories, etc. Many things cause the cessation of a machine when doing production either due to damage by the device itself or because of something else [2]. Therefore we need a way or method to maintain the function of the machine itself, keeping it able to operate to produce a lot of quality products. If a machine cannot function normally, the company will suffer a lot of losses resulting in a decline in the productivity of the company itself. In dealing with this, the company usually carries out certain types of maintenance so that all of its machines can operate normally [3].

According to Reyes et al.[4], maintenance means improving pre-existing capabilities, whereas repairs are defined as making things better and operating optimally. In general, the problem that occurs on the production floor is the damage to the opinion machine [5]. A common problem that is often found on the production floor is that production equipment does not operate properly so that it affects the next process. In a sugar mills, there are several workstations in the production process, including the preparation station, grinding station, purification station, evaporation station, cooking station, Lutheran station, and finishing station. In the production process activities, all stations must have their own constraints, but at the grinding station, there are more obstacles due to broken machines or other things. Therefore in this study, measurement of engine performance will be carried out at the grinding station, whether the performance or productivity of the machine within six months of operation is said to be useful [6].

2. Literature Review

Maintenance is an effort that is done intentionally and systematically on the equipment to achieve an acceptable or desirable result or condition. According to Charles et al. [7], support is a primary function in a company that can be defined as an activity to take care of facilities so that the equipment is in a condition that is ready to use as needed. Total Productive Maintenance (TPM) is one of the maintenance process methods developed to improve productivity in the work area by making the process more reliable and less wasteful [8]. This method is part of lean manufacturing.

Measurement of the success of TPM in evaluating and measuring the extent to which the successful implementation of TPM, the primary measurement tool used is the Overall Equipment Effectiveness (OEE) [9]. Mathematically, OEE formula is as follows:

$$\text{OEE} = \text{Availability x Performance Rate x Quality} \quad (1)$$

Where:

Availability = Machine readiness or willingness to operate
Performance = Number of product units produced by the machine in the time available
Quality = Comparison of the number of functional units with the number of units produced

The FMEA is a systematic method in identifying and preventing problems that occur in products and processes. There are two uses of FMEA, namely in the field of design (FMEA Design) and in the process (FMEA Process). FMEA Design will help eliminate design-related failures, such as failures due to improper strength, inappropriate material, and so on. The Risk Priority Number (RPN) methodology is a technique for analyzing risks related to potential problems that have been identified during the creation of FMEA [10]. This RPN is a mathematical product of the seriousness of the effects (Severity), the possibility of the cause will cause failure related to effects (Occurrence), and the ability to detect faults before they occur in the customer (Detection) [11].

Fault Tree Analysis (FTA) is a technique for identifying the failures of a system. FTA is function-oriented or better known as the 'top-down approach' because this analysis starts at the system level (top) and continues downward [12]. FTA is a technique used to identify risks that contribute to failure. This method is carried out with a top-down approach, which begins with the assumption of failure or loss from a top event and then details the causes of a top event to a root cause [13].

The age replacement model or the method of age replacement is a method of preventive maintenance that is carried out by setting the interval of preventive maintenance based on the time intervalof damage that requires the existence of replacement repair measures with minimization criteria [14]. Replacement maintenance model where the time interval for component replacement is done by taking into account the service life of the component, so as to avoid replacing newly installed equipment, it will be replaced in a relatively short time [15][16].

3. Research Methodology

3.1 Problem identification and problem formulation

This activity is carried out to identify things that will be the beginning of problems that occur in the field or company and the formulation of problems in research are things that are considered to be resolved from some of the problems that have been identified at the time of the survey so that what is the purpose of the research this can be achieved.

3.2 Research objectives

The purpose of the study was carried out using material from various sources, later the purpose was made based on the specified material. In this discussion is a continuation of the formulation of problems that have been determined, and will be used as a goal to be resolved in this study.

3.3 Literature study

Literature study activities carried out together during learning activities in the classroom or individually by using material from various sources. And collection of references from various research journals or books related to this research. 4. Field Study Field study activities carried out at PT. PG Candi Baru Sidoarjo, using material obtained from various sources. And gathering information from various parties within the company to support the research conducted. Information is collected in the field with interviews with staff or maintenance, and data collection is also carried out on installation staff at the installation department.

3.4 Data collection

Stage In this research activity, the data collection process was carried out by conducting observations, interviews, and direct observations in the field. The data that has been collected include: engine type and damage time, data on the amount of sugar product production, machine working hours, breakdown time, power cutoff, scrap, planned downtime, the time between engine failures, when repairing the engine, number of employees, price of components, the cost of sugar production

3.5 Data processing

Stage In this stage the data that has been obtained from data collection activities will be processed using OEE method starting from the availability calculation, performance calculation, quality calculation and then the measurement results are obtained OEE, from the effects of processing later, it will also be found six big losses, namely six big losses that occurred at the mill station. After that, identify the causes of engine damage using the FMEA method, which is to identify the types of machine damage that occur at the grinding station and then calculate the RPN. Then find the root of the problem of engine damage using the FTA; after that, it will be calculated again with the Age Replacement method to determine the optimal maintenance costs.

3.6 Analysis phase

After all data and information are collected and processed, the results of the processing are then analyzed so that conclusions can be drawn. At this stage, the OEE value will be obtained from the calculation results and get a factor of 6 significant losses or six big loss that occurs at the grinding station. After that, it will get the most prominent engine damage factor from the results of the RPN calculation, then FTA analysis to find out the root cause of the problem of damage and analyze the Age Replacement calculation to get optimal cost results and scheduling repairs.

3.7 Conclusion and suggestion stage

At this stage contains conclusions obtained from the data analysis stage, which answers the purpose of the problem in this study. And the advice in this study is to contain things that can be used as a reference to be better or better the company or researchers in the future.

4. Results and Discussion

4.1 OEE Calculation

Overall Equipment Effectiveness (OEE) Calculation Analysis in 2015 and 2016 for six months of operation of PT. PG Candi Baru Sidoarjo has increased and decreased. This is due to the availability value, performance efficiency, and unstable rate of quality. The benefits are as follows in Table 3.

From the results of the OEE value in 2015, the highest value was found in May amounted to 84.115 %, while the OEE value in 2016 was found in June amounted to 82.389%. The smallest OEE value in 2015 was in July with a value of 38,250 %, while the lowest OEE value in 2016 was in July with an amount of 14,219 %. The results of the above benefits show that the comparison of OEE values in 2015 with 2016 is better in 2015; this happens because, in 2016, the amount of availability, performance efficiency, and rate of quality are smaller.

Table 1. Calculation of OEE in 2015

No.	Period	Availability (%)	Performance Efficiency (%)	Rate of Quality Product (%)	OEE (%)
1.	May	98,95	100.00	85	84.12
2.	June	99.97	100.00	85	84.98
3.	July	94.32	47.71	85	38.25
4.	August	97.97	100.00	85	83.28
5.	September	99.11	84.62	85	71.29
6.	October	98.12	100.00	85	83.41
7.	November	90.26	100.00	85	76.72
8.	December	-	-	-	-

Table 2. Calculation of OEE in 2016

No.	Period	*Availability* (%)	*Performance Efficiency* (%)	*Rate of Quality Product* (%)	OEE (%)
1.	May	-	-	-	-
2.	June	96.93	100.00	85	82.39
3.	July	94.79	17.65	85	14.22
4.	August	95.47	100.00	85	81.15
5.	September	97.45	84.62	85	70.09
6.	October	96.51	100.00	85	82.03
7.	November	95.23	100.00	85	80.95
8.	December	90.55	100.00	85	76.97

Table 3. Comparison of OEE values

No.	Period	2015 (%)	2016 (%)
1.	May	84.11	-
2.	June	84.98	82.39
3.	July	38.25	14.22
4.	August	83.28	81.15
5.	September	71.29	70.09
6.	October	83.41	82.03
7.	November	76.72	80.95
8.	December	-	76.97

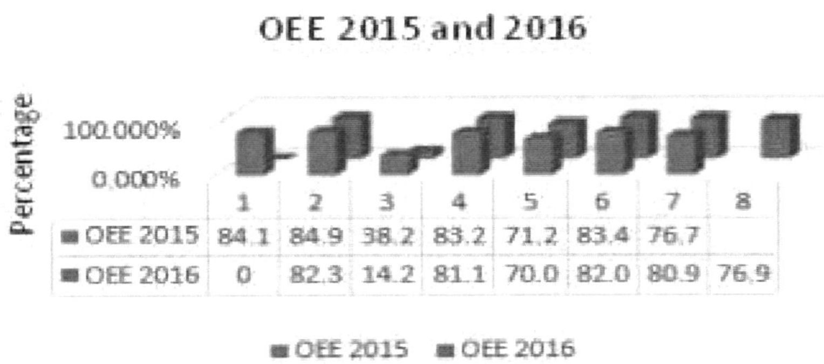

Figure 1. Comparison of OEE Value at Glingan Station in 2015 and 2016

4.2 Six big loses calculation

Downtime losses consist of equipment failure loss and set-up and adjustment. Equipment failure, namely damage to the machine or equipment that occurs suddenly, this unwanted damage will cause losses because the production process will stop. After all, the machine can not operate so much time is wasted and even experienced product defects.

$$\text{Equipment Failure} = \text{Total Breakdown} / \text{Loading Time X } 100\% \quad (2)$$

Table 4. Equipment failure loss calculation

No.	Period	2015			2016		
		Total Break Down (hour)	Loading Time (hour)	Equipment Failure (%)	Total Break Down (hour)	Loading Time (hour)	Equipment Failure (%)
1.	May	1.00	96	1.04	-	-	-
2.	June	0.21	720	0.03	15.48	504	3.07
3.	July	13.0	229	5.68	3.75	72	5.21
4.	August	15.09	744	2.03	33.71	744	4.53
5.	September	4.68	528	0.89	13.49	528	2.56
6.	October	13.96	744	1.88	25.99	744	3.49
7.	November	23.37	240	9.74	34.33	720	4.77
8.	December	-	-	-	68.05	720	9.45

Table 5. Set-Up and adjustment calculation

No.	Period	Years in 2015			Years in 2016		
		Set-Up and Adjustment (hour)	Loading Time (hour)	Set-Up and Adjustment (%)	Set-Up and Adjustment (hour)	Loading Time (hour)	Set-Up and Adjustment (%)
1.	May	0	96	0.000%	-	-	-
2.	June	0	720	0.000%	0	504	0.000%
3.	July	13	229	5.677%	0	72	0.000%
4.	August	0	744	0.000%	0	744	0.000%
5.	September	0	528	0.000%	0	528	0.000%
6.	October	0	744	0.000%	0	744	0.000%
7.	November	0	240	0.000%	13	720	1.839%
8.	December	-	-	-	0	720	0.000%

Set-up and adjustment is the time of installation and adjustment of equipment or machine set-up, including the time of adjustment and adjustment or replacement of one type of product to another product.

In the analysis of this improvement is to determine the type of distribution pattern in accordance with the data that already exists, the process is done with the help of software Minitab to determine the type of distribution that is appropriate or approaching, the following results of processing values in Minitab.

$$\text{Set-up and adjustment} = \text{total set up and adjustment} / \text{loading time X } 100\% \quad (3)$$

Table 6. Recapitulation of damage distribution types (TTF)

No.	Damage type	Jenis Distribusi	AD
1.	The IMC chain is stuck.	Lognormal	0.300
2.	IMC chain broken /dropped / off track	Weibull	0.585
3.	Rolling Mill	Lognormal	0.854

After the distribution parameter is known, the treatment is proposed to be done with the age replacement method with the type of damage to the IMC, and the treatment is performed on the 12th day with a value of IDR. 3,806,425.92 so that the optimal maintenance interval is once every 12 days. Then for the type of broken IMC damage, plummeted/exhausted the line on treatment on the 13th day with a value of IDR 5,614,466.88 so that the optimal interval of treatment is once every 13 days and for the type of rolling mill damage the treatment is carried out on the 10th day with a value of IDR. 34,272,482.34 so that the optimal maintenance interval is once every ten days.

Whereas the current treatment is IDR. 137,902,216 x 6 times damage = IDR. 827,413,296 for the IMC crash, for the current treatment IMC damage is broken, dropped / exited the line IDR. 138,892,556 x 7 times the damage = IDR. 972,247,892 and the mill slip IDR 727,243,926 x 9 times damage = IDR 6,545,195,334.

Table 7. Current and Proposed Treatments

Damage type	Current Treatments (IDR)	Proposed Treatments (IDR)
The IMC chain is stuck	827,413,296	57,096,389
IMC chain broken / dropped / off track	972,247892	77,738,772
Rolling Mill	6,488,631,684	616,904,682

From the results of the comparison of the table above, it shows that the proposed maintenance costs are far less so that it is more optimal to minimize the maintenance costs that exist during the sugar cane grinding period.

5. Conclusions

1. From the results of the OEE value in 2015, the highest value was found in May amounted to 84.115 %, while the OEE value in 2016 was found in June amounted to 82.389 %. The smallest OEE value in 2015 was in July with a value of 38,250 %, while the lowest OEE value in 2016 was in July with a value of 14,219 %. With a total OEE average value in 2015 of 74.576 % while the total OEE average value in 2016 amounted to 69.684 %.
2. From the results of the comparison of six big losses in 2015 and 2016 shows that the increase in 2015 amounted to 1,040.368 hours to 1,141609 hours in 2016.

3. From the results of the RPN value that appears that the type of damage that often occurs is the IMC chain crash with an RPN value of 48, the IMC chain is broken, dropped/exited the path with an RPN of 48 and the slip mill with an RPN value of 32.
4. From the results of the FTA analysis, the IMC damage is abrupt because the poor quality of the chain cannot withstand the juice of the pulp of water and lots of soil dirt and sugar cane, broken IMC breaks, dropped / off the track because the quality of the bad chain cannot stand the juice of the juice and slip mill due to dirty sugarcane untreated so that there is a lot of sugarcane droppings such as soil and curds which accumulate in the rolling mill.
5. From the results of the calculation of the proposed treatment with the type of damage, the IMC is done on treatment on the 12th day with a value of IDR 3,806,425.92 so that the optimal maintenance interval is once every 12 days. Then for the type of broken IMC breakage, dropped/exited the line on treatment on the 13th day with a value of IDR 5,614,466.88 so that the optimal maintenance interval is once every 13 days and for the type of rolling mill damage the treatment is carried out on the 10th day with a value of IDR. 34,272,482.34 so that the optimal maintenance interval is once every ten days.

References

[1] Wakjira M W and Singh A P 2012 Total productive maintenance: A case study in manufacturing industry *Glob. J. Res. Eng.* **12** 25–32
[2] Lazim H M, Salleh M N, Subramaniam C and Othman S N 2013 Total Productive Maintenance and Manufacturing Performance : Does Technical Complexity in the Production Process Matter ? *Int. Journl Trade, Econ. Financ.* **4** 380–3
[3] Patidar L, Soni P K and Naik B D 2015 Relationship of 5S and MP with Mediator of TPM and TQM *Int. Res. J. Eng. Technol.* **2** 1217–22
[4] Reyes J, Alvarez K, Martínez A and Guamán J 2018 Total Productive Maintenance for the Sewing Process in Footwear *J. Ind. Eng. Manag.* **11** 814–22
[5] Ihueze C and Ebisike S 2018 Optimization of Production System : An Application of Total Productive Maintenance (TPM) *Arch. Curr. Res. Int.* **13** 1–9
[6] Hąbek P and Molenda M 2017 Using the FMEA method as a support for improving the social responsibility of a company *6th International Conference on Operations Research and Enterprise Systems. ICORES* p 39
[7] Charles E, Okpala C and Onyekachi E E 2016 Benefits and Challenges of Total Productive Maintenance Implemention *Int. J. Adv. Eng. Technol* **7** 196–200
[8] Eissa A E M 2019 The Use Failure Mode and Effects Analysis as Quantitative Risk Analysis Tool *J. Appl. Sci.* **2019** 1–7
[9] Habidin F, Hudin N S, Salmuni W, Mustaffa W, Asma S, Rosli M, Yong S, Ong Y, Fuzi N M and Zulkifle N N 2017 A Proposed Total Productive Maintenance (TPM) Tool for Lean Dashboard , Statistics and Performance Efficiency in Malaysian Modern Manufacturer *Int. J. Acad. Res. Bus. Soc. Sci.* **7** 1012–8
[10] Rajput H S and Jayaswal P 2012 A Total Productive Maintenance (TPM) Approach To Improve Overall Equipment Efficiency *IJMER* **2** 4383–6
[11] Thakore R, Dave R and Parsana T 2015 Research Article A Case Study : A Process FMEA Tool to Enhance Quality and Efficiency of Bearing Manufacturing Industry *SJET* **3** 413–8
[12] Afefy I H 2013 Implementation of Total Productive Maintenance and Overall Equipment Effectiveness Evaluation *Int. J. Mech. Mechatronics Engg.* **13** 69–75
[13] Prabowo R, Singgih M L, Karningsih P D and Widodo E 2020 New Product Development from Inactive Problem Perspective in Indonesian SMEs to Open Innovation *J. Open Innov. Technol. Mark. Complex.* **6** 20
[14] Almeanazel O T R 2010 Total Productive Maintenance Review and Overall Equipment *J. Mech.*

Ind. Eng. **4** 517–22
[15] Wang Y, Chin K, Ka G, Poon K and Yang J 2009 Risk Evaluation in Failure Mode and Effects Analysis Using Fuzzy Weighted Geometric Mean q *Expert Syst. Appl.* **36** 1195–207
[16] McKone K E, Schroeder R G and Cua K O 2001 The impact of total productive maintenance practices on manufacturing performance *J. Oper. Manag.* **19** 39–58

Food safety assurance through thermal process on canned corned beef

M M Rohaman and N C Siregar

Researcher at Center for Agro-Based Industry, Ministry of Industry, Republic of Indonesia

E-mail: nobel_7@yahoo.com

Abstract. As one of critical control point (CCP) of canned processed corned beef, thermal process application needs to be investigated to ensure that the product get enough heat to destruct all bacteria and spores to get commercial sterile condition while saving energy to produce heat. The result showed that at retort temperature of 121^0C the sterilization value or minimum F_0 value for corned beef 340 grams was 12.08 minutes and maximum value of 15.72 minutes. This value is above of F_0 for processed meat of beef which is about 7 minutes. The observation of the process showed that the sterilization was over process and the holding time needs to be reduced by 10 minutes from 72 minutes. This value was similar for canned size of 198 grams which gave minimum F_0 of 12.07 minutes and-maximum value of 16.69 minutes but with different holding time. The later was obtained by sterilizing it with holding time of 57 minutes. The determination of optimum heat adequacy time (sterilization) is very useful for the industrial canning of meat, because with the optimum sterilization time, safe sterilization results can be obtained and the nutritional value can be maintained with efficient operating costs.

1. Introduction

In the current era of globalization, the concept of quality assurance and product safety must be applied in every activity / process related to the food production chain, including in the development of processed meat products. Meat processors should understand the basic principles of correct processing, and be accustomed to doing so, so that a quality assurance and product safety system can be applied. Product quality assurance and safety systems are based on a critical control point hazard analysis (HACCP) which has now become a necessity for export-oriented products to America, Europe, Asia and others. In the implementation of HACCP, what needs to be considered is to focus attention on CCP (Critical Control Point) which is applied to each process specifically such as ensuring adequate heat analysis on thermally processed / sterilized food. The International Organization for Standardization define that critical control point (CCP) is step in the process at which control measure are applied to prevent or reduce a significant food safety hazard to an acceptable level, and defined critical limit and measurement enable the application of corrections [1].

The HACCP system in principle is the application of risk management in controlling food safety hazards that are preventive and proactive. Food safety hazard have meaning that is biological, chemical, or physical agent in food with the potential to cause an adverse health effect [1]. Therefore, hazard analysis of critical control point (HACCP) is analysis of hazard which is conducted based on preliminary information to determine the hazards that need to be controlled on critical control point [1]. Hazard control starts from raw materials, processing, packaging, labelling, storage and use of food products including product distribution to consumers. The implementation of the HACCP system in the processed meat industry will guarantee product quality and safety so that it is expected to provide wider market opportunities, both national and international markets that require the application of HACCP as one of the requirements for product acceptance. This is in line with the demands of equivalence and harmonization of regulations and standards for the food safety system implemented in Indonesia with the system implemented by importing countries and of course also aimed at protecting domestic consumers.

The data of National Socio-Economic Survey conducted by Central Bureau of Statistics of Indonesia *(Biro Pusat Statistik or BPS)* revealed that the average beef consumption in Indonesian in 2017 was estimated at 3.00 kg/capita/year. The BPS stated that the population Indonesia in 2017 was 261.36 million people. If multiplied, then National beef demand in 2017 reached 784 thousand tons. Meanwhile, beef production in 2017 was 486,320 tons [2]. Therefore, to meet the needs of beef demand, the additional beef supply was needed by importing. The very high consumption of beef in Indonesia need serious attention in its quality and safety aspects.

Publications that integrate from the analysis of HACCP implementation to the thermal process application (sterilization) of the process are very few. One of the publications related to cold chain development on the process of supply to production processed meat products to increase transparency needs in the effectiveness of control at all the stages of the process in the cold supply chain The system can apply RFID, GPS, GIS and other technologies can support real time supervision on meat supply chain [3][4]. However, those papers did not explain enough regarding sterilization process on corned beef.

Retort temperatures conducted in this research also different from previous study which is useful as a reference for further research. Sterilized products have been organoleptically tested and have been accepted by consumers. However, this study is limited to two things namely HACCP analysis and thermal process. Hence, the aims of this research are to analyze and determine CCP from HACCP Implementation of Corned Beef production and to obtain sterilization value of corned beef on two different sizes which was conducted on horizontal retort, still, and saturated steam system.

2. Materials and Methods

2.1. Materials and instruments

Materials used in this research are canned corned beef on two different sizes of 340 g and 198 g which was produced and canned in one of processed meat factory in Bali, Indonesia. Retort used was horizontal retort, still, and using saturated-steam system. The dimension of can 340 g was 95.80 mm x 47.50 mm x 91.30 mm with rectangular shape, while for can 198 g was 95.80 mm x 47.50 mm x 56.40 mm. The full capacity of retort with can size of 340 g and 198 g was about 3000 pieces and 4900 pcs respectively. The sterilization was conducted at full capacity to make worst case scenario. The retort and length of retort was 1080 mm and 4100 mm respectively. The equipment to conduct the temperature measurement was Data Logger and Sensor from Ellab, Hillerod, Denmark with resolution of 0.01°C. The data was interpreted using Valsuite Software by Ellab installed on laptop. The data logger and sensor were attached to the product at the coldest zone of the retort.

2.2. Method

The determination of Critical Control Point was carried out using analysis of HACCP decision tree [5][6]. The hazard analyzed was biological, physic, and chemical hazard. After CCP was determined, then the next step was to ensure that the CCP can be controlled. However, the scope of this article limited to one of the two CCP that was sterilization which was one of thermal process application.

Sterilization value (F_0) values was calculated by using General Method with Trapezoidal rule approach which was still golden standard to determine heating value (F-value) [7][8]. The parameter used was product temperature measured by Ellab Data Logger and Sensor. The reference temperature is 121.11°C and z value of 10°C which is value to decrease one log cycle of number of microorganism of *Clostridium botulinum*. To ensure that the heat application is enough to destroy such microorganism, then this value was compared to D value of Clostridium botulinum at 121.11°C which is about 0.21 – 0.26 minutes. The commercial sterilization was met when the sterilization value give 12 log cycle or more toward destruction of *Clostridium botulinum* [9][10].

3. Result and Discussion

3.1. Analysis of Fresh Beef Meat and Canned Corned Beef

Analysis of fresh beef meat and canned corned beef was carried out to see the metal contamination in the fresh meat and quality parameter in canned corned beef and compared them to Indonesian National Standard (SNI). The Table of 1 showed the analysis of fresh beef meat and Table 2 showed the analysis of canned corned beef [11][12].

Table 1. Analysis of fresh beef meat

Parameter	Unit	Result	Method
Metal contamination:			SNI. 01-2896-1998, point 5 & 6
1. Lead (Pb)	mg/kg	< 0,059	-
2. Copper (Cu)	mg/kg	0,24	-
3. Zinc (Zn)	mg/kg	5,69	-
4. Tin (Sn)	mg/kg	< 0,8	-
5. Mercury (Hg)	mg/kg	< 0,0005	-
Arsenic	mg/kg	< 0,0002	SNI 01-4866-1998
Colorant	-	Negative	SNI 01-2895-1992
Borax	-	Negative	SNI 01-2894-1992 point 3.1
Sodium nitrite	-	Negative	SNI 01-2894-1992 point 2.4

Table 2. Analysis of canned corned beef (Compared to SNI 01-3775-1995 [12])

Parameter		Result	SNI	Testing method
Form		Normal	Normal	
Taste		Normal	Normal	
Odor		Normal	Normal	
Color		Normal	Normal	
Moisture	%	66.6		SNI 01-2891-1992 point 5.1
Ash	%	2.62		SNI 01-2891-1992 point 6.1
Protein (Nx6,25)	%	11.4	min. 17	SNI 01-2891-1992 point 7.1
Fat	%	7.75	max. 12	SNI 01-2891-1992 point 8.2
Carbohydrate	%	11.6	max. 5	Reducing
Sodium (Na)	mg/100 g	703		AOAC, 15 ed. 1990.985.35
Calcium (Ca)	mg/100 g	29.7		AOAC, 15 ed. 1990.985.35
Sodium Nitrite	mg/kg	10.8		SNI 01-2896-1998 point 5&6
Lead (Pb)	mg/kg	<0.07	max. 2,0	
Copper (Cu)	mg/kg	1.24	max. 20.0	
Zinc (Zn)	mg/kg	16.5	max. 40.0	
Tin (Sn)	mg/kg	<0.6	max. 250.0	
Mercury (Hg)	mg/kg	<0.0005	max. 0.03	
Arsenic (As)	Mg.kg	-	max. 1.0	SNI 01-2896-1992 point 6.
- Spores forming thermophilic bacteria	Colony/g	<10	max. 102	SNI 01-2897-1992
- Coliform	APM/g	<3	<3	
- C. perfringens*	Colony/g	0	0	

Based on the result of analysis toward beef that would be processed, the value has met the requirements of Indonesian National Standard SNI No.01-6366-2000 on Maximum Value of Microbial Contamination and SNI No. 01-3775-1995 Canned Corned Beef.

3.2. Determination of Critical Control Point (CCP)

Determination or identification of critical control point (CCP) was carried out in processed meat factory and analyzed by using HACCP decision tree [5][6]. The decision tree was interpreted was showed in table form for the sake of brevity. The table to identify CCP was displayed in Table 3.

3.3. Thermal Process (Sterilization) on Canned Corned Beef

The implementation of sterilization which was one of critical control point (CCP) was carried out in canned corned beef factory where retort located. The retort used was saturated steam system, still basket, and horizontal. The process condition controller was manual and operated by retort operator. The product packaging used was rectangular shape and made of tin plate. The retort itself consist of four baskets.

The equipment to measure product temperature was data logger and sensor made by Ellab company with resolution of 0.01°C. To calculate the sterilization value, the equation from Bigelow [13] was used.

$$L = \frac{T - T_{ref}}{z} \quad (1)$$

with: L = lethality rate
T = product temperature (°C)
T_{ref} = reference temperature (°C)
z = temperature needed to decrease 1 log cycle of decimal time

and sterilization value can be obtained from this formula:

$$F_0 = \int_0^t 10^{\frac{T - T_{ref}}{z}} \quad (2)$$

with: F_0 = sterilization value (minutes)
T = product temperature (°C)
T_{ref} = reference temperature (°C)
z = temperature to decrease 1 log cycle of decimal time
t = total time of sterilization process include pre-heating, holding, and cooling.

The equation (2) can be solved using general method (graphical integration) using trapezoidal rule approach of equation (1). It is because the product temperature is not constant during sterilization process. This is different with retort temperature which is expected is constant during sterilization (holding time).

Table 3. Determination of CCP of Corned Beef Production

PROCESS STEP	HAZARD Biological (B) Chemical (C) Physical (P)	1. Do control action exist?? *If **not** :Not CCP, show where and how the hazard is controlled.* *If **Yes** : next step.*	2. Are these action specifically designed to reduce the occurrence of hazards until they are acceptable? *If **not** : next step* *Jika **Yes** : CCP*	3. Are the hazard of product contamination comes from personnel, tools, and the process environment? If ***not*** : next step. If ***Yes*** *: Not CCP (controlled by SOP & GMP)*	4. Can hazard cross the acceptable level? *If **not** : Not CCP.* *If **Yes** : next step.*	5. Is the next step can reduce the hazard to an acceptable level? *If **not** : CCP.* *If **Yes** : Not CCP*	CCP
Acceptance of frozen meat raw materials	B : Growth of pathogenic microbes (*Salmonella* sp.) P : Filth, bone pieces, the rest of the packaging. C : Heavy metal contamination	Yes	No	Yes			Not CCP
Acceptance of fresh meat	B : Growth of pathogenic microbes (*Salmonella* sp., *Staphylococcus sureus*). P : Filth, sand, soil C : Heavy metal contamination, pesticide	Yes	No	Yes			Not CCP
Cutting	B : - P : Knife shards of frozen meat C : -	Yes	No	Yes			Not CCP
Thawing	B : Growth of pathogenic microbes (*Salmonella* sp.) P : - C : -	Yes	No	Yes			Not CCP
Trimming	B : Growth of pathogenic microbes (*Salmonella* sp.). P : Filth, hair C : -	Yes	No	Yes			Not CCP
Grinding	B : - P : Knife shards C : -	Yes	No	Yes			Not CCP
Curing (Salter)	B : - P : Filth/sand C : Nitrite doses that exceed the rules	Yes	No	Yes			Not CCP
Curing	B : Growth of pathogenic microbes (*Staphylococcus aureus*) - P : - C :	Yes	No	Yes			Not CCP

Table 3. Determination of CCP of Corned Beef Production (*cont.*)

PROCESS STEP	HAZARD Biological (B) Chemical (C) Physical (P)	1. Do control action exist?? *If **not** :Not CCP, show where and how the hazard is controlled.* *If **Yes** : next step.*	2. Are these action specifically designed to reduce the occurrence of hazards until they are acceptable? *If **not** : next step* *Jika **Yes** : CCP*	3. Are the hazard of product contamination comes from personnel, tools, and the process environment? If ***not*** : next step. If ***Yes*** *: Not CCP (controlled by SOP & GMP)*	4. Can hazard cross the acceptable level? *If **not** : Not CCP.* *If **Yes** : next step.*	5. Is the next step can reduce the hazard to an acceptable level? *If **not** : CCP.* *If **Yes** : Not CCP*	CCP
Heating	B : -						
	P : Filth (thread, lice from herbs)	Yes	No	Yes			Not CCP
	C :						
Filling	B : Microbial contamination from workers (*Staphylococcus aureus*). Filth (hair)	Yes	No	Yes			Not CCP
	P : -						
	C :						
Seaming	B : -						
	P : Can defect on double seam.						
	C : -	Yes	No	Yes			Not CCP
Sterilization	B : Mesophilel and thermophile microbial growth, microbial contamination from cooling water. Can deformation.	Yes	Yes				**CCP**
	P :						
	C :						
Quarantine	B : Thermophilic microbial growth -						
	P : -	Yes	Yes				**CCP**
	C : -						
Sortation	B : -	ob					
	P : Damaged cans come in the packaging section.	Yes	No	Yes			Not CCP
	C : -						
Packaging	B : -						
	P : Damage to the packaging						
	C : -	Yes	No	Yes			Not CCP
Sortation	B : -						
	P : Rust, dent, package damage-	Yes	No	Yes			Not CCP
	C :						
Distribution	B :						
	P : Dented / damaged / wet packaging items	Yes	No	Yes			Not CCP
	C :						

IOP Conf. Series: Materials Science and Engineering **885** (2020) 012064 doi:10.1088/1757-899X/885/1/012064

Based on Table 3, it can be seen that the critical control point (CCP) in canned corned beef production was sterilization and quarantine. However, this article was limited to sterilization which was commonly as critical control point for low acid food production. Low acid foods itself have pH > 4.6 and aw > 0.85 and categorized as potential hazardous foods e.g. canned corned beef.

The sterilization in this research was carried out at the one of processed meat factory. The sterilization was carried out in saturated steam retort with full capacity of products to make worst case scenario. The result of sterilization of F_0 value (sterilization value) was showed below:

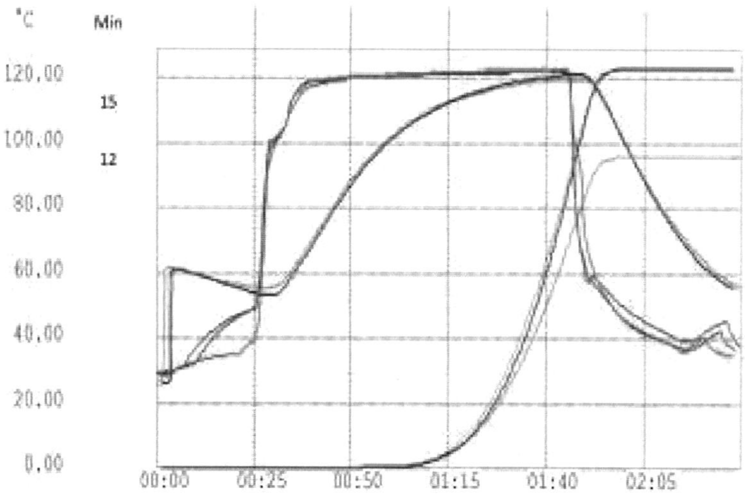

Figure 1. Graph of sterilization process on canned corned beef 340 g

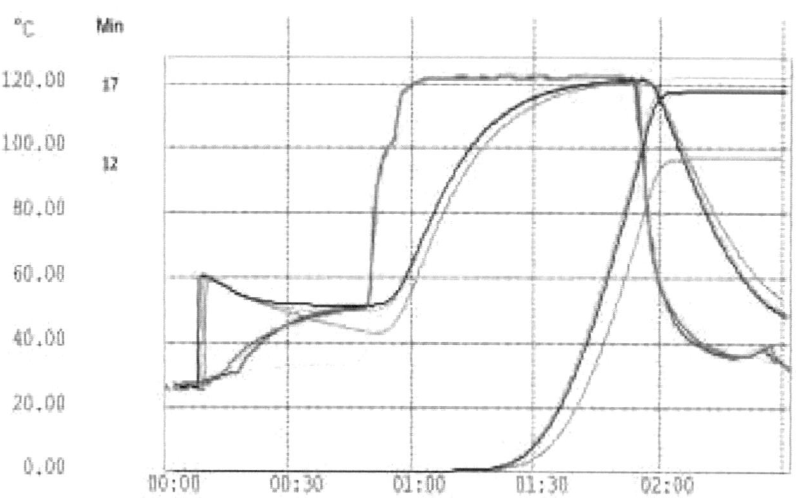

Figure 2. Graph of sterilization process on canned corned beef 198 g

The result above show that the sterilization value (F_0) minimum and maximum for canned corned beef product with size of 340 g was 12.08 minutes and 15.72 minutes respectively. This value was reached when product was sterilized for 72 minutes at retort temperature of 121°C. The company set the standard for 7 minutes for corned beef. This value was higher than the company requirements and can cause overheating and loss of much energy. However, this value met the requirements of the Regulation of The Head of The National Agency of Drug and Food Control of Republic of Indonesia No. 24 of 2016 on Requirement of Commercially Sterile Food which require that the sterilization value was minimum of 3.00 or in other words, the sterilization value must met at least 12 cycle log destruction of mesophilic bacteria of *Clostridium botulinum.* However, more severe processes equivalent to $F_0 = 6$ min or longer usually applied to control spoilage organism as many companies did [9]. This case also similar with can size of 198 g which gave minimum sterilization value of 12.07 minutes and maximum value of 16.69 minutes. This value was reached when product was sterilized for 57 minutes. The determination of optimum heat adequacy time (sterilization) is very useful for the industrial canning of meat, because with the optimum sterilization time, safe sterilization results can be obtained and the and the nutritional value can be maintained with efficient operating costs.

Some studies conduct the sterilization process on corned beef and give recommended process condition, those are: No.1 Can ($2\frac{11}{16}\ x\ 4$ inches) 120 minutes at retort temperature of 113°C and 75 minutes at 115°C. This study also investigated and recommended that for No.2 Can ($2\frac{11}{16}\ x\ 2\frac{11}{16}$ inches), the sterilization (holding) time should be 150 minutes at retort temperature 113°C and 90 minutes at 115°C. The last, for 6lb Can, the recommended time for sterilization was 300 minutes at 113°C and no information at retort temperature of 115°C [14]. However, there is no information on sterilization value (F_0) toward recommended processes suggested. The information about F_0 is very important to make a model using Formula Method to recalculation process time to obtain sterilization value (F_0) target.

The other studies recommended that for can size of 76 mm (diameter) x 51 mm (height) or in other word with can volume of 180 mL, the sterilization value that was recommended was 4.5 – 6.6 minutes [13], 5 min , 4.0 – 4.9 min , and ≥ 4.0 minutes [9]. However, this study did not explain the schedule process or operating condition how to achieve that value. The other differences were the shape of can used in this research was rectangular instead of cylinder. Therefore, this research (article) explain the operation condition, holding time, and F_0 obtained with others factor involved.

4. Conclusion

The result showed that sterilization was one of the critical control point (CCP) in canned corned beef production and need to be control to acceptable level. The sterilization process showed at retort temperature of 121°C the sterilization value or minimum F_0 value for canned corned beef 340 grams was 12.08 minutes and maximum value of 15.72 minutes. The value for canned size of 198 grams gave minimum F_0 of 12.07 minutes and-maximum value of 16.69 minutes but with different holding time of 57 minutes. This value is more than requirement and need to be adjusted for saving energy and to maintain the quality of the product from overheating. The determination of optimum heat adequacy time (sterilization) is very useful for the industrial canning of meat, because with the optimum sterilization time, safe sterilization results can be obtained and the and the nutritional value can be maintained with efficient operating costs.

Acknowledgment

The authors would like to thank Center for Agro-Based Industry for funding this research activity and PT. CIP, Bali, Indonesia for the opportunity and collaboration to conduct the research.

IOP Conf. Series: Materials Science and Engineering **885** (2020) 012064 doi:10.1088/1757-899X/885/1/012064

References

[1] ISO 2018 *ISO 22000: Food safety management systems — Requirements for any organization in the food chain*
[2] BPS 2019 *Beef Production by Province: 2009-2018* (Biro Pusat Statistik)
[3] Wei J and Lv S 2019 Research on the Distribution System of Agricultural Products Cold Chain Logistics Based on Internet of Things *IOP Conference Series: Earth and Environmental Science* vol 237 (IOP Publishing) p 52036
[4] Raab V, Petersen B and Kreyenschmidt J 2011 Temperature monitoring in meat supply chains *Br. Food J.*
[5] CODEX 2009 *Hazard Analysis and Critical Control Point (HACCP) system*
[6] CODEX 1997 *Hazard analysis and critical control point (HACCP) system and guidelines for its application* vol 3
[7] Bigelow W D, Bohart G S, Richardson A C and Ball C O Heat Penetration in Processing Canned Foods. Research Laboratory, National Canners Association; Washington, DC, USA: 1920 *Bulletin*
[8] Patashnik M 1953 A simplified procedure for thermal process calculation *Food Technol.* **7** 1–6
[9] Holdsworth S D, Simpson R and Barbosa-Cánovas G V 2008 *Thermal processing of packaged foods* vol 284 (Springer)
[10] BPOM 2016 Regulation of The Head of The National Agency of Drug and Food Control of Republic of Indonesia No. 24 of 2016: Requirement of Commercially Sterile Food
[11] SNI 2000 Indonesian Nasional Standard (SNI) No.01-6366-2000 on Maximum Value of Microbial Contamination
[12] SNI 2000 Indonesian Nasional Standard (SNI) No. 01-3775-1995 on Canned Corned Beef
[13] Bigelow W D 1921 The logarithmic nature of thermal death time curves *J. Infect. Dis.* 528–36
[14] Featherstone S 2015 *A complete course in canning and related processes: Volume 3 Processing Procedures for Canned Food Products* (Woodhead Publishing)

IOP Conf. Series: Materials Science and Engineering **885** (2020) 012065 doi:10.1088/1757-899X/885/1/012065

Identification of production planning and supply of glass bottles at the bottled tea company

Nurulinzany and Amrin

Department of Agro Industrial Engineering, Polytechnic ATI Makassar, Indonesia

E-mail: nurulinzany@gmail.com

Abstract. Sinar Sosro Limited (SS Ltd) is a company that produces a variety of packaged teas with superior products of bottled tea with 220 ml glass bottles. Based on the preliminary research, the glass bottles still can meet the production but with the demand. In their production planning, the company applies a method of estimating or predicting so that production results cannot meet production demand. This is the background of the researchers to make production planning based on forecasting methods and determine the quantity of glass bottle orders to meet demand. In forecasting the demand for production of the 220 ml glass bottle packaging, the methods used are Moving Average, Exponential Smoothing, and Linear Regression by comparing the size of forecasting errors using Mean Absolute Deviation (MAD), Mean Square Error (MSE), and Mean Absolute Percentage Error (MAPE). To determine the supply of glass bottles used the EOQ (Economy Order Quantity) method to determine the purchase of the most economical glass bottles. Base on the data, the Moving Average method (n=9) was the best forecasting used to predict the 220 ml glass bottle packaging demand for the next period of 233,862 crates with MAD forecasting error of 13,680, MSE of 561,420,907, and MAPE 5.49% which is the smallest value of the three methods. The application of the EOQ method results in the most economical order of glass bottles that is 136,234 crates for one message with a total repurchase frequency of 21 times a year.

1. Introduction

Economic growth in the era of globalization requires every company to compete in the industry or service. One of them is the beverage industry which has very fierce competition because drinks are a daily necessity, especially of packaged tea [1][2].

Sinar Sosro Limited (SS Ltd) is a company engaged in the beverage industry as a market leader who controls the market share with the mainstay product namely Sosro Tea Bottles packaging 220 ml glass bottles. However, in the production of sosro bottled tea, the 220 ml glass bottles do not match the number of requests. This can be seen from the market demand data of 2,967,644 crates in 2018 while the company was only able to cover the demand of 1,692,046 crates for 2018.

From this issue arises the problem of inaccurate demand forecasting which has an impact on production planning. Even though the company has forecasted the production demand, the amount produced is still unable to meet the actual demand [3].

Forecasting is generally done based on past data and analyzed using certain methods to get the smallest forecast error value. Several methods can be used to solve these problems, including the method of forecasting Linear Regression, Moving Average, and Exponential Smoothing [4]. These three methods are used to predict future demand using past data.

Then to avoid any shortage or excess supply of glass bottles, it is also necessary to control the supply of glass bottles. Inventory control is intended to predict the amount of supply of glass bottles needed in the production process in the coming period so as not to experience shortages or excess. This inventory is also to avoid deficiency or excess glass bottles intended to anticipate an increase in demand resulting in shortages of inventory.

Economic Order Quantity (EOQ) is one of the inventory control methods that aims to determine the optimum amount of ordering the most economical goods or materials for each order. This method can determine the optimal frequency of ordering to minimize production costs without reducing the target or profit to be achieved [5][6].

2. Methods

The data obtained is then observed and processed to obtain the desired results, and the data processing analysis stages are [7][4]:

1. Forecasting of the production requests for 220 ml glass bottle size packaging using forecasting models as follows:
   a. Linear regression

$$\hat{y} = a + bx$$

$$b = \frac{n.\Sigma xy - \Sigma x.\Sigma y}{n.\Sigma x^2 - (\Sigma x)^2}$$

$$a = \frac{\Sigma y - b.\Sigma x}{n}$$

   b. Exponential Smoothing

$$F_t = F_{t-1} + \alpha\,(A_{t-1} - F_{t-1})$$

   c. Moving Average

$$\textit{Moving Average} = \frac{\Sigma \text{ Previous periods request}}{s}$$

2. Calculation of the forecast error from the production of 220 ml glass bottles:
   a. Mean Absolute Deviation (MAD)

$$\text{MAD} = \frac{\Sigma\, |actual - forecasting|}{n}$$

   b. Mean Square Error (MSE)

$$\text{MSE} = \frac{\Sigma\, (actual - forecasting)^2}{n}$$

   c. Mean Absolute Percentage Error (MAPE)

$$\text{MAPE} = \frac{100\ \sum_{i=1}^{n} |actual_i - forecasting_i|\ /actual}{}$$

3. Determination of the best method to predict demand for the production of the 220 ml glass bottle size by comparing the smallest error value.
4. Calculation of the number of orders in a single message by using the EOQ method

$$EOQ = \sqrt{\frac{2.D.S}{H}}$$

5. Calculation of order frequencies based on the EOQ method

$$F = \frac{D}{EOQ}$$

6. Calculation of the total cost (TIC) of inventory based on the EOQ method

$$TIC = S \times \frac{D}{Q} + H \times \frac{Q}{2}$$

7. Calculation of Safety Stock

$$SS = SD \times Z$$

8. Calculation of the point of reordering inventory (ROP / *Reorder Point*)

$$ROP = (d \ \times L) + SS$$

3. Result

Following the formulation of the problem, the data collected from the company include data on the demand for production of 220 ml bottles of sosro tea bottles with glass bottles from January 2018 to December 2018, data on the inventory of glass bottles, ordering costs, and storage costs. Table 1 was obtained from the production and logistics department of SS Ltd.

Table 1. Production data of the 220 ml glass bottle packaging and the glass bottle supply (2018)

Period	Production (crate)	Demand (crate)	Bottle Supplies (crate)
January	130.115	150.183	140.152
February	150.755	184.952	160.246
March	140.456	257.755	150.457
April	150.286	214.589	160.429
May	150.655	255.435	160.289
June	180.522	230.187	185.835
July	170.247	200.125	177.902
August	143..049	221.226	150.289
September	132.342	262.908	140.235
October	145.688	233.286	155.206
November	150.635	227.410	156.234
December	190.345	259.588	198.213
Total	1.692.046	2.697.644	1.935.487

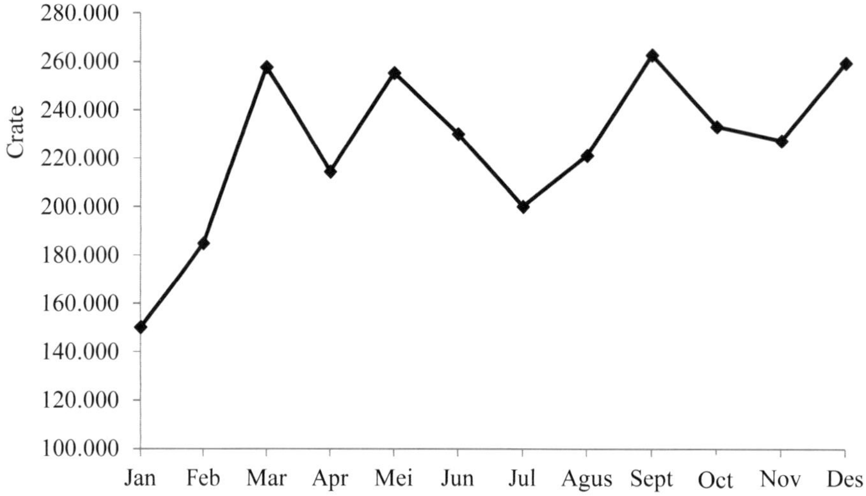

Figure 1. Demand of the 220 ml glass bottles in 2018

From Table 1, it can be seen that the demand for the 220 ml glass bottle size packaging did not meet with the production amount. Therefore, proper production planning is needed by doing the forecasting method. For this reason, the methods used in this forecasting are Linear Regression, Moving Average, and Exponential Smoothing. These three methods are used because they use data that occurs within a certain period and is past or past. Costs incurred to order goods at SS Ltd. can be seen in Table 2. Ordering costs do not depend on the number ordered but depend on the number of times an order is placed.

Table 2. Costs of glass bottles

Cost Components	(IDR/order)	
Telephone Charges	IDR	200,000
Administrative Costs	IDR	200,000
Loading and unloading coasts	IDR	2,500,000
Transportation costs	IDR	2,000,000
Total Booking Fee	IDR	4,900,000
Storage Fee (Price per crate IDR 15,000 x 10%)	IDR	1,500/crate
Lead Time	7 days	

Forecasting can be done by looking at the demand data in the previous period. Forecasting is done in this study to provide an overview of companies on how to determine alternative policies towards demand in the coming period. There are three methods used in this forecasting research namely Moving Average, Exponential Smoothing and Linear Regression.

By comparing the forecasting results of the three methods is expected to obtain the smallest forecast error value, so that it can be used as a guideline for companies to determine to forecast in the future.

To get the correct forecasting method so that it can be applied to the company, it is important to compare the accuracy of forecasting errors in each method. The accuracy of forecasting errors consists of MAD MSE, and MAPE.

Table 3. Comparison of forecasting error for demand for the 220 ml glass bottles

No.	Method	Forecasting	Size of Forecasting Error		
			MAD	MSE	MAPE
1	Moving Average n = 9	233,862	13,680	283,370,239	5.49%
2	Exponential Smoothing α = 0,5	254,910	27,179	1,366,058,662	11.46%
3	Linear Regression	259,422	21,940	718,339.387	10.25%

Based on the results of the comparison, the appropriate method used to apply to forecast to the demand for the 220 ml glass bottles is the Moving Average method n = 9. This means forecasting can begin from the tenth month of October 2018, because in January 2018 until September 2018 is used as a basis for determining the forecast of production demand for the 220 ml glass bottle size packaging for the next 1 year. The forecasting is shown in Table 4.

Table 4. Results of forecasting production demand with the Moving Average Method n = 9

Period (2019)	Forecast (crate)
January	233,862
February	236,003
March	233,844
April	234,250
May	238,042
June	239,910
July	237,355
August	237,807
September	238,962
October	236,671
November	236,983
December	237,092
Total	2,840,789

Based on the data processing above, it can be seen the forecasting of demand for the production of 220 ml glass bottles for one year later in 2019, with the number of demand forecasting of 2,840,789 crates.

The calculation of the analysis of shard bottle inventory control can be done using the EOQ method with the assumption that 1 cate contains as many as 24 glass bottles. The EOQ method allows companies to determine the most economical quantity of glass bottle orders with a constant number of requests and lead time.

a. EOQ calculation

This economical glass bottle purchase is based on the calculation in Table 5.

Table 5. EOQ formulation data

Data Type	2019
Glass Bottle Total Needs (D)	2.840.789 crates
One-Time Booking Fee (S)	Rp4.900.000/order
Storage Fee (H)	Rp1.500/crate
EOQ	136.234 crates

b. Frequency of purchase of glass bottle
Using the EOQ method, the number of order frequencies in one year can be calculated as follows:

$$F = \frac{D}{EOQ} = \frac{2.840.789}{136.234} = 21$$

Orders in 2019 are made 21 times a year precisely every 18 days.

c. Total inventory cost
To calculate the total cost of inventory, what you need to know is as follows:

$$TIC = S \times \frac{D}{EOQ} + H \times \frac{EOQ}{2}$$

$$= 4{,}900{,}000 \times \frac{2.840.789}{136.234} + 1.500 \times \frac{136.234}{2}$$
$$= 102{,}176{,}154 + 102{,}175{,}500$$
$$= \text{IDR } 204{,}351654$$

Based on the above calculation, it is known that the total inventory cost that must be borne by the company in 2019 is IDR 204,351,652.-

d. Safety stock
The calculation of standard deviation can be seen in Table 6.

Table 6. Calculation of standard deviations

Period	Glass Bottle Need (Y)	X	(Y-X)	$(Y\text{-}X)^2$
January	233,862	236,733	-2,871	8,242,641
February	236,003	236,733	-730	532,900
March	233,844	236,733	-2,889	8,346,321
April	234,250	236,733	-2,483	6,165,289
May	238,042	236,733	1,309	1,713,481
June	239,910	236,733	3,177	10,093,329
July	237,355	236,733	622	386,884
August	237,807	236,733	1,074	1,153,476
September	238,962	236,733	2,229	4,968,441
October	236,671	236,733	-62	3,844
November	236,983	236,733	250	62,500
December	237,092	236,733	359	128,881
Total	2,840,789	2,840,796	-15	41,797,987

$$X = \frac{\text{Glass Bottle Needs 1 year}}{n}$$
$$= \frac{2{,}840{,}789}{12}$$
$$= 236{,}732{,}417 = 236{,}733$$

$$SD = \sqrt{\frac{\Sigma\,(Y-X)^2}{n}}$$
$$= \sqrt{\frac{41{,}797{,}987}{12}}$$
$$= 1{,}866{,}32 = 1{,}866$$

Assuming that the company uses a 5% significant deviation is not tolerated and uses one side of normal, the value of which can be seen in the table that is 1,65, Then the quantity of safety stock is:

$$SS = SD \times Z$$
$$= 1{,}866 \times 1{,}65$$
$$= 3{,}078{,}9 \text{ crates} = 3{,}079 \text{ crates}$$

e. Reorder Point
SS Ltd has a lead time of 7 days with an average number of working days of 300 days for a year. Then the calculation of the glass bottle ordering point is as follows:

$$ROP = (d \times L) + SS$$
$$=\left(\left(\frac{2.840.789}{300}\right) \times 7\right) + 3{,}079$$
$$= (9{,}469 \times 7) + 3{,}079$$
$$= 69{,}362 \text{ crates}$$

Based on the reorder point calculation, the company must order a glass bottle when it reaches the amount of 69,362 crates.

4. Conclusion

Based on the calculation of the results of forecasting errors, the most appropriate method used for forecasting requests for the 220 ml glass bottle packaging is the Moving Average method n = 9 with the lowest forecast error rate compared to other methods. Where the MAD value is 13,680, MSE is 283,370,239, and MAPE is 5.49%. The most economical number of orders for purchasing glass bottles is 136,234 crates with ordering frequency 21 times a year, total inventory costs are IDR 204,351,654 safety supplies are 3,079 crates of glass bottles and re-ordering points when reaching 69,362 crates of glass bottles.

References

[1] Vicentini A, Liberatore L and Mastrocola D 2016 Functional Foods: Trends And Development Of The Global Market *Ital. J. Food Sci.* **28** 338–51
[2] Corbo M R, Bevilacqua A, Petruzzi L, Casanova F P and Sinigaglia M 2014 Functional beverages: the emerging side of functional foods: commercial trends, research, and health implications *Compr. Rev. food Sci. food Saf.* **13** 1192–206
[3] Govindan K and Soleimani H 2017 A review of reverse logistics and closed-loop supply chains: a Journal of Cleaner Production focus *J. Clean. Prod.* **142** 371–84

[4] Santoso and Rainisa M H 2017 *Perencanaan dan Pengendalian Produksi* (Bandung: Alfabeta)
[5] Kumar R 2016 Economic Order Quantity (EOQ) Model *Glob. J. Financ. Econ. Manag.* **5** 1–5
[6] Cárdenas-Barrón L E, Chung K-J and Treviño-Garza G 2014 Celebrating a century of the economic order quantity model in honor of Ford Whitman Harris *Int. J. Prod. Econ.* **155** 1–7
[7] Gaspersz V 2011 *Production Planning And Inventory Control* (Jakarta: PT. Gramedia)

IOP Conf. Series: Materials Science and Engineering **885** (2020) 012066 doi:10.1088/1757-899X/885/1/012066

Basic properties evaluation of alkyl palmitate to be used as oil insulation

A Rajab, J S Ramadhani, R Kurniawan, K Qibran, Melda L, and M. I. Hamid

Electrical Engineering, Andalas University, Kampus Limau Manis, Padang, Indonesia

Email: a.rajab@eng.unand.ac.id

Abstract. This paper discusses the evaluation of alkyl palmitate to be used as the oil insulation in an oil-filled transformer. Three kinds of alkyl palmitates, namely, Methyl-, Ethyl-, and Isopropyl-palmitates, were prepared through the esterification of methyl-, ethyl-, and iso-propyl-alcohols, respectively, and palmitate acid. The resulted alkyl palmitates were then evaluated for their basic properties such as breakdown voltage, viscosity, density, acidity, and water content based on the ASTM D 6871 standard specification. The peroxide value of the alkyl palmitates was also examined to indicate their oxidation stability and compared the result with that of mineral oil. The results show that viscosity, water content, and density of all alkyl palmitates fulfill the values specified by the standard, whereas, the breakdown voltage and acidity do not. The peroxide value of all alkyl palmitates is still much higher than that of mineral oil.

1. Introduction

Oil insulation plays an important role in the operation of an oil-filled transformer. In addition to serving as insulation, the oil also works as a cooling medium and provides information on the health condition of the transformer [1,2]. Mineral oil has been using for more than a century and still the most widely used oil today [3–5]. It possesses some advantages like high oxidation stability, high dielectric strength, low kinematic viscosity, and low melting point [6]. However, the oil has a low fire point property makes it risky to be used in transformers located in an area like a hospital, populated area, etc. Besides, the oil is also recognized as a toxic substance, and poorly biodegradable. Therefore, the oil is unfriendly to the environment. Since the mineral oil derived from petroleum resources, it belongs to the non-renewable oil group [7]. Other oils like silicone oils, halogenated oils, and askarel have also been implemented as oil insulation due to their good characteristics. Their use is limited to some specific applications due to their high cost, low biodegradation level, and non-renewable nature. The existence of PCB constituent in Askarel oil leads to the banning of the oil in the USA and other countries having a strict environmental regulation [8].

The need for renewable and environmentally friendly oil insulation in distribution and power transformers is emerging during the last three decades due to the rising concern on environmental consideration. The oil insulation derived from natural esters or vegetable oils is found to meet these criteria. The oils are non-toxic and fully biodegradable [9]. The natural ester or vegetable oils-based oil insulation are now commercially available as alternatives for mineral-based oils in transformer applications [10]. In addition to its traditional advantages like those mentioned earlier, natural esters have more benefits since the oils have a higher flash/fire points and low thermal expansion coefficient. The higher flash/fire points promote a low fire risk oil, whereas the lower thermal expansion coefficient allows a more compact design of a natural ester oil-filled transformer compared to the mineral oil ones [11–14].

Nevertheless, the viscosity and pour point values of natural esters of triglyceride (tri-ester) type are generally much higher than those of mineral oils. These physical properties are essential and cannot be ignored. The viscosity is essential to serve as the cooling medium to dissipate heat generated in the transformer [15]. A lower viscosity oil allows a more efficient cooling capability [16]. The pour point is the lowest temperature at which oil could still flow. A low pour point is important, particularly in cold climates, to ensure an easy circulation of the oil to serve its function as a coolant. Even with addition pour point depressants, the pour point of a natural ester in its original form (tri-ester) cannot match that of a typical transformer mineral oil [6]. These obstacles motivate the development of low viscosity oil insulation from the natural ester, namely, monoester. This paper presents the basic properties evaluation of monoesters of alkyl palmitates. The properties like breakdown voltage, viscosity, relative density, acidity, water content, and oxidation stability are measured. The measurement results are compared with values specified by ASTM D6871 standard, the specification standard for as-received new natural ester used for transformer insulation.

2. Experiment

2.1. Sample Preparation

The sample used in the experiment was monoester of alkyl palmitate type. Three kinds of alkyl palmitate samples were prepared through the esterification of methyl-, ethyl-, and isopropyl-alcohols and palmitate acids based on the equation (1)[17].

(1)

The resulted samples are expected to have relatively high oxidation stability since the palmitate acid belongs to a saturated fatty acids group, whose oxidation stability is typically higher than those of the unsaturated ones. The use of monoester, instead of tri-ester, is intended to have an oil derived from natural oil resources having a lower viscosity.

2.2. Procedure

The standard procedure for testing an as-received new natural ester used for transformer insulation and the required value of each oil properties are listed in Table 1 [18]. For instance, the testing procedure for breakdown voltage follows the standard procedure ASTM D1816, and the required value for the breakdown voltage is not less than 35 kV.

Table 1. As-received new natural ester property requirements for transformer use

Property	Limit	Test Method
Fire Point, min, °C	300	D 92
Pour Point, max, °C	-10	D 97
Relative Density, 15 °C/15 °C, max	0.96	D 1298
Viscosity (40 °C), max, cSt	50	D 445
Breakdown voltage, min, kV	35	D 1816
Dissipation factor at 60 Hz, 25 °C, max, %	0,2	D 924
Acid number, max, mg KOH/g	0,06	974
Water content, max, mg/kg	200	D 1533

3. Results and Analysis

3.1 Breakdown Voltage

The breakdown voltage of all alkyl palmitate samples is shown in Figure 2. It seems that the difference in the number of carbon atom in a molecular structure does not correlate with the breakdown voltage of the alkyl palmitates. It should be noticed that methyl-, ethyl- and isopropyl-components have a different carbon atom number, which are 1, 2, and 3, respectively. It can also be seen from Figure 2 that the breakdown voltage of all alkyl palmitates is in the range of 31 to 32 kV, which are less than that specified by the ASTM D 6871 standard. These breakdown voltage values could be improved to the acceptable level. Our previous work on methyl ester shows that the breakdown voltage of the methyl ester can be enhanced by filtration and water reduction [19]. The improvement of the breakdown voltage would be conducted in the subsequent investigation.

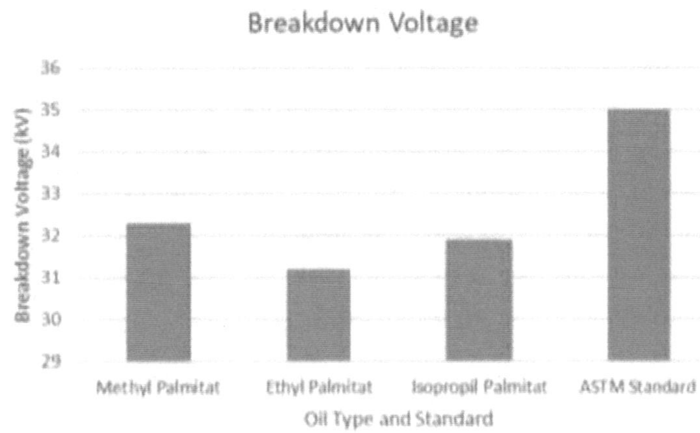

Figure 1. The Breakdown voltage of alkyl palmitates and ASTM D 6871 standard specification.

3.2 Kinematic Viscosity

Monoester typically has a much lower viscosity compared to the tri-ester. The kinematic viscosity of all alkyl palmitate samples and the value specified by the ASTM D 6871 standard are shown in Figure 3. It is clear that the viscosity of all tested samples is much less than that specified by the standard. The kinematic viscosity of all tested oils is in the range of 0.8 – 2.0. These values are even lower than that of mineral oil, which means that the cooling efficiency of monoester is better than that of mineral.

3.3 Oxidation Stability

The oil insulation is usually used in the oil-filled transformer for a long time up to 40 years. One of the determinant factors in evaluating the long-term performance of oil insulation is the oxidation stability [7]. However, the oxidation stability property is not listed in the ASTM D 6871 standard, as shown in Table 1. Hence, the oxidation stability of mineral oil is used for comparison. In this study, the evaluation of the oxidation stability of all oil samples was performed by measure their peroxide value. The higher peroxide value means a higher amount of oil sample oxidized during the test. Figure 4 shows the peroxide value of all oil samples and mineral oil. It can be perceived from Figure 4 that the peroxide value of all alkyl palmitate samples is much higher than that of mineral oil, indicating that the alkyl palmitates are more vulnerable to the oxidation than the mineral oil.

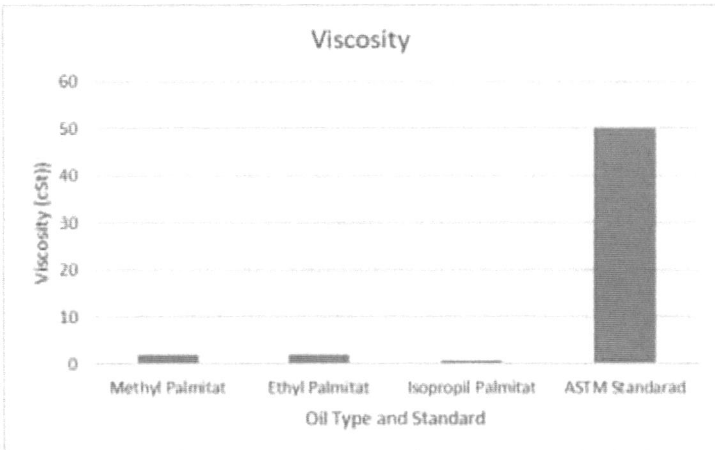

Figure 2. The kinematic viscosity of alkyl palmitates and the ASTM D 6871 standard specification.

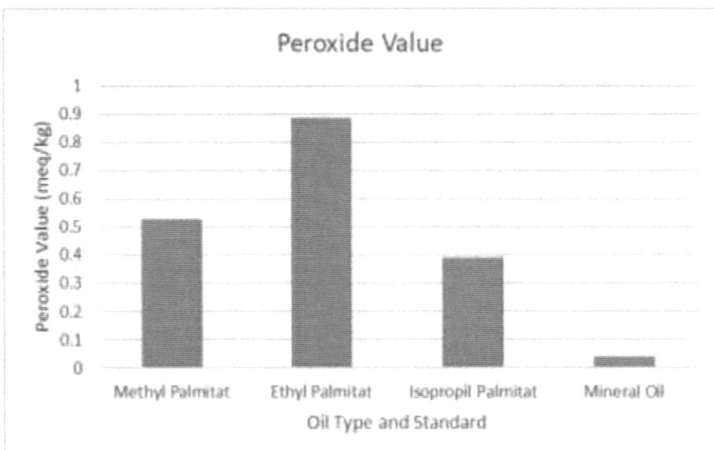

Figure 3. Peroxide value of alkyl palmitates and that of mineral oil.

3.4 . Water Content

It is crucial to keep the water content of an insulation oil at a low level since the electrical properties of the insulation oil are susceptible to the water. The tolerable levels of relative water content are 30 % and 10 % for clean and unclean insulation oils, respectively [20]. For monoesters, these saturation levels correspond to the absolute water contents of approximately 300 ppm and 100 ppm, respectively [16].

The existence of water in the alkyl palmitate samples cannot be avoided. Water (H_2O) is another product resulted from the esterification process, in addition to alkyl palmitates, as indicated by equation (1). An attempt to reduce the water content of oil samples was conducted by filtering the oils using an anhydrate natrium sulfate ($NaSO_4$). This technique successfully reduces the water content of all tested samples to acceptable levels. Figure 5 shows the water content of all tested oils and the specification value from ASTM D 6871. It is clear that the water content of all alkyl palmitate samples is less than that specified by the standard.

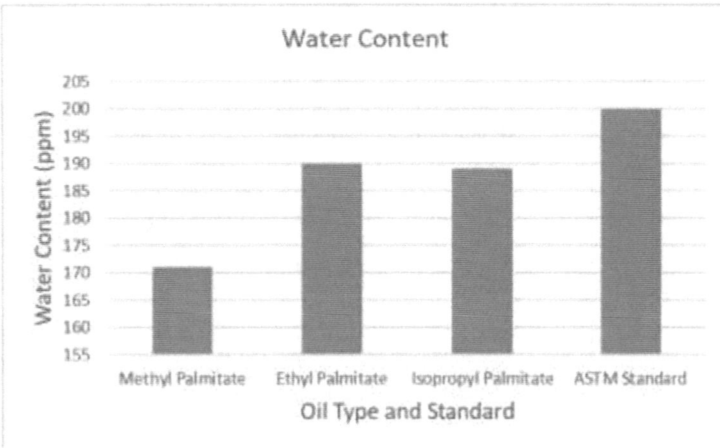

Figure 4 The water content of alkyl palmitates and ASTM D 6871 Standard specification.

3.5 . Acid Number

As mentioned in section 2.1, the alkyl palmitates used in the experiment are resulted from the esterification reactions between alcohols and palmitate acid. During the esterification process, the acid palmitate does not completely react with the alcohols to form alkyl palmitates. Therefore, the existence of acid constituent in the resulted oils cannot be avoided, and it is reflected by the higher acid number owned by the oils, as depicted in Figure 6. The increase in the number of C atom from methyl- to iso-propyl palmitates does not seem to affect the change in the acid number of the tested oils. Figure 6 also shows that the acid number of all tested samples is much higher than the ASTM D 6871 standard specification. Treatment needs to be conducted to decrease the acid constituent of the oil since the presence of acid of a long carbon chain type with the acid number larger than 9 mg KOH/g can affect the breakdown voltage of the ester oil [21].

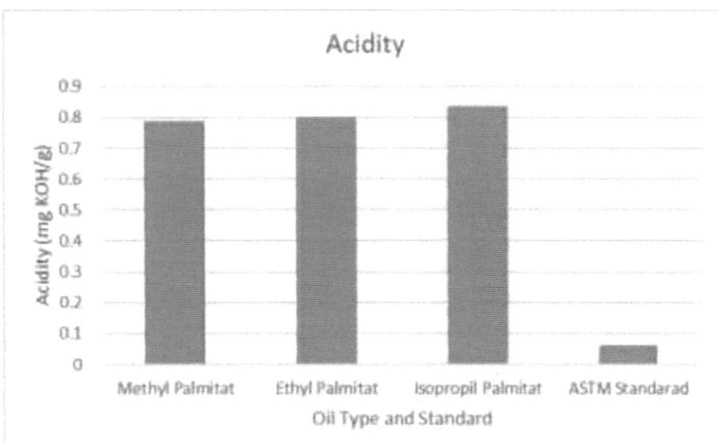

Figure 5. The acid number of alkyl palmitates and ASTM D 6871 Standard specification.

3.6 . Relative Density

Figure 7 shows the measurement results on the relative density of methyl-, ethyl- and isopropyl palmitates and the specified value from the ASTM D 6871. According to the standard, the relative density for unused natural ester should not less than 0.96. Figure 7 shows that the relative density of all alkyl palmitates is well below that of the standard, indicating the suitability of the oils to be used as oil insulation from the relative density point of view.

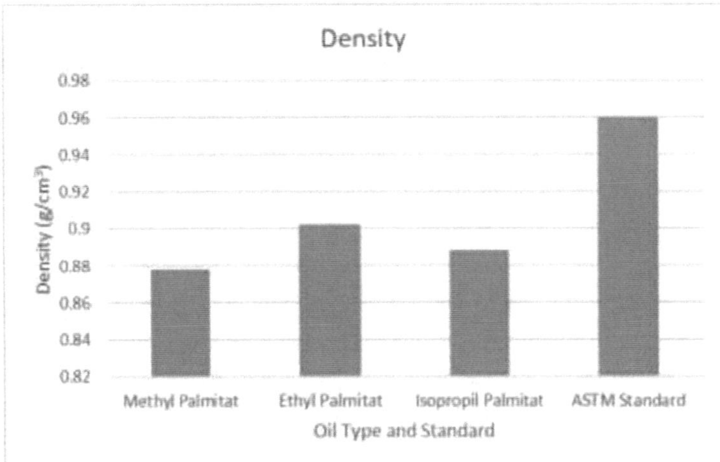

Figure 6. The density of alkyl palmitates and ASTM D 6871 Standard specification

4. Conclusion

The basic properties of alkyl palmitates have been evaluated for possible use as oil insulation. Among six basic properties that have been studied, it is found that the viscosity, water content, and the relative density have complied with the specifications of the ASTM D 6871 standard, whereas the breakdown voltage and acidity have not fulfilled the standard. The oxidation stability of the oils is also less than that of the mineral oil. The breakdown voltage almost reaches the standard values, and it could be further enhanced. In addition, the acidity and oxidation stability will be also improved in the coning studied.

Acknowledgment

The authors would like to thank Lembaga Penelitian dan Pengabdian Masyarakat (LPPM) Universitas Andalas for granting this research under the Klaster Riset-Publikasi Percepatan ke Guru (KRP2GB-PDU-Unand) scheme, with the contract No. T/60/UN.16.17/PT.IS-KRP2GB/LPPM/2019.

References

[1] Meira M, Alvarez R, Catalano L, Ruschetti C and Verucchi C 2018 Gases Production in Mineral and Vegetable Dielectric Oils: a Critical Comparison *2018 IEEE Biennial Congress of Argentina (ARGENCON)* (IEEE) pp 1–8

[2] Mengounou G M and Imano A M 2016 Effect of Processing on Dielectric Strength of Palm Kernel Methyl Esters *Afrique Sci.* **12** 157–65

[3] Wilhelm H M, Tulio L, Jasinski R and Almeida G 2011 Aging markers for in-service natural ester-based insulating fluids *IEEE Trans. Dielectr. Electr. Insul.* **18** 714–9

[4] Perrier C and Beroual A 2009 Experimental investigations on insulating liquids for power transformers: mineral, ester, and silicone oils *IEEE Electr. Insul. Mag.* **25** 6–13

[5] Wilhelm H M, Stocco M B C, Tulio L, Uhren W and Batista S G 2013 Edible natural ester oils

as potential insulating fluids *IEEE Trans. Dielectr. Electr. Insul.* **20** 1395–401
[6] Alahmer A, Mayyas A, Mayyas A A, Omar M A and Shan D 2011 Vehicular thermal comfort models; A comprehensive review *Appl. Therm. Eng.* **31** 995–1002
[7] Oommen T V 2002 Vegetable oils for liquid-filled transformers *IEEE Electr. Insul. Mag.* **18** 6–11
[8] McShane C P 2001 Relative properties of the new combustion-resist vegetable-oil-based dielectric coolants for distribution and power transformers *IEEE Trans. Ind. Appl.* **37** 1132–9
[9] McShane C P, Corkran J, Rapp K and Luksich J 2006 Natural ester dielectric fluid development *IEEE/PES Transmission and Distribution Conference and Exposition*
[10] Rafiq M, Lv Y Z, Zhou Y, Ma K B, Wang W, Li C R and Wang Q 2015 Use of vegetable oils as transformer oils–a review *Renew. Sustain. Energy Rev.* **52** 308–24
[11] Trnka P, Mentlík V and Svoboda M 2014 Ecologically acceptable insulating liquids for electrical appliances *2014 IEEE 18th International Conference on Dielectric Liquids (ICDL)* (IEEE) pp 1–4
[12] Oommen T V, Claiborne C C and Walsh E J 1998 Introduction of a new fully biodegradable dielectric fluid *1998 IEEE Annual Textile, Fiber and Film Industry Technical Conference (Cat. No. 98CH36246)* (IEEE) pp 1–3
[13] Claiborne C C, Walsh E J and Oommen T V 1999 An agriculturally based biodegradable dielectric fluid *1999 IEEE Transmission and Distribution Conference (Cat. No. 99CH36333)* vol 2 (IEEE) pp 876–81
[14] Oommen T V, Claiborne C C, Walsh E J and Baker J P 2000 A new vegetable oil based transformer fluid: development and verification *2000 annual report conference on electrical insulation and dielectric phenomena (Cat. No. 00CH37132)* vol 1 (IEEE) pp 308–12
[15] Viertel J, Ohlsson K and Singha S 2014 Studies of the improvement of the viscosity of natural ester liquids *2014 IEEE 18th International Conference on Dielectric Liquids (ICDL)* (IEEE) pp 1–4
[16] Kanoh T, Iwabuchi H, Hoshida Y, Yamada J, Hikosaka T, Yamazaki A, Hatta Y and Koide H 2008 Analyses of electro-chemical characteristics of palm fatty acid esters as insulating oil *2008 IEEE International Conference on Dielectric Liquids* (IEEE) pp 1–4
[17] Bertrand Y and Lauzevis P 2013 Development of a low viscosity insulating liquid based on natural esters for distribution transformers
[18] ------------ 2003 *Standard Specification for Natural (Vegetable Oil) Ester Fluids Used in Electrical Apparatus. Annual Book of ASTM Standard*
[19] Rajab A, Putra F E, Ramadhani J S, Silitonga M S I, Kurniawan R, Qibran K, Melda L, Hamid M I, Kiasatina A and Akbar M F 2019 Filtration and water reduction of methyl ester for insulation purpose. *Telkomnika* **17** 3183–90
[20] Martin R, Athanassatou H, Duart J C, Perrier C, Sitar I, Walker J, Claiborne C, DE T B, US D C and DE E G 2010 *Experiences in service with new insulating liquids*
[21] Azis N 2012 *Ageing assessment of insulation paper with consideration of in-service ageing and natural ester application* (The University of Manchester (United Kingdom))

IOP Conf. Series: Materials Science and Engineering **885** (2020) 012067 doi:10.1088/1757-899X/885/1/012067

Analysis of social demographic factors and its effect on work-related musculoskeletal disorders in front shovel operators of a mining company

R A Imran[1], I Bakri[2], Mulyadi[2], A E U Husain[2]

[1]Industrial Engineering Department, Universitas Jenderal Soedirman, Indonesia
[2]Industrial Engineering Department, Universitas Hasanuddin, Makassar, Indonesia

Email: rani.aulia.imran@unsoed.ac.id

Abstract. Worker comfort is one important factor in the production operation. The attention to comfort in the workplace will be able to reduce the occurrence of complaints. Facilities and workplace conditions that are not ergonomic increase worker's complaints, such as musculoskeletal disorders are often experienced. Many factors affect the occurrence of work-related musculoskeletal disorders. In this research, social demographic factors retrieved and the influence its effect on musculoskeletal experienced by front shovel operators at a mining company in South Sulawesi was analyzed. This research was based on subjective questionnaries and data obtained were processed by using bivariate statistical method. Dependent and independent variables are musculoskeletal disorder, social demography factor, handgrip strength, subjective physical complaint and environmental factors (temperature, noise, vibration, lighting, and dust). Results concluded from 28 front shovel operators, they have a senior high school education background (86%), with over 3 years work (3-5 years 32%,> 5 years 39%), generally slept in a day for 5-8 hours (71%), and most operators are smokers (75%) with the number of cigarettes consumed in an average a day 4-12 sticks (57%), generally operators are having a breakfast (89%) by consuming rice and side dishes (88%). Statistical tests showed no significant relationship between social-demographic factors to muscle complaints. The influence of environmental factors has a significant relationship to subjective physical complaints but the results obtained inconsistent because the number of samples is still limited.

1. Background

Comfort at work is one important factor in the production process, and by paying attention to it will be able to reduce the occurrence of worker complaints. The complaints that arise are due to the absence of ergonomic work facilities, in accordance with the awkward posture of the worker causing them to feel less comfortable [1]. In industrial activities, exposure and risks in the workplace tend to be around the workplace and workers. Sometimes these conditions cannot always be avoided due to work demands [2]. Comfort in working is strongly influenced by the place and condition of the work environment. Uncomfortable places and working conditions can cause loss and even accidents for workers, which one of the complaints is musculoskeletal disorders.

Heavy equipment is a large-sized machine designed to carry out construction functions, such as earthwork, road construction, building construction, mining, and plantations. In the operation of heavy equipment many aspects must pay attention to, starting from tool operation procedures, aspects of work safety, operator expertise and knowledge, as well as inspection and maintenance aspects [3]. Mining company in this study as one of the largest mining companies in Indonesia has to facilitate its operational activities using a number of heavy equipment. Several heavy equipment is used to excavate soil and rocks, for example front shovel to dig up material that is located on the surface where the machine is located. The use of heavy equipment is to facilitate humans in supporting their work, so the expected results can be achieved more easily with a relatively shorter time. Technology has brought convenience to human life, however, interactions between workers with the environment and work tools can have a negative impact on workers [4]. The most dominant factor influencing the occurrence of low back pain in mining operators operating heavy equipment is the use of static vehicles. The more often operators use static vehicles, the more likely it is to get low back pain.

The amount number and type of heavy equipment used will results in having operators to work in the area in large numbers. The number of operators who work to operate heavy equipment at the company are approximately 700 workers. The partition of operators work time are divided into 3 shifts; each work shift has 8 of work hours. Research by [5] conclude that heavy equipment operators in nickel mining may be exposed to a higher low back pain risk due to the land condition of the overburden overlying nickel deposits. There are many factors that influence the occurrence of muscloskeletal disorders due to work, one of the factors is social-demographic factors. A large number of operators certainly have different backgrounds and socio-demographic conditions. Social demographic factors pay more attention to various individual and group characteristics which include social-demographic, educational, and economic characteristics. Social and demographic characteristics include gender, age, marital status, and religion. Educational characteristics include the level of education. Economic characteristics include occupation, economic status, and income.

Dominant factors not related to work that have an impact on indications of failure are work environment factors such as dusty air, hot air, noise, and workplace or vibrating work tools. Most respondents indicated experiencing severe fatigue felt that they were disturbed by the work environment [6]. The quality of a good work environment and in accordance with the human condition as workers will support the work performance and productivity produced. Control and handling of work environment factors such as noise, temperature, vibration, and lighting are problems that must be dealt with seriously and continuously. Noise, hot temperatures, vibrations and poor lighting in the workplace are sources that cause work pressure and decrease work productivity [7][8]. This has become one of the factors that influence the chances of muscloskeletal disorders, due to the workload provided is not fit with the capabilities of the workers. It will not benefit for the company and either for the workers.

2. Methods

This research was conducted to the front shovel operator by purposive sampling, a mining company at South Sulawesi, Indonesia. Material handling for the excavation using seven units of front shovels and handled by 28 operators with eight working hours of each operator, excluding the rest time and departure time to and from the location.

Figure 1. Front Shovel

Several methods of data collection are carried out, namely questionnaires, interviews and field observations. This questionnaire was given to respondents to get data on individual factors (age, years of service, smoking habits), then added factors such as rest periods and breakfast habits. The level of subjective complaints of skeletal muscle disorders felt by the operator as well as conditions around the work environment also collected. Interview by conducting questions and answer, meanwhile direct measurement of grip strength from front shovel operators were measured. Physical condition before and after doing work measured directly by the strength on the dominant hand using the handgrip dynamometer on the operator at each 1st Shift, 2nd Shift, and 3rd Shift. Measurement tool used in this study is subjective physical fatique the Borg Scale. The Borg Scale was introduced by Gunnar Borg on a scale of 6 – 20 usually used for the diagnosis of shortness of breath, dyspnea, chest pain, and bone or muscle pain. The Borg scale, follows the heartbeat of a healthy adult by multiplying with 10. The larger the scale means the greater the energy expended [9].

3. Results and discussion

The company studied is a nickel production company in Indonesia. To meet the demand requirements, the company manufacture in 24 hours to produce ± 100,000 tons of nickel. In order to run the production process to continuously, the worker's schedule is divided into 3 work shifts, i.e. Shift 1 starts from 06.45 to 14.45, Shift 2 starts from 14.45 to 22.45, and Shift 3 starts from 22.45 to 06.45 WITA. The nickel production process starts from exploration, mining, processing to mine closure. Every shift is given a target of ± 35,000 tons production a day, with adjustment in weekly basis.

Table 1. Participants characteristics (n=28)

Characteristics		Total	Percent
Gender	Man	28	100%
	Women	0	0%
Age	≤40 years	9	45%
	>40 years	11	55%
Education	Grade School	1	4%
	Junior High	0	0%
	Senior High	24	86%
	Diploma	0	0%
	University	3	11%
Employment	< 1 year	0	0%
	1-3 years	8	29%
	3-5 years	9	32%
	> 5 years	11	39%
Slept	< 3 hours	1	4%
	3-5 hours	5	18%
	5-8 hours	20	71%
	> 8 hours	2	7%

Heavy equipment operators especially front shovel machine was total of 28 people, all of them are man (100%). As can be seen in Table 1 that 3% of operators have an elementary education background, 11% have university degrees, and 86% have a high school education background. Based on the length of work, 29% of operators have worked for 1-3 years, 32% for 3-5 years and 39% for> 5 years. The data shows that more than all operators have worked for more than 1 years in operating the front shovel. Operators were having rest periods <3 hours (4%), 7% of operator slept for> 8 hours, 18% of operator slept for 3-5 hours, and most operators have a rest periods ranging from 5-8 hours (71%).

Table 2. Participants lifestyle characteristics (n=28)

Characteristics		**Total**	**Percent**
Smoking	Yes	21	75%
	No	7	25%
Number of Cigarettes	4-12 sticks	12	43%
	>12 sticks	9	32%
Breakfast	Yes	25	89%
	No	3	11%
Breakfast menu	Rice and side dishes	22	79%
	Bread/bakery	3	11%
Sickness Prior	Yes	24	86%
	No	4	14%
Discomfort	Yes	16	57%
	No	12	43%

As many as 75% of operators consume cigarettes every day shown in Table 2, also that 43% of operators who smoke consume >12 cigarettes per day and 57% of operators consume 4-12 cigarettes/day. Most operator (89%) having a breakfast and 11% of operators did not, while most of them (88%) were consume rice when having breakfast and 12% consume bread/bakery. Most of operator (86%) have experienced sickness/pain prior this research and 57% felt discomfort in work.

Table 3. Operator front shovel handgrip stregth (n=20)

Descriptive	Age	1st Shift		2nd Shift		3rd Shift	
		Pre	Post	Pre	Post	Pre	Post
Mean	40,50	50,18	40,01	48,17	37,62	46,62	29,14
Min	37,00	42,40	32,90	40,60	30,20	40,10	22,60
Max	43,00	59,20	48,30	55,80	44,60	56,40	37,60
Stdv	1,96	5,36	4,03	5,33	4,74	5,20	4,47

Based on the handgrip strength data, it can be seen in Table 3 and Figure 2 that there is a decrease in the grip strength of each operator before (pre) and after work (post). On the average, operator subjective physical fatigue experience was increased in 2nd Shift and 3rd Shift compared to 1st Shift, which the biggest difference on the 3rd Shift. Overall, the handgrip strength category was decrease from level good (46,5 - 55,5 kg) to moderate (36,5 - 46,0 kg) [10] after work.

Subjective physical fatigue is tiredness that felt by the operator after doing work or activity presented in a scale, as seen in Figure 3. The result accordance with handgrip strength which the 3rd Shift is the most cause fatigue in front shovel operator, with the highest Borg Scale of all.

Risk factors that are not related to work and are associated with indications of fatigue are environmental influences [6]. The environmental factors subjectively studied were temperature, noise, vibration, lighting, and dust of workplace; whom the operator fills in the questionnaire based on what they felt about their environmental conditions.

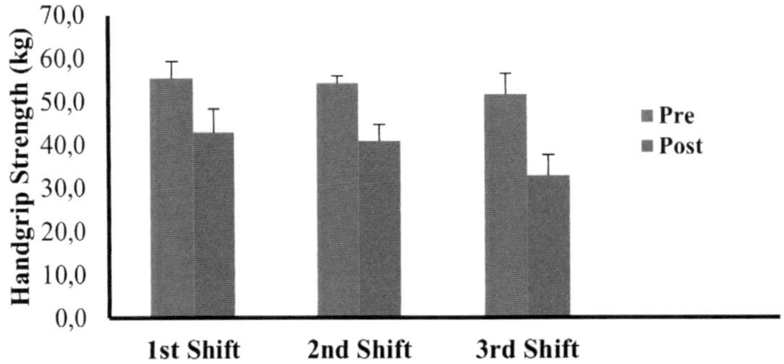

Figure 2. Handgrips strength each shift of front shovel operator

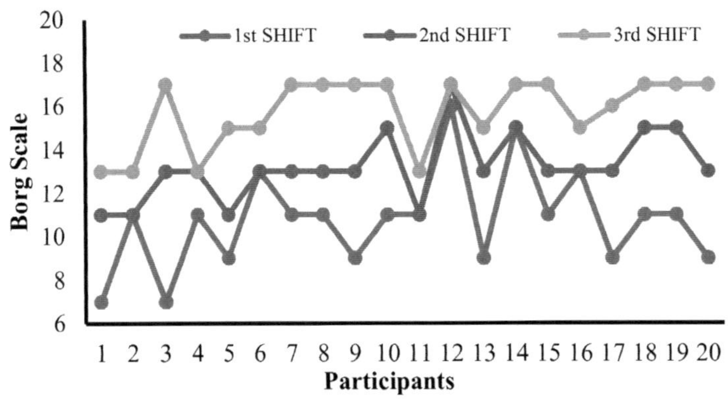

Figure 3. The subjective operator fatigue measured with Borg Scale

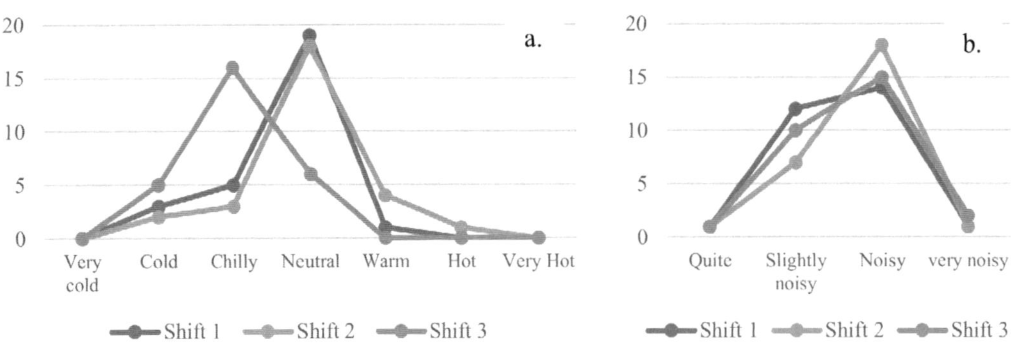

Figure 4. Subjective measure of environment conditions (a) Temperature and (b) Noise

The operator's work environment is in the front shovel cabin, where each unit is equipped with air conditioning. As seen in Figure 4a, the cabin was in neutral level for two Shifts and get chilly at 3rd Shift, eventhough some participant felt warm at 2nd Shift. Figure 4b shown the sounds surround workplace felt noisy at all Shifts and more operators perceived noisier at 2nd Shift.

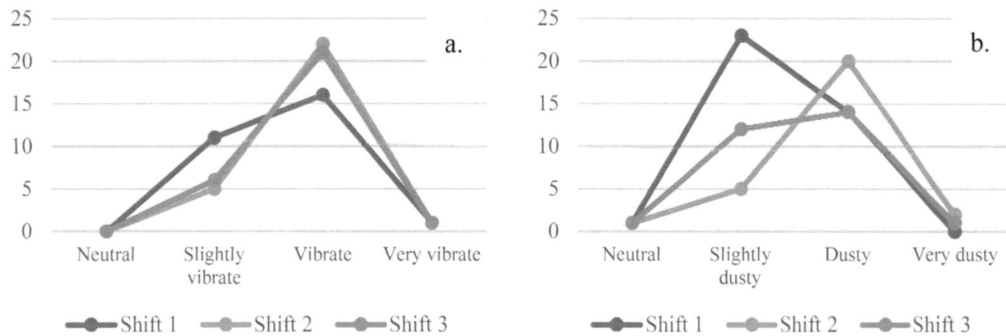

Figure 5. Subjective measure of environment conditions (a) Vibration and (b) Dust

Figure 5a shown the vibration felt at all shifts comply with the study by [11], suggest measuring vibration levels to identify and assessment the risk of equipment is needed. On the other hand, seen in Figure 5b that the work environment was slightly dusty in 1st Shift but became dusty in 2nd Shift.

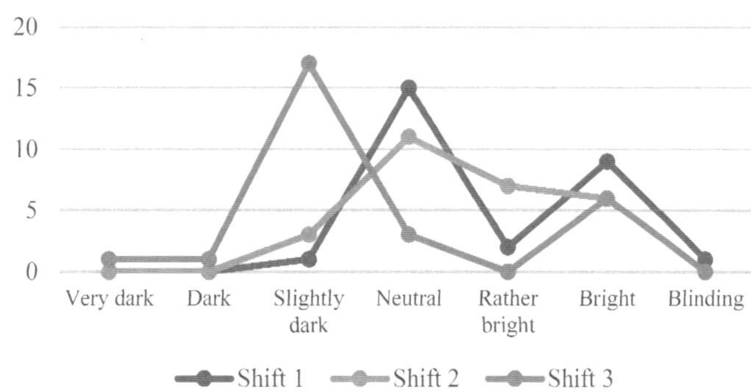

Figure 6. Subjective measure of environment conditions of lighting/lumination

Far away terrain, straight, and saturated requires operators to always be in high concentration when working (vigilance), especially when working at night. In addition to the risk terrain, continuous high concentrations are also caused by a lack of lighting at the worksite [6]. The result corresponded those study as seen in Figure 6, shown that 3rd Shift at the cabin was slightly dark. This could be one of the reasons the physical fatigue increased in 3rd Shift.

Table 4. Comparison of environment subjective measurement of the front shovel operators

Variable	1st Shift	2nd Shift	3rd Shift
Temperature	68% neutral	64% neutral	55% slightly cold
Noise	50% noisy	66% noisy	63% noisy
Vibration	57% vibrate	78% vibrate	75% vibrate
Lighting	53% neutral	41% neutral	61% slightly dark
Dust	60% slightly dusty	72% dusty	50% dusty

Overall measurement of subjective environmental conditions can be seen in Table 4 to compare the results of the commonly perceived by operators. Based on [12], the condition of moderate workload and 8 hours of work is recommended to have a temperature of 28°C, a noise threshold of 85 dB, a whole-body vibration threshold of 0.8661 m/sec^2, with 100 lux of lighting. This results in Table 4 are accordingly with [6], most of the respondents who indicated experiencing severe fatigue felt that they were disturbed by the work environment. Work environment factors such as dusty air, hot air, noise, and workplace or vibrating work tools have an impact on indications of fatigue; especially related to the distance and breadth of eyesight if the road is not visible due to dust will interfere with driving ability. Operators who are exposed to noise more than 85 dB have a risk of 0.198 times higher of fatigue [13], and there is a positive association between whole-body vibration and low back disorders [11].

Table 5. Binary logistic regression analysis of variable handgrip strength and demographics

Variables	Coef.	Std. Err.	z	P>|z| Sig.
Employment	-.0988	.5006	-0.20	0.844
Sleep time	-.9412	. 7951	-1.18	0.236
Smoking Habit	-.2426	.9363	-0.26	0.796
Breakfast	.7612	.6753	1.13	0.260
Constant	2.9332	2.9875	0.98	0.326

Regression analysis of handgrip strength and demographics variable can be seen in Table 5. The results of bivariate statistical analysis showed that there was no significant relationship between handgrip strength and social demographic factors of the operators. The smoking habit result is comply the study by [4], there was no significant relation with muscle pain. However, on the contrary employment and sleep time have the indication of fatigue on other studies [6][13]. Participants who were indicated to have severe fatigue were work within 1 - 3 years, with a risk of 1.195 times higher for fatigue. Based on the results of those study, whose work duration was 1 - 3 years did not yet have a good rest period. If an adult has less than 5 hours sleep it will decrease the mental ability and have an impact on work performance, with a risk of 4.09 times greater for fatigue.

Table 6. Correlation of operator demograpics, Borg Scale and subjective fatigue measurement

Correlation	Edu-cation	Employ ment	Slept	Smok-ing	Number of Cigarettes	Break-fast	Break-fast Menu	Age	Borg Scale
Borg Scale	-0,384	0,097	-0,035	-0,126	0,168	0,035	0,064	-0,132	
Handgrip difference	-0,190	0,327	0,013	-0,237	0,151	-0,144	-0,024	0,195	0,426

Eventhough the regression analysis did not conclude a significant relationship, there were very weak correlations between numbers of cigarettes and Borg scale (16.8%) and numbers of cigarettes and handgrip difference (15.1%) The positive correlation between numbers of cigarettes with both is possibly due to fatigue, because there are correlations between stress level with smoking behavior [14], psycologycal distress with smoking habit [15], and stress level with risk of musculoskeletal pain [4]. There was also weak correlation between age and handgrip difference (19.5%). The bigger handgrip differences mean the strength between pre-post work is significantly decreased; thus, the older operators will be higher strength differences. Low correlation between Borg scale and handgrip differences (42.6%) also be seen in the study; the higher strength differences will be higher perceived fatigue. Participants mean age were 40,5 years, and [16] concluded that after about the age of thirty the chances of developing a musculoskeletal disorder remain fairly consistent regardless of age. Nonetheless, according to [13] operators who have more than 10 years of work experience or who have aged over 40 years have a smaller incidence rate.

Table 7. Results of the handgrip strength and environmental factors linear regression

Variable	1st Shift Sig.	2nd Shift Sig.	3rd Shift Sig.
(Constant)	0	0	0
Temperature	0,703	0,726	0,672
Noise	0,434	0,565	0,689
Vibration	0,690	0,271	0,196
Lighting	0,238	0,750	0,352
Dust	0,598	0,554	0,651

Statistical test results shown by Table 7, working environment and their effect on the handgrip strength of the front shovel operators on every shift showed there were no significant relationship on the influence of the work environment on the handgrip strength of the operators. Statistical testing linear regression also conducted to determine the relationship between working environment conditions and their effects on subjective physical fatigue experienced by operators.

Table 8. Results of the Borg Scale and environmental factors linear regression

Variable	1st Shift Sig.	2nd Shift Sig.	3rd Shift Sig.
(Constant)	0,096	0	0
Temperature	0,020	0,291	0,087
Noise	0,728	0,895	0,523
Vibration	0,813	0,614	0,040
Lighting	0,067	0,321	0,706
Dust	0,020	0,526	0,825

The results obtained from analysis shown on Table 8 that in 1st Shift which are significant and have an effect on subjective physical fatigue were temperature (p value=0.02) and dust (p value= 0.02). This could be due to the first shift at 06.45 to 14.45 WITA, with known temperature in the morning was 27°C, afternoon 29°C, night 23°C, and maximum temperature was 31°C [17]. A hot condition above 30°C causes decreased consciousness and generally experience fatigue [13]. However, in 2nd Shift can be seen that there is no influence of the five factors on subjective physical fatigue. The 3rd Shift shows that only vibrations that affect subjective physical fatigue felt by workers after doing work (p value=0.04). Driving these heavy machines in mine workings involve uneven terrain, changing slopes on haul road, needs careful speed control etc. all of which can influence level of exposures [11]. Bio-mechanical factors, vibration and heat stress were highlighted contributed to the health risks caused by poor ergonomics [18].

Muskuloskeletal disorder place a signifcant burden on mining company finances, but in many cases largely preventable through workplace design, usage of correct tools, proper housekeeping and equipment modifcations [16]. Based on this study, improvements need to be made so that operator's comfort will increase. The results obtained from the overall data are still visible inconsistent results between environmental conditions against subjective physical fatigue, most likely due to the limited number of samples and differences in operator's perceived their work condition. For further research, it is advisable to increase the number of samples, posture analysis, body discomfort/pain mapping, and measure the environmental conditions with quantitative data to improve a mining workers quality of life.

4. Conclusion

The results of this study indicate that in 28 front shovel operators have a high school education background (86%), with a work period of > 5 years (39%), the rest time is generally within 5-8 hours a day (71%), and 75% is smokers with the average number of cigarettes consumed in a day is 4-12 cigarettes (57%), generally having breakfast in the morning (89%) by consuming rice and side dishes (88%). The average strength of the operator's handgrip decreased from 48,32 kg (good) then after work was 35,58 kg (moderate); with a Borg scale after work was 13.25 (slightly heavy). The results of bivariate statistical analysis showed that there was no significant relationship between social demographic factors and muscle complaints experienced by the operator. However, there is a low correlation between Borg scale and handgrip strength (42.6%).

The environmental factors mostly felt by operators were neutral temperature, noisy, vibrate, neutral lighting, and dusty categories. The results of statistical analysis show that environmental factors at 1st Shift temperature and dust have a significant effect on the Borg scale, whereas on 3rd Shift only vibrations have a significant effect on the Borg scale. It can be concluded that environmental factors have a significant relationship with subjective physical fatigue, eventhough there was no significant effect on handgrip strength.

References

[1] Nazlina, Buchari and Ria S I 2008 Usulan Perancangan Postur Kerja dengan Menggunakan Pendekatan Biomekanika dan Fisiologi pada Aktivitas Pencetakan Batu Bata *Seminar Nasional Teknik Industri dan Kongres BKSTI V*

[2] Raliby, Widodo E M and Aman M 2008 Studi Intervensi Ergonomi dan Penilaian Tingkat Resiko Terhadap Pengrajin Pahat Batu di Sentra Industri Pahat Batu Prumpung *Semin. Nas. Tek. Ind. dan Kongr. BKSTI V*

[3] PT. Karya Master Mandiri Indonesia 2019 Pelatihan dan Sertifikasi Operator Front Shovel

[4] Basuki K 2009 Faktor Risiko Kejadian Low Back Pain Pada Operator Tambang Sebuah Perusahaan Tambang Nickel Di Sulawesi Selatan *Indones. J. Heal. Promot.* **4** 115–21

[5] Yassierli 2017 Implementation of ergonomic programs to reduce sick leave due to low back pain among nickel mining operators *Int. J. Ind. Ergon.* **61** 81–7

[6] Susilowati I H, Syaaf R Z, Satrya C, Hendra H and Baiduri B 2013 Pekerjaan, Nonpekerjaan, dan Psikologi Sosial sebagai Penyebab Kelelahan Operator Alat Berat di Industi Pertambangan Batu Bara *Kesmas Natl. Public Heal. J.* **8** 91

[7] Suma'mur P K 1987 *Hiperkes, keselamatan kerja dan ergonomi* (Jakarta: Muara Agung Dharma Bhakti)

[8] Tarwaka, Bakri S and Sudiajeng L 2004 *Ergonomi untuk keselamatan, kesehatan kerja, dan produktivitas* (Indonesia: UNIBA Press)

[9] Bridger R S 2008 *Introduction to Ergonomics, Third Edition.* (CRC Press)

[10] Fenanlampir A and Faruq M M 2015 *Tes dan Pengukuran Dalam Olah Raga* (Yogyakarta: CV. Andi Offset (Penerbit ANDI))

[11] Mandal B B and D. Manwar V 2017 Prevalence of musculoskeletal disorders among heavy earth moving machinery operators exposed to whole-body vibration in opencast mining *Int. J. Community Med. Public Heal.* **4** 1566

[12] Indonesia M K R 2018 *Peraturan Menteri Ketenagakerjaan Republik Indonesia Nomor 5 Tahun 2018 Tentang Keselamatan Dan Kesehatan Kerja Lingkungan Kerja* (Indonesia)

[13] Mulyani T 2012 *Faktor-Faktor Yang Berhubungan Dengan Keluhan Fatigue Pada Operator Unit Hauling Coal Dan Overburden Di Pt Buma, Mitra Kerja Pt Berau Coal Lati Tanjung Redeb Kalimantan Timur Tahun 2012* (Universitas Indonesia)

[14] Arniati L N 2019 *Hubungan Antara Tingkat Stres Dengan Perilaku Merokok Perawat Pria di RSUD Sukoharjo* (Universitas Muhammadiyah Surakarta)

[15] Considine R, Tynan R, James C, Wiggers J, Lewin T, Inder K, Perkins D, Handley T and Kelly B 2017 The contribution of individual, social and work characteristics to employee mental health in a coal mining industry population *PLoS One* **12** 1–15

[16] Weston E, Nasarwanji M F and Pollard J P 2016 Identification of Work-Related Musculoskeletal Disorders in Mining. *J. safety, Heal. Environ. Res.* **12** 274–83

[17] Anon Prediksi BMKG, Cuaca Malili Pagi dan Siang Cerah Berawan, Malam Hujan - Tribun Timur

[18] Schutte P C 2005 Ergonomics in the South African mining industry *J. South African Inst. Min. Metall.* **105** 369–72

IOP Conf. Series: Materials Science and Engineering **885** (2020) 012068 doi:10.1088/1757-899X/885/1/012068

Quality Improvement Using PDCA Methodology in the Beverage Industry

Raodah, W Astutik, A A Aris and S Bahri

Department of Industrial Engineering, Universitas Hasanuddin, Makassar, Indonesia

Email: rararaodah@gmail.com

Abstract. This paper has an objective to apply PDCA methodology as a way to improve the quality of production processes in the beverage industry. As part of it, a prior analysis has been done and found the highest breakdown was in the conditioning area of the bottle making production. It has been proofed by the secondary data which showedthe average breakdown in the conditioning area of the bottle making production was about 40% from July to September 2019. A preliminary brainstorming involved a manager and a supervisor of the production department to identify the possible causes of the problem. From brainstorming, there were several potential causes of the breakdown problem have been identified using the cause and effect analysis. They were categorised from machine, man and method factors. For machine factors the potential causes came from the faulty conditioning capacitors, the cable of power condition was short, semi-finished bottle burns in conditioning and low of reliability of conditioning sensors. Meanwhile, from the man factor the potential causes came from the lack of skill from operators in centering the conditioning of the collect. Therefore, from the method factor, the cause was because lack of method with relation to the centering conditioning of the collect. The next stage of this research will apply PDCA methodology to overcome the problem and improve the quality of processes in this company.

1. Introduction

To produce good products in terms of defect-free, it is necessary for the industry to improve their processes continuously and having people who commit to achieve company's quality goals. PT. XYZ is a manufacturing company that produces various types of beverages. Before the product is made, the bottle packaging must be produced first. This research has an objective to improve company performance. Currently, the company is facing a situation where machine breakdown contribute to unsuccesful in meeting the target.

One of the biggest contributors to the breakdown was the breakdown in the conditioning area. The conditioning process is the process of reheating semi-finished bottle. It can be seen in Table 1 the average breakdown frequency in the conditioning area for three months as follows:

IOP Conf. Series: Materials Science and Engineering **885** (2020) 012068 doi:10.1088/1757-899X/885/1/012068

Table 1. Frequency of bottle maker breakdown periods in July – September 2019

	Frequency (Hour)
Semi-finished bottle counts incorrect RF generator conditioning	507
Semi-finished bottle presence on injection mold B1	365
Board NR 13	244
Semi-finished bottle presence on injection mold B2	113
Injection press not open	90
Air high pressure not present blowing phase	57
Dryer not working	51

From Table 1, the most often contributors to the breakdown was the semi-finished bottle count incorrect RF generator conditioning. So, it becomes a critical factor for the improvement in the area.

Table 2. The average frequency of breakdown in the conditioning area

		July	August	September
Breakdown (Hour)		154	211	142
The average frequency			169	
Production planning	Hour	413	392	475
	Average hours per month		427	
Percentage	%per month	37%	54%	30%
	Average % per month		40%	

Based on Table 2 above, the average breakdown frequency per month is 169 hours with a percentage of 40%. The application of PDCA approach can reduce breakdown in the conditioning area so that it can improve the production process so that the machine runs according to the expected planning. Further, it can reduce inefficient costs due to time wasted due to breakdown.

2. Literature Review

The products or services produced must be affordable and good quality, so that customers are satisfied and remain loyal to the products or services produced. Based on this, the company has to control the process so they meets the customer expectations. One of the most common use of methods to quality improvement is PDCA. The PDCA methodology was developed in 1930 where products were considered easy to find, ordinary and startfacing competition in the market so that they were increasingly directed towards implementing quality management [1][2].

The creator of PDCA was William Edward Deming in 1950 known as one of the most famous improvement methods in the world. This method was successfully implemented in Japanese companies and commonly known as Deming Cycle. The PDCA cycle was originally used as a tool to control product quality, but to date it has been recognized as a method to improve organizational processes. The cycle is characterized by a focus on continuous improvement to improve the quality of products and processes [3][4]. According to Prashar [5], the PDCA cycle as a tool to maintain quality or standards set by the company and to improve processes and enable managers to monitor every steps.

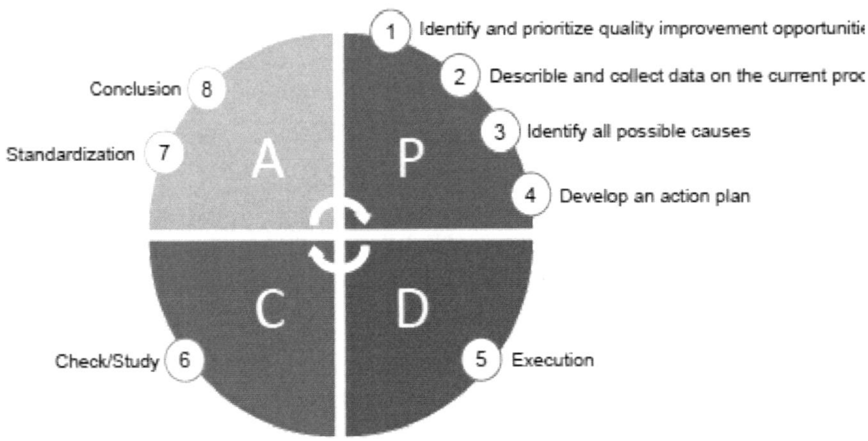

Figure 1. PDCA cycle [3]

The PDCA cycle divided into Plan, Do, Check and Action, and can be explained as follows [3]:

a) Plan
In this step, opportunities for improvement are identified and prioritized; the process is consistently investigated in the data; the cause of the problem is determined; and possible actions to mitigate the issues to be acted upon.

b) Do
The purpose of this step is to implement an action plan; document data; and record the unexpected events, lessons, and knowledge that have been obtained.

c) Check
At this stage, the results of the action are analyzed. The new situation compared to the old one, verifies whether there have been improvements and whether the goal has been achieved. At this stage, various supporting tools such as graphics should be used.

d) Action
At this stage, method development is carried out. The team involved in developing the method standardizes (if the results have been achieved); repeat tests to collect new data and re-evaluate interventions (if the data collected is inadequate or circumstances have changed); or start over from the initial stage (if the action taken has not resulted in an effective improvement). In order for the steps presented in Figure 1 to be carried out effectively, it may be necessary to use other quality tools. This tool helps primarily to analyze problems and determine the actions that must be carried out.

IOP Conf. Series: Materials Science and Engineering **885** (2020) 012068 doi:10.1088/1757-899X/885/1/012068

3. Methodology

The data collected in this study were secondary that was the engine breakdown data for three months (July-September 2019). This research will use the PDCA as methodology to solve problems and improve the quality of processes in this beverage industry. A set of tools are listed below and some of them will be used for every stages of PDCA for this research.

Table 3. Tools for PDCA cycle [6]

Seven basic quality tools (7QC tools)	Steps of PDCA-cycle				
	Plan	Do	Plan, Check	Plan, Act	Check
	Problem identification	Implement solutions	Process analysis	Solutions development	Result evaluation
Flow chart	✓			✓	
Cause-and-effect diagram	✓		✓		
Check sheet	✓		✓		✓
Pareto diagram	✓		✓		✓
Histogram	✓				✓
Scatter plot			✓	✓	✓
Control chart	✓		✓		✓

4. Analysis

Figure 2 shown a process flowchart of the bottle making and the focus of this research will be an improvement in conditioning area.

No	Process	Work Method
	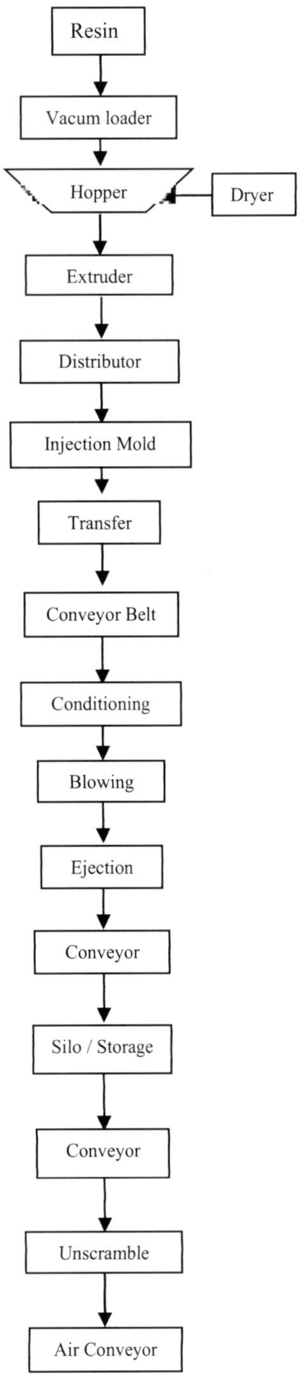	1. The resin is inserted into the hopper by using a vacuum loader by the time set.
		2. The resin is dried using a dryer until the recommended temperature is reached, then the resin is added to the Extruder for the melting process.
		3. After PET resin is melting inside the extruder, then the screw pushes the material toward the distributor center.
		4. In the distributor section, the material is distributed for 4 parts, namely are A1 - A2 - B1 - B2 by passing hot channel 1, hot channel 2, Pet valve, injector chamber, pet valve, pet inlet.
		5. After the material enters the injection mold, then the semi-finished bottle will be forming based on the process that occurs in the injection mold.
		6. After the semi-finished bottle is formed, the transfer will be performed and then place it on the conveyor belt and then carry it to the conditioning area
		7. After the semi-finished bottle arrived at the conditioning area, the conditioning system will work to reheat the semi-finished bottle.
		8. When the semi-finished bottle is in the blowing area, the blowing system will work according to the process to form a semi-finished bottle into a bottle in the form of mold on the blowing system.
		9. After the blowing process is complete and the bottle is formed or printed, then the conveyor belt will bring the bottle to the ejector area. The ejector will press the bottle from the conveyor and fall onto the conveyor.
		10. Next, the finished bottles will be transferred to the Silo by using a conveyor onto the silo to be put into the Silo.
		11. Then the bottle is removed from the Silo by using the Belt conveyor Feeder and then transferred to the Unscramble.
		12. After entering the Unscramble, the bottle is set and positioned so that when it comes out of the unscrambled, the bottle is already standing and then directed to the air conveyor.
		13. After the bottle is in a standing position Air conveyor carries the bottle to the filling area by blowing on the bottle.

Figure 2. Process flowchart of bottle making

IOP Conf. Series: Materials Science and Engineering **885** (2020) 012068 doi:10.1088/1757-899X/885/1/012068

As previously explained, the red dotted line in the flowchart shows the area that is the focus of research, specifically to the RF generator conditioning when breakdown existed. At this stage, interviews were conducted with supervisors and managers in the bottle production section related to factors causing a breakdown in the conditioning area. The results of the interviews can be seen in the Cause and Effect Diagram (Figure 3).

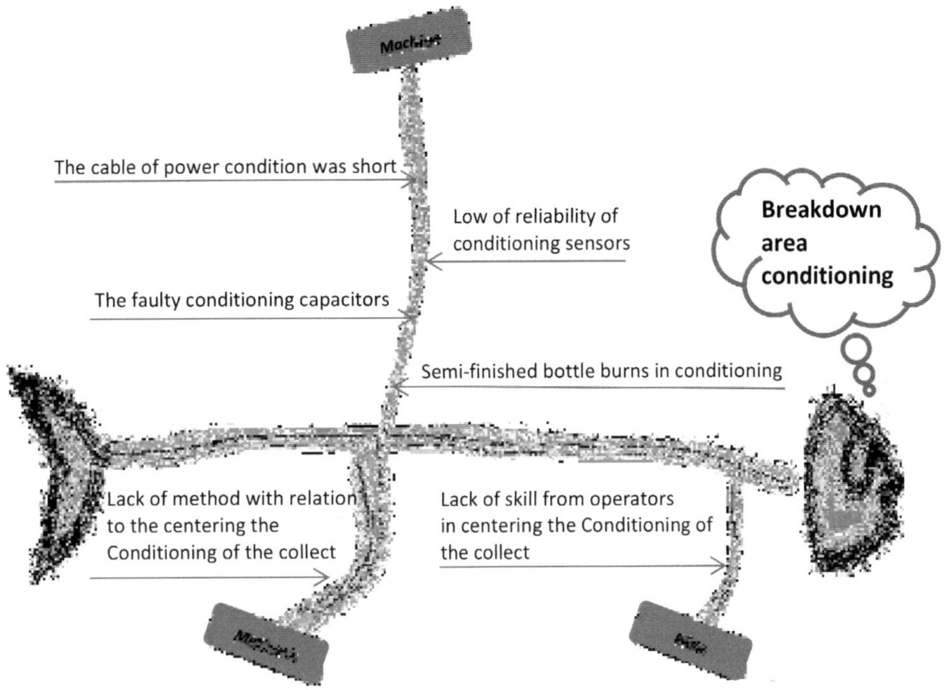

Figure 3. Cause and effect diagram of conditioning breakdown

5. Discussion and Conclusion

Preliminary results of this study will use as a basis for further research. The results obtained in the form of factors that cause the occurrence of breakdown engine in the conditioning area which categorize under man, methods and machines factors. From the man factor, it found that the lack of skill from operators in centering the conditioning of the collect was potentially caused to the machine breakdown [7]. While from the method, found the lack of method with relation to centering of the conditioning and the collect. Another potential cause ofthe machine breakdown was from machine itself i.e. the cable of power condition was short, the faulty conditioning capacitors, semi-finished bottle burns in conditioning and low of reliability of conditioning sensors. Future research will use the PDCA methodology to solve problem in the conditioning area using the initial results reported in this paper.

References

[1] De Souza J M 2016 PDCA and lean manufacturing: Case study in appliance of quality process in alpha graphics *J. Leg. Bus. Sci* **17** 11–7

[2] Khusaini N S, Jaffar A and Noriah Y 2014 A survey on lean manufacturing tools implementation in malaysian food and beverages industry using rasch model *Advanced Materials Research* vol 845 (Trans Tech Publ) pp 642–6

[3] Silva A S, Medeiros C F and Vieira R K 2017 Cleaner Production and PDCA cycle: Practical application for reducing the Cans Loss Index in a beverage company *J. Clean. Prod.* **150** 324–38

[4] Jagusiak-Kocik M 2017 PDCA cycle as a part of continuous improvement in the production company-a case study *Prod. Eng. Arch.* **14** 19–22

[5] Prashar A 2017 Adopting PDCA (Plan-Do-Check-Act) cycle for energy optimization in energy-intensive SMEs *J. Clean. Prod.* **145** 277–93

[6] Soković M, Jovanović J, Krivokapić Z and Vujović A 2009 Basic quality tools in continuous improvement process *J. Mech. Eng.* **55** 1–9

[7] Rasmussen J 1982 Human errors. A taxonomy for describing human malfunction in industrial installations *J. Occup. Accid.* **4** 311–33

IOP Conf. Series: Materials Science and Engineering **885** (2020) 012069 doi:10.1088/1757-899X/885/1/012069

Identifying Cybersickness When Wearing a Head-Mounted Display through Heart Rate Variability Data

N O Setiowati, T Wijayanto and F Trapsilawati

Department of Mechanical and Industrial Engineering, Universitas Gadjah Mada, Yogyakarta, Indonesia.

E-mail: nonioktianas@gmail.com

Abstract. Cybersickness is one of the common problems related to Virtual Reality (VR). The symptoms are classified into three categories: nausea, oculomotor, and disorientation. The purpose of this study is to identify cybersickness symptoms based on simulator sickness questionnaire (SSQ) and physiological signal using heart rate variability (HRV) when wearing a head-mounted display. Fifteen participants performed a 15-min driving simulation task using head-mounted display. From the SSQ results, the most dominant category of cybersickness symptoms induced by participants are nausea and disorientation. The results showed a correlation between the nausea score and the high frequency (HF)/component of HRV ($r=0.618$, $p=0.032$) as well as the ratio of low and high frequency component (LF/HF) ($r=-0.718$, $p=0.009$). These findings confirm that the occurrence of nausea when wearing a head-mounted display is influenced by the activation of sympathetic nervous system.

1. Introduction

Virtual Reality (VR) and Head-Mounted Display (HMD) have been increasingly popular for professional use and entertainment use over the past years. VR technology has been used, for example, for military training simulations [1][2], medical procedures [3], and gaming or entertainment. Apart from various technological improvements, one of the common problems that is constantly associated with VR activities is dizziness and uncomfortable feelings. This phenomenon is called cybersickness [4].

Cybersickness is a common side effect of VR related activities. It is often indicated with nausea, dizziness and vomiting as the symptoms [5]. A study by Cobb *et al.* [6] stated that 80% of their participants showed cybersickness symptoms within 10 minutes during VR activities. Another study by Regan and Price [7] found that 60% of their participants showed cybersickness symptoms within 20 minutes during VR activities. These symptoms can last for more than five hours after they are no longer in the Virtual Environment [7].

The cybersickness symptoms are classified into three categories, namely nausea, disorientation, and oculomotor [8]. Those symptoms are usually measured using Simulator Sickness Questionnaire (SSQ), a subjective measurement developed by Kennedy *et al.* [8]. Measurement using physiological signal such as Heart Rate Variability (HRV) to obtain more objective results have received lack of attention. There are a few similar studies about HRV to identify cybersickness [9][1] [10][11][12]. But a more detailed study about the correlation between HRV parameters and cybersickness symptoms need to be done to validate the findings of those studies.

Heart rate variability describes changes in heart rate intervals associated with automatic nerve regulation such as the sympathetic and parasympathetic nervous system [13]. Parasympathetic activity decelerates the HR and on the other hand, sympathetic activity elevates it. The dynamic balance between parasympathetic and sympathetic activity causes a continuous oscillation of the HR called HR variability [14]. Under resting conditions, parasympathetic activity is dominant [15]. The advantages of using physiological indices such HRV parameters is to understand participants' physical condition. HRV parameters can be used to examine the participants' conditions based on parasympathetic and sympathetic activities, so the most correlated HRV parameters to cybersickness need to be investigated.

From the aforementioned background, the purpose of this study was to identify the dominant categories of cybersickness symptoms when wearing HMD in a driving simulator. This study also sought to investigate whether HRV correlates with cybersickness occurrence when wearing HMD. In this study, cybersickness was measured objectively using HRV and subjectively using SSQ. HRV parameters used in this study are the time domain components (SDNN and RMSSD) and the frequency domain components (LF, HF, and LF/HF).

2. Methods

2.1. Participants

Fifteen students (Mean of ages=23.5 (± SE=0.57)) volunteered in this study. Participants had healthy eyes or had corrected to normal vision. All participants reported that they were infrequent game players (less than 5 times a month). Three from fifteen participants discontinued the experiment prematurely due to severe cybersickness and were excluded from the data analysis.

2.2. Apparatus

The driving simulation task was performed using a PC-based driving simulator (City Car Driving v 1.5.6, Forward Development, Russia) run on a PC with a quad-core Intel® Core™ i7-6700K@ 4.00 GHz, 16GB of RAM, and a NVIDIA® Geforce® GTX 980 with 12GB of GDDR5. The virtual environment of the driving simulator was projected to an HMD (Oculus Rift, Oculus VR, Irvine, CA). The driving simulator was equipped with a driving chair, a manual gear controller. and a three-pedal set.

2.3. Experiment Procedure

Before starting the experiment, the participants received the details of this study and were informed that they were free to abort the experiment at any time without being penalized. All participants gave written consent prior to the experiment and stated that they were in a normal state of health (no dysfunctional olfactory). Polar heart rate sensor (Polar H7, Polar, Finland) was attached to every participant's chest during the experiment to record the inter beat interval (IBI). After the fifteen minutes driving task was completed, the participant was asked to fill in SSQ to indicate the cybersickness symptoms they experienced. All procedures in this study had followed the ethical clearance for human subjects.

2.4. Subjective Measurements

Participants' cybersickness was measured using Simulator Sickness Questionnaire (SSQ) developed by Kennedy *et al.* [8]. It consists of 16 symptoms related to the simulator sickness. Participants were asked to rate these symptoms on a 4-point Likert scales (0 – not at all, 1 – slight, 2 – moderate, and 3 -severe) after the driving simulation task. The SSQ symptoms were categorized into three subscales, nausea (SSQ-N), disorientation (SSQ-D), and oculomotor (SSQ-D) and finally the cybersickness score (SSQ-Total) was calculated using the pre-defined weighting factor as explained in Kennedy *et al.* [8].

2.5. Heart Rate Variability Measurements

HRV was measured using a polar heart rate sensor (H10, Polar, Finland). The heart rate sensor was attached to participant's chest before participants starting the experiment. Inter beat interval data (IBI) was recorded using the Polar Beat application in a smartphone. IBI data were then imported to Kubios HRV software (Kubios HRV Standard 3.2.0) to pre-process the data and to obtain the frequency and time domain components. The frequency domain data used in this study were Low Frequency (LF, 0.04 – 0.15 Hz) and High Frequency (HF, 0.15 – 0.40 Hz) and LF/HF ratio. The time domain data used in this study were Standard Deviation of Normal to Normal R-R intervals (SDNN) and Root Mean Square of Successive Heartbeat Interval Difference (RMSSD) as indicators.

2.6. Statistical Data Analysis

The data are presented in mean ± standard error (SE). Prior to statistical analysis, a normality test was used to check the parametric assumption of the data. If normality assumption is violated, then nonparametric test will be performed. RM ANOVA test was employed to test the significant difference among the SSQ subscales. If there is a significant difference, then post hoc test with Bonferroni correction would be administered. Maximum value of SSQ subscales are different so the data have to be normalized. HRV analysis with frequency domain were expressed in original units or as the natural logarithm (ln) of original units to achieve a more normal distribution [16]. Correlation test (Spearman's rho test) was administered to find the relationship between physiological indices (HRV parameters) and cybersickness level. The significance level was set $\alpha = 0.05$. All statistical tests were performed using JASP software.

3. Results

3.1. Simulator Sickness Questionnaire

The mean SSQ score are shown in Figure 1, the results showed that Nausea and Disorientation had the highest score among all symptoms. Nausea mean score is 53.26 ± 13.46, Oculomotor mean score is 48.00 ± 9.43, and Disorientation mean score is 53.36 ± 12.57. However, the RM ANOVA test showed that there was no significance difference among these three subscales (F=3.093, p=0.066).

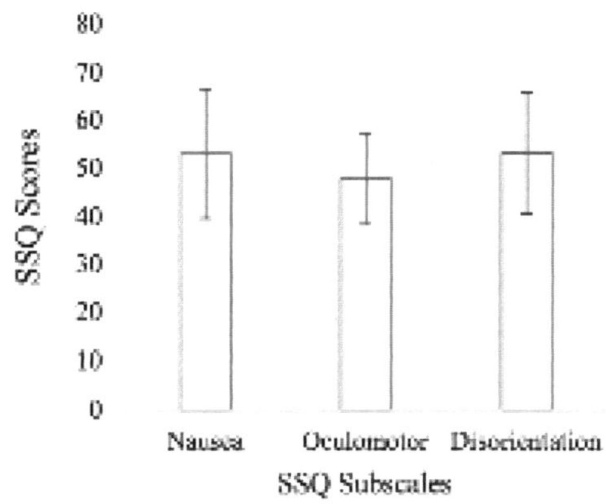

Figure 1. Mean Ratio SSQ subscales scores

3.2. Heart Rate Variability (HRV)

The mean HRV scores for Frequency Domain and Time Domain are shown in Figure 2, the results showed that the lnLF mean score is 4.55 ± 0.13, lnHF mean score is 1.95 ± 0.15, and lnLF/HF mean score is 2.46 ± 0.16. The mean HRV scores for Time Domain showed that SDNN mean score is 12.6 ± 0.69, and RMSSD mean score is 7.07 ± 0.44. High LF power means that sympathetic activity is more dominant than parasympathetic activity, sympathetic activity governs the "fight or flight" response while the parasympathetic activity controls the "rest and digest" response. The main overall end effect of the sympathetic activity is to prepare the body for physical activity [20].

3.3. Correlation between SSQ and HRV

Correlation test (Spearman's rho test) was performed to find the correlation between SSQ subscales (Nausea, Disorientation, and Oculomotor) and HRV subscale (LF, HF, LF/HF, SI, SDNN, and RMSSD). The result showed that there is a strong positive correlation (r=0.618, p=0.032) between HF and Nausea score. A strong negative correlation is found between LF/HF ratio and Nausea score (r=-0.718, p=0.009). But there was no significance difference between the other HRV parameters and SSQ subscales. The correlation plots between HRV (HF and LF/HF) and SSQ nausea score, and the correlation coefficient are shown in Figure 3.

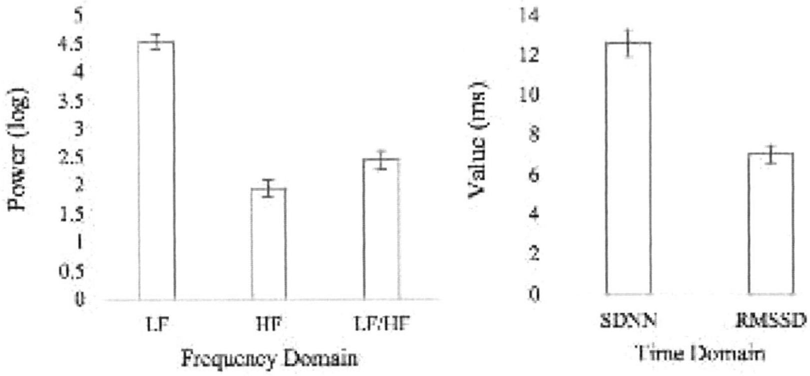

Figure 2. Frequency Domain and Time Domain HRV Analysis

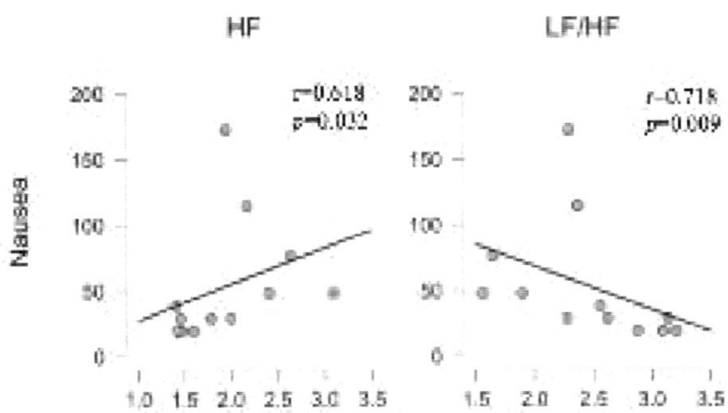

Figure 3. Correlation Plot between HRV and SSQ Subscales

4. Discussion

This study used subjective and objective measurements to identify cybersickness symptoms induced by participant through participant's Heart Rate Variability. Subjective responses measured by Simulator Sickness Questionnaire (SSQ) and objective responses measured by Heart Rate Variability (HRV).

The SSQ mean score showed that Nausea and Disorientation had the highest score compared to Oculomotor. It means that Nausea and Disorientation are the strongest symptoms induced by participants that caused Cybersickness. But there was no significance difference among these three subscales ($p>0.05$).

HRV analysis using Frequency Domain method showed the high score of LF power and the low score of HF power. Psychological stress was significantly associated with an increase in the LF/HF ratio [17]. The LF power is associated with sympathetic activities and the HF power is associated with parasympathetic activities [18]. It means that cybersickness induced by participant caused the increased of LF power and the decreased of HF power. The increased of LF power means that sympathetic nervous system (SNS) is dominant. The SNS governs the "fight or flight" response while the PNS (parasympathetic nervous system) controls the "rest and digest" response. The main overall end effect of the SNS is to prepare the body for physical activity [19].

Correlation test (Spearman's rho) was performed to find the coefficient correlation between SSQ subscales (Nausea, Disorientation, Oculomotor and Total SSQ) and HRV subscales (LF, HF, LF/HF, SDNN, and RMSSD). The Coefficient Correlations showed that there is a strong positive correlation between HF and Nausea score. And a strong negative correlation was found between LF/HF ratio and Nausea score. The results are statistically significant. But there was no significance difference between the other HRV variables and SSQ subscales.

A strong positive correlation between HF power and Nausea score showed that increasing Nausea induced by participants caused the increase of HF power. HF or High Frequency range of HRV is considered to be representative of vagal activity [20]. Previous study [20] also has the same findings that there is a transient trending increase in HF power right before strong nausea. Findings confirm gradual sympathetic activation with increasing nausea and further evidence of increase in vagal tone before strong nausea. Vagal activity results in various effects such as heart rate reduction [20].

A strong negative correlation was found between LF/HF ratio and Nausea score. The increased of LF/HF ratio showed the decreased of nausea experienced by participant. It means that the balance of autonomic nervous system corelated to nausea induced by participants. From Correlation test performed in this study, we can conclude that there are relationships between HRV parameters and SSQ subscales. Further research with more participants needs to be done.

5. Conclusion

The purpose of this study is to find the correlation between the induced cybersickness and participant's heart rate variability. From the SSQ results, Nausea and Disorientation had the highest score among all symptoms. The Coefficient Correlations showed that there is a strong positive correlation between HF power and Nausea score, also a strong negative correlation between LF/HF ratio and Nausea score. The results are statistically significant. But there was no significance difference between the other HRV parameters and SSQ subscales. These findings confirm that the occurrence of nausea when wearing a head-mounted display is influenced by the activation of sympathetic nervous systems. Further research with more participants needs to be done.

Acknowledgment

We thank all participants for participating in this experiment and the assistants of Ergonomic Lab of Universitas Gadjah Mada for the technical supports. Special thanks to Indonesia Endowment Fund for Education (LPDP), the sponsor of this research.

References

[1] Garcia-Agundez A, Reuter C, Caserman P, Konrad R and Göbel S 2019 Identifying cybersickness through heart rate variability alterations *Int. J. Virtual Real.* **19** 1–10

[2] Wu D, Courtney C G, Lance B J, Narayanan S S, Dawson M E, Oie K S and Parsons T D 2010 Optimal arousal identification and classification for affective computing using physiological signals: Virtual reality stroop task *IEEE Trans. Affect. Comput.* **1** 109–18

[3] Rizzo A, Parsons T D, Lange B, Kenny P, Buckwalter J G, Rothbaum B, Difede J, Frazier J, Newman B and Williams J 2011 Virtual reality goes to war: A brief review of the future of military behavioral healthcare *J. Clin. Psychol. Med. Settings* **18** 176–87

[4] Mittelstaedt J, Wacker J and Stelling D 2018 Effects of display type and motion control on cybersickness in a virtual bike simulator *Displays* **51** 43–50

[5] LaViola Jr J J 2000 A discussion of cybersickness in virtual environments *ACM Sigchi Bull.* **32** 47–56

[6] Cobb S V G, Nichols S, Ramsey A and Wilson J R 1999 Virtual reality-induced symptoms and effects (VRISE) *Presence Teleoperators Virtual Environ.* **8** 169–86

[7] Regan E C and Price K R 1994 The frequency of occurrence and severity of side-effects of immersion virtual reality. *Aviat. Space. Environ. Med.*

[8] Kennedy R S, Lane N E, Berbaum K S and Lilienthal M G 1993 Simulator sickness questionnaire: An enhanced method for quantifying simulator sickness *Int. J. Aviat. Psychol.* **3** 203–20
[9] Malińska M, Zużewicz K, Bugajska J and Grabowski A 2015 Heart rate variability (HRV) during virtual reality immersion *Int. J. Occup. Saf. Ergon.* **21** 47–54
[10] Watanabe H, Teramoto W and Umemura H 2007 Effect of predictive sign of acceleration on heart rate variability in passive translation situation: preliminary evidence using visual and vestibular stimuli in VR environment *J. Neuroeng. Rehabil.* **4** 36
[11] Ohyama S, Nishiike S, Watanabe H, Matsuoka K, Akizuki H, Takeda N and Harada T 2007 Autonomic responses during motion sickness induced by virtual reality *Auris Nasus Larynx* **34** 303–6
[12] Zużewicz K, Saulewicz A, Konarska M and Kaczorowski Z 2011 Heart rate variability and motion sickness during forklift simulator driving *Int. J. Occup. Saf. Ergon.* **17** 403–10
[13] Billman G E, Huikuri H V, Sacha J and Trimmel K 2015 An introduction to heart rate variability: methodological considerations and clinical applications *Front. Physiol.* **6** 55
[14] Renkin E M 1984 *Handbook of physiology: Section 2, The cardiovascular system. Microcirculation: pt. 2* vol 4 (American Physiological Soc.)
[15] Mathew N L 1988 Sympathetic parasympathetic interactions in the heart *Neurocardiology* 85–98
[16] Tarvainen M P, Lipponen J, Niskanen J P and Ranta-Aho P 2017 *Kubios HRV Version 3 - User's Guide* (Kuopio: University of Eastern Finland)
[17] Sloan R P, Shapiro P A, Bagiella E, Boni S M, Paik M, Bigger Jr J T, Steinman R C and Gorman J M 1994 Effect of mental stress throughout the day on cardiac autonomic control *Biol. Psychol.* **37** 89–99
[18] Rawenwaaij-Arts C, Kallee L and Hopman J 1993 Task Force of the European Society of Cardiology and the North American Society of Pacing and Electrophysiology. Heart rate variability. Standards of measurement, physiologic interpretation, and clinical use. Circulation 1996; 93: 1043-1065. *Intern. Med* **118** 436–47
[19] McCorry L K 2007 Physiology of the autonomic nervous system *Am. J. Pharm. Educ.* **71**
[20] LaCount L T, Napadow V, Kuo B, Park K, Kim J, Brown E N and Barbieri R 2009 Dynamic cardiovagal response to motion sickness: a point-process heart rate variability study *2009 36th Annual Computers in Cardiology Conference (CinC)* (IEEE) pp 49–52

AUTHOR INDEX

Agus, M 26, 165
Ainuddin, A 22
Aisyah, Siti 172, 177
Akbar, Z 146
Akil, Y S 297
Al Adawiyah, Syardah Ugra 244
Amar, K 297, 398
Amin, Widya Ashari 72
Amrin 423
Apollo 92
Aras, Neny Rasnyanti M 76
Arbain, P 7
Arifin, Abdul Nasser 273
Aris, A A 267, 448
Arminas 227, 309
Asmal, S 267, 324
Asqarye, A 344
Astutik, W 448
Aulia, I R 227
Awaliyah, Rizqa 256
Azis, M F 22, 43
Azis, Muhammad Fadli 56
Bachtiar, E 109
Baharuddin, A Velahyati 198
Baharuddin, A V 293
Bahri, S 448
Bakri, I 250, 382, 438
Banjarsanti, SSN. 232
Basri, M 227
Bustan, M. Djoni 61
Cahyadi, Dwi 316
Darmawan, A 344
Dewarani, R F 172, 177
Djonny, M 83
Eltahan, Muhammed 15
Erdawaty 109
Fatmawati, A 104
Fitriani, Anisa 184
Ganesa, D 289
Griseldis, G L 392
habibi, Baso 139
Habibuddin, Julianti 56
Habibuddin, J 22, 43
Haddade, Jalaluddin 132
Hakim, AR 151
Hamid, M I 431
Haming, Puadi 256
Hamzah, Ainun Jariah 219
Hanafi, R 344
Hariyati, H 87
Harnaningrum, R N 353
Hartanto, E S 369
Haryati, Sri 61
Hasanah, F 369
Hasanuddin, Z B 146
Hendrawan, R 309
Herlinah 139
Heryani, S 369
Hidayanto, Andi Farid 316
Hidayati, Qory 34
Hidayat 104
Husain, A E U 438
Ihsan, Muhammad 132
Imansuri, F 26, 159, 165
Imran, R A 438
Indah, A B R 329
Indah, N 289
Irwan, M 281
Islam, Sri S 184
Jalaluddin 123
Jayanti, Regna Tri 61
Jufri 123
Junaedy, M F 109
Kasa, Agustina 49
Kelah, N Ester 250
Khaerani, N B 309
Kukuh, P. 232
Kurniawan, R 431
Lando, Asiyanthi Tabran 273
Langkong, J 87
Lestari, Tika 184
Lukmandono 405
Madjid, M 92
Mahendradatta, M 87
Mangngenre, S 297, 329
Mansyur 67
Margaritifera, Theta 232
Marit, I Y 377
Marzuki, I 109
Masing 281
Melawaty, L 83
Melda, L 431
Mellolo, A 83
Metandi, F 151
Miyauchi, H 297
Moharm, Karim 15

Muchtar, T ... 43
Mudiastuti, R D ... 329
Mukhsen, M. Iqbal ... 118
Mulyadi ... 203, 240, 250, 438
Mustafa ... 281
Muthalib, I S. ... 392
Muthalib, I S ... 324
Mutmainnah ... 22, 43, 56
Nadiyah, K ... 398
Nainggolan, E A ... 304
Nilda ... 240, 329
Nizaora, Ditha ... 316
Nur, Rusdi ... 118
Nur, R ... 92, 104
Nur, Taufik ... 34
Nursanti, E ... 377
Nurulinzany ... 423
Nurwahidah, Andi ... 203, 240
Nurwahidah, A ... 250
Paroka, Daeng ... 123
Pasaribu, M ... 304
Patoding, Hestikah Eirene ... 1, 49
Prabowo, R ... 405
Pramudi, R ... 339
Purba, F S P ... 304
Purba, Humiras Hardi ... 172, 177
Purwanto, D ... 43
Qibran, K ... 431
Rachim, F ... 109
Rajab, A ... 431
Rakka, Christo ... 118
Ramadhan, D G ... 211
Ramadhan, F ... 109
Ramadhani, J S ... 431
Ramdhani, I ... 382
Randis ... 97
Raodah ... 448
Rapi, Amrin ... 273
Rapi, A ... 293, 324
Rinamurti, M ... 360
Rochman, Abd ... 139
Rohaman, M M ... 414
Rombe ... 1
Rusman, M ... 392
Safril ... 26, 159, 165
Salam, A E U ... 146
Salam, Nasaruddin ... 132
Santoso, A ... 353
Sarungallo, Rosalia Sira ... 83
Satria, Dedi Nismar ... 203
Sau, Matius ... 1, 49
Sawal, Ahmad ... 219
Setiawan, A M ... 109
Setiawan, F ... 339
Setiawan, H ... 360
Setiowati, N O ... 455
Shabrina, M P ... 293
Sihombing, R ... 104
Simatupang, Dimas Frananta ... 67
Siregar, N C ... 414
Sitompul, C ... 339
Sitompul, E Y R ... 304
Soeprapto, Etwin Fibrianie ... 316
Sofiyanurriyanti ... 211
Subakdo, W A ... 263
Suhaebri, T ... 22
Sukardin, M Setiawan ... 132
Sulistyowati, E ... 405
Sumasto, F ... 26, 159, 165
Suminto ... 7
Suparno ... 353
Susanti, I ... 369
Suyuti, M A ... 92
Syamsul, N I ... 329
Tarakka, Rustan ... 123, 132
Tarigan, Jenny ... 67
Tawali, A B ... 87
Topadang, Anton ... 189
Trapsilawati, F ... 455
Triyono, Agus ... 189
Tyassena, Flaviana Yohanala Prista ... 72
Utomo, Karyo Budi ... 7
Vitasari, P ... 377
Vojisavljevic, V ... 398
Wardani, Dilla Ayu ... 184
Widodo, E ... 353
Wijaya, H ... 369
Wijaya, S H ... 97
Wijayanto, T ... 455
Wirandi, M ... 26, 159, 165
Yanti, Nur ... 34
Yusuf, M. Alif AF ... 118
Zainal ... 289